VOLUME SIX HUNDRED AND TWENTY THREE

METHODS IN ENZYMOLOGY

RNA Recognition

METHODS IN ENZYMOLOGY

Editors-in-Chief

ANNA MARIE PYLE
Departments of Molecular, Cellular and Developmental Biology and Department of Chemistry
Investigator, Howard Hughes Medical Institute
Yale University

DAVID W. CHRISTIANSON
Roy and Diana Vagelos Laboratories
Department of Chemistry
University of Pennsylvania
Philadelphia, PA

Founding Editors

SIDNEY P. COLOWICK and NATHAN O. KAPLAN

VOLUME SIX HUNDRED AND TWENTY THREE

METHODS IN ENZYMOLOGY

RNA Recognition

Edited by

AMANDA E. HARGROVE
Department of Chemistry,
Duke University,
Durham, NC, United States

ELSEVIER

ACADEMIC PRESS
An imprint of Elsevier

Academic Press is an imprint of Elsevier
50 Hampshire Street, 5th Floor, Cambridge, MA 02139, United States
525 B Street, Suite 1650, San Diego, CA 92101, United States
The Boulevard, Langford Lane, Kidlington, Oxford OX5 1GB, United Kingdom
125 London Wall, London, EC2Y 5AS, United Kingdom

First edition 2019

ISBN: 978-0-12-816831-8
ISSN: 0076-6879

For information on all Academic Press publications
visit our website at https://www.elsevier.com/books-and-journals

Publisher: Zoe Kruze
Acquisition Editor: Ashlie M. Jackman
Editorial Project Manager: Shellie Bryant
Production Project Manager: James Selvam
Cover Designer: Miles Hitchen

Typeset by SPi Global, India

Contents

Contributors

Rebecca L. Adams
Department of Molecular, Cellular and Developmental Biology, Yale University, New Haven, CT, United States

Dev P. Arya
Laboratory of Bioorganic and Medicinal Chemistry, Department of Chemistry, Clemson University, Clemson, SC, United States

Lauren M. Aufdembrink
Department of Genetics, Cell Biology, and Development, University of Minnesota, Minneapolis, MN, United States

Benjamin F. Bachan
Department of Genetics, Cell Biology, and Development, University of Minnesota, Minneapolis, MN, United States

Dennis Bong
Department of Chemistry and Biochemistry, The Ohio State University, Columbus, OH, United States

David R. Calabrese
Chemical Biology Laboratory, National Cancer Institute, Frederick, MD, United States

Jessica L. Childs-Disney
Department of Chemistry, The Scripps Research Institute, Jupiter, FL, United States

Christine S. Chow
Wayne State University, Detroit, MI, United States

Colleen M. Connelly
Chemical Biology Laboratory, National Cancer Institute, Frederick, MD, United States

Matthew G. Costales
Department of Chemistry, The Scripps Research Institute, Jupiter, FL, United States

Yumin Dai
Department of Chemistry and Center for Drug Discovery, Virginia Tech, Blacksburg, VA, United States

Rhiju Das
Department of Biochemistry, Stanford University School of Medicine; Biophysics Program; Department of Physics, Stanford University, Stanford, CA, United States

Matthew D. Disney
Department of Chemistry, The Scripps Research Institute, Jupiter, FL, United States

Danielle N. Dremann
Wayne State University, Detroit, MI, United States

Aaron E. Engelhart
Department of Genetics, Cell Biology, and Development; Department of Biochemistry, Molecular Biology and Biophysics, University of Minnesota, Minneapolis, MN, United States

Christopher S. Eubanks
Department of Chemistry, Duke University, Durham, NC, United States

Adrian R. Ferré-D'Amaré
Biochemistry and Biophysics Center, National Heart, Lung and Blood Institute, Bethesda, MD, United States

Erin E. Gallagher
Department of Medicinal Chemistry, College of Pharmacy, University of Michigan, Ann Arbor, MI, United States

Amanda L. Garner
Department of Medicinal Chemistry, College of Pharmacy; Program in Chemical Biology, University of Michigan, Ann Arbor, MI, United States

Aaron C. Goldstrohm
Department of Biochemistry, Molecular Biology and Biophysics, University of Minnesota, Minneapolis, MN, United States

Michael Gotrik
Department of Biochemistry, Stanford University, Stanford, CA, United States

Amanda E. Hargrove
Department of Chemistry, Duke University, Durham, NC, United States

Joseph M. Heili
Department of Genetics, Cell Biology, and Development; Department of Biochemistry, Molecular Biology and Biophysics, University of Minnesota, Minneapolis, MN, United States

Dziyana Hnedzko
Department of Chemistry, Binghamton University, Binghamton, NY, United States

Tanner G. Hoog
Department of Genetics, Cell Biology, and Development, University of Minnesota, Minneapolis, MN, United States

Kaiyi Huang
Life Sciences Institute, Zhejiang University, Hangzhou, Zhejiang, China

Nicholas C. Huston
Department of Molecular, Cellular and Developmental Biology, Yale University, New Haven, CT, United States

Christopher P. Jones
Biochemistry and Biophysics Center, National Heart, Lung and Blood Institute, Bethesda, MD, United States

Stuart F.J. Le Grice
Basic Research Laboratory, National Cancer Institute, Frederick, MD, United States

Yue Li
Department of Chemistry, The Scripps Research Institute, Jupiter, FL, United States

Yufeng Liang
Department of Chemistry and Biochemistry, The Ohio State University, Columbus, OH, United States

Daniel A. Lorenz
Program in Chemical Biology, University of Michigan, Ann Arbor, MI, United States

Jie Mao
Department of Chemistry and Biochemistry, The Ohio State University, Columbus, OH, United States

John D. McAnany
Department of Chemistry, University of Rochester, Rochester, NY, United States

Benjamin L. Miller
Department of Dermatology, University of Rochester, Rochester, NY, United States

Giacomo Padroni
Department of Chemistry, Duke University, Durham, NC, United States

Subrata Panja
T.C. Jenkins Department of Biophysics, Johns Hopkins University, Baltimore, MD, United States

Matthew R. Pawlak
Department of Genetics, Cell Biology, and Development, University of Minnesota, Minneapolis, MN, United States

Ashley N. Peralta
Department of Chemistry and Center for Drug Discovery, Virginia Tech, Blacksburg, VA, United States

Anna M. Pyle
Department of Molecular, Cellular and Developmental Biology; Department of Chemistry, Yale University, New Haven, CT; Howard Hughes Medical Institute, Chevy Chase, MD, United States

Chen Qiu
Epigenetics and Stem Cell Biology Laboratory, National Institute of Environmental Health Sciences, National Institutes of Health, Research Triangle Park, NC, United States

Ramya Rangan
Biophysics Program, Stanford University, Stanford, CA, United States

Nihar Ranjan
Department of Medicinal Chemistry, National Institute of Pharmaceutical Education and Research (NIPER)-Raebareli, New Transit Campus, Lucknow, India

Aiming Ren
Life Sciences Institute, Zhejiang University, Hangzhou, Zhejiang, China

Eriks Rozners
Department of Chemistry, Binghamton University, Binghamton, NY, United States

Webster L. Santos
Department of Chemistry and Center for Drug Discovery, Virginia Tech, Blacksburg, VA, United States

John S. Schneekloth Jr.
Chemical Biology Laboratory, National Cancer Institute, Frederick, MD, United States

Chringma Sherpa
Basic Research Laboratory, National Cancer Institute, Frederick, MD, United States

Aiai Sun
Life Sciences Institute, Zhejiang University, Hangzhou, Zhejiang, China

Traci M. Tanaka Hall
Epigenetics and Stem Cell Biology Laboratory, National Institute of Environmental Health Sciences, National Institutes of Health, Research Triangle Park, NC, United States

Rafael C.A. Tavares
Department of Molecular, Cellular and Developmental Biology; Department of Chemistry, Yale University, New Haven, CT, United States

Gabriele Varani
Department of Chemistry, University of Washington, Seattle, WA, United States

Sai Pradeep Velagapudi
Department of Chemistry, The Scripps Research Institute, Jupiter, FL, United States

Matthew J. Walker
Department of Chemistry, University of Washington, Seattle, WA, United States

Andrew M. Watkins
Department of Biochemistry, Stanford University School of Medicine, Stanford, CA, United States

Hannah Wayment-Steele
Department of Chemistry, Stanford University, Stanford, CA, United States

Sarah A. Woodson
T.C. Jenkins Department of Biophysics, Johns Hopkins University, Baltimore, MD, United States

Michelle Wu
Program in Biomedical Informatics, Stanford University, Stanford, CA, United States

Luqian Zheng
Life Sciences Institute, Zhejiang University, Hangzhou, Zhejiang, China

Preface

RNA molecules are increasingly recognized both for their regulatory roles and as potential therapeutic targets in a range of microbial and viral infections and human diseases. Such RNAs include bacterial riboswitches that control translation, viral RNA structures critical to replication, microRNAs that control mRNA levels and expression, and long noncoding RNAs (lncRNAs) that regulate complex gene expression networks.*

Central to these RNA functions, and any related efforts to target them, is the specific recognition of RNA molecules. The mysteries of RNA recognition are particularly fascinating but difficult to unlock given the increased dynamics and reduced chemical diversity of RNA relative to protein and the more complex structure of RNA relative to DNA. Indeed, our current understanding suggests that it is the combination of sequence, structure, and dynamics that ultimately leads to specific recognition. Given these complexities, it is not surprising that a diverse set of methods have been and are continuously being developed to create a deep understanding of RNA recognition.

This volume includes a sampling of both traditional and novel methods in the study of RNA recognition, including the lessons we learn from natural systems, from engineering biomolecules, and from studying how small molecule probes can selectively recognize RNA. In each case, the authors illustrate how the methods described reveal insights into RNA recognition.

Looking at natural systems, several chapters investigate riboswitches, both as important biological regulators and as excellent model systems to understand RNA recognition. Ren and coauthors detail riboswitch characterization by X-ray diffraction and NMR methods while Ferré-D'Amaré, Woodson, and coauthors gain insights from cotranscriptional folding of riboswitches by combining super-helicases with single molecule fluorescence measurements. Das and coauthors reveal methods to model and push towards riboswitch optimality in both natural and synthetic systems as well as methods to model 3D RNA structures, which have been benchmarked using riboswitch aptamers. At the same time, such detailed characterization of structure and folding is generally more difficult for long RNAs, and Pyle

*Given the many excellent reviews and primary literature reports on these topics, readers are referred to individual chapters for the most relevant and up-to-date respective references.

and coauthors outline how to combine a suite of probing methods for the accurate determination of the natively folded secondary structure of long RNA molecules, using the Hepatitis C virus genome as a model system. Importantly, the majority of RNAs are thought to be bound by proteins, and Hall, Goldstrohm, and coauthors illuminate how crystallization and structure determination of protein:RNA complexes can reveal intricate details of cooperativity in RNA recognition.

Insights gained from studying native RNA structure, folding, and interactions directly contribute to, and are complemented by, efforts to design ligands for RNA. Ligand approaches include the engineering of peptide and nucleic derivatives as well as natural products and synthetic small molecules. The power of peptide design is demonstrated by Santos and coauthors through the study of branched peptides with RNA-targeting modifications and by Varani and coauthors through the design of conformationally constrained cyclic peptides. Peptide nucleic acids offer a distinct approach by taking advantage of nucleobase hydrogen bonding patterns, as illustrated by Rozners and coauthors in the recognition of RNA through triplex formation and by Bong and coauthors who generate allosteric triggers of RNA-driven catalysis.

Finally, small-molecule targeting of RNA is an area of intense interest but with the need for improved methods to evaluate recognition and selectivity, the latter of which can be particularly challenging for these lower molecular weight molecules in comparison to peptide or oligonucleotide derivatives. In addition, small molecules are particularly suited for the recognition of RNA structure rather than sequence, and high-resolution structures remain a challenge for many RNAs. Naturally occurring small-molecule RNA-ligands such as aminoglycosides offer unique insights into RNA recognition and also serve as good model systems, as revealed by Chow and coauthors in the use of mass spectrometry to monitor RNA:aminoglycoside interactions and by Arya and coauthors who use fluorescently-labeled aminoglycosides to screen for other RNA ligands. Hargrove and coauthors describe a method using pattern recognition to understand determinants of RNA structural differentiation and selectivity via aminoglycoside binding.

These and other methods can be extended to a range of small molecule classes. Garner and coauthors present a high-throughput method to screen natural product extracts against multiple pre-microRNAs for both discovery and selectivity analysis, and Disney and coauthors depict a computational and experimental pipeline to identify and validate small molecule ligands for specific RNA secondary structures. Yet another approach to the challenge of selectivity is the generation of multivalent small molecules, and Miller and coauthors detail the use of dynamic combinatorial chemistry to identify multivalent ligands without the need for high-resolution structure. In addition to screening techniques, new biophysical methods for determining small molecule binding properties are also critical to progress in the field. Schneekloth and coauthors describe ligand-observed NMR techniques to rapidly assess binding and selectivity of small molecules to RNA and Engelhart and coauthors report a method to easily assess stability of small molecule:RNA complexes via thermal denaturation with fluorogenic dyes.

We hope that this compilation of select methods will serve as a valuable resource to those entrenched in the study of RNA and an inspiration to those interested in starting their own RNA adventure.

Finally, I express my sincere gratitude to the authors who contributed their expertise, time, and wisdom to this volume as well as to the editorial and production team for their valuable assistance.

Amanda E. Hargrove
Department of Chemistry
Duke University

CHAPTER ONE

Preparation of cooperative RNA recognition complexes for crystallographic structural studies

Chen Qiu[a], Aaron C. Goldstrohm[b], Traci M. Tanaka Hall[a,*]

[a]Epigenetics and Stem Cell Biology Laboratory, National Institute of Environmental Health Sciences, National Institutes of Health, Research Triangle Park, NC, United States

[b]Department of Biochemistry, Molecular Biology and Biophysics, University of Minnesota, Minneapolis, MN, United States

*Corresponding author: e-mail address: hall4@niehs.nih.gov

Contents

Abstract

It is essential that mRNA-binding proteins recognize specific motifs in target mRNAs to control their processing, localization, and expression. Although mRNAs are typically targets of many different regulatory factors, our understanding of how they work together is limited. In some cases, RNA-binding proteins work cooperatively to regulate an mRNA target. A classic example is *Drosophila melanogaster* Pumilio (Pum) and Nanos (Nos). Pum is a sequence-specific RNA-binding protein. Nos also binds RNA, but interaction with some targets requires Pum to bind first. We recently determined crystal structures of complexes of Pum and Nos with two different target RNA sequences. A crystal structure in complex with the *hunchback* mRNA element showed how Pum and Nos

ISSN 0076-6879
https://doi.org/10.1016/bs.mie.2019.04.001
2019 Published by Elsevier Inc.

together can recognize an extended RNA sequence with Nos binding to an A/U-rich sequence 5′ of the Pum sequence element. Nos also enables recognition of elements that contain an A/U-rich 5′ sequence, but imperfectly match the Pum sequence element. We determined a crystal structure of Pum and Nos in complex with the *Cyclin B* mRNA element, which demonstrated how Nos clamps the Pum-RNA complex and enables recognition of the imperfect element. Here, we describe methods for expression and purification of stable Pum-Nos-RNA complexes for crystallization, details of the crystallization and structure determination, and guidance on how to analyze protein-RNA structures and evaluate structure-driven hypotheses. We aim to provide tips and guidance that can be applied to other protein-RNA complexes. With hundreds of mRNA-binding proteins identified, combinatorial control is likely to be common, and much work remains to understand them structurally.

1. Introduction

Post-transcriptional regulation, mediated by hundreds of mRNA-binding proteins (mRBPs), plays an essential role in controlling gene expression. mRBPs recognize their target mRNAs through specific sequence or structural motifs often in the 5′ or 3′ untranslated regions (UTRs), acting as an initial step leading to regulation of RNA processing, localization, translation and degradation (Mitchell & Parker, 2014; Singh, Pratt, Yeo, & Moore, 2015). An individual mRBP is capable of recognizing hundreds of mRNAs. A second RBP can add an additional layer of selectivity when its binding site is in the vicinity of the first binding site. Cooperative binding of the two RBPs yields a more stable complex while each individual protein might bind only weakly or transiently. X-ray crystallography is a valuable technique to visualize the protein-RNA complexes and provide insights into combinatorial target RNA recognition and regulation by multiple RBPs. Hundreds of structures of individual RBPs in complex with their cognate RNA are available. However, structural knowledge of cooperative binding by multiple RBPs to RNA is limited. It is more challenging to obtain a stable protein-RNA complex consisting of multiple proteins for crystallographic studies. Here, we describe methods for preparing a ternary complex of *Drosophila* Pumilio (Pum) and Nanos (Nos) proteins and their target RNA for biochemical and structural studies. We also discuss some techniques and strategies that might be helpful for crystallizing protein-RNA complexes in general.

Drosophila Pum is a founding member of the eukaryotic PUF (Pum and *fem-3* mRNA-binding factor, FBF) protein family (Zamore, Williamson, &

Lehmann, 1997). PUF proteins are characterized by a sequence-specific RNA-binding domain known as the PUM-homology domain (PUM-HD), and the domain is evolutionarily conserved across eukaryotes (Fig. 1A) (Wickens, Bernstein, Kimble, & Parker, 2002; Zhang et al., 1997). The prototypical PUM recognition/response element (PRE) contains 5′-UGUAnAUA-3′ (n = any base) (Wang, McLachlan, Zamore, & Hall, 2002). Yeast and worm encode multiple PUF proteins exhibiting different sequence specificities with varying length of *n* from 1 to 5 in the middle region (Gerber, Herschlag, & Brown, 2004; Koh et al., 2009). Crystal structures of PUM-HDs in complex with RNA reveal modular RNA recognition by eight PUM repeats arranged in a crescent shape (Lu & Hall, 2011; Miller, Higgin, & Hall, 2008; Wang et al., 2002; Wang, Opperman, Wickens, & Hall, 2009; Wilinski et al., 2015; Zhu, Stumpf, Krahn, Wickens, & Hall, 2009). Typically, one PUM repeat comprising three α helices uses three amino acid residues in the second α helix to contact one RNA base in a sequence-specific manner. PUF proteins bind to PREs usually found in the 3′-UTR of target mRNAs and recruit partner proteins leading to repression of protein expression and mRNA degradation (Miller & Olivas, 2011). PUF proteins function as broad-spectrum regulators in higher eukaryotes for mRNAs involved in embryonic, germline and neuronal development (Gerber, Luschnig, Krasnow, Brown, & Herschlag, 2006; Goldstrohm, Hall, & McKenney, 2018; Kershner & Kimble, 2010).

A classic example of PUF-mediated regulation is demonstrated by the essential role of *Drosophila* Pum in establishing the embryonic body pattern through repression of maternal *hunchback* (*hb*) mRNA (Forbes & Lehmann, 1998; Lehmann & Nusslein-Volhard, 1987). Pum protein and *hb* mRNA are distributed throughout the *Drosophila* embryo (Macdonald, 1992). Nos protein forms a gradient, emanating from the posterior to the anterior end (Barker, Wang, Moore, Dickinson, & Lehmann, 1992; Wang & Lehmann, 1991). At the posterior, *hb* mRNA is repressed by Pum in concert with Nos, whereas at the anterior, Hb protein is produced due to the absence of Nos (Lehmann & Nusslein-Volhard, 1991; Murata & Wharton, 1995; Sonoda & Wharton, 1999). The proper spatial distribution of Hb protein is critical for proper body pattern formation: no abdominal segments are formed in embryos lacking *pum* or *nos*. Extensive genetic and biochemical studies have revealed the importance of two copies of Nos Response Elements (NREs) in the 3′-UTR of *hb* mRNA (Wharton & Struhl, 1991). Both NRE1 and NRE2 contain two conserved elements termed Box A and Box B. The Box B sequence corresponds to a

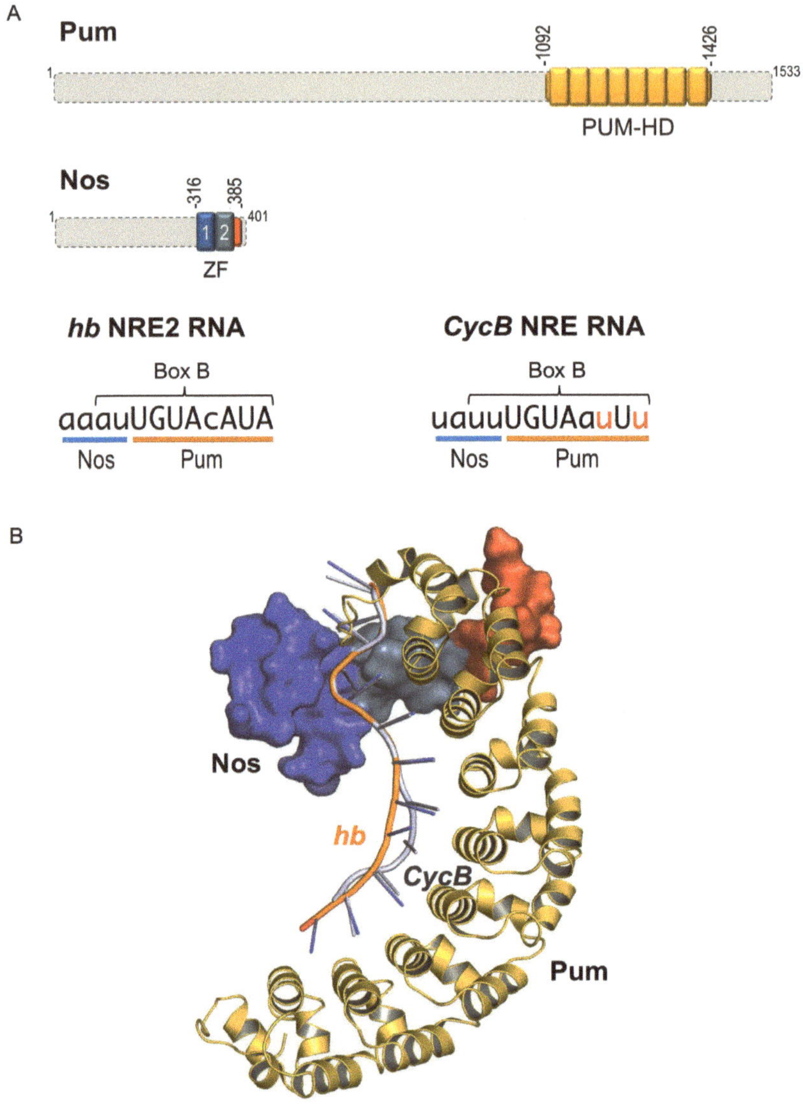

Fig. 1 Crystal structures of Pum-Nos-RNA ternary complexes. (A) Schematic drawings of Pum and Nos structural domains and RNA sequences. The eight repeats forming the PUM-HD RNA-binding domain are indicated by yellow rectangles, and the two ZF domains of Nos are represented by blue rectangles. The C-terminal tail of Nos is colored red. NRE RNA sequences bound by Pum and Nos are underlined orange and blue, respectively. The PRE consensus sequences are shown in black uppercase letters at highly conserved positions and in the *CycB* sequence, nucleotides that do not match the consensus are red. The Nos binding sequence is A/U-rich and shown in lowercase. (B) Superposition of crystal structures of Pum and Nos in complex with *hb* (orange, PDB ID: 5KL1) and *CycB* (light blue, PDB ID: 5KL8) RNAs. The two RNAs are shown with the protein structures from the Pum-Nos-*hb* RNA complex. Pum is shown as a ribbon diagram, and Nos is shown as a surface representation. The PUM-HD and ZF domains are colored as in the schematic drawings.

PRE (5′-UGUAHAUA, H=C/U), which is bound by Pum. RNA-bound Pum recruits Nos, which binds RNA upstream of the PRE *via* its C-terminal zinc-finger (ZF) domain (Fig. 1A) (Curtis et al., 1997). Nos subsequently recruits the CNOT deadenylase complex through its N-terminal Nos effector domain, resulting in deadenylation and decapping of the mRNA (Raisch et al., 2016). Box A has been shown recently to be a binding site for a third mRBP called Brain Tumor (Brat), which also synergistically inhibits *hb* mRNA translation (Arvola, Weidmann, Tanaka Hall, & Goldstrohm, 2017; Loedige et al., 2014; Sonoda & Wharton, 2001).

Our crystal structures of Pum-Nos-NRE ternary complexes illuminated the mechanisms of Pum-Nos cooperativity at the atomic level (Fig. 1B) (Weidmann et al., 2016). Within the ternary complex, Pum recognizes the PRE sequence in a manner similar to that also observed in our Pum-PRE binary complex. Additionally, Nos embraces both Pum and RNA, effectively clamping them together. The Nos tandem ZFs bind three nucleotides immediately upstream of the PRE, and the Nos C-terminal tail after the ZF domain interacts with a loop region between repeats seven and eight of the PUM-HD. Nos alone does not bind the NRE in the absence of Pum. Nevertheless, Nos enhances Pum binding to RNA. Another known Pum-Nos target *Cyclin* B (*CycB*) mRNA contains an imperfect PRE: 5′-UGUAAUUU (Asaoka-Taguchi, Yamada, Nakamura, Hanyu, & Kobayashi, 1999) (Fig. 1A). Pum alone does not form a stable complex with the *CycB* mRNA element, but addition of Nos results in a stable ternary complex. A crystal structure of a ternary complex of Pum and Nos with the *CycB* RNA showed how the imperfect PRE sequence is tolerated to allow Pum binding (Fig. 1B). We also used SEQRS (*in vitro* selection, high-throughput sequencing of RNA, and sequence specificity landscapes) analysis, which revealed that Nos alters the binding specificity of Pum. A 4-nt Nos-binding site (NBS) with A/U rich sequences was identified upstream of the 5′-UGUA of the PRE, while sequence selection at the 3′-end of the PRE became relaxed.

2. Expression and purification of Pum-Nos-RNA ternary complexes

2.1 Expression system

cDNAs encoding *Drosophila* Pum (residues 1091–1426) and Nos (289–401) were subcloned into the pSMT3 vector, and plasmids were transformed into *E. coli* strain BL21-CodonPlus (DE3)-RIL (Agilent). The pSMT3 vector (kindly provided by Christopher Lima, Sloan Kettering Institute) has become our first choice when working with a new protein (Mossessova & Lima, 2000). The N-terminal fused His_6-SUMO tag improves protein solubility and is cleavable by the SUMO protease, Ulp1. Ulp1 recognizes the tertiary structure of the ubiquitin-like protein SUMO rather than an amino acid sequence so the cleavage is highly specific. When the BamHI restriction site is used for the 5′ end cloning, only a serine residue is left at the N-terminus of the expressed protein after Ulp1 cleavage. The cleaved His_6-SUMO tag runs at ~17 kDa on SDS-PAGE with a calculated MW of 13.7 kDa. The Pum and Nos proteins were purified using the protocol below (Fig. 2). Steps 1–11 are the same for Pum and Nos and can be performed in parallel.

For this work, we focused on the structures of truncated Pum and Nos because the full-length proteins expressed poorly. The optimal truncated proteins were determined in part by analysis of sequence homology, specifically the well-conserved PUM-HD of Pum and the tandem ZFs of Nos. In addition, functional assays for RNA-binding and translational repression were employed to delineate the regions of Pum and Nos necessary for RNA recognition and synergistic mRNA regulation.

2.2 Protein expression

1. On day 1, inoculate a small volume (10–50 mL) culture of Luria-Bertani (LB) medium containing 50 μg/mL of chloramphenicol and 50 μg/mL kanamycin with *E. coli* from freshly transformed colonies or frozen glycerol stock. Inclusion of chloramphenicol maintains the pACYC plasmid containing extra copies of tRNA genes in the CodonPlus strain. The culture should contain at least 5 mL for each liter of expression culture. Grow overnight at 37 °C with shaking at 220–250 rpm.
2. The next morning, day 2, add 5 mL overnight culture to inoculate a 1 L culture of LB medium containing 50 μg/mL kanamycin. Include 0.5 mM $Zn(OAc)_2$ in the media for Nos expression since Nos contains tandem ZFs. Incubate cultures at 37 °C with shaking at 220–250 rpm until the OD_{600} reaches ~0.6. Lower the incubator temperature to 16 °C. Induce protein expression with 0.2 mM IPTG and incubate the culture overnight at 16 °C with shaking.
3. On day 3, spin down the cultures. Wash the cell pellets with PBS and store the pellets at −20 °C for later use or continue the purification.

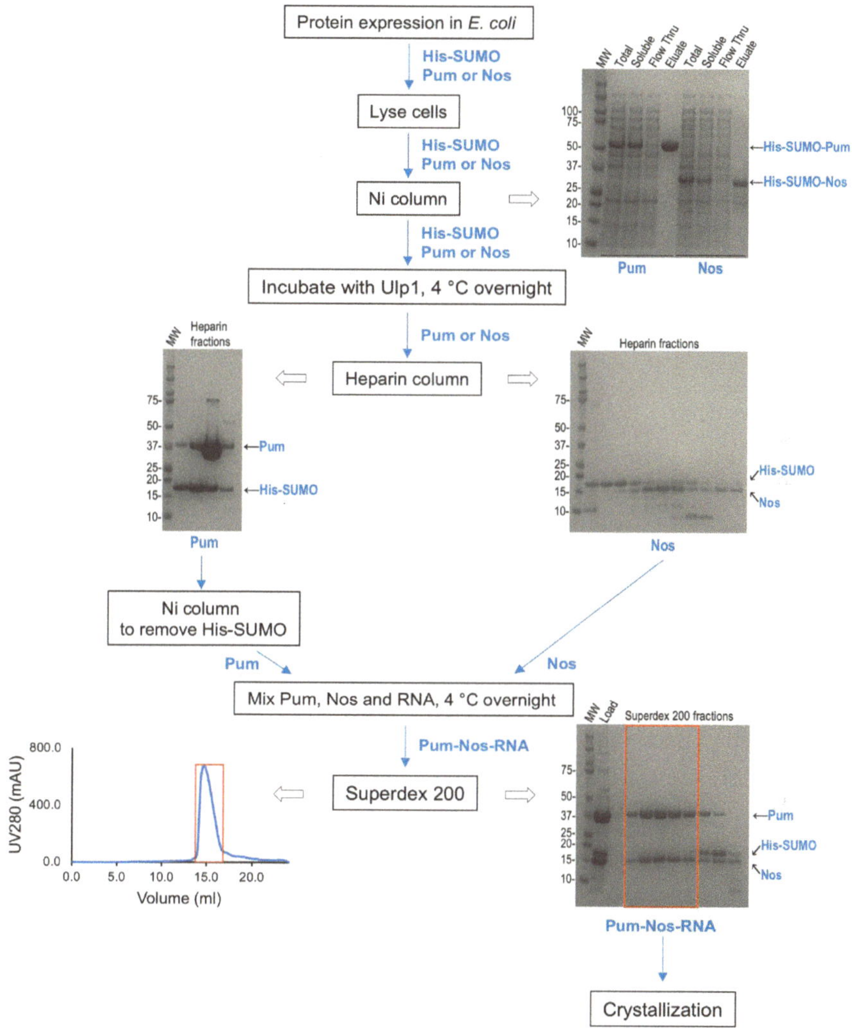

Fig. 2 Flowchart of protein purification and Pum-Nos-RNA ternary complex formation. Coomassie-stained SDS-PAGE gels showing samples at different stages of complex preparation are shown. In the top gel, the lanes are aliquots of total cell lysate (Total), soluble supernatant fraction (Soluble), remaining proteins flowing through the Ni column (Flow Thru), and imidazole eluate from the Ni column (Eluate). A chromatogram of the Superdex 200 gel filtration purification of the Pum-Nos-RNA ternary complex is shown, and the pooled peak fractions are indicated by the red box. Fractions from this peak are shown in the bottom SDS-PAGE gel.

2.3 Protein purification

4. Resuspend the cell pellet from 1 L culture in 40 mL pre-chilled lysis buffer containing 20 mM Tris (pH 8.0), 0.5 M NaCl, 20 mM imidazole, 5% (v/v) glycerol, and 0.1% (v/v) β-mercaptoethanol. If the cell pellet is freshly prepared, do a freeze-thaw to improve cell lysis efficiency by sonication.
5. Sonicate to disrupt the cells for 30 s at 60% amplitude, cycling 1 s on and 2 s off. Repeat three times. Spin the lysate at 28,000 × *g* for 35 min to pellet and remove the insoluble material. Save a small aliquot (~50 μL) of the total cell lysate and supernatant fraction for SDS-PAGE analysis (Fig. 2).
6. Mix the supernatant with 5 mL Ni-NTA resin (Qiagen) pre-equilibrated with lysis buffer in a 50-mL Falcon tube and rotate for 1 h at 4 °C to bind His-tagged protein. To prepare the resin, we transfer 10 mL of the 50% slurry to a Bio-Rad Econo-Pac column, elute the storage solution, wash the resin with lysis buffer, and transfer the resuspended resin to the Falcon tube.
7. After protein binding, spin the lysate and resin mixture at 500 × *g* for 5 min at 4 °C. Remove the supernatant containing unbound proteins. Save a small aliquot (~50 μL) of the supernatant solution for SDS-PAGE analysis (Fig. 2). Add 50 mL lysis buffer to the tube and transfer the resuspended resin to a gravity column. We use Bio-Rad Econo-Pac columns.
8. Wash the resin with lysis buffer until no protein is detected in the washes. Usually, 200–300 mL wash is sufficient. We monitor this by performing a rapid protein detection assay. We mix 9 μL of wash elution with 50 μL diluted Bio-Rad Protein Assay Dye Reagent (one part dye concentrate to four parts distilled water). The yellow dye solution turns blue in the presence of protein. The high NaCl concentration in the lysis and wash buffer helps to reduce nucleic acid contamination.
9. Elute protein with a buffer containing 20 mM Tris (pH 8.0), 50 mM NaCl, 0.2 M imidazole (a 2 M imidazole stock solution is pH adjusted to 8.0), and 1 mM DTT. Monitor by rapid protein detection assay and stop eluting when the dye solution no longer changes color. Usually, 50 mL eluent is collected. Note the NaCl concentration is reduced from 500 mM in the lysis and wash buffer to 50 mM in the elution buffer. Using the low-salt elution buffer avoids dialysis or dilution to prepare the sample for subsequent ion-exchange or heparin-affinity

chromatography. Imidazole has no effect on protein binding to a heparin or Q column. It is typically good practice to check the purity of the eluted protein and the progression of the purification by SDS-PAGE (Fig. 2).

10. To cleave the fusion protein, add 100 μL Ulp1 (1 mg/mL) to the eluent. Incubate at 4 °C overnight. We obtained the Ulp1 expression plasmid from Dr. Christopher Lima at Sloan Kettering Institute. The His-tagged Ulp1 was purified by one-step Ni-chelating column. One liter of *E. coli* culture yields >20 mg Ulp1. The Ulp1 protease is aliquoted and stored at −80 °C. The incubation time for Ulp1 cleavage varies for different proteins. Some fusion proteins can be completely cleaved within 1–2 h at 4 °C. For those proteins, it is better to incubationincubate for 2 h and then move to the next purification step on the same day. Some proteins require more cleavage time, such as overnight incubation; however, we have had proteins precipitate during overnight Ulp1 cleavage reactions. This step should be assessed by SDS-PAGE the first time protein is prepared and cleaved.

11. Filter the protein solution using Millex-GP syringe filters with 0.22 μm pore size. Apply the protein solution to a 5-mL Hi-Trap Heparin column (GE Healthcare) equilibrated in Buffer A containing 20 mM Tris (pH 8.0) and 1 mM DTT. Elute the protein with a linear NaCl gradient from 5% to 100% Buffer B containing 20 mM Tris (pH 8.0), 1 mM DTT, and 1 M NaCl. Pum co-elutes with His-SUMO in a single peak at 260 mM NaCl from the Heparin column (Fig. 2). Nos separates from the majority of His-SUMO. However, an SDS-PAGE gel of the Nos preparation shows some faster migrating fragments that are likely to be degradation products (Fig. 2).

12. To remove contaminating His-SUMO from Pum, pooled peak fractions are applied to 1 mL Ni-NTA resin. His-SUMO binds to the Ni-NTA resin and Pum flows through.

13. Concentrate purified Pum (after removal of His-SUMO) and Nos (eluted from the Heparin column) separately using Amicon Ultra Centrifugal Filter Units (EMD Millipore), 30,000 and 10,000 NMWL (nominal molecular weight limit), respectively. Measure the OD_{280} with a NanoDrop or other spectrophotometer. Stop concentrating when the OD_{280} is in the range of 1–2. Determine the protein concentration using a calculated protein extinction coefficient based on the numbers of tryptophan and tyrosine residues. The online tool

ProtParam in ExPASy can be used to calculate the extinction coefficient (Gasteiger et al., 2005).

14. One liter of bacterial culture yields about 30 mg of Pum or 1 mg of Nos. The yield of Nos is lower, and we do not typically have excess protein to store. The concentrated Pum protein that is not used immediately can be flash frozen in liquid nitrogen and stored at −80 °C. After thawing the protein, a gel filtration column will prepare the protein for use.

2.4 Formation and purification of a Pum-Nos-RNA complex

15. A 16-nt *hb* NRE2 RNA (5′-AAAUUGUACAUAAGCC) and a 14-nt *CycB* NRE RNA (5′-UAUUUGUAAUUUAU) were custom synthesized by Dharmacon. Deprotect the RNAs as directed using the supplied buffer and dry using a SpeedVac. Resuspend the pellet in a buffer of 10 mM Tris, pH 7.5 and 1 mM EDTA. The RNA concentration is determined by OD_{260} using a NanoDrop or other spectrophotometer (for a 1 cm pathlength, the $OD_{260} = 1$ for a solution of 40 μg/mL RNA). RNA is stored at −20 °C.

16. Mix Pum, Nos, and RNA in a molar ratio of approximately 1:1.1:1.1 and incubate at 4 °C overnight. Since a minor amount of the Nos protein was partially degraded, the calculated protein concentration based on UV_{280} may not be accurate. To compensate, we included excess Nos in the binding reaction. Any unbound Nos protein will be separated, along with excess RNA, in the subsequent gel filtration step. With OD_{280} values of 1–2 for the two proteins, the concentration of the complex is ~100 μM.

17. Run the assembled complex mixture on a Superdex 200 10/300 GL column (GE Healthcare) equilibrated in 20 mM Hepes (pH 7.4), 0.15 M NaCl, and 2 mM DTT. The maximum sample loading volume for the 10/300 GL column is 500 μL. If the sample volume is >500 μL, concentrate using an Amicon Ultra Centrifugal Filter Unit with a 30,000 NMWL. In this case, the molecular weight of the ternary complex of Pum-Nos-RNA is about 54 kDa. The extra Nos protein with a MW of 12 kDa, including the degraded products that lack the C-terminus and fail to interact with Pum, can pass through the 30,000 NMWL filter membrane. Thus the concentration step can contribute to purification of the ternary complex. Always centrifuge the sample in a microcentrifuge at maximum speed (~16,000 × *g*) for

10 min to remove possible precipitation before loading onto the column.

18. The Pum:Nos:RNA complex elutes from the Superdex 200 10/300 GL column at 14.7 mL. Pool the peak fractions and concentrate to OD_{280} ~7. The ratio of OD_{260}/OD_{280} is about 1.7, indicating the presence of RNA. Spin down the sample in a microcentrifuge at maximum speed (~16,000 × *g*) for 10 min. Set up the crystallization experiment immediately.

3. Crystallization and structure determination of a Pum-Nos-RNA complex

3.1 Crystallization

3.1.1 Crystallization screening and optimization

Initial crystallization screening was carried out using commercial sparse-matrix screens. 250 nL sample was mixed with 250 nL reservoir solution in a 96-well sitting drop plate using a Mosquito® robot (TTP Labtech). Crystals were observed in multiple conditions with $(NH_4)_2SO_4$ for the sample of Pum-Nos-*hb* RNA. A grid screen of varying $(NH_4)_2SO_4$ concentration and pH was performed in a 24-well hanging-drop vapor diffusion format. Single crystals were grown in the condition of 1.1 M $(NH_4)_2SO_4$, 0.1 M MES, pH 5.6.

During co-crystallization of a complex of multiple macromolecules (with more than one protein and/or with RNA), it is always unknown whether the crystals contain all of the components. Verification that crystals have the desired content at an early stage can avoid the waste of time and effort in optimizing hits with some missing components. One way to determine the content of crystals is to analyze them on a polyacrylamide gel. Crystals are retrieved and washed carefully before dissolving in water or directly in gel-loading buffer. The presence of RNA can be checked by TBE-urea gel, staining with an RNA dye such as methylene blue. Alternatively, diffraction data can be collected if the initial crystals are of diffraction quality. The unit cell content of the crystal can be analyzed for probable protein residue number in terms of the Matthews coefficient. With the structures of one or two components available, one can perform molecular replacement to see whether there is extra density for the third component even with low-resolution data. After the protein-RNA complex composition is confirmed, further optimization can be carried out to obtain crystals diffracting to higher resolution.

3.1.2 Crystallization of the Pum-Nos-CycB RNA complex

Crystals of Pum-Nos-*CycB* RNA were obtained in the condition of 0.9 M $(NH_4)_2SO_4$, 0.1 M MES, pH 5.6 by microseeding with the Pum-Nos-*hb* RNA complex crystals.

1. Transfer a few Pum-Nos-*hb* RNA crystals to a 1.5-mL microcentrifuge tube with a seed bead (Hampton Research) and 50 μL of 1.1 M $(NH_4)_2SO_4$, 0.1 M MES, pH 5.6. Vortex the tube for 2 min.
2. Make 10-fold serial dilutions of the seed stock with 1.1 M $(NH_4)_2SO_4$, 0.1 M MES, pH 5.6; (1, 10^{-1}, 10^{-2}, 10^{-3}, 10^{-4}).
3. Set up a 24-well plate with varying concentrations of $(NH_4)_2SO_4$ from 0.8 to 1.2 M in the reservoir. Pipet 2 μL Pum-Nos-*CycB* RNA sample and 2 μL diluted seed stock on a cover slide for hanging drop vapor diffusion crystallization.

3.1.3 Crystallization of the Pum-RNA complex

To address the question whether Nos induces any structural change in Pum, we also grew crystals of the Pum-RNA binary complex.

1. Thaw a tube of frozen Pum protein from step 13 in Section 2.4 on ice. Centrifuge the protein at 16,000 × *g* for 10 min to remove any precipitation.
2. Run a Superdex 75 10/300 GL column (GE Healthcare) with buffer of 20 mM Hepes (pH 7.4), 0.15 M NaCl, and 2 mM DTT. Pool the peak fractions and concentrate to OD_{280} around 3–4.
3. Mix Pum protein with RNA at a molar ratio of 1:1.2. Keep on ice for 2 h. We usually make the RNA stock concentration 0.5–1 mM, so the protein will not be too dilute after mixing with RNA.
4. Centrifuge the sample at 16,000 × *g* for 10 min. Set up crystallization screens.

PUF proteins bind to short single-stranded RNAs. Different from the ternary complex preparation, purification of the PUF-RNA binary complex by a gel filtration column is not necessary. One advantage is that a variety of RNA oligonucleotides can be screened simply by mixing purified protein with different RNA. Indeed the sequence and length of RNA affect the outcome of crystallization. In this case, two RNAs were used for screening: an 8-nt PRE sequence (5′-UGUACAUA) and a 12-nt sequence including 4 nt upstream of the PRE (5′-AAAUUGUACAUA). Only the 8-nt RNA produced crystals, while the 12-nt did not crystallize. The initial crystals of the Pum-RNA complex appeared in clusters of rods. To grow single crystals, the microseeding method was used as described above. The optimized condition contained 15% (*w*/*v*) PEG 3350, 0.1 M bis-tris, pH 6.5, and 0.2 M NH_4OAc.

3.2 Structure determination

Crystals were transferred to a cryoprotective-solution containing the crystallization solution with an additional 20% (v/v) glycerol and flash frozen in liquid nitrogen. X-ray diffraction data were collected at the SER-CAT Beamline 22-ID at the Advanced Photon Source, Argonne National Laboratories.

3.2.1 The Pum-Nos-RNA ternary complex

Crystals of the Pum-Nos-*hb* RNA complex belong to the $P6_522$ space group with unit cell dimensions: $a = b = 137.0$ Å, $c = 221.4$ Å, $\alpha = \beta = 90°$, $\gamma = 120°$. There is one complex in an asymmetric unit, corresponding to 78% solvent and a Matthews coefficient of 5.7. The high solvent content correlates with the low-resolution diffraction limit of the crystals. The best crystal diffracted to 3.7 Å Bragg spacings, although with a relatively large crystal size of $0.2 \times 0.2 \times 0.4$ mm. Crystal dehydration has been used as a post-crystallization treatment to remove excess solvent, tighten crystal packing, and thus improve diffraction. We tried dehydration by increasing the precipitant concentration in either the drop or the reservoir from 1.1 M $(NH_4)_2SO_4$ to 2.4 M and equilibrating overnight or for 4 days. However, crystal diffraction did not improve.

The dataset was processed and scaled using HKL2000 (Otwinowski & Minor, 1997). The structures of *Drosophila* Pum (PDB: 3H3D) (Edwards, Pyle, Wharton, & Aggarwal, 2001) and zebrafish Nos ZF domain (PDB: 3ALR) (Hashimoto et al., 2010) were used as the search model for molecular replacement with Phenix Phaser (Adams et al., 2010). One round of Phenix. refine revealed the density of the bound RNA. Iterative rounds of manual model building with COOT (Emsley & Cowtan, 2004) and refinement with Phenix led to the final structure of Pum-Nos-*hb* RNA with $R_{work} = 0.264$ and $R_{free} = 0.30$. In the structure only residues 316–385 of Nos were modeled with visible density, while the Nos construct contained residues 289–401. More than 20 residues at both the N- and C-termini were disordered. Removing disordered regions from protein constructs often increases the likelihood of crystallization and improves crystal quality. The Nos protein sequence contains poly-serine and poly-asparagine in the region preceding the first ZF: $_{292}$SNSSNNNNNNN$_{302}$. We truncated the Nos construct to 307–401. However, the shortened Nos protein aggregated. Taken together, the 3.7 Å structure of the Pum-Nos-*hb* RNA ternary complex was the best we could achieve. Despite the modest resolution, the structure provides molecular insights into the Pum-Nos cooperative RNA recognition.

Crystals of the Pum-Nos-*CycB* RNA complex diffracted even more weakly than the *hb* crystals, with the resolution of 4.0 Å. The structure was refined with the high-resolution Pum protein structure as the reference model.

3.2.2 The Pum-RNA binary complex

Crystals of the Pum-RNA binary complex belong to the C2 space group with unit cell dimensions: $a = 194.9$ Å, $b = 29.5$ Å, $c = 62.0$ Å, $\alpha = \gamma = 90°$, $\beta = 101.2°$. There is one complex in an asymmetric unit, corresponding to 44% solvent and a Matthews coefficient of 2.2. The crystal diffracted to 1.14 Å resolution. The structure was determined by molecular replacement using the *Drosophila* Pum structure (PDB: 3H3D) as the search model.

4. Analysis of protein:RNA crystal structures

With a beautiful crystal structure of a protein:RNA complex in hand, it is satisfying to admire the outcome of weeks, months, or perhaps years of work. Yet harvesting the fruit of that labor lies in the in-depth analyses of the crystal structure. It is helpful to make a list of all the questions that you have and what to look for in the structures. Some common questions for all structures:

1. What is the overall fold of the protein? Is it similar to other proteins? What features are conserved? What are unique features?
2. Using the structure to make a structure-based sequence alignment, are the typical functional residues conserved? What are the roles of well-conserved amino acid residues?
3. Does the structure of the protein change upon ligand binding? What might be the importance of conformational changes for function?
4. Do observed protein-RNA contacts provide information about RNA-binding sequence or structure specificity?

4.1 Characterizing protein:RNA interactions

Of course, one of the first things we do when we have a new protein:RNA structure is to look at all of the protein:RNA contacts. One can generate a list of sets of protein atoms and solvent molecules that are within ~3.5 Å of the RNA. However, for structures like our PUF protein:RNA complexes, where the RNA is of a manageable length, it is best to look at the protein: RNA interactions manually. We typically follow along the RNA and look at the protein and solvent atoms that are within ~3.5 Å of the RNA and determine whether the interaction is chemically reasonable. This practice is also good for quality control of the structure. We use MolProbity (Davis et al., 2007) to assess our structure, and sometimes we find an Arg/Gln/His side chain flip that produces better contacts with the RNA. We usually draw a diagram that illustrates the contacts. We classify interactions into those that appear to be sequence specific because they interact with the RNA bases and those that interact with the RNA backbone. We examine which bases are recognized specifically and which are not. We also determine whether RNA structures are recognized, such as a set of RNA backbone interactions that can only occur if the RNA folds in a particular way. MolProbity's characterization of the RNA structure is also helpful for identifying RNA structural elements.

In the specific case of the Pum:Nos:RNA and Pum:RNA complexes, we examined interactions of both Pum and Nos with the *hb* RNA and noted differences in interactions with the *CycB* RNA. Our ternary complex structures were of modest resolution, so we were limited in our ability to identify water-mediated interactions. However, our Pum:*hb* NRE RNA binary complex structure was of very high resolution, allowing us to use this structure to provide details.

*4.1.1 Pum-*hb *RNA interactions*

We found that Pum recognizes the PRE within the *hb* NRE RNA similar to what had been observed in previous Pumilio:RNA complexes.

a. Pum uses characteristic tripartite recognition motifs (TRMs) to bind to specific RNA bases using the PUF RNA recognition code: NQ/Y for uracil, SQ/R or CQ/R for adenine, and SE/N for guanine.

b. Pum did not specifically recognize the nucleotide at position 5 (+5C), instead this base directly stacked with the 4th base. We had previously identified this type of RNA recognition, the "base omission" recognition mode, for some human Pum protein and RNA combinations (Lu & Hall, 2011).

c. Recognition of the *hb* RNA PRE by Pum was the same in the binary and ternary complexes.

*4.1.2 Pum-*CycB *RNA interactions*

We found that the interactions of Pum with the *CycB* RNA included 5′ interactions that were similar to those with the *hb* RNA and 3′ interactions that were distinct (Fig. 1B).

a. In our structures of ternary complexes of Pum and Nos with *hb* or *CycB* RNAs, the structure and recognition of the common 5′-UGUA sequence was the same.

b. In contrast, the 3′ divergent sequence of the *CycB* NRE RNA required adjustments in the RNA structure for Pum to bind to "mismatched" +5A and +6U bases. The +8U base of CycB RNA was not resolved, but interaction with +7U was the same for *hb* and *CycB* RNAs.

4.1.3 Nos-RNA interactions

We found that Nos interacted with three nucleotides immediately upstream of the PRE in the *hb* and *CycB* RNAs (Fig. 3).

a. Nos forms a binding pocket for the −1U nucleotide.

b. Nos binds to the −2A and −3A nucleotides on the surface of the two zinc fingers.

4.1.4 Pum protein conformational changes

We also discovered changes in the Pum structure with formation of the ternary complex that resulted in additional RNA interaction. If multiple structures of a protein are available, aligning the structures and assessing structural changes may turn up interesting differences. A previous structure of Pum alone required truncation of the C-terminus for crystallization, and in the crystal lattice, the C-terminal regions of two independent Pum molecules interacted (Edwards et al., 2001). Because of this truncation, the Pum protein in our binary complex contained 22 additional residues that formed two α helices at the C-terminus. In the presence of Nos, the final α helix unfolded and the peptide formed part of the RNA interaction surface in the ternary complex.

4.2 Characterizing protein:protein interactions

The crystal structures of the ternary complexes revealed crucial interactions between the C-terminal tail of Nos and the non-RNA-binding surface of Pum (Fig. 3). Just as we examined the structures for all contacts between the protein and RNA, we also looked at interactions between the two proteins. We looked specifically at the roles of residues that had previously been shown to be important for Pum and Nos function. A *nos* mutant with a strong phenotype resulted from a deletion of seven residues in the C-terminal region of Nos (Arrizabalaga & Lehmann, 1999), and our structures showed that the deleted region includes residues that interact with

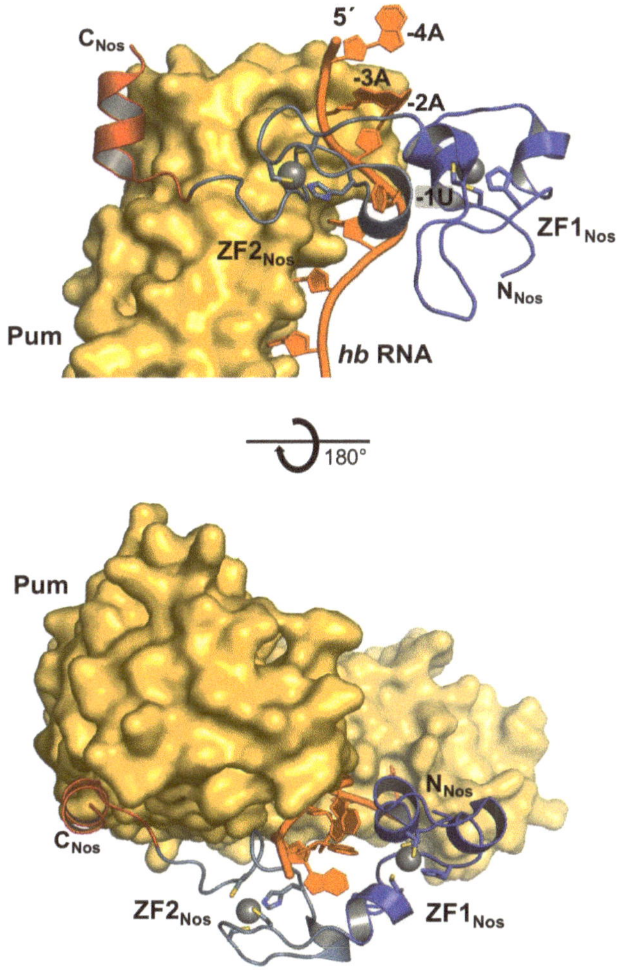

Fig. 3 Nos clasps the RNA nucleotides 5′ of the Pum binding site with its ZF domains and wraps around Pum with its C-terminal tail. Two orientations of the Pum-Nos-*hb* RNA ternary complex are shown. Pum (yellow) is shown as a surface representation, and Nos is shown as a ribbon diagram. The ZFs and C-terminal tail are colored as in Fig. 1.

Pum. In addition, our structures revealed that Nos interacts with Pum between PUM repeats seven and eight. This finding indicates why an insertion in this loop disrupted Pum function and why Nos could only interact with human Pum1 when three residues in the equivalent loop are deleted (Sonoda & Wharton, 1999). Other mutations that disrupt complex formation were also found at this interface (Wharton, Sonoda, Lee, Patterson, & Murata, 1998).

4.3 Methods to evaluate structure-driven hypotheses

We took advantage of several different methods to evaluate complex formation, translational repression, and RNA sequence specificity (Weidmann et al., 2016).

a. Electrophoretic mobility shift assays (EMSAs) were used to assess complex formation (Black, Chan, Min, Wang, & Bell, 1998; Ryder, Recht, & Williamson, 2008). Although EMSAs have their limitations as a non-equilibrium method, they have the advantage of visualizing unbound RNAs and distinct complexes. The ability to observe a supershifted RNA band when both Pum and Nos bound to the PRE RNA was helpful. We could manipulate the two proteins or RNA using this *in vitro* method. This approach revealed critical elements for complex formation and clearly demonstrated that Nos enables binding of Pum to the *CycB* RNA.

b. Cell-based reporter assays to assess translational repression were used to determine the effects of Nos variations (Van Etten, Schagat, & Goldstrohm, 2013). We took advantage of the fact that *Drosophila* D. mel-2 cells do not express Nos and the level of Pum expression is not sufficient to repress a reporter mRNA in the absence of exogenous Nos. Since we relied upon the endogenous Pum, we could alter the reporter RNA sequence or Nos protein.

c. An *in vitro* selection technique called SEQRS (Campbell et al., 2012) was used to identify the consensus RNA motif for Pum alone or in the presence of Nos. This method is thoroughly described in a recent review (Lou et al., 2017). The strength of this method is in the ability to analyze the sequences as enrichment advances through rounds of selection.

5. Concluding remarks

The breadth of information about RNA-binding proteins and their motifs continues to expand with the development of novel genomics methods. These data provide the opportunity to explore the range of RNA recognition strategies by mRBPs. X-ray crystallography remains an effective method to determine the molecular mechanisms of protein-RNA interaction. It provides high-resolution information and is particularly well suited for medium-to-large sized complexes (50–200 kDa). Collaborative mRBP complexes have been identified more slowly. Often they begin with genetic studies that indicate regulation of a common pathway followed by careful biochemical studies to firm up the relationships. The story of Pum and Nos cooperative regulation arose from *Drosophila* genetics and developed through a series of studies by multiple groups. The crystal structures of Pum-Nos-RNA complexes have added a satisfying picture to illustrate new principles of RNA recognition by mRBPs working together to dynamically control gene expression.

Acknowledgments

We thank members of the Goldstrohm lab and Zachary Campbell's lab for their work on other biochemical and cell-based assays that completed the evaluation of the structural findings. We are grateful to our colleagues, Robin Stanley and Huanchen Wang, for their critical comments on the manuscript. We thank Lars Pedersen and the staff of the Southeast Regional Collaborative Access Team beamlines for assistance with X-ray data collection that led to the structures described. This work was supported in part by National Institute of General Medical Sciences, National Institutes of Health grant R01GM105707 (A.C.G.) and the Intramural Research Program of the National Institutes of Health, National Institute of Environmental Health Sciences (T.M.T.H.).

References

Adams, P. D., Afonine, P. V., Bunkoczi, G., Chen, V. B., Davis, I. W., Echols, N., et al. (2010). PHENIX: A comprehensive python-based system for macromolecular structure solution. *Acta Crystallographica. Section D, Biological Crystallography*, *66*(Pt. 2), 213–221. https://doi.org/10.1107/S0907444909052925.

Arrizabalaga, G., & Lehmann, R. (1999). A selective screen reveals discrete functional domains in Drosophila Nanos. *Genetics*, *153*(4), 1825–1838.

Arvola, R. M., Weidmann, C. A., Tanaka Hall, T. M., & Goldstrohm, A. C. (2017). Combinatorial control of messenger RNAs by Pumilio, Nanos and brain tumor proteins. *RNA Biology*, *14*(11), 1445–1456. https://doi.org/10.1080/15476286.2017.1306168.

Asaoka-Taguchi, M., Yamada, M., Nakamura, A., Hanyu, K., & Kobayashi, S. (1999). Maternal Pumilio acts together with Nanos in germline development in Drosophila embryos. *Nature Cell Biology*, *1*(7), 431–437. https://doi.org/10.1038/15666.

Barker, D. D., Wang, C., Moore, J., Dickinson, L. K., & Lehmann, R. (1992). Pumilio is essential for function but not for distribution of the Drosophila abdominal determinant Nanos. *Genes & Development*, *6*(12A), 2312–2326.

Black, D. L., Chan, R., Min, H., Wang, J., & Bell, L. (1998). The electrophoretic mobility shift assay for RNA binding proteins. In *RNA: Protein interactions* (pp. 109–136). Oxford University Press.

Campbell, Z. T., Bhimsaria, D., Valley, C. T., Rodriguez-Martinez, J. A., Menichelli, E., Williamson, J. R., et al. (2012). Cooperativity in RNA-protein interactions: Global analysis of RNA binding specificity. *Cell Reports*, *1*(5), 570–581. https://doi.org/10.1016/j.celrep.2012.04.003.

Curtis, D., Treiber, D. K., Tao, F., Zamore, P. D., Williamson, J. R., & Lehmann, R. (1997). A CCHC metal-binding domain in Nanos is essential for translational regulation. *EMBO Journal*, *16*(4), 834–843. https://doi.org/10.1093/emboj/16.4.834.

Davis, I. W., Leaver-Fay, A., Chen, V. B., Block, J. N., Kapral, G. J., Wang, X., et al. (2007). MolProbity: All-atom contacts and structure validation for proteins and nucleic acids.

Nucleic Acids Research, 35(Web Server issue), W375–W383. https://doi.org/10.1093/nar/gkm216.

Edwards, T. A., Pyle, S. E., Wharton, R. P., & Aggarwal, A. K. (2001). Structure of Pumilio reveals similarity between RNA and peptide binding motifs. *Cell, 105*(2), 281–289.

Emsley, P., & Cowtan, K. (2004). Coot: Model-building tools for molecular graphics. *Acta Crystallographica. Section D, Biological Crystallography, 60*, 2126–2132. https://doi.org/10.1107/S0907444904019158. Pt. 12 Pt. 1.

Forbes, A., & Lehmann, R. (1998). Nanos and Pumilio have critical roles in the development and function of Drosophila germline stem cells. *Development, 125*(4), 679–690.

Gasteiger, E., C, H., Gattiker, A., Duvaud, S., Wilkins, M. R., Appel, R. D., et al. (2005). Protein identification and analysis tools on the ExPASy server. In J. M. Walker (Ed.), *The proteomics protocols handbook* (pp. 571–607).

Gerber, A. P., Herschlag, D., & Brown, P. O. (2004). Extensive association of functionally and cytotopically related mRNAs with Puf family RNA-binding proteins in yeast. *PLoS Biology, 2*(3), E79. https://doi.org/10.1371/journal.pbio.0020079.

Gerber, A. P., Luschnig, S., Krasnow, M. A., Brown, P. O., & Herschlag, D. (2006). Genome-wide identification of mRNAs associated with the translational regulator PUMILIO in Drosophila melanogaster. *Proceedings of the National Academy of Sciences of the United States of America, 103*(12), 4487–4492. https://doi.org/10.1073/pnas.0509260103.

Goldstrohm, A. C., Hall, T. M. T., & McKenney, K. M. (2018). Post-transcriptional regulatory functions of Mammalian Pumilio proteins. *Trends in Genetics*. https://doi.org/10.1016/j.tig.2018.09.006.

Hashimoto, H., Hara, K., Hishiki, A., Kawaguchi, S., Shichijo, N., Nakamura, K., et al. (2010). Crystal structure of zinc-finger domain of Nanos and its functional implications. *EMBO Reports, 11*(11), 848–853. https://doi.org/10.1038/embor.2010.155.

Kershner, A. M., & Kimble, J. (2010). Genome-wide analysis of mRNA targets for Caenorhabditis elegans FBF, a conserved stem cell regulator. *Proceedings of the National Academy of Sciences of the United States of America, 107*(8), 3936–3941. https://doi.org/10.1073/pnas.1000495107.

Koh, Y. Y., Opperman, L., Stumpf, C., Mandan, A., Keles, S., & Wickens, M. (2009). A single C. elegans PUF protein binds RNA in multiple modes. *RNA, 15*(6), 1090–1099. https://doi.org/10.1261/rna.1545309.

Lehmann, R., & Nusslein-Volhard, C. (1987). hunchback, a gene required for segmentation of an anterior and posterior region of the Drosophila embryo. *Developmental Biology, 119*(2), 402–417.

Lehmann, R., & Nusslein-Volhard, C. (1991). The maternal gene nanos has a central role in posterior pattern formation of the Drosophila embryo. *Development, 112*(3), 679–691.

Loedige, I., Stotz, M., Qamar, S., Kramer, K., Hennig, J., Schubert, T., et al. (2014). The NHL domain of BRAT is an RNA-binding domain that directly contacts the hunchback mRNA for regulation. *Genes & Development, 28*(7), 749–764. https://doi.org/10.1101/gad.236513.113.

Lou, T. F., Weidmann, C. A., Killingsworth, J., Tanaka Hall, T. M., Goldstrohm, A. C., & Campbell, Z. T. (2017). Integrated analysis of RNA-binding protein complexes using in vitro selection and high-throughput sequencing and sequence specificity landscapes (SEQRS). *Methods, 118–119*, 171–181. https://doi.org/10.1016/j.ymeth.2016.10.001.

Lu, G., & Hall, T. M. (2011). Alternate modes of cognate RNA recognition by human PUMILIO proteins. *Structure, 19*(3), 361–367. https://doi.org/10.1016/j.str.2010.12.019.

Macdonald, P. M. (1992). The Drosophila pumilio gene: An unusually long transcription unit and an unusual protein. *Development, 114*(1), 221–232.

Miller, M. T., Higgin, J. J., & Hall, T. M. (2008). Basis of altered RNA-binding specificity by PUF proteins revealed by crystal structures of yeast Puf4p. *Nature Structural & Molecular Biology, 15*(4), 397–402. https://doi.org/10.1038/nsmb.1390.

Miller, M. A., & Olivas, W. M. (2011). Roles of Puf proteins in mRNA degradation and translation. *Wiley Interdisciplinary Reviews: RNA, 2*(4), 471–492. https://doi.org/10.1002/wrna.69.

Mitchell, S. F., & Parker, R. (2014). Principles and properties of eukaryotic mRNPs. *Molecular Cell, 54*(4), 547–558. https://doi.org/10.1016/j.molcel.2014.04.033.

Mossessova, E., & Lima, C. D. (2000). Ulp1-SUMO crystal structure and genetic analysis reveal conserved interactions and a regulatory element essential for cell growth in yeast. *Molecular Cell, 5*(5), 865–876.

Murata, Y., & Wharton, R. P. (1995). Binding of pumilio to maternal hunchback mRNA is required for posterior patterning in Drosophila embryos. *Cell, 80*(5), 747–756.

Otwinowski, Z., & Minor, W. (1997). Processing of X-ray diffraction data collected in oscillation mode. *Methods in Enzymology, 276*, 307–326. https://doi.org/10.1016/S0076-6879(97)76066-X [20].

Raisch, T., Bhandari, D., Sabath, K., Helms, S., Valkov, E., Weichenrieder, O., et al. (2016). Distinct modes of recruitment of the CCR4-NOT complex by Drosophila and vertebrate Nanos. *EMBO Journal, 35*(9), 974–990. https://doi.org/10.15252/embj.201593634.

Ryder, S. P., Recht, M. I., & Williamson, J. R. (2008). Quantitative analysis of protein-RNA interactions by gel mobility shift. *Methods in Molecular Biology, 488*, 99–115. https://doi.org/10.1007/978-1-60327-475-3_7.

Singh, G., Pratt, G., Yeo, G. W., & Moore, M. J. (2015). The clothes make the mRNA: Past and present trends in mRNP fashion. *Annual Review of Biochemistry, 84*, 325–354. https://doi.org/10.1146/annurev-biochem-080111-092106.

Sonoda, J., & Wharton, R. P. (1999). Recruitment of Nanos to hunchback mRNA by Pumilio. *Genes & Development, 13*(20), 2704–2712.

Sonoda, J., & Wharton, R. P. (2001). Drosophila brain tumor is a translational repressor. *Genes & Development, 15*(6), 762–773. https://doi.org/10.1101/gad.870801.

Van Etten, J., Schagat, T. L., & Goldstrohm, A. C. (2013). A guide to design and optimization of reporter assays for 3′ untranslated region mediated regulation of mammalian messenger RNAs. *Methods, 63*(2), 110–118. https://doi.org/10.1016/j.ymeth.2013.04.020.

Wang, C., & Lehmann, R. (1991). Nanos is the localized posterior determinant in Drosophila. *Cell, 66*(4), 637–647.

Wang, X., McLachlan, J., Zamore, P. D., & Hall, T. M. (2002). Modular recognition of RNA by a human Pumilio-homology domain. *Cell, 110*(4), 501–512.

Wang, Y., Opperman, L., Wickens, M., & Hall, T. M. (2009). Structural basis for specific recognition of multiple mRNA targets by a PUF regulatory protein. *Proceedings of the National Academy of Sciences of the United States of America, 106*(48), 20186–20191. https://doi.org/10.1073/pnas.0812076106.

Weidmann, C. A., Qiu, C., Arvola, R. M., Lou, T. F., Killingsworth, J., Campbell, Z. T., et al. (2016). Drosophila Nanos acts as a molecular clamp that modulates the RNA-binding and repression activities of Pumilio. *eLife, 5*. https://doi.org/10.7554/eLife.17096.

Wharton, R. P., Sonoda, J., Lee, T., Patterson, M., & Murata, Y. (1998). The Pumilio RNA-binding domain is also a translational regulator. *Molecular Cell, 1*(6), 863–872.

Wharton, R. P., & Struhl, G. (1991). RNA regulatory elements mediate control of Drosophila body pattern by the posterior morphogen nanos. *Cell, 67*(5), 955–967.

Wickens, M., Bernstein, D. S., Kimble, J., & Parker, R. (2002). A PUF family portrait: 3'UTR regulation as a way of life. *Trends in Genetics, 18*(3), 150–157.

Wilinski, D., Qiu, C., Lapointe, C. P., Nevil, M., Campbell, Z. T., Tanaka Hall, T. M., et al. (2015). RNA regulatory networks diversified through curvature of the PUF protein scaffold. *Nature Communications, 6*, 8213. https://doi.org/10.1038/ncomms9213.

Zamore, P. D., Williamson, J. R., & Lehmann, R. (1997). The Pumilio protein binds RNA through a conserved domain that defines a new class of RNA-binding proteins. *RNA*, *3*(12), 1421–1433.

Zhang, B., Gallegos, M., Puoti, A., Durkin, E., Fields, S., Kimble, J., et al. (1997). A conserved RNA-binding protein that regulates sexual fates in the C. elegans hermaphrodite germ line. *Nature*, *390*(6659), 477–484. https://doi.org/10.1038/37297.

Zhu, D., Stumpf, C. R., Krahn, J. M., Wickens, M., & Hall, T. M. (2009). A 5′ cytosine binding pocket in Puf3p specifies regulation of mitochondrial mRNAs. *Proceedings of the National Academy of Sciences of the United States of America*, *106*(48), 20192–20197. https://doi.org/10.1073/pnas.0812079106.

CHAPTER TWO

Methods for thermal denaturation studies of nucleic acids in complex with fluorogenic dyes

Lauren M. Aufdembrink[a,†], Tanner G. Hoog[a,†], Matthew R. Pawlak[a], Benjamin F. Bachan[a], Joseph M. Heili[a,b], Aaron E. Engelhart[a,b,*]

[a]Department of Genetics, Cell Biology, and Development, University of Minnesota, Minneapolis, MN, United States

[b]Department of Biochemistry, Molecular Biology and Biophysics, University of Minnesota, Minneapolis, MN, United States

*Corresponding author: e-mail address: enge0213@umn.edu

Contents

Abstract

Thermal denaturation is a common technique in the biophysical study of nucleic acids. These experiments are typically performed by monitoring the increase in absorbance (hyperchromism) of a sample at 260nm with temperature (Mergny & Lacroix, 2003; Puglisi & Tinoco, 1989). This wavelength is chosen as nucleic acids of mixed sequence typically exhibit their maximum absorbance here. Exceptions exist, however, some non-canonical nucleic acid structures exhibit differing spectral changes with temperature,

† These authors contributed equally to this work.

Methods in Enzymology, Volume 623
ISSN 0076-6879
https://doi.org/10.1016/bs.mie.2019.05.029

resulting in other wavelengths being convenient reporters of secondary structure. In the case of nucleic acids that bind visible light-absorbing ligands, such as fluorogenic aptamers, another wavelength can be a convenient reporter of secondary structure stability and RNA-ligand recognition. As it can be difficult, if not impossible, to know which wavelength to employ a priori, we have developed a system for obtaining the full UV-visible spectrum of a sample at each wavelength, allowing for the subsequent extraction of the absorbance-temperature profile at the desired wavelength. Here, we describe the apparatus and software used to do so. We also describe another technique for the use of a qPCR instrument for measuring secondary structure stability of fluorescent nucleic acid-ligand complexes.

1. Introduction

The hyperchromicity associated with the folded-to-single stranded transition of a nucleic acid has long been used as a means of measuring secondary structure stability. This technique formed the basis for the determination of the nearest-neighbor biophysical parameters used routinely in molecular biology labs for T_M prediction of, e.g., PCR primers (SantaLucia, Allawi, & Seneviratne, 1996; Schroeder & Turner, 2009). Typically, this technique is performed by placing a DNA sample in a quartz cuvette, which is held in a temperature-controlled holder, with bidirectional temperature control performed via a Peltier-effect device. The sample's absorbance is typically monitored at 260 nm, since this is the wavelength at which a mixed-sequence DNA or RNA strand typically exhibits maximum absorbance (Fig. 1A; Puglisi and Tinoco, 1989).

This is not always the case, however—G-quadruplex containing nucleic acids, for example, typically exhibit minimal change in absorbance at 260 nm. These structures exhibit *hypochromism* with increasing temperature at 295 nm (Mergny & Lacroix, 2009). In the case of nucleic acids in complex with a ligand, such as a secondary structure-stabilizing drug, or a fluorogenic dye, another wavelength is often desired (Dolgosheina et al., 2014; Paige, Wu, & Jaffrey, 2011; Waring, 1965). This can be a visible wavelength associated with the dye, or another UV wavelength, with the A_{260} transition obscured or shifted due to the absorbance of the ligand (Fig. 1B).

We have found that the optimal wavelength for observing a secondary structure transition is often best empirically determined. Thus, we have found it useful to collect the full absorbance spectrum of samples at each temperature, enabling later analysis of the data. We have found most commercially available instruments have the option to collect only one wavelength or a few discrete wavelengths. A diode array instrument that

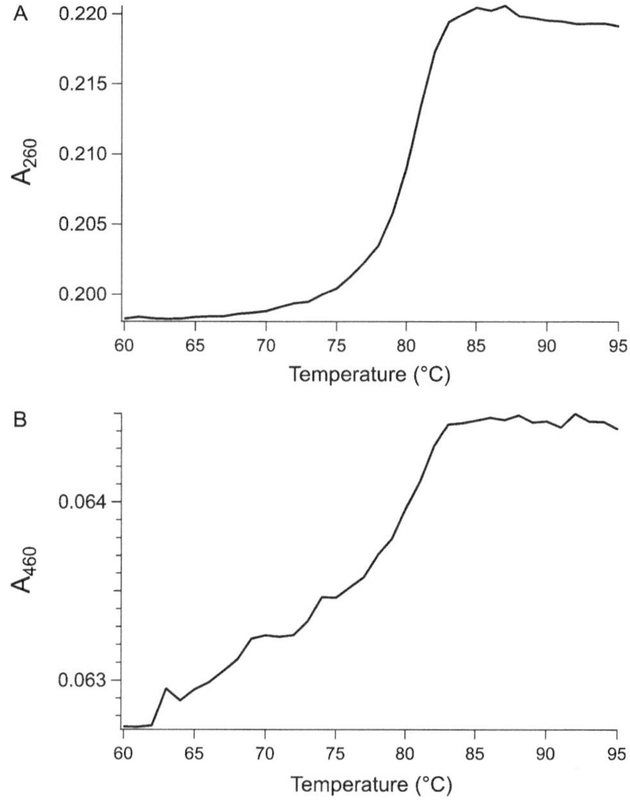

Fig. 1 UV-visible monitored melts of DNA-ethidium bromide complex. A 40 bp DNA duplex complexed with ethidium bromide at nearest-neighbor stoichiometry was subjected to thermal denaturation. Extraction of a melt curve at both the DNA-associated wavelength (260 nm, (A)) and at the ligand-associated wavelength (460 nm, (B)) demonstrate that the melting transition can be monitored at both wavelengths. Data are from the first cooling trace. Samples were in 100 mM Tris-HCl pH 8, 100 mM KCl, 10 mM $MgCl_2$.

collects full spectra exists (the Agilent 8453/8454) with a Peltier holder, but only in a single cell format. The Agilent Cary 60 is among the more inexpensive cuvette-based UV-visible spectrometers available (ca. $11K new as of 2017) but is unique in that it employs a flashlamp source with a 80 Hz repetition rate, enabling scanning a full UV-visible spectrum in a few seconds. It does not have a multi-cell Peltier module, but one is commercially available from a third-party vendor (qCHANGER 6 from Quantum Northwest, ca. $7K new as of 2017). Here, we describe the use of the combination of these two instruments for data collection in full-spectrum UV-visible thermal denaturation.

Another common method of monitoring the thermal denaturation of a DNA sample is via fluorescence. This is routinely performed in qPCR experiments, which often take advantage of the presence of a dye that becomes fluorescent in the presence of a DNA duplex, such as SYBR Green (Dorak, 2006). By measuring fluorescence vs temperature, a thermal denaturation curve is obtained. We routinely employ this technique to monitor the secondary structure stability of fluorescent aptamers (Fig. 2). The use of

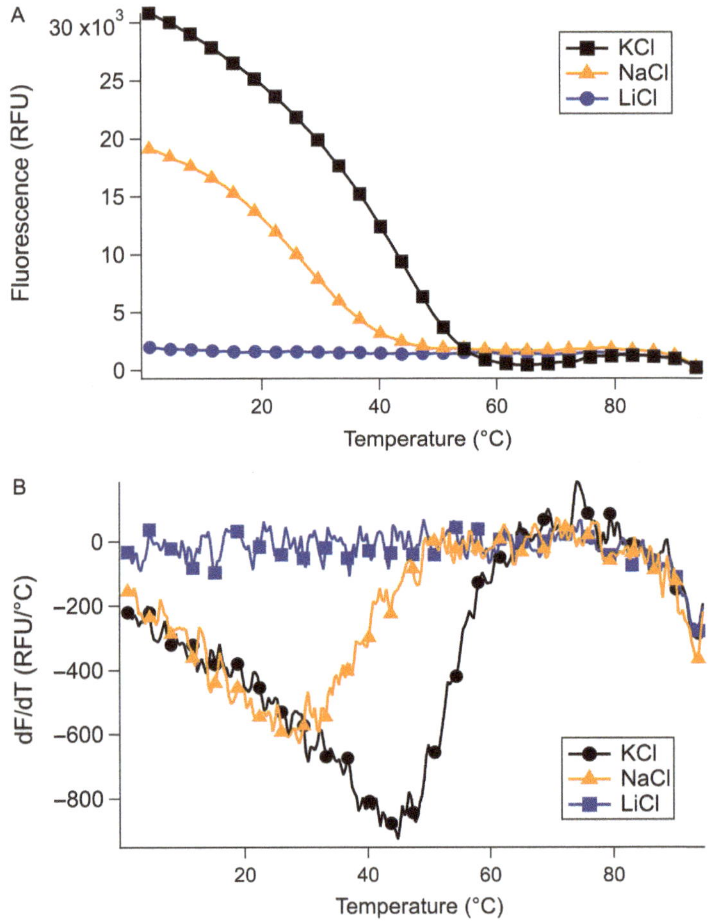

Fig. 2 Fluorescence-monitored aptamer melts. The Broccoli aptamer has a closely related core sequence to the Spinach aptamer and the engineered Baby Spinach aptamer, both of which are known to form G-Quadruplexes (Filonov, Moon, Svensen, & Jaffrey, 2014; Huang et al., 2014; Warner et al., 2014). Fluorescence-monitored melts of the Broccoli-DFHBI-1T complex show the expected cation dependence of T_M for a G-Quadruplex ($K^+ > Na^+$ ⊠ Li^+). (A) Fluorescence data. (B) Data processed to show dF/dT. Data are from the first cooling trace. Samples were in 100 mM Tris-HCl pH 8, 100 mM MCl with M^+ as specified in the legend, 5 mM $MgCl_2$.

fluorescence and/or ligand-associated absorbance affords a means by which to observe both secondary structure stability and RNA-ligand recognition within the same experiment.

2. Sample preparation

Careful sample preparation is critical for nucleic acid samples to be studied by thermal denaturation. Folded RNAs can bind divalent cations avidly, and they are synthesized (in live cells or in vitro) in the presence of these ions (Bray et al., 2018; Milligan, Groebe, Witherell, & Uhlenbeck, 1987; Petrov, Bowman, Harvey, & Williams, 2011). G-Quadruplexes require a monovalent cation to fold, and their stability varies depending on the identity of the cation present, with $K^+ > Na^+ \approx NH_4^+ \ggg Li^+$ (Hud, Smith, Anet, & Feigon, 1996). Sodium is ubiquitous in the environment and is a common adduct observed in mass spectrometry of oligonucleotides (Greig & Griffey, 1995). Aqueous ammonia is commonly used for deprotection of synthetic oligonucleotides, which can result in the presence of residual ammonium ion (Caruthers et al., 1987). G-Quadruplex structures can exhibit remarkable stability in the presence of potassium, with some structures not denaturing even near the boiling point of water (Engelhart, Plavec, & Hud, 2009). Some G-Quadruplexes have been reported to fold in the presence of even submicromolar concentrations of potassium (Wu et al., 2016). Accordingly, great care must be taken in the preparation of samples for analysis by thermal denaturation.

2.1 In vitro transcription

2.1.1 Preparation of RNA samples

For longer (>30 nt) RNAs, samples are typically prepared by in vitro transcription with T7 RNA polymerase (Milligan et al., 1987). The buffer employed is a modification of the buffer used with the commercially available enzyme from New England Biolabs (40 mM Tris-HCl pH 7.9, 1 mM DTT, 2 mM spermidine). The buffer also contains a higher concentration of $MgCl_2$ (42 mM) to compensate for a higher concentration of NTPs than employed in the commercial kit (4 mM each NTP). The buffer used also contains added (100 mM) KCl; this is to enable on-line monitoring of fluorescent aptamers, many of which are G-Quadruplexes and require a suitable monovalent cation to fold correctly (Huang et al., 2014; Trachman et al., 2017; Warner et al., 2014, 2017).

To achieve high yields, concentrations of polymerase in excess of that used in commercial single-enzyme products are used (1 μM). *E. coli* inorganic pyrophosphatase is also added (2 μM) to the transcription reactions to suppress the formation of magnesium pyrophosphate, which precipitates and removes magnesium ion from the reaction mixture. T7 RNA polymerase is overexpressed in-house from the plasmid pT7-911Q; inorganic pyrophosphatase is commercially available in suitable concentrations (Ichetovkin, Abramochkin, & Shrader, 1997; Rio, 2013).

1. The optimal procedure for preparation of the template is empirical and depends on how much secondary structure the RNA to be transcribed contains. For RNAs that do not contain highly stable secondary structures, simply annealing the strand corresponding to the promoter to the antisense template strand will suffice. For most transcripts, a fully double-stranded template is required, with an annealing from 95 °C to 25 °C over 10 min yielding a satisfactory template. This is readily performed on a thermal cycler. For highly-structured templates, a fully double-stranded template can be prepared by either PCR, or overlap extension of two fragments corresponding to the full-length transcript and promoter region.
2. Set up small pilot-scale (10 or 20 μL) reactions. In our experience, we have found these enzyme, NTP, and Mg^{++} concentrations to be optimal or near-optimal for transcription of typical aptamer-length (50–200 nt) RNAs. Optimization of template concentration can be performed at this stage. The optimal concentration will vary with template length; for RNAs of the lengths used here, 50–200 nM is a reasonable starting point. For some transcripts, additional GTP (4–10 mM total) produces higher yields due to the consumption of GTP by abortive cycling from the T7 RNAP Class I promoter (Martin, Muller, & Coleman, 1988).
3. For transcriptions of fluorescent aptamers, monitoring of transcription can be performed on a plate reader using suitable wavelengths for the aptamer (Heili et al., 2018; Höfer, Langejürgen, & Jäschke, 2013).
4. Analyze RNA yield and purity by electrophoresis. If purity is not suitable, adjust the method of template preparation. As reported aptamers generally transcribe well, we have found that when heterogeneous transcripts are present, they are most commonly caused by issues with template preparation, and this is typically the best step to optimize.
5. Using optimal conditions, set up a reaction of scale appropriate for the experiment required (typically 1–10 mL).
6. Purify the desired RNA by large-scale PAGE, or one of the purification techniques given in Section 2.2 (Wyatt, Chastain, & Puglisi, 1991).

A representative optimized set of buffer conditions for a transcription reaction of a fluorescent aptamer is 40 mM Tris-HCl pH 7.9, 1 mM DTT, 2 mM spermidine, 8 mM GTP, 4 mM ATP, 4 mM CTP, 4 mM UTP, 100 mM KCl (can be omitted if not monitoring a G-quadruplex-forming aptamer), 200 nM dsDNA template.

2.2 Purification of samples

2.2.1 HPLC purification

Purification of nucleic acids of lengths typical of thermal denaturation experiments (10–100 nt) in suitable quantities can be performed on a nominally analytical-scale column (5 × 250 mm DNAPAC PA-100, referred to in older literature as NucleoPac PA-100) (Ausserer & Biros, 1995). Following dry-down of deprotection solution after solid-phase synthesis, the oligonucleotide can be resuspended in a minimum volume of water (100–200 μL for a 1 μmol scale synthesis), filtered to remove insoluble protecting groups and dissolved silicates from the CPG support, and used directly in this procedure.

1. Perform a pilot scale injection of 1–5 μg nucleic acid. A linear gradient of 1–100% 2 M NH_4OAc from 1 to 30 m, 100–1% from 30 to 35 m, 1% from 35 to 40 m, will separate most oligonucleotides of this length, with single-nucleotide resolution for sequences on the short end of this range (ca. 30 nt or less). *Note*: Exceptionally stable secondary structures, such as some G-Quadruplexes, may migrate anomalously in this solvent system and need to be separated using a different eluent system employing high pH (Ausserer & Biros, 1995).
2. Inject at larger scale, taking care not to inject >50% of the volume of the loop. Loss of resolution will occur at higher column loading, but this column can typically accommodate ca. 100 μg per injection with ca. 95% product purity for 10–15 nt oligonucleotides.
3. Collect fractions manually or on a fraction collector, concentrate by lyophilization or vacuum centrifugation, desalt by Sep-Pak (Section 2.2.2) or Sephadex (Section 2.2.3), and reinject for verification of purity. Ion exchange (Section 2.2.4) can be performed at this stage if required.

2.2.2 Sep-Pak purification

Sep-Pak purification with the C_{18} Sep-Pak Plus can produce desalted oligonucleotides with ca. 3–4 eq. (as measured by ^1H NMR) of triethylammonium counterion. Recoveries are typically ca. 90–95% with this protocol.

1. Equilibrate the Sep-Pak with 5 mL 100% acetonitrile, 5 mL 50% acetonitrile, 50% 100 mM triethylammonium acetate, then 5 mL 100 mM triethylammonium acetate. These flow-through can be discarded and flow rate is not critical at this step.
2. Load the nucleic acid, supplemented to 100 mM triethylammonium acetate, at a rate of 1 drop per second or lower. Save the flow-through at this step, in case of failed loading. A syringe pump, if available, is a convenient means of regulating flow rate. Take the UV spectrum of the flow-through; A_{260} should be negligible. The most common problem at this step is flowing nucleic acid through the cartridge too quickly. Loading a dilute aqueous solution of a dye with visible absorption (e.g., ethidium bromide) onto the cartridge is a convenient means of showing researchers unfamiliar with the technique the acceptable range of flow rates, as a band of dye will accumulate at the top of the cartridge if it is transferring to the resin.
3. Wash the cartridge with two 5 mL portions of water, flowing the water through at ca. 1 mL/s. A small amount of sample breakthrough (5–10% of loaded nucleic acid) may be observed by UV spectroscopy at this stage.
4. Elute the nucleic acid with two 1 mL portions of 40% acetonitrile, 60% water. *Note*: The unmixed eluent, particularly from the first portion, will not be homogeneous. When the acetonitrile/water solution is flowed through the Sep-Pak, the initial flow-through will be the pure water held up in the cartridge, and the acetonitrile/water solution will follow. To ensure accurate quantitation of the eluted nucleic acid, vortex and spin down the tube prior to measuring concentrations.
5. The eluted nucleic acid can be dried on a vacuum concentrator. Alternately, if the sample is to be lyophilized, remove ca. 50% of the volume of the eluent on a vacuum concentrator to remove acetonitrile before freezing the sample.

2.2.3 Sephadex desalting

Sep-Pak desalting is typically adequate for thermal denaturation studies, particularly when coupled with the ion exchange method (Section 2.2.4). Prior to ion exchange, several equivalents of triethylammonium typically remain. For scrupulously desalted nucleic acids, size-exclusion resins are useful. In contrast to spin columns containing Sephadex G-15, large columns (e.g., ca. 3 cm × 1 m) containing the same resin can give baseline separation between salt and nucleic acid.

1. Pack the column by making a slurry of the resin with mobile phase. The typical mobile phase employed is pure water of the highest grade available—typically 18.2 MΩ water treated with UV light to destroy organics (e.g., DNAse and RNAse). Do not stir the resin excessively to avoid generation of fines.
2. Equilibrate the column. *Note*: Sephadex resins have a small amount of cation exchange capacity (Eaker & Porath, 1967). In sensitive applications, this can result in unintended exchange of cations, including from previously loaded samples. For sensitive applications, charge the column with a large excess of the desired counterion (5–10 mL of a 1 M solution), then wash extensively with water. The silver chloride method, discussed in Step 7, can be used to determine when salt elutes.
3. Load the sample. A typical volume for this size column is ca. 1 mL.
4. As the sample loads, continue applying mobile phase to the top of the column until 4–5 1 mL fractions have loaded onto the column. Let the column get as low as possible without drying out. Columns that dry out may crack, in which case the resin can be removed (pushing air into the bottom of the column with a 50-mL syringe is a useful means of extruding the resin from the top of the column), a slurry generated, and the column repoured.
5. Leave 1–2 cm of mobile phase on top of the column, and apply a cap connected to tubing routed to a bottle of mobile phase. Allow gravity and siphon action to apply mobile phase to the column.
6. Collect fractions, either manually or by employing a fraction collector. The presence of DNA and RNA can be monitored by UV absorbance.
7. A useful way to monitor when salt elutes off the column is by generation of silver chloride. If a chloride salt is not present in the sample, columns of this size have sufficient salt removal capacity that salt can be added to the sample at 0.1–1 M. Use the chloride salt of the desired counterion. Following purification, take 1–2 μL aliquots of fractions and mix with an equivalent volume of 100 mM $AgNO_3$ on a microscope slide. The white precipitate of AgCl is readily visible on a black benchtop or piece of paper. A dilution series of the soluble chloride salt present can be used to generate a standard array of dots and determine the visual detection limit, which is typically ca. 1 μM. Baseline separation of 10 mL or more between nucleic acid and salt is typical.
8. Pool nucleic acid-containing fractions and lyophilize.
9. Wash the column extensively (3–4 column volumes).

10. Samples prepared by this method are typically of the desired counterion, particularly if purified from a solution predominantly containing that ion and if care is taken to account for the cation exchange capacity of Sephadex. If desired, however, the sample can be resuspended, and the ion exchange procedure (Section 2.2.4) can be performed at this point.

2.2.4 Ion exchange of nucleic acids

In some cases, adventitious cations, such as sodium from glass or residual ammonium ion from DNA deprotection, can interfere with experiments, particularly those employing G-Quadruplexes, which fold in the presence of sodium, ammonium, or potassium ions (Engelhart et al., 2009). In these experiments, the use of an ion exchange resin to produce an oligonucleotide containing one predominant counterion is helpful.

1. Prepare an ion-exchange resin of the desired counterion by treating Dowex 50WX8, hydrogen form, 100–200 mesh with ca. 1.5 eq. (relative to the sulfonic acid loading of the resin) of the hydroxide salt of the desired ion (e.g., lithium hydroxide, tetramethylammonium hydroxide) in a 1 M solution. The resin will become warm and a color change will occur. The supernatant will be yellow to orange-colored.
2. Wash the resin extensively with 18.2 MΩ water, stopping when the pH is no longer basic. Many washes (10 or more) will be required. Water from purification instruments will typically be pH 5.5–6 due to absorption of CO_2 from the atmosphere. The resin can be stored in the refrigerator under water, but it will leach colored impurities over time. It should be washed several times to clarity immediately before use. Resin stored and prepared in this fashion and used as described here does not introduce impurities observable by ^{1}H NMR.
3. Using an extensively desalted oligonucleotide (by Sep-Pak, Section 2.2.2, is suitable, or using size-exclusion desalting, Section 2.2.3), add a solution of the oligonucleotide to 100 eq. of resin (the ratio of sulfonate groups on the resin to phosphate groups in the oligonucleotide). Equilibration is rapid. After ca. 5 min, remove the supernatant, wash with 2–3 additional volumes of water, and lyophilize the combined sample.

3. Thermal denaturation

3.1 UV-visible absorbance-monitored

Thermal denaturation with the Cary 60 and qCHANGER 6 relies on both the Cary 60's connection to the controlling PC (via USB) and the qCHANGER 6's connection (via a serial port). Most commercially available USB-to-serial converters are suitable; we have obtained the most consistent results with those employing an FTDI chipset. Off-brand serial converters can cause runs to fail in an unpredictable way.

The qCHANGER 6 also requires a connection to heat transfer fluid to remove or supply heat to the Peltier device as required. Several means of doing this are possible. Quantum Northwest can supply the device with one of two solutions. The first, the BATH 10, consists of a modified plastic kitchen tub and an aquarium pump. In our experience, this configuration does not work well. No active cooling is supplied, and the water heats up over the course of an experiment, resulting in inconsistent ramp rates. The second employs a liquid cooling system designed for computer processors, which sinks heat to a radiator/fan combination. This device, the Koolance EXT-440CU, is sold at or near cost by Quantum Northwest as the KOOL 440 and is also available from other vendors for ca. $300 (as of 2019). It typically produces consistent ramp rates going to 10–15 degrees below ambient. For lower temperatures, a recirculating refrigerated chiller, operated at 10–15 °C, is also suitable. In this configuration, it is possible to bring samples to at least −5 °C. Samples in scrupulously clean cuvettes can be supercooled to this temperature without ice crystallization occurring often.

Control of the Cary 60/qCHANGER 6 relies on the ADLs_QNW/Version_201 software available from Quantum Northwest.

1. Clean the inside of the cuvette thoroughly. In order to remove both salt solutions and oil, using 50% ethanol, 50% water, or water followed by ethanol, is helpful. Dry in a low-temperature (75 °C) oven.
2. Prepare samples with the desired oligonucleotide, buffer, and salt concentration. Typical volumes are 400 μL for a 1 mm pathlength cuvette or 3 mL for a 10 mm pathlength cuvette.
3. Filling 1 mm cuvettes can be difficult due to the surface tension of water. A thin-stemmed plastic transfer pipette can be stretched and cut, resulting in a very narrow stem that can fit within the narrow cuvettes. If water does not flow to the bottom of the cuvette readily or beads within the cuvette, this usually indicates residues have adhered to the side of the cuvette. Cuvettes are typically cleaned with water and ethanol. If this does not suffice, elevated heat, acid (1–5 M HCl, 1–5 M nitric acid, or, as a last resort, aqua regia) can be used. Ultrasonic cleaners should not be used to avoid damaging cuvettes.
4. Cuvettes with slip-fit Teflon stoppers will often not seal reliably at elevated temperatures, resulting in evaporation. Evaporation can be kept to a minimum by employing Teflon tape, which is more deformable

than the stopper. Thick Teflon tape is optimal for this purpose; it is sold for gas line use and is yellow in color. Push the stopper through the Teflon tape as it is inserted into the cuvette; deformation and sealing can be verified by observing the color change in the tape.

5. Clean the outside of the cuvette thoroughly. In order to remove both salt solutions and oil, using 50% ethanol, 50% water, or water followed by ethanol, is helpful.
6. Insert the cuvette into the sample holder with a suitable spacer. Cuvette positions are numbered 1–6 from front to back.
7. Ensure that all programs communicating with the spectrophotometer are closed. Start the Quantum Northwest ADL script that communicates with the Cary 60 and qCHANGER 6.
8. Select the COM port to ensure the controlling PC is communicating with the qCHANGER 6. Set the COM port with options 1–4. The PC will then check if the connection is OK. If the wrong COM port is selected, an error message will appear on the screen. If the correct COM port is selected, then you will be returned to the main interface.
9. Set the "Instrument and Graph Parameters" as desired. The Cary 60 flashlamp pulses at 80 Hz, so a "Signal Averaging Time (SAT)" of 1 s corresponds to 80 reads. The maximum step size is 5 nm. For monitoring of samples not containing ligand, scanning from 350 to 200 nm is typically sufficient. The "Y-min" and "Y-max" values only impact the visual presentation of the data as it is collected. "Temperature Monitor" will default to "QNW Sample Holder" if no external probe is present. For "Method," select "StepT." "HoldT" will hold a static temperature, while "RampT" will attempt to collect data as the cuvette holder temperature ramps; scanning is typically slower than the cuvette holder temperature is ramped.
10. Set the "Method Parameters" as desired. Typical values employed are "Start T" = 20 °C, "Stop T" = 95 °C, "Return T" (the temperature to which the block is set after the run is complete) = 20 °C, "T Step" = 1 °C, "Equilibration Time" (the initial equilibration, only performed before the first datapoint is collected) = 0.1 min, and "Stir Speed" (RPM of the qCHANGER 6's internal stirrer) = 0 (off).
11. Select the cells that contain samples in "Cell Selection and Sample ID Entry," and give them descriptive names.
12. Set "Cycle Modes" as desired; typical parameters are "Step Up/Down" selected, number of cycles = 2 (at least two are recommended, as many samples are not at equilibrium and exhibit a different melting transition

following the annealing afforded by the first heat/cool cycle), "When Done: Ramp to Return T" and "Controller Off" selected, "Data Simulation = None," and "Report" only "Data" Selected.

13. Initialize the melt. Once the temperature equilibrates, the spectra will appear in split-screen with one subwindow for each cuvette in real time.
14. Runs can take 24–36 h or more, depending on spectral width, data width (in nm), signal averaging, collection interval, etc. Once the run completes, save the data in CSV format. For large runs, export may take several minutes.

3.2 Fluorescence-monitored

Fluorescence-monitored melts are also a convenient means of observing denaturation of secondary structures. This typically is employed in our laboratory for monitoring of fluorescent aptamer or fluorescent aptamer-based sensor folding, but it also can be used for monitoring of FRET pairs, fluorescent stains, etc. The assay employed in our laboratory uses the Open qPCR from Chai Bio. This device is several-fold less expensive than other qPCRs ($4299 for a single-channel instrument and $5999 for a dual-channel instrument as of 2019), enabling its use for experiments such as slow ramp rates that are not possible in heavily used core instruments with limitations on appointment time.

The device is open hardware and the software and schematics are available online. The excitation source is a CREE C503B-BAN-CZ0A0452 LED, with excitation centered around 470 nm and usable excitation of ca. 40% of max between 450 and 500 nm. The instrument's melt program only monitors the green channel (specified as 513–555 nm for the single channel version and 508–532 nm for the dual channel version). These filters allow for fluorescence-monitored thermal denaturation of green aptamers (e.g., Broccoli, with $E_{max} = 507$ nm) as well as the Corn aptamer, a YFP mimic, with $E_{max} = 545$ nm (Fig. 3).

Setup of experiments on the Open qPCR is performed via its web interface. A PC, phone, tablet, etc. with a web browser on the same local network can be used to access the instrument, set up, and collect data.

1. Using a web browser, go to the IP address displayed on the LCD screen of the instrument. In the instrument's default configuration, this IP will be in the local 192.168.0.0 Class C IP address block.
2. Log in through the web interface, and click the green "Create a New Experiment" button. Enter an experiment name when prompted.

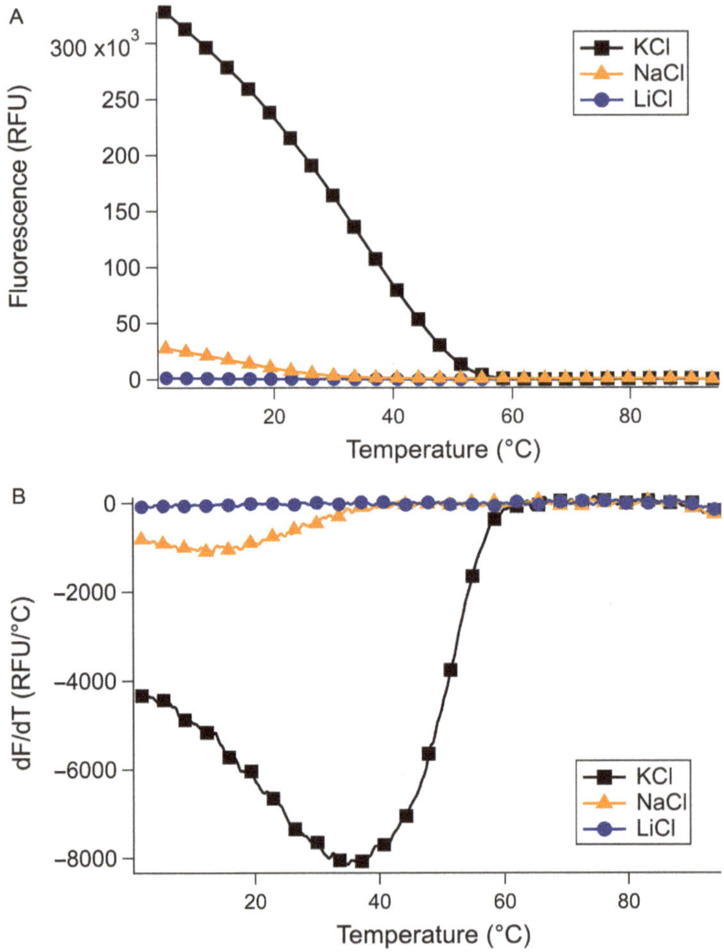

Fig. 3 Fluorescence-monitored aptamer melts using a yellow aptamer. The Corn aptamer contains a G-Quadruplex (Song et al., 2017; Warner et al., 2017). While the Open qPCR only monitors melt fluorescence in the green channel, the spectral bandwidth is sufficient to detect Corn fluorescence. Fluorescence-monitored melts of the Broccoli-DFHBI-1T complex show the expected cation dependence of T_M for a G-Quadruplex ($K^+ > Na^+$ ⊠ Li^+). (A) Fluorescence data. (B) Data processed to show dF/dT. Data are from the first cooling trace. Samples were in 100 mM Tris-HCl pH 8, 100 mM MCl with M^+ as specified in the legend, 5 mM $MgCl_2$.

3. At the assay setup screen, enter the experiment parameters. The melt curve function of the instrument is designed to perform a single melt trace at the end of an experiment to provide qualitative information about the purity of an amplification reaction; it therefore

collects only one melt trace per sample. To collect multiple heat/cool cycles, define individual programs for each cycle by creating separate "experiments."

4. Typical parameters for melts are two heat/cool traces from 5 °C to 95 °C. In a typical experiment, the samples are placed in the instrument, and the instrument is set to hold at 5 °C, then after user intervention, ramp from 5 °C to 95 °C. Following each trace, begin the next program (e.g., after the first heating trace, begin the first cooling trace program, which consists of a 95–5 °C ramp). This should be done rapidly, before the instrument begins to cool or warm up from the previous cycle.
5. Ramp rates can be determined empirically based on the system. The instrument has a wide range of possible ramp rates, with a maximum of 5 °C/s and a claimed minimum of 0.00001 °C/s (i.e., 0.864 °C/day). Most experiments with aptamers involve intramolecular folding, and their secondary structures often refold readily as a result of the original selection conditions. Thus, fairly high ramp rates are feasible in these experiments; a typical ramp rate is 0.1 °C/s.

   The ramp rate chosen will depend on the needs of the experiment. Typically, several ramps are performed to ensure the sample is properly annealed (Fig. 4A). The ramp rate and temperature range chosen must balance two competing phenomena. A sample ramped too quickly will exhibit hysteresis (Fig. 4A). However, slow ramping and/or time spent at high temperatures can result in sample degradation—a particular concern for RNA (Fig. 4B). In many cases, a ca. 5 °C bounding range obtained from an (artificially low) apparent T_M from a cooling trace and an (artificially high) apparent T_M from a heating trace is sufficient to determine the operational conditions for a given aptamer system. Conversely, if detailed biophysical characterization of a system is desired, measuring T_Ms under as close to equilibrium conditions as possible is critical.

4. Data analysis

4.1 Data extraction

4.1.1 UV-visible

The CSV data files obtained from the Quantum Northwest program contain a pair of columns for each cell/temperature pair corresponding to wavelength and absorbance. Extraction of a desired wavelength/temperature/cell can be performed in Excel for smaller files.

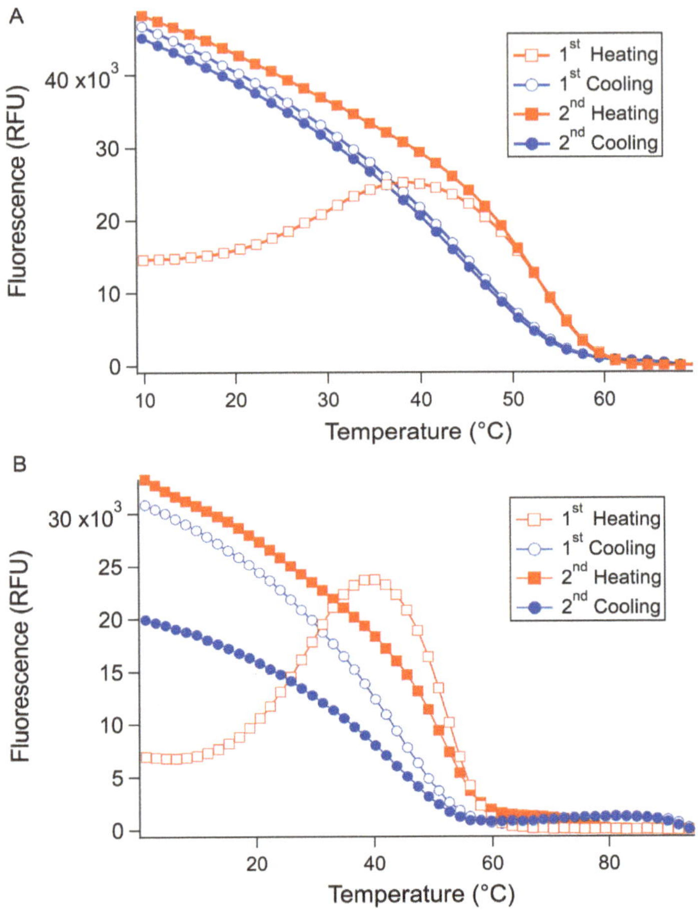

Fig. 4 Information obtained from multiple temperature ramps. A sample of Broccoli-DFHBI-1T complex was subjected to fluorescence-monitored thermal denaturation over two heat-cool cycles. Traces are reported using the 260 nm absorbance values. (A): (1) The first heating trace exhibits a markedly different shape than the other three traces. When this is observed, the possibility of the sample not existing in its equilibrium fold as prepared should be considered. (2) Hysteresis is observed. The heating traces exhibit a higher apparent T_M than the cooling traces. This is a hallmark of a sample that has been ramped at a temperature rate high enough that it is not at equilibrium. (B): (3) A sample subjected to a longer ramp going to higher temperatures exhibits loss of fluorescence in later ramps due to heat-induced degradation. This is of particular concern for RNA (Adamala & Szostak, 2013) and can limit the minimum ramp rate and maximum temperature accessible for a given system. Samples were in 100 mM Tris-HCl·pH 8, 100 mM KCl, 5 mM $MgCl_2$.

For experiments containing many lines or more datapoints per spectra, Excel will often freeze or crash, even on computers with 16 GB or more of RAM. These files are not typically large (up to ca. 10 MB for a single experiment), but they contain many columns. In these cases, the use of the commercially available program Delimit (Delimitware) readily allows for extraction of a specific wavelength.

4.1.2 Fluorescence

Data from the Open qPCR instrument are supplied in CSV format and are readily extracted. The instrument provides data in two modes: (1) absolute fluorescence and (2) the first derivative of the fluorescence trace. Take care to extract data from the absolute fluorescence curve.

4.2 Plotting

Data are typically plotted in Igor Pro. Owing to the command line nature of the program, modular scripts are easily generated for display of data. For example, if a data set from a single sample containing alternating columns with temperature and absorbance data for two each of heating and cooling traces (i.e., eight columns) is pasted into an Igor table from the program used to extract the data (e.g., Excel), the below code snippet will rename the waves and display the trace. "SampleName" can then be changed by using search and replace for each sample.

```
•Rename wave0,'SampleName-H1-Temp'; Rename wave1,'SampleName-H1-Abs'; ;DelayUpdate
•Rename wave2,'SampleName-C1-Temp'; Rename wave3,'SampleName-C1-Abs'; ;DelayUpdate
•Rename wave4,'SampleName-H2-Temp'; Rename wave5,'SampleName-H2-Abs'; ;DelayUpdate
•Rename wave6,'SampleName-C2-Temp'; Rename wave7,'SampleName-C2-Abs';
•Display 'SampleName-H1-Abs' vs 'SampleName-H1-Temp'
•AppendToGraph 'SampleName-C1-Abs' vs 'SampleName-C1-Temp'
•AppendToGraph 'SampleName-H2-Abs' vs 'SampleName-H2-Temp'
•AppendToGraph 'SampleName-C2-Abs' vs 'SampleName-C2-Temp'
•ModifyGraph mode=0,rgb('SampleName-H1-Abs')=(65280,0,0);DelayUpdate
•ModifyGraph rgb('SampleName-C1-Abs')=(0,0,65280);DelayUpdate
•ModifyGraph rgb('SampleName-H2-Abs')=(65280,0,0);DelayUpdate
•ModifyGraph rgb('SampleName-C2-Abs')=(0,0,65280)
```

```
•Label left "\\Z24A\\B260\\M"
•ModifyGraph mode=4,mskip=10,marker('SampleName-C1-
Abs')=8;DelayUpdate
•ModifyGraph marker('SampleName-H2-Abs')=16,marker('SampleName-
C2-Abs')=5
•ModifyGraph msize=8
•Legend/C/N=text0/J/S=3/Z=1/A=LT "\\s('SampleName-H1-Abs')
SampleName - H1\r\\s('SampleName-C1-Abs') SampleName -
C1";DelayUpdate
•AppendText "\\s('SampleName-H2-Abs') SampleName -
H2\r\\s('SampleName-C2-Abs') SampleName - C2"
```

The optimum wavelengths for analysis are typically those where the most change in absorbance is observed over a given temperature range. For a canonical DNA or RNA duplex, for example, this is usually 260 nm. Similarly, the ligand's maximum absorbance is typically used for ligand-monitored reactions. This can occasionally vary, however: for instance, a ligand with significant UV absorbance can shift the optimum wavelength for the nominally DNA-associated absorbance band, or a ligand that exhibits solvatochromism may have an optimum monitoring wavelength that differs from its low-temperature maximum. A detailed description of the analysis of thermal denaturation data is beyond the scope of this chapter. The reader is directed to a review by Mergny and Lacroix for more comprehensive information (Mergny & Lacroix, 2003).

5. Outlook

The rapid scanning and capability to collect full spectra using the UV-visible instrument configuration described have enabled multiwavelength melts. This configuration is also capable of performing kinetic analyses, and this will enable full-spectrum kinetic analysis of colorimetric reactions, providing richer datasets and more detail about reaction progress. Similarly, the ability to perform temperature-dependent fluorescence experiments in low-volume (20 μL) in an inexpensive instrument in medium-throughput (16 samples/run) format has enabled a range of experiments that were otherwise infeasible. Several groups have employed the melt curve capability of qPCR for thermofluorimetric analysis (TFA)-based determination of binding interactions, including one using the Open qPCR (Damase, Miura, Parent, & Allen, 2018; Hu, Kim, & Easley, 2015). The low cost and flexibility of the platform suggest a proliferation of new qPCR-based (or simply low-volume, fluorescence-based) assays will be observed in the not-too-distant future.

Acknowledgments

The authors thank Louis Libertini (Quantum Northwest) and Nicholas V. Hud (Georgia Institute of Technology) for helpful discussions. This work was supported by NASA Contract 80NSSC18K1139 under the Center for Origin of Life (to A.E.E.).

References

Adamala, K., & Szostak, J. W. (2013). Nonenzymatic template-directed RNA synthesis inside model protocells. *Science (New York, N.Y.)*, *342*(6162), 1098–1100. https://doi.org/10.1126/science.1241888.

Ausserer, W. A., & Biros, M. L. (1995). High-resolution analysis and purification of synthetic oligonucleotides with strong anion-exchange HPLC. *BioTechniques*, *19*, 136–139.

Bray, M. S., et al. (2018). Multiple prebiotic metals mediate translation. *Proceedings of the National Academy of Sciences of the United States of America*, *115*(48), 12164–12169. https://doi.org/10.1073/PNAS.1803636115.

Caruthers, M. H., et al. (1987). Chemical synthesis of deoxyoligonucleotides by the phosphoramidite method. *Methods in Enzymology*, *154*, 287–313. https://doi.org/10.1016/0076-6879(87)54081-2.

Damase, T. R., Miura, T. A., Parent, C. E., & Allen, P. B. (2018). Application of the open qPCR instrument for the in vitro selection of DNA aptamers against epidermal growth factor receptor and *Drosophila* C virus. *ACS Combinatorial Science*, *20*(2), 45–54. https://doi.org/10.1021/acscombsci.7b00138.

Dolgosheina, E. V., et al. (2014). RNA mango aptamer-fluorophore: A bright, high-affinity complex for RNA labeling and tracking. *ACS Chemical Biology*, *9*(10), 2412–2420. https://doi.org/10.1021/cb500499x.

Dorak, M. T. (2006). In M. T. Dorak (Ed.), *Real-time PCR*. Taylor and Francis.

Eaker, D., & Porath, J. (1967). Sorption effects in gel filtration: I. A survey of amino acid behavior on sephadex G-10. *Separation Science*, *2*(4), 507–550. https://doi.org/10.1080/01496396708049718.

Engelhart, A. E., Plavec, J., & Hud, N. V. (2009). Metal ion interactions with G-Quadruplex structures. In *Nucleic acid–metal ion interactions* (pp. 118–153). RSC Publishing. https://doi.org/10.1039/9781847558763-00118.

Filonov, G. S., Moon, J. D., Svensen, N., & Jaffrey, S. R. (2014). Broccoli: Rapid selection of an RNA mimic of green fluorescent protein by fluorescence-based selection and directed evolution. *Journal of the American Chemical Society*, *136*(46), 16299–16308. https://doi.org/10.1021/ja508478x.

Greig, M., & Griffey, R. H. (1995). Utility of organic bases for improved electrospray mass spectrometry of oligonucleotides. *Rapid Communications in Mass Spectrometry*, *9*(1), 97–102. https://doi.org/10.1002/rcm.1290090121.

Heili, J. M., et al. (2018). Real-time visualization of in vitro transcription of a fluorescent RNA aptamer: An experiment for the upper-division undergraduate or first-year graduate laboratory. *Journal of Chemical Education*, *95*, 1867–1871. https://doi.org/10.1021/acs.jchemed.7b00735.

Höfer, K., Langejürgen, L. V., & Jäschke, A. (2013). Universal aptamer-based real-time monitoring of enzymatic RNA synthesis. *Journal of the American Chemical Society*, *135*(37), 13692–13694. https://doi.org/10.1021/ja407142f.

Hu, J., Kim, J., & Easley, C. J. (2015). Quantifying aptamer-protein binding via thermofluorimetric analysis. *Analytical Methods: Advancing Methods and Applications*, 7(17), 7358–7362. https://doi.org/10.1039/c5ay00837a.

Huang, H., et al. (2014). A G-quadruplex-containing RNA activates fluorescence in a GFP-like fluorophore. *Nature Chemical Biology*, *10*(8), 686–691. https://doi.org/10.1038/nchembio.1561.

Hud, N. V., Smith, F. W., Anet, F. A., & Feigon, J. (1996). The selectivity for K^+ versus Na^+ in DNA quadruplexes is dominated by relative free energies of hydration: A thermodynamic analysis by ^1H NMR. *Biochemistry*, *35*(48), 15383–15390. https://doi.org/10.1021/bi9620565.

Ichetovkin, I. E., Abramochkin, G., & Shrader, T. E. (1997). Substrate recognition by the leucyl/phenylalanyl-tRNA-protein transferase. Conservation within the enzyme family and localization to the trypsin-resistant domain. *The Journal of Biological Chemistry*, *272*(52), 33009–33014. https://doi.org/10.1074/JBC.272.52.33009.

Martin, C. T., Muller, D. K., & Coleman, J. E. (1988). Processivity in early stages of transcription by T7 RNA polymerase. *Biochemistry*, *27*, 3966–3974. https://doi.org/10.1021/bi00411a012.

Mergny, J.-L., & Lacroix, L. (2003). Analysis of thermal melting curves. *Oligonucleotides*, *13*(6), 515–537. https://doi.org/10.1089/154545703322860825.

Mergny, J. L., & Lacroix, L. (2009). UV melting of G-quadruplexes. *Current Protocols in Nucleic Acid Chemistry*, *37*, 17.1.1–17.1.15. https://doi.org/10.1002/0471142700.nc1701s37.

Milligan, J. F., Groebe, D. R., Witherell, G. W., & Uhlenbeck, O. C. (1987). Oligoribonucleotide synthesis using T7 RNA polymerase and synthetic DNA templates. *Nucleic Acids Research*, *15*(21), 8783–8798. https://doi.org/10.1093/nar/15.21.8783.

Paige, J. S., Wu, K. Y., & Jaffrey, S. R. (2011). RNA mimics of green fluorescent protein. *Science (New York, N.Y.)*, *333*(6042), 642–646. https://doi.org/10.1126/science.1207339.

Petrov, A. S., Bowman, J. C., Harvey, S. C., & Williams, L. D. (2011). Bidentate RNA-magnesium clamps: On the origin of the special role of magnesium in RNA folding. *RNA (New York, N.Y.)*, *17*(2), 291–297. https://doi.org/10.1261/rna.2390311.

Puglisi, J. D., & Tinoco, I. (1989). Absorbance melting curves of RNA. *Methods in Enzymology*, *180*, 80108–80109. https://doi.org/10.1016/0076-6879(89)80108-9.

Rio, D. C. (2013). Expression and purification of active recombinant T7 RNA polymerase from E. coli. *Cold Spring Harbor Protocols*, *2013*(11). https://doi.org/10.1101/pdb.prot078527.

SantaLucia, J., Allawi, H. T., & Seneviratne, P. A. (1996). Improved nearest-neighbor parameters for predicting DNA duplex stability. *Biochemistry*, *35*(11), 3555–3562. https://doi.org/10.1021/bi951907q.

Schroeder, S. J., & Turner, D. H. (2009). Optical melting measurements of nucleic acid thermodynamics. *Methods in Enzymology*, *468*, 371–387. https://doi.org/10.1016/S0076-6879(09)68017-4.

Song, W., et al. (2017). Imaging RNA polymerase III transcription using a photostable RNA-fluorophore complex. *Nature Chemical Biology*, *13*(11), 1187–1194. https://doi.org/10.1038/nchembio.2477.

Trachman, R. J., et al. (2017). Structural basis for high-affinity fluorophore binding and activation by RNA mango. *Nature Chemical Biology*, *13*, 807–813. https://doi.org/10.1038/nchembio.2392.

Waring, M. J. (1965). Complex formation between ethidium bromide and nucleic acids. *Journal of Molecular Biology*, *13*, 269–282. https://doi.org/10.1016/S0022-2836(65)80096-1.

Warner, K. D., Chen, M. C., Song, W., Strack, R. L., Thorn, A., Jaffrey, S. R., et al. (2014). Structural basis for activity of highly efficient RNA mimics of green fluorescent protein. *Nature Structural & Molecular Biology*, *21*(8), 658–663. https://doi.org/10.1038/nsmb.2865.

Warner, K. D., et al. (2017). A homodimer interface without base pairs in an RNA mimic of red fluorescent protein. *Nature Chemical Biology*, *13*(11), 1195–1201. https://doi.org/10.1038/nchembio.2475.

Wu, T., et al. (2016). Tuning the sensing range of potassium ions by changing the loop size of G-quadruplex sensors. *New Journal of Chemistry*, *40*(11), 9285–9290. https://doi.org/10.1039/C6NJ02136K.

Wyatt, J. R., Chastain, M., & Puglisi, J. D. (1991). Synthesis and purification of large amounts of RNA oligonucleotides. *BioTechniques*, *213*, 776–783.

CHAPTER THREE

Identifying and validating small molecules interacting with RNA (SMIRNAs)

Matthew D. Disney*, Sai Pradeep Velagapudi, Yue Li, Matthew G. Costales, Jessica L. Childs-Disney
Department of Chemistry, The Scripps Research Institute, Jupiter, FL, United States
*Corresponding author: e-mail address: disney@scripps.edu

Contents

Abstract

High throughput sequencing has revolutionized our ability to identify aberrant RNA expression and mutations that cause or contribute to disease. These data can be used directly to design oligonucleotide-based modalities using Watson-Crick pairing to target unstructured regions in an RNA. A complementary, although more difficult, strategy to deactivate a malfunctioning RNA is to target highly structured regions with small molecules. Indeed, RNA structures are directly causative of disease. Herein, we discuss emerging strategies to design high affinity, selective, bioactive ligands targeting RNA, or small molecules interacting with RNA (SMIRNAs), and target validation and profiling methods. An experimental foundation is required for a lead identification strategy

Methods in Enzymology, Volume 623
ISSN 0076-6879
https://doi.org/10.1016/bs.mie.2019.04.027

for RNA structures, constructed from a library-vs.-library screen that probes vast libraries of small molecules for binding RNA three dimensional folds. Dubbed 2-dimensional combinatorial screening (2DCS), the resulting data can be mined against transcriptomes or the composite of RNAs that are produced in an organism to define folded RNA structures that can be targeted. By applying SMIRNAs to cells and using target validation tools such as Chemical Cross-Linking and Isolation by Pull-down (Chem-CLIP) and Small Molecule Nucleic Acid Profiling by Cleavage Applied to RNA (RiboSNAP), all targets engaged in cells can be defined, along with rules for molecular recognition to affect RNA biology. This chapter will describe lessons learned in applying these approaches in vitro, in cells, and in pre-clinical animal models of disease, enabling SMIRNAs to capture opportunities in chemical biology.

1. Introduction

Affecting biology by targeting RNA has been historically focused on the bacterial ribosome. One of the first drugs to target the bacterial ribosome was streptomycin, isolated from *Streptomyces* bacteria in soil (Jones, Metzger, Schatz, & Waksman, 1944). Ironically, many consider Streptomycin as not "drug-like," despite it being a transformative, first-in-class treatment for tuberculosis infections at a time when many lived out their last days quarantined in sanitariums. Indeed, much work has been completed to elucidate how antibacterials inhibit translation—by binding highly folded regions in ribosomal RNA (Poehlsgaard & Douthwaite, 2005; Tenson & Mankin, 2006). The discovery of a suite of compounds that differentially affect ribosome function has not only provided life-saving medicines but also invaluable chemical probes that enabled in-depth studies on translation, both biochemical and structural.

Another advance to target RNA was the development of antisense oligonucleotides (ASOs) that recognize sequence. Paul Zamecnik, who discovered tRNA, and coworkers used oligonucleotides complementary to mRNAs found in Rous sarcoma virus (RSV) to inhibit viral replication, transcription and translation (Stephenson & Zamecnik, 1978; Zamecnik & Stephenson, 1978). They subsequently demonstrated ASOs work against other infectious diseases, laying the foundation for treatment of human diseases. Amongst many applications, an ASO was developed to treat spinal muscular atrophy (SMA) by targeting sites in survival of motor neuron 2 (*SMN2*) pre-mRNA, facilitating inclusion of exon 7 and hence synthesis of a full length and functional SMN protein. This ASO is a transformative medicine for children with SMA (Morrow, 2017; Stein & Castanotto, 2017).

The above examples demonstrate that targeting RNA can improve human disease even when the therapeutic modalities are not "drug-like" (streptomycin and other antibiotics) or are of large molecular weights (ASOs). Indeed, there are exemplar cases of small molecules developed to purposefully target human RNAs despite the perception that SMIRNAs cannot be sufficiently selective or potent to affect biology or disease outcome. Herein, we describe: (i) methods to define experimentally the folded RNA structures that can be targeted by small molecules; (ii) a computational lead identification strategy for an RNA target of interest using the data defined in (i); (iii) deploying SMIRNAs to affect RNA biology and study molecular recognition; and (iv) various target validation approaches in cells to establish compound mode of action by defining the RNAs that are targeted by SMIRNAs and their binding sites within them. Collectively, these studies define principles for molecular recognition that can drive SMIRNA activity in vitro, in cells, and in pre-clinical animal models, affording chemical probes and lead medicines targeting RNA.

2. Two-dimensional combinatorial screening (2DCS), a library-vs.-library screening approach to select the RNA three dimensional folds that bind small molecules

In the broadest sense, RNAs have unstructured regions and regions of highly defined structure that define their function (Rouskin, Zubradt, Washietl, Kellis, & Weissman, 2014; Spitale et al., 2015; Wan et al., 2014). This delineation has major implications for targeting RNA. For example, an ASO is tailored for the former, typically coding sequences, although regions of defined structure have been identified in some mRNAs. The latter represent ideal target sites for SMIRNAs (Angelbello et al., 2018; Disney, Dwyer, & Childs-Disney, 2018). The challenge to enable design of SMIRNAs is to define the optimal RNA three dimensional folds for a particular small molecule and chemotypes that confer avid, selective binding to RNA. As these types of data are generally lacking, a small molecule library-vs.-RNA fold library screen was developed, dubbed 2-Dimensional Combinatorial Screening (2DCS; Fig. 1). A detailed description of how to implement this approach can be found in (Aminova & Disney, 2010).

The 2DCS approach requires encoding ofboth the small molecules and the RNA three dimensional folds that bind them. The small molecules are spatially

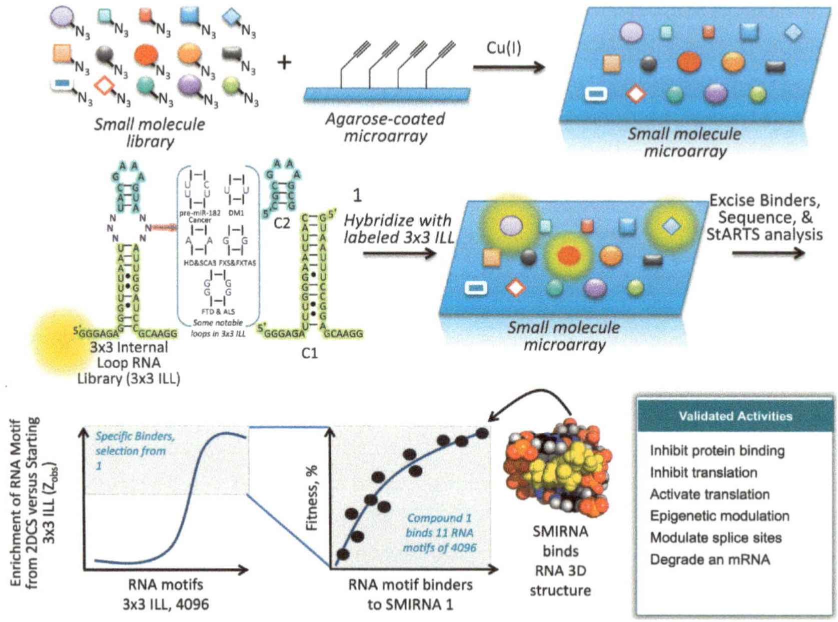

Fig. 1 Discovering the three-dimensional RNA folds that bind small molecules using a library-vs.-library screen named 2-Dimensional Combinatorial Screening (2DCS). A small molecule library is site-selectively immobilized (shown) or absorbed onto agarose-coated microarrays. The arrays are incubated with a labeled RNA fold library that displays discrete structural elements (such as the 3×3 internal loop library (ILL)) in the presence of competitor oligonucleotides that mimic regions common to all library members (C1 and C2, for example). Bound RNAs are excised from the surface, and their identities are determined by RNA-seq. Statistical analysis of RNA-seq data reveals privileged RNA fold-small molecule interactions that inform design of SMIRNAs that modulate RNA biology. Along with 2DCS, we developed an approach named Structure-Activity Relationship Through Sequencing (StARTS), which allows for facile annotation of 2DCS data by scoring the relative affinity and selectivity of RNA motifs that bind to a small molecule (a metric of small molecule "fitness").

arrayed and either covalently attached (Childs-Disney, Wu, Pushechnikov, Aminova, & Disney, 2007) or non-covalently absorbed (Velagapudi et al., 2018) onto microarrays and thus are spatially encoded via their position on the surface (Fig. 1). To allow for isolation of RNAs bound to the small molecules displayed on the surface, microarrays use an agarose substrate, vide infra. Arrays have been constructed by using a wide variety of compounds ranging from known nucleic acid binders to compounds with unknown RNA binding capacity to known drugs.

The arrays are hybridized with a library of three dimensionally folded RNA structural motifs that are embedded in a unimolecular hairpin that can be amplified and sequenced to decode the bound RNA folds (Fig. 1). Small molecule microarrays are hybridized with labeled RNA motif libraries in the presence of unlabeled competitor RNAs that mimic the constant regions in the RNA fold libraries; these competitor oligonucleotides are engineered such that they cannot be amplified via RT-PCR with primers that amplify the RNA fold libraries. This approach is general for many types of RNA fold libraries (internal loops, hairpins, bulges, etc.) and likely has an extended applicability beyond what has been published to date.

The members of the RNA fold library that bind to small molecules on the array are manually excised from the agarose surface, amplified, and sequenced. When sequence analyses are completed by RNA-seq with at least six-fold coverage of the selection, affinity landscapes between small molecules and the various RNA folds can be rapidly assigned (Velagapudi et al., 2017). Assigning affinity landscapes is accomplished by measuring the frequency of enrichment of a specific RNA fold selected by a small molecule to the frequency of this same fold in the starting library using a pooled population comparison. This analysis affords a Z_{obs} score for each RNA fold, a metric of statistical significance, which correlates directly with relative affinity. A Z_{obs} value >8 assigns a SMIRNA (RNA fold-small molecule pair) as binding selectively (Velagapudi et al., 2017). Z_{obs} within a selection can be normalized to provide a Fitness Score for binders with $Z_{obs} > 8$, with a score of 100 indicating the optimal, most fit interaction. Indeed, 2DCS in conjunction with this statistical analysis has defined avid, selective RNA fold-small molecule interactions, chemotypes that confer RNA binding capacity, and the beginning of rules or framework that govern molecular recognition by SMIRNAs. Additionally, it was found that known drug classes target RNA (Velagapudi et al., 2018). It is thus provocative to think that RNA should be considered in toxicological optimization of small molecule drug candidates.

3. Inforna: A sequence-based approach to define RNAs that are druggable from the output of 2DCS

3.1 Overview of Inforna

Decades of research has enabled annotation of folded structures within an RNA sequence by various methods including phylogenic comparison, free energy minimization with or without experimental constraints

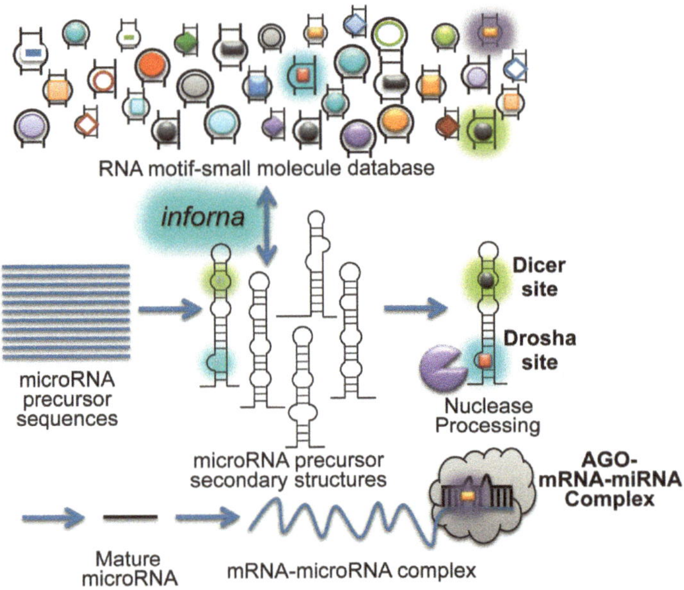

Fig. 2 Design of SMIRNA using Inforna, a lead identification strategy for RNA targets. Privileged RNA fold-small molecule interactions identified by 2DCS are annotated and deposited into a database. Inforna designs lead small molecules by comparing the folds in the database to folds in a cellular RNA target. Illustrated is implementation of Inforna to identify SMIRNAs for miRNA targets.

(Mathews et al., 2004), and a combination of conservation and free energy minimization (Mathews & Turner, 2002). Protocols to annotate RNA structure from sequence using free energy minimization are available (Mathews, 2014). By combining these advances with the RNA fold-small molecule interactions identified by 2DCS, we developed a lead identification strategy for structured RNAs named Inforna (Fig. 2) (Disney et al., 2016; Velagapudi, Gallo, & Disney, 2014).

Inforna was validated by identifying folded structures in microRNA (miRNA) precursors that overlap with privileged interactions discovered by 2DCS (Velagapudi et al., 2014). MiRNAs comprise a class of noncoding (nc)RNAs that regulate gene expression and hence many biological processes including disease. They are produced as primary transcripts (pri-miRNA) that undergo two processing steps, one by the nuclear nuclease Drosha, which liberates a precursor (pre-) miRNA that is translocated to the cytoplasm, and the second by the cytoplasmic nuclease Dicer. This final mature miRNA regulates gene expression by binding to complementary 3′ untranslated regions present in a subset of mRNAs, suppressing the amount of protein that is translated (Bartel, 2004, 2009).

An analysis of the RNA structures that are presented by pri- and pre-miRNAs has provided answers to fundamental questions in RNA folding and motif redundancy (Liu et al., 2016). Among >1500 human miRNA precursors, both unique motifs, found only in a singular miRNA, and redundant motifs, found in many miRNAs, are present. A database of these and other folded structures in the human transcriptome are available in an on-line database (Liu et al., 2016).

In this first iteration of Inforna, a target agnostic approach was taken. That is, each human miRNA precursor, with a three dimensionally folded structure, was analyzed by Inforna for overlap with privileged RNA fold-small molecule interactions from 2DCS studies (Velagapudi et al., 2014). To refine the lead small molecules designed by Inforna, only those motifs present in the Drosha or Dicer processing sites in disease-associated miRNAs were further investigated. Over two dozen disease-associated miRNA hairpin precursors were found to be targetable with small molecules, and 44% of the lead small molecules reduced levels of its cognate miRNA in cells without lead optimization (Velagapudi et al., 2014). Importantly, Inforna provides insight into potential off-targets. That is, Inforna reports if: (i) a motif is present in multiple targets (and their location within the target); and (ii) a less optimal motif for the small molecule (binds with lower affinity) is present in another target.

Among the lead interactions designed by Inforna in this inaugural study, an optimal interaction with a Fitness Score of 100% (highest affinity interaction amongst 4096 RNA folds in the selection) was identified between a benzimidazole and the Drosha site in pri-miR-96 (Velagapudi et al., 2014). In vitro, the small molecule bound to the target with low micromolar affinity, bound to the predicted site as determined by nuclease protection assays, and inhibited Drosha processing of a model pri-miR-96 RNA (Velagapudi et al., 2014). In breast cancer cells, the compound increased the levels of the pri-miR-96 and diminished levels of pre-miR-96 and miR-96 at micromolar concentration, bolstering compound mode of action as inhibition of Drosha processing as designed (Velagapudi et al., 2014). The pro-apoptotic transcription factor Forkhead Box O1 (FOXO1) is silenced by miR-96 in many cancers. This protein functions to transactivate Bcl-2-like protein 11 (Bim), a member of the B-cell lymphoma 2 (Bcl-2) protein family that promotes apoptosis. Application of the small molecule to breast cancer cells increased FOXO1 levels and triggered apoptosis. To dissect the mechanisms by which the small molecule triggered apoptosis, an siRNA against FOXO1 was delivered; reduction of FOXO1 ablated the activity of the pri-miR-96-targeting small molecule. Notably, a control siRNA had no effect on compound activity. Full miRNA profiling in cells showed that the compound is selective as only miR-96 levels were reduced to a statistically significant extent while an antagomir directed at miR-96 reduced levels of multiple miRNAs (Velagapudi et al., 2014).

To increase the potency of our lead pri-miR-96 targeting compound such that it could be applied in vivo, Inforna was used to define a fragment that could bind to a second motif within pri-miR-96 (Velagapudi et al., 2016). Thus, the lead small molecule could be tethered to the new fragment to target two sites simultaneously thereby enhancing affinity. The compound, named TargaprimiR-96 ("Targ" indicates targets and primiR-96 indicates the RNA target) has a low nanomolar affinity, selectively triggered apoptosis in triple negative breast cancer but not healthy breast (do not express miR-96) cells, and, importantly, decreased tumor load in a mouse model (Velagapudi et al., 2016). Thus, rational design can provide lead compounds that can be optimized and reduce tumor load in a preclinical mouse model.

Using Inforna, we have successfully designed a small molecule (TargapremiR-210) that inhibits processing of another cancer-associated miRNA, miR-210 (Costales et al., 2017; Velagapudi et al., 2014). In hypoxic cancers, miR-210 targets glycerol-3-phosphate dehydrogenase 1-like (*GPD1L*) mRNA to repress its translation. In a normoxic environment, GPD1L protein binds prolyl hydroxylase (PHD) to promote hyperhydroxylation of hypoxia inducible factor 1-alpha (HIF-1α) to mediate its polyubiquitination and subsequent degradation by the proteasome (Huang et al., 2009; Mathew & Simon, 2009). At low oxygen concentrations, such as in solid breast cancer tumors, miR-210 represses *GPD1L* mRNA. This, in turn, decreases PHD activity, stabilizing cytoplasmic HIF-1α levels, allowing for its dimerization with hypoxia inducible factor 1-beta (HIF-1β) in the nucleus, to form the active HIF-1 transcription factor and turn on hypoxia-associated genes. As miR-210 is among the genes upregulated by HIF-1, overexpressed miR-210 triggers a positive feedback loop to drive further miR-210 expression (Kelly, Souza, Clish, & Puigserver, 2011).

Application of TargapremiR-210 to hypoxic, but not normoxic, breast cancer cell lines silenced miR-210 specifically and short-circuited the miR-210-HIF-1α pathway. In particular, it increased expression of GPD1L, decreased expression of HIF-1α, and triggered apoptosis only in hypoxic breast cancer (no effect on normoxic breast cancer cells). Further, TargapremiR-210's activity was reduced in hypoxic cells upon forced overexpression of miR-210. Notably, TargapremiR-210 inhibited hypoxic tumor growth in a mouse model (Costales et al., 2017).

Altogether, these studies show that organic ligands can be designed to target RNA using only its sequence; the only other manner in which sequence-based design had been previously possible is via oligonucleotide-based approaches. These designed compounds can provide lead Targeted Therapeutics against previously thought undruggable RNA targets both in cells and in animal models. Notably, Inforna-designed compounds can be lead optimized using chemical similarity searching (Kumar et al., 2012; Parkesh et al., 2012), traditional medicinal chemistry approaches including defining structure-activity relationships (SAR), and targeting multiple structures within the same RNA with a single small molecule, or multivalency (Rzuczek et al., 2017, 2013; Velagapudi et al., 2016; Yang, Gao, Southern, Sarkar, & Disney, 2016).

3.2 Materials

1. Sequence of the desired RNA target or its .CT file.
2. Software to generate a .CT file of the RNA target's secondary structure, which can be generated online by mFold (http://mfold.rna.albany.edu), RNAstructure (https://rna.urmc.rochester.edu/RNAstructureWeb/) or ViennaFold (http://rna.tbi.univie.ac.at/). RNAstructure can also be downloaded and run locally (https://rna.urmc.rochester.edu/register.html).
3. To use Inforna, visit https://disney.florida.scripps.edu/software/ and submit a software license agreement. Upon submission, log-in credentials will be provided.

3.3 Equipment

1. A computer with internet access and a web browser

3.4 Procedure

A detailed procedure for using Inforna to generate lead small molecules for an RNA target of interest is available in (Disney et al., 2016) in the Supporting Information. In brief:

1. Generate a .CT file, a connectivity table that describes the RNA's secondary structure. Ensure that the sequence does not contain Ts (Us only).
2. Log in to Inforna using the credentials provided after submission of a software license agreement.
3. Upload the .CT file.

4. Choose search options, whether to force loop nucleotides and closing base pairs to match an RNA motif-small molecule interaction in the database (most stringent) or loop nucleotides only (least stringent).
5. Search the database and export the results as a.csv file. The results will include the motif in the RNA target, motif in the database, simplified molecular-input line-entry system (SMILES) notations for the lead small molecules, and Fitness Scores and related meta data.

4. Validating the targets of SMIRNAs in cells

4.1 Overview of target validation

Target validation is critical to define SMIRNA mode of action and to define occupied off-targets. Target validation methods for RNA, however, have generally not been developed (Childs-Disney & Disney, 2016) and have not been considered broadly when developing SMIRNAs. In the protein world, activity-based protein profiling (Cravatt, Wright, & Kozarich, 2008) and other cross-linking methods (Parker et al., 2017) have been used to validate targets. The most common method to identify RNA-small molecule binding sites is to monitor sites of protection from nuclease cleavage or chemical modification reagents. For example, in vitro mapping studies identified sites in the ribosome that bound antibiotics (Moazed & Noller, 1987; Stern, Moazed, & Noller, 1988). Some binding sites, however, can go undetected: (i) if nucleotides within a binding site do not react with a chemical modifying reagent; (ii) the irreversible nature of the reaction with the probe, which would require long SMIRNA residence time. These are not trivial issues and significantly complicate analysis and detection.

In this section, we describe the development and implementation of two target validation and profiling approaches for SMIRNAs that rely on cross-linking and cleavage, named Chemical Cross-Linking and Isolation by Pull-down (Chem-CLIP; Fig. 3) (Guan & Disney, 2013a; Yang, Wilson, Velagapudi, & Disney, 2015) and Small Molecule Nucleic Acid Profiling by Cleavage Applied to RNA (RiboSNAP; Fig. 4) (Li & Disney, 2018; Rzuczek et al., 2017), respectively. Indeed, both approaches have been further developed to map the binding sites of small molecules in cells. Because Chem-CLIP and RiboSNAP probes undergo a proximity-based reaction with the RNA target in a live cell, it mitigates complications observed with competitive chemical modification or nuclease protection discussed above.

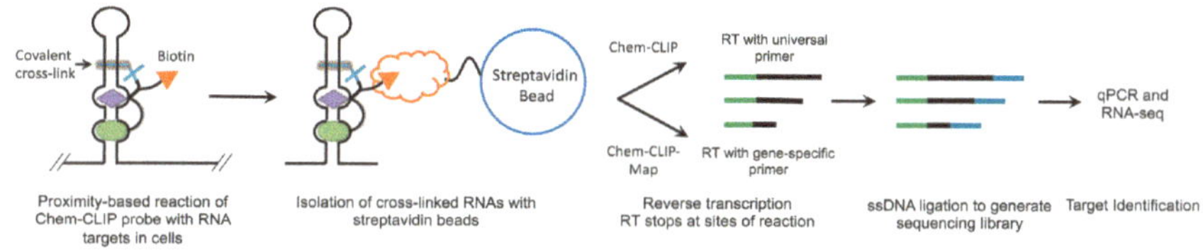

Fig. 3 Overview of the target validation and profiling method Chemical Cross-Linking and Isolation by Pull-down (Chem-CLIP). A Chem-CLIP probe is an SMIRNA appended with a cross-linking module (typically the nitrogen mustard chlorambucil) and a biotin purification module. Thus, RNA targets that bind the SMIRNA are brought into close proximity with the cross-linking module, forming a covalent adduct upon reaction. In brief, an appropriate cell line is incubated with the Chem-CLIP probe, followed by isolation of total RNA. RNAs that reacted with the Chem-CLIP probe are isolated and purified via its biotin tag and streptavidin beads. After elution of the adducts, the RNAs are reverse transcribed with either a universal primer (Chem-CLIP to identify all targets) or a gene-specific primer (Chem-CLIP-Map to identify the SMIRNA binding site). After RT, a single stranded DNA adaptor is ligated to enable generation of a sequencing library; qPCR and sequencing (RNA-seq) are then completed.

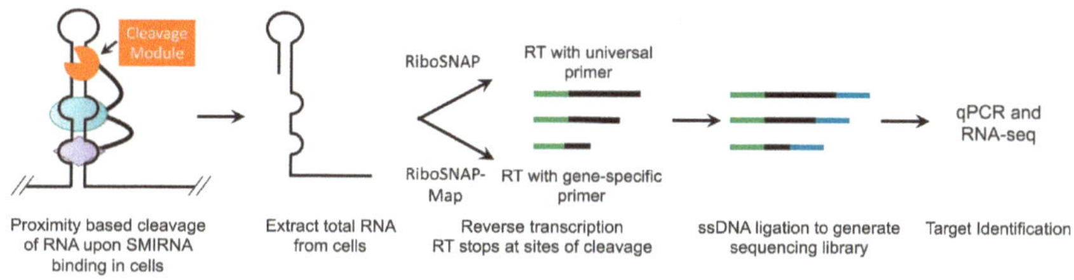

Fig. 4 Overview of the target validation and profiling method Small Molecule Nucleic Acid Profiling by Cleavage Applied to RNA (RiboSNAP). A RiboSNAP probe is an SMIRNA appended with a cleavage module, which can be activated by light (such as N-HPT), recruit a nuclease to effect cleavage at the target site (RIBOTACs), or cleave the target directly (bleomycin). RNA targets that bind the SMIRNA bring the cleavage module into close proximity such that the RNA is cleaved. Briefly, cells are treated with a RiboSNAP probe and total RNA is isolated. In contrast to Chem-CLIP studies, total RNA can be analyzed directly, avoiding the pull-down step. Reverse transcription is either completed with a universal primer to identify all targets (RiboSNAP) or with a gene-specific primer to localize the SMIRNA binding site (RiboSNAP-Map). A sequencing library is then generated by ligation of a single stranded DNA adaptor and subsequent qPCR and RNA-seq analysis.

4.2 Chemical cross-linking and isolation by pull-down (Chem-CLIP)

The Chem-CLIP approach is a chemical method of cross-linking that is operationally similar to the technique described by Darnell and co-workers to define protein-RNA binding sites via Cross-linking and immunoprecipitation (CLIP) (Licatalosi et al., 2008; Ule et al., 2003).

In Chem-CLIP, two functionalities are installed onto a SMIRNA; (i) a proximity-based chemical crosslinking unit; and (ii) a tag that allows for purification of cross-linked material (Fig. 3) (Guan & Disney, 2013a; Yang et al., 2015). These units must be installed in positions that do not significantly affect molecular recognition. The position used for covalent attachment to array surfaces in 2DCS studies or the scaffold used to link the fragments (i.e., a peptidomimetic such as a peptoid). After incubating cells with a Chem-CLIP probe, biotin is used to isolate and purify RNA targets that reacted with the cross-linker (pull-down) (Fig. 3). The targets are identified by RT-qPCR or RNA-seq.

The molecular recognition of TargapremiR-210 was studied in cells by using Chem-CLIP. In these studies, the interaction the SMIRNA with hundreds of cellular RNAs of different types was investigated (Costales et al., 2017). Indeed, the TargapremiR-210 Chem-CLIP probe bound the desired target, pre-miR-210, in cells to the greatest extent, as determined by measuring its enrichment in the pulled down fractions vs. lysate. Important control experiments included Competitive Chem-CLIP (C-Chem-CLIP) in which cells are co-treated with the Chem-CLIP probe and parent SMIRNA; the SMIRNA competes for binding to the RNA target and prevents reaction with the Chem-CLIP probe. C-Chem-CLIP studies with TargapremiR-210 showed that the parent SMIRNA bound pre-miR-210 in breast cancer cells. Furthermore, C-Chem-CLIP allows the relative occupancy of a series of SMIRNAs to a target (Costales et al., 2017; Disney et al., 2012; Yang et al., 2015).

By analysis of the pulled down material from a Chem-CLIP study, differential selectivity of a SMIRNA in cells can be assessed. As mentioned above, TargapremiR-210 engaged the desired target to the greatest extent, demonstrating specific binding. These studies revealed that other RNA targets were occupied, albeit to lesser extent, yet their biological function was unaffected. For example, the motif in pre-miR-210's Dicer site is also present in pre-miR-497, and both are occupied by the SMIRNA. However, only miR-210 abundance is affected because the binding site is located in a *functional*, Dicer processing site.

Chem-CLIP was also applied to fragment assembled compounds that target pri-miR-96 (Velagapudi et al., 2016) and to RNA repeat expansions including the r(CUG) repeat that causes myotonic dystrophy type 1 (Rzuczek et al., 2017), the r(CGG) repeat that causes fragile X-associated tremor ataxia syndrome (Yang et al., 2015, 2016), and the r(G_4C_2) repeat that causes C9orf72 amyotrophic lateral sclerosis and frontotemporal dementia (c9ALS/FTD) (Su et al., 2014; Wang et al., 2018). Interestingly, these studies showed that SMIRNAs selectively recognize mutant alleles containing the RNA repeat expansion. For the case of r(CUG)-targeting ligands, the SMIRNA targets disease driving repeats over other transcripts with shorter, non-toxic repeats of r(CUG), including the wild type allele (Rzuczek et al., 2017). Chem-CLIP has also been used to show that known drugs target microRNA precursors in cancer cells, supporting the notion that anti-cancer drugs target RNA and affect their biology (Velagapudi et al., 2018). Perhaps other drugs target RNA and affect its biology.

As mentioned above, Chem-CLIP has been further developed to map the binding sites of SMIRNAs in cells, or Chem-CLIP-Map (Fig. 3). In the first iteration, binding sites are revealed by completing a traditional Chem-CLIP experiment, digesting the captured RNAs with antisense oligonucleotides and RNase H, and analyzing the fragments by RT-qPCR. This technique was developed for RNA repeat expansions as it is difficult for reverse transcriptase (RT) to read through the repeats and to PCR amplify. In multiple examples, this method localized the binding of an SMIRNA to its cognate repeat expansion (Disney et al., 2012; Rzuczek et al., 2017; Yang et al., 2015; Yang, He, et al., 2016). These binding sites gratifyingly are in accordance with Inforna-based design.

For other RNA targets, RT-PCR, including RNA-seq analysis, can be used to define cellular binding sites. In these experiments, the site that the probe reacts with in an RNA target can be detected as an "RT stop" (Sexton, Wang, Rutenberg-Schoenberg, & Simon, 2017; Velagapudi, Li, & Disney, 2019). Thus, pulled down RNA can be reverse transcribed with a gene-specific RT primer, followed by addition of adaptors and amplification via RT-PCR. Chem-CLIP-Map-Seq was validated with Targaprimir-96, demonstrating that the reaction site for the Chem-CLIP probe is proximal to the SMIRNA binding site and in accordance with Inforna design and modeling (Velagapudi et al., 2019).

4.2.1 Materials

1. Chem-CLIP probe
2. Disease-relevant cell line and corresponding growth medium
3. TRIzol reagent (Ambion)
4. miRNeasy Kit (Qiagen)
5. Quick-RNA Miniprep Kit (Zymo Research)
6. Dynabeads™ MyOne™ Streptavidin C1 (Thermo Fisher)
7. Ethanol and isopropanol

8. 1 × Wash Buffer: 10 mM Tris-HCl, pH 7.0, 1 mM EDTA, 4 M NaCl, 0.2% (v/v) Tween-20
9. 1 × Elution Buffer: 10 mM Tris-HCl pH 8.0, 5 mM EDTA, 1 mM biotin, and 1 μL SUPERase In RNase Inhibitor (20 units/μL; Thermo Fisher), 2.5 μL Proteinase K
10. Gene specific primers
11. qScript cDNA Synthesis Kit (Quantabio)
12. miScript II RT kit (Qiagen)
13. Power SYBR Green Master Mix (Thermo Fisher Scientific)
14. Superscript™ III Reverse Transcription Kit (Thermo Fisher Scientific)
15. RNase H (2 units/μL; New England Biolabs)
16. RNase A (10 mg/mL; ThermoFisher Scientific)
17. Agencourt RNAClean XP (Beckman Coulter)
18. ssDNA adaptor: 5′phosphate-NNNAGATCGGAAGAGCG-TCGTGTAG-3C spacer
19. T4 RNA Ligase I (NEB)
20. 1 mM ATP
21. 50% PEG-8000
22. Phusion polymerase (NEB)
23. AMPure XP beads (Beckman Coulter)
24. Agencourt XP beads (Beckman Coulter)
25. PCR Cloning Kit (such as NEB® PCR Cloning Kit, catalog #: E1202)

4.2.2 Equipment

1. Clinical centrifuge
2. Microcentrifuge
3. qPCR thermocycler (such as an Applied Biosystem's 7900HT Fast Real Time PCR System)
4. PCR thermocycler
5. Heat block, water bath, incubator
6. Nanodrop UV spectrophotometer
7. Magnetic rack
8. Gel electrophoresis equipment
9. Phosphorimager (such as Molecular Dynamics Typhoon; GE Healthcare Life Sciences)

4.2.3 Protocol (Fig. 3)

1. Culture the appropriate cell line in complete growth medium (100 mm dishes) to about 70–80% confluency. Treat cells with appropriate Chem-CLIP probe in growth medium for 18–24 h. Although the concentration

of Chem-CLIP probe is likely unique for each target, we typically use the concentration that reduces levels of the mature miRNA by 50% in an appropriate cell line, as determined by RT-qPCR. In the case of miR-96, the concentration of the Chem-CLIP probe used was 1 μM, where levels of mature miR-96 were reduced by 50% in MDA-MB-231 cells.

2. Extract total RNA using either TRIzol reagent, Quick-RNA Miniprep Kit, or miRNeasy Kit as appropriate per manufacturer's protocol. Quantify the amount of RNA isolated using a Nanodrop spectrophotometer.
3. Using approximately 30 μg of total RNA, pull down the RNAs that reacted with the Chem-CLIP probe by incubating with 50 μL of Dynabeads MyOne Streptavidin C1 beads in 1 × PBS and gently shaking for 30 min at room temperature.
4. Separate the beads from the supernatant containing unbound RNAs by using a magnetic rack. Remove the supernatant.
5. Wash the beads containing RNA-compound adducts three times with 1 × Wash Buffer and then twice with nuclease free water.
6. Elute bound RNA in 200 μL of 1 × Elution Buffer by heating at 37 °C for 30 min. Repeat this step, combining the fractions.
7. Purify the eluted RNA using Zymo's Quick-RNA Miniprep Kit per the manufacturer's protocol.
8. Amplification
   a. Chem-CLIP:
      i. Reverse transcribe approximately 200–1000 ng of total RNA, which can be completed using a miScript II RT Kit (for miRNAs) or a qScript cDNA Synthesis Kit (for longer RNAs) per the manufacturer's protocol.
      ii. Complete qPCR using Power SYBR Green Master Mix per the manufacturer's recommended protocol and a Real Time PCR amplification thermocycler such as an Applied Biosystem's 7900HT Fast Real Time PCR System. Normalize expression levels to a housekeeping gene or to 18S rRNA.
   b. Chem-CLIP-Map-Seq:
      iii. Reverse transcription: Reverse transcribe ~500 ng of pulled down RNA with a gene-specific primer with an adapter sequence at the 3′ end (GATCTCGCAGCTGCGGGTCCT) using Superscript III (SSIII) per the manufacturer's protocol. Next, add 0.5 μL of 2 units/μL RNase H and 0.5 μL of 10 mg/mL RNase A and incubate at 37 °C for 30 min to degrade RNA. Clean up the RT reaction using an Agencourt

RNAClean XP Kit per manufacturer's protocol for small RNAs by adding 1.8 volume Agencourt RNAClean XP beads, 3 volumes 2.5 M NaCl, 20% PEG-8000, and 1 volume isopropanol.

iv. ssDNA ligation: Ligate the purified RT reaction to a ssDNA adaptor (5′phosphate-NNNAGATCGGAAGAGCGTCGT GTAG-3C spacer) using T4 RNA ligase I. In a 20 μL total volume, add together 2 μL 10 × T4 RNA ligase buffer, 1 μL 1 mM ATP, 10 μL 50% PEG-8000, 5 μL cDNA (from step i), 1 μL of 20 μM ssDNA adaptor, and 1 μL of T4 RNA ligase and incubate overnight at room temperature. After overnight incubation, clean up using RNA Clean XP as described above in step i.

v. Amplification: PCR amplify the cDNA from step ii using Phusion polymerase using the following primers: 5′-CTACACGACGCTCTTCCGATCT-3′ and 5′-CAGA-CGTGCTCTTCCGAT-3′. Clean up the PCR reaction using Ampure XP beads by adding 1.8 volume Agencourt Ampure XP beads, 3 volumes of 2.5 M NaCl, 20% PEG-8000, and 1 volume isopropanol.

vi. Generation of sequencing library: Generate a sequencing library using NEBNext® Multiplex Oligos for Illumina® (NEB) per the manufacturer's recommended protocol.

vii. Sequence the library: Either clone the library generated in (iv) into a vector for sequencing or sequence directly with an Illumina® high throughput sequencer.

viii. Analyze the sequencing data by comparing probe-treated and vehicle-treated samples. In particular, calculate the percentage of sequencing reads that terminate at a given nucleotide in both data sets and then determined statistical significance using a two-tailed Student *t*-test or the like. Statistically significant differences identify the cross-linked region and hence the binding site.

4.3 Small molecule nucleic acid profiling by cleavage applied to RNA (Ribo-SNAP)

In the Chem-CLIP studies, target profiling is accomplished by pulling down cross-linked RNA targets. Pull-down and isolation of the targets can be laborious. Thus, we sought to develop an approach to profile the binding of SMIRNAs directly from total RNA harvested from cells by using small molecules that cleave an RNA target. This RiboSNAP approach can be implemented in three different ways, each of which has a subset of ideal targets (Fig. 4). Importantly, this approach has been used to demonstrate exquisite selectively for SMIRNAs in animal models. Compounds have cleaved a designed target and have not cleavage off-targets as assessed via RNA-seq. from mouse tissues (Angelbello et al., 2019).

In the first RiboSNAP method reported, a SMIRNA was appended with a cleavage module that is activated by light (Guan & Disney, 2013b). Briefly, 1-hydroxy-6-thioxo-1,6-dihydropyridine-2-carboxylic acid (N-HPT) was appended to an RNA binder. Upon irradiation with light, N-HPT produced hydroxyl radicals that cleaved the desired RNA (Guan & Disney, 2013b). As light is required for targeted cleavage, the tissues to which this approach can be applied are limited.

The second approach uses a small molecule to recruit an endogenous nuclease to the desired target to induce its cleavage. As the cleavage-inducing probe is comprised of RNA-binding and nuclease-recruiting modules, they were dubbed ribonuclease targeting chimeras, or RIBOTACs (Costales, Matsumoto, Velagapudi, & Disney, 2018). Because cleavage is affected by a recruited nuclease and not the small molecule itself, the sites of cleavage can be complex and distal to the SMIRNA binding site.

To alleviate issues encountered with light-activated cleavage modules (N-HPT) and complex cleavage patterns (RIBOTACs), we employed bleomycin A5 as cleaving module for RiboSNAP probes (Fig. 4). Bleomycin A5, a derivative of the well-known DNA cleavage anti-tumor agent bleomycin, has a free amine that drives DNA-binding affinity and cleavage. Fortuitously, this amine provides a conjugation site to a SMIRNA while reducing DNA affinity and cleavage. Indeed, conjugation of the cleaving module at this site allows for RNA specific cleavage in vitro and in cells (Li & Disney, 2018; Rzuczek et al., 2017).

We have validated using bleomycin A5 as the cleavage module, showing in both cases that conjugation to an RNA binder allows for selective cleavage. That is, it appears that conjugation of any RNA-binder to bleomycin A5's free amine direct cleavage away from DNA and to the RNA target to which the SMIRNA binds. For example, when bleomycin A5 was appended to a dimeric SMIRNA that selectively recognizes r(CUG) repeat expansion, cleavage was not only selective for the mutant allele over wild type but also over transcripts with short, non-disease-causing r(CUG) repeats, mirroring the results obtained from the corresponding Chem-CLIP probe (Rzuczek et al., 2017). Likewise, conjugation of bleomycin A5 to Targaprimir-96 led to selective cleavage of pri-miR-96 in vitro and in cells. Further, profiling of all miRNAs expressed in triple negative breast cancer cells showed that only miR-96 was significantly affected (Li & Disney, 2018).

The exact binding sites of bleomycin RiboSNAP probes, or RiboSNAP-Map, can be determined similarly as described for Chem-CLIP-Map (Fig. 4). After incubation with the RiboSNAP probe, total RNA is harvested and subjected to gene-specific RT for the RNA target that was depleted in profiling experiments (e.g., pri-miR-96), followed by ligation of DNA adaptors to which PCR products are complementary. After PCR amplification, the products are sequenced, and binding sites are revealed where amplification stops. Akin to the results observed for Chem-CLIP-Map experiments, the binding site of Targaprimir-96 was proximal to the ligand binding site, in corroboration with Inforna design and three-dimensional modeling (Li & Disney, 2018). Below, we describe implementation of RiboSNAP using bleomycin A5 as the cleavage module to map SMIRNA binding sites.

4.3.1 Materials

Please see Section 4.2.1, except item #1 is a RiboSNAP probe

4.3.2 Equipment

Please see Section 4.2.2

4.3.3 Protocol (Fig. 4)

1. Culture the appropriate cell line in complete growth medium in 100 mm dishes to ~70% confluency. Treat with a bleomycin RiboSNAP probe, prepared in growth medium, for 6 h (optimized to obtain specific cleavage sites without destroying the target RNA and is likely specific to each target). Akin to Chem-CLIP studies described above, the concentration of RiboSNAP probe typically used reduces the cellular levels of the RNA of interest by ~50%, as determined by RT-qPCR. In the case of miR-96, the concentration of bleomycin RiboSNAP probe was chosen based on the above criterion, or 500 nM in MDA-MB-231 cells.
2. Extract total RNA using TRIzol reagent per the manufacturer's recommended protocol and quantify the amount present by using a Nanodrop spectrophotometer.
3. Reverse transcription and PCR amplification

a. RiboSNAP
  i. Reverse transcribe approximately 1 μg of total RNA using a miScript II RT Kit (for miRNAs) or a qScript cDNA Synthesis Kit (for longer RNAs) per the manufacturer's protocol.
  ii. Complete qPCR using Power SYBR Green Master Mix per the manufacturer's recommended protocol and Real Time PCR amplification thermocycler. Normalize expression levels to a housekeeping gene or to 18S rRNA.

b. RiboSNAP-Map
  i. Reverse transcribe total RNA using a gene specific primer containing a 5′ adapter and SuperScript™ III. Anneal the primer by incubating the following at 65 °C for 5 min followed by ice for 5 min: 10 μg of total RNA, 2 pmol primer, and 1 μL of 10 mM dNTP mix in a total volume of 13 μL. Then, add 4 μL 5 × First-Strand Buffer, 1 μL of 0.1 M DTT, 1 μL of RNaseOUT, and 1 μL of SSIII RT. Incubate the samples at at 50 °C for 1 h and then at 85 °C for 10 min to inactivate the RT.
  ii. Add 0.5 μL of 2 units/μL RNase H and 0.5 μL of 10 mg/mL RNase A to the RT reaction and incubate at 37 °C for 30 min to degrade the RNA.
  iii. Follow steps (ii)–(iv) in Section 4.2.3, Chem-CLIP-Map-Seq to construct libraries for sequencing and hence to identify binding sites.

5. Summary

Herein, we have provided an overview for the design of small molecules that target RNA structures from sequence (Figs. 1 and 2). Indeed, many RNA structures are implicated in disease, particularly neurological diseases (Bernat & Disney, 2015). As oligonucleotides are best suited for targeting unstructured regions, small molecules provide an important complementary strategy for deactivating disease-causing RNAs. Essential to the development of SMIRNAs are target validation methods to confirm compound mode of action. Although these methods have been developed and implemented in the protein-targeting field, they have generally been lacking for RNA. A detailed description of two methods, Chem-CLIP (based on covalent cross-linking upon binding of the SMIRNA to the target; Fig. 3) and RiboSNAP (based on cleavage of the RNA target upon SMIRNA binding; Fig. 4), including advantages and limitations have been provided along with protocols for their implementation.

Acknowledgment

This work was funded by the National Institutes of Health Grants R01-GM097455, DP1-NS096898, and P01-NS099114 to M.D.D.

References

Aminova, O., & Disney, M. D. (2010). A microarray-based method to perform nucleic acid selections. *Methods in Molecular Biology*, *669*, 209–224.

Angelbello, A. J., Chen, J. L., Childs-Disney, J. L., Zhang, P., Wang, Z. F., & Disney, M. D. (2018). Using genome sequence to enable the design of medicines and chemical probes. *Chemical Reviews*, *118*(4), 1599–1663.

Angelbello, A. J., Rzuczek, S. G., Mckee, K. K., Chen, J. L., Olafson, H., Cameron, M. D., et al. (2019). Precise small-molecule cleavage of an r(CUG) repeat expansion in a myotonic dystrophy mouse model. *Proceedings of the National Academy of Sciences of the United States of America*, *116*(16), 7799–7804. https://doi.org/10.1073/pnas.1901484116.

Bartel, D. P. (2004). MicroRNAs: Genomics, biogenesis, mechanism, and function. *Cell*, *116*(2), 281–297.

Bartel, D. P. (2009). MicroRNAs: Target recognition and regulatory functions. *Cell*, *136*(2), 215–233.

Bernat, V., & Disney, M. D. (2015). RNA structures as mediators of neurological diseases and as drug targets. *Neuron*, *87*(1), 28–46.

Childs-Disney, J. L., & Disney, M. D. (2016). Approaches to validate and manipulate RNA targets with small molecules in cells. *Annual Reviews of Pharmacology and Toxicology*, *56*, 123–140.

Childs-Disney, J. L., Wu, M., Pushechnikov, A., Aminova, O., & Disney, M. D. (2007). A small molecule microarray platform to select RNA internal loop-ligand interactions. *ACS Chemical Biology*, *2*(11), 745–754.

Costales, M. G., Haga, C. L., Velagapudi, S. P., Childs-Disney, J. L., Phinney, D. G., & Disney, M. D. (2017). Small molecule inhibition of microRNA-210 reprograms an oncogenic hypoxic circuit. *Journal of the American Chemical Society*, *139*(9), 3446–3455.

Costales, M. G., Matsumoto, Y., Velagapudi, S. P., & Disney, M. D. (2018). Small molecule targeted recruitment of a nuclease to RNA. *Journal of the American Chemical Society*, *140*(22), 6741–6744.

Cravatt, B. F., Wright, A. T., & Kozarich, J. W. (2008). Activity-based protein profiling: From enzyme chemistry to proteomic chemistry. *Annual Reviews of Biochemistry*, *77*, 383–414.

Disney, M. D., Dwyer, B. G., & Childs-Disney, J. L. (2018). Drugging the RNA world. *Cold Spring Harbor Perspectives in Biology*, *10*(11), a034769.

Disney, M. D., Liu, B., Yang, W. Y., Sellier, C., Tran, T., Charlet-Berguerand, N., et al. (2012). A small molecule that targets $r(CGG)^{exp}$ and improves defects in fragile X-associated tremor ataxia syndrome. *ACS Chemical Biology*, 7(10), 1711–1718.

Disney, M. D., Winkelsas, A. M., Velagapudi, S. P., Southern, M., Fallahi, M., & Childs-Disney, J. L. (2016). Inforna 2.0: A platform for the sequence-based design of small molecules targeting structured RNAs. *ACS Chemical Biology*, *11*(6), 1720–1728.

Guan, L., & Disney, M. D. (2013a). Covalent small-molecule-RNA complex formation enables cellular profiling of small-molecule-RNA interactions. *Angewandte Chemie International Edition in English*, *1*(10), 201301639.

Guan, L., & Disney, M. D. (2013b). Small molecule-mediated cleavage of RNA in living cells. *Angewandte Chemie International Edition in English*, *52*(5), 1462–1465.

Huang, X., Ding, L., Bennewith, K. L., Tong, R. T., Welford, S. M., Ang, K. K., et al. (2009). Hypoxia-inducible mir-210 regulates normoxic gene expression involved in tumor initiation. *Molecular Cell*, *35*(6), 856–867.

Jones, D., Metzger, H. J., Schatz, A., & Waksman, S. A. (1944). Control of gram-negative bacteria in experimental animals by streptomycin. *Science*, *100*(2588), 103–105.

Kelly, T. J., Souza, A. L., Clish, C. B., & Puigserver, P. (2011). A hypoxia-induced positive feedback loop promotes hypoxia-inducible factor 1alpha stability through miR-210 suppression of glycerol-3-phosphate dehydrogenase 1-like. *Molecular and Cellular Biology*, *31*(13), 2696–2706.

Kumar, A., Parkesh, R., Sznajder, L. J., Childs-Disney, J. L., Sobczak, K., & Disney, M. D. (2012). Chemical correction of pre-mRNA splicing defects associated with sequestration of muscleblind-like 1 protein by expanded r(CAG)-containing transcripts. *ACS Chemical Biology*, 7(3), 496–505.

Li, Y., & Disney, M. D. (2018). Precise small molecule degradation of a noncoding RNA identifies cellular binding sites and modulates an oncogenic phenotype. *ACS Chemical Biology*, *13*, 3065–3071.

Licatalosi, D. D., Mele, A., Fak, J. J., Ule, J., Kayikci, M., Chi, S. W., et al. (2008). HITS-CLIP yields genome-wide insights into brain alternative RNA processing. *Nature*, *456*(7221), 464–469.

Liu, B., Childs-Disney, J. L., Znosko, B. M., Wang, D., Fallahi, M., Gallo, S. M., et al. (2016). Analysis of secondary structural elements in human microRNA hairpin precursors. *BMC Bioinformatics*, *17*, 112.

Mathew, L. K., & Simon, M. C. (2009). Mir-210: A sensor for hypoxic stress during tumorigenesis. *Molecular Cell*, *35*(6), 737–738.

Mathews, D. H. (2014). RNA secondary structure analysis using RNAstructure. *Current Protocols in Bioinformatics*, *46*, [12.16.11-12.16.25].

Mathews, D. H., Disney, M. D., Childs, J. L., Schroeder, S. J., Zuker, M., & Turner, D. H. (2004). Incorporating chemical modification constraints into a dynamic programming algorithm for prediction of RNA secondary structure. *Proceedings of the National Academy of Sciences of the United States of America*, *101*(19), 7287–7292.

Mathews, D. H., & Turner, D. H. (2002). Dynalign: An algorithm for finding the secondary structure common to two RNA sequences. *Journal of Molecular Biology*, *317*(2), 191–203.

Moazed, D., & Noller, H. F. (1987). Interaction of antibiotics with functional sites in 16S ribosomal-RNA. *Nature*, *327*(6121), 389–394.

Morrow, T. (2017). New therapy for spinal muscular atrophy offers modest bang for pharamaceutical buck. *Managed Care*, *26*(2), 36–37.

Parker, C. G., Galmozzi, A., Wang, Y., Correia, B. E., Sasaki, K., Joslyn, C. M., et al. (2017). Ligand and target discovery by fragment-based screening in human cells. *Cell*, *168*(3), 527–541. [e529].

Parkesh, R., Childs-Disney, J. L., Nakamori, M., Kumar, A., Wang, E., Wang, T., et al. (2012). Design of a bioactive small molecule that targets the myotonic dystrophy type 1 RNA via an RNA motif-ligand database and chemical similarity searching. *Journal of the American Chemical Society*, *134*(10), 4731–4742.

Poehlsgaard, J., & Douthwaite, S. (2005). The bacterial ribosome as a target for antibiotics. *Nature Reviews Microbiology*, *3*(11), 870–881.

Rouskin, S., Zubradt, M., Washietl, S., Kellis, M., & Weissman, J. S. (2014). Genome-wide probing of RNA structure reveals active unfolding of mRNA structures in vivo. *Nature*, *505*(7485), 701–705.

Rzuczek, S. G., Colgan, L. A., Nakai, Y., Cameron, M. D., Furling, D., Yasuda, R., et al. (2017). Precise small-molecule recognition of a toxic CUG RNA repeat expansion. *Nature Chemical Biology*, *13*(2), 188–193.

Rzuczek, S. G., Gao, Y., Tang, Z. Z., Thornton, C. A., Kodadek, T., & Disney, M. D. (2013). Features of modularly assembled compounds that impart bioactivity against an RNA target. *ACS Chemical Biology*, *8*(10), 2312–2321.

Sexton, A. N., Wang, P. Y., Rutenberg-Schoenberg, M., & Simon, M. D. (2017). Interpreting reverse transcriptase termination and mutation events for greater insight into the chemical probing of RNA. *Biochemistry*, *56*(35), 4713–4721.

Spitale, R. C., Flynn, R. A., Zhang, Q. C., Crisalli, P., Lee, B., Jung, J. W., et al. (2015). Structural imprints in vivo decode RNA regulatory mechanisms. *Nature*, *519*(7544), 486–490.

Stein, C. A., & Castanotto, D. (2017). FDA-approved oligonucleotide therapies in 2017. *Molecular Therapy*, *25*(5), 1069–1075.

Stephenson, M. L., & Zamecnik, P. C. (1978). Inhibition of Rous sarcoma viral RNA translation by a specific oligodeoxyribonucleotide. *Proceedings of the National Academy of Sciences of the United States of America*, *75*(1), 285–288.

Stern, S., Moazed, D., & Noller, H. F. (1988). Structural analysis of RNA using chemical and enzymatic probing monitored by primer extension. *Methods in Enzymology*, *164*, 481–489.

Su, Z., Zhang, Y., Gendron, T. F., Bauer, P. O., Chew, J., Yang, W. Y., et al. (2014). Discovery of a biomarker and lead small molecules to target r(GGGGCC)-associated defects in c9FTD/ALS. *Neuron*, *83*(5), 1043–1050.

Tenson, T., & Mankin, A. (2006). Antibiotics and the ribosome. *Molecular Microbiology*, *59*(6), 1664–1677.

Ule, J., Jensen, K. B., Ruggiu, M., Mele, A., Ule, A., & Darnell, R. B. (2003). CLIP identifies Nova-regulated RNA networks in the brain. *Science*, *302*(5648), 1212–1215.

Velagapudi, S. P., Cameron, M. D., Haga, C. L., Rosenberg, L. H., Lafitte, M., Duckett, D. R., et al. (2016). Design of a small molecule against an oncogenic noncoding RNA. *Proceedings of the National Academy of Sciences of the United States of America*, *113*(21), 5898–5903.

Velagapudi, S. P., Costales, M. G., Vummidi, B. R., Nakai, Y., Angelbello, A. J., Tran, T., et al. (2018). Approved anti-cancer drugs target oncogenic non-coding RNAs. *Cell Chemical Biology*, *25*(9), 1086–1094.

Velagapudi, S. P., Gallo, S. M., & Disney, M. D. (2014). Sequence-based design of bioactive small molecules that target precursor microRNAs. *Nature Chemical Biology*, *10*(4), 291–297.

Velagapudi, S. P., Li, Y., & Disney, M. D. (2019). A cross-linking approach to map small molecule-RNA binding sites in cells. *Bioorganic & Medical Chemistry Letters*, *29*(12), 1532–1536.

Velagapudi, S. P., Luo, Y., Tran, T., Haniff, H. S., Nakai, Y., Fallahi, M., et al. (2017). Defining RNA-small molecule affinity landscapes enables design of a small molecule inhibitor of an oncogenic noncoding RNA. *ACS Central Science*, *3*(3), 205–216.

Wan, Y., Qu, K., Zhang, Q. C., Flynn, R. A., Manor, O., Ouyang, Z., et al. (2014). Landscape and variation of RNA secondary structure across the human transcriptome. *Nature*, *505*(7485), 706–709.

Wang, Z. F., Ursu, A., Childs-Disney, J. L., Guertler, R., Yang, W. Y., Bernat, V., et al. (2018). The hairpin form of $r(G_4C_2)^{exp}$ in c9ALS/FTD is repeat-associated non-ATG translated and a target for bioactive small molecules. *Cell Chemical Biology*, *26*, 179–190.

Yang, W. Y., Gao, R., Southern, M., Sarkar, P. S., & Disney, M. D. (2016). Design of a bioactive small molecule that targets r(AUUCU) repeats in spinocerebellar ataxia 10. *Nature Communications*, 7 11647.

Yang, W. Y., He, F., Strack, R. L., Oh, S. Y., Frazer, M., Jaffrey, S. R., et al. (2016). Small molecule recognition and tools to study modulation of $r(CGG)^{exp}$ in fragile X-associated tremor ataxia syndrome. *ACS Chemical Biology*, *11*(9), 2456–2465.

Yang, W. Y., Wilson, H. D., Velagapudi, S. P., & Disney, M. D. (2015). Inhibition of non-ATG translational events in cells via covalent small molecules targeting RNA. *Journal of the American Chemical Society*, *137*(16), 5336–5345.

Zamecnik, P. C., & Stephenson, M. L. (1978). Inhibition of Rous sarcoma virus replication and cell transformation by a specific oligodeoxynucleotide. *Proceeding of the National Academy of Sciences of the United States of America*, *75*(1), 280–284.

CHAPTER FOUR

Dynamic combinatorial chemistry as a rapid method for discovering sequence-selective RNA-binding compounds

John D. McAnany[a], **Benjamin L. Miller**[b,*]
[a]Department of Chemistry, University of Rochester, Rochester, NY, United States
[b]Department of Dermatology, University of Rochester, Rochester, NY, United States
*Corresponding author: e-mail address: benjamin_miller@urmc.rochester.edu

Contents

Abstract

The ever-growing number of RNA species that are recognized as having a role in human disease is driving a demand for novel molecular probes and therapeutics. Producing sequence-selective RNA-binding molecules remains a substantial challenge, however. One approach that has been successful in producing molecules with high affinity and specificity for disease-relevant RNAs is the use of dynamic combinatorial chemistry, a fragment-based method in which fragments combine reversibly in the presence of the target. We describe methods for the design, synthesis, and screening of dynamic combinatorial libraries targeting RNA.

Methods in Enzymology, Volume 623
ISSN 0076-6879
https://doi.org/10.1016/bs.mie.2019.05.012

Abbreviations

DCL	dynamic combinatorial library
DCM	dichloromethane
DIPEA	*N,N*-diisopropylethylamine
DMF	dimethylformamide
Fmoc	fluorenylmethoxycarbonyl
Fmoc-Ahx-OH	6-(fluorenylmethoxycarbonyl-amino)hexanoic acid (CAS: 88574-06-5)
Fmoc-Ala-OH•2H$_2$O	N_α-(fluorenylmethoxycarbonyl)-L-alanine dihydrate (CAS: 35661-39-3)
Fmoc-Anp-OH	3-(fluorenylmethoxycarbonyl-amino)-3-(2-nitrophenyl) propanoic acid (CAS: 171778-06-6)
Fmoc-Asn(Trt)-OH	N_α-(fluorenylmethoxycarbonyl)-N_γ-triphenylmethyl-L-asparagine (CAS: 132388-59-1)
Fmoc-Cys(S*t*Bu)-OH	N_α-(fluorenylmethoxycarbonyl)-*S*-*tert*-butylthio-L-cysteine (CAS: 73724-43-3)
Fmoc-Cys(Trt)-OH	N_α-(fluorenylmethoxycarbonyl)-*S*-triphenylmethyl-L-cysteine (CAS: 103213-32-7)
Fmoc-His(Trt)-OH	N_α-(fluorenylmethoxycarbonyl)-$N_{(im)}$-triphenylmethyl-L-histidine (CAS: 109425-51-6)
Fmoc-Lys(Boc)-OH	N_α-(fluorenylmethoxycarbonyl)-N_ε-*tert*-butyloxycarbonyl-L-lysine (CAS: 71989-26-9)
Fmoc-Met-OH	N_α-(fluorenylmethoxycarbonyl)-L-methionine (CAS: 71989-28-1)
Fmoc-Phe-OH	N_α-(fluorenylmethoxycarbonyl)-L-phenylalanine (CAS: 35661-40-6)
Fmoc-Pro-OH	N_α-(fluorenylmethoxycarbonyl)-L-proline (CAS: 71989-31-6)
Fmoc-Ser(Trt)-OH	N_α-(fluorenylmethoxycarbonyl)-*O*-triphenylmethyl-L-serine (CAS: 111061-56-4)
Fmoc-Thr(Trt)-OH	N_α-(fluorenylmethoxycarbonyl)-*O*-triphenylmethyl-L-threonine (CAS: 133180-01-5)
Fmoc-Val-OH	N_α-(fluorenylmethoxycarbonyl)-L-valine (CAS: 68858-20-8)
FSS	frameshift stimulatory signal
HBTU	2-(1H-benzotriazole-1-yl)-1,1,3,3-tetramethyluronium hexafluorophosphate (CAS: 94790-37-1)
HIV-1	human immunodeficiency virus type 1
ITC	isothermal titration calorimetry
LC-MS	liquid chromatography-mass spectrometry
MALDI-MS	matrix assisted laser desorption ionization-mass spectrometry
RBDCC	resin-bound dynamic combinatorial chemistry
RB-DCL	resin-bound dynamic combinatorial library
SPPS	solid-phase peptide synthesis
SPR	surface plasmon resonance
***t*Bu**	*tert*-butyl

1. Introduction

Interest in the production of sequence-specific RNA-binding compounds has dramatically increased in recent years. This has occurred in tandem with an increasing understanding that complex RNA structures perform specific functions in human cells, and these functions have relevance to both human health and human disease. Targeting RNA sequences with synthetic molecules as probes for understanding biology or as candidate therapeutics is therefore recognized as an important goal. While the characteristics that define ideal RNA-binding compounds are not well understood, efforts by many laboratories are allowing the field to make rapid progress (Shortridge et al., 2018).

The generation and screening of large libraries of compounds is often the first step in producing a small molecule that will interact with a biological target. Fragment-based approaches are an attractive alternative to traditional library methods (Erlanson, McDowell, & O'Brien, 2004; Keserű et al., 2016). In particular, use of a dynamic library of fragments aids in the process of library generation by decreasing the complexity and labor needed to produce a given number of library species. A Dynamic Combinatorial Library (DCL) is assembled from a group of small molecules that can reversibly form more complex species, and either break apart or interconvert (Corbett et al., 2006; Li, Nowak, & Otto, 2013). Introduction of a target biomolecule causes the composition of this equilibrating mixture to shift, based on the affinity of library components for the biomolecule. Entirely solution-phase DCLs are common, but there are substantial benefits to immobilizing monomer species onto a solid support. Resin-Bound Dynamic Combinatorial Chemistry (RBDCC) libraries (RB-DCLs) were developed to aid analysis of results of a dynamic library selection experiment, as attachment of library components to the solid phase permits their spatial segregation and analysis by analytical techniques such as mass spectrometry (McNaughton & Miller, 2006).

Once the dynamic library is assembled, a screening procedure is performed. The library is allowed to come to equilibrium with and without a target molecule. Library species that are enriched in the presence of the target most likely have an affinity for that target. The structures of DCL-identified molecules can then be refined using standard medicinal chemistry methods, yielding compounds useful as molecular probes or therapeutic agents. As discussed in reviews referenced above, DCLs have been employed successfully in the identification of protein-binding compounds, receptors for small molecules, and even in the context of developing new materials. Here, we focus on the use of DCLs (and in particular, RB-DCLs) for the rapid identification of RNA-binding compounds.

By way of example, we describe a library that has been successfully screened against several RNA sequences. In particular, we will focus on its use in discovering compounds binding the Frameshift Stimulatory Signal (FSS) RNA of HIV-1. The FSS is a region of the Gag-Pol mRNA that forms a thermodynamically stable hairpin and induces a −1 ribosomal frameshift. How often this frameshift occurs is critical to the infectivity of the virus (Telenti et al., 2002). Hit compounds that were identified from this exemplar library have since been refined and shown to decrease the infectivity of HIV-1 by altering the degree of frameshifting, consistent with direct target RNA engagement in human cells (Ofori et al., 2014).

2. Dynamic combinatorial library production and use

Aspects of DCL design and use that are not in common with the synthesis of a more traditional library are discussed herein.

2.1 Selection of molecular fragments for RNA binding

The DCL consists of fragment molecules (monomers) and the structures formed dynamically from the fragments. All DCL components including monomers have the potential of interacting with the biological target. As such, the monomers and the moieties that make them up need to be designed carefully to not impart affinity without specificity. For example, in the case of RNA, use of aminoglycoside derivatives would clearly impart affinity for RNA, but the polycationic nature of aminoglycosides also makes it less likely that compounds identified will bind sequence-specifically (although there are also examples of the successful use of aminoglycosides to obtain sequence-selective RNA binders, so this is not absolute) (Childs-Disney, Parkesh, Nakamori, Thornton, & Disney, 2012; Luedtke, Liu, & Tor, 2003). Knowledge is slowly being acquired about which moieties provide this type of specific affinity for RNA to a compound, but this knowledge is in its infancy (Morgan, Forte, & Hargrove, 2018).

2.2 Selection of exchange chemistry

The exchange chemistry used in a DCL experiment must (1) be compatible with a broad range of functionality; (2) occur under conditions compatible with the biological target; (3) have sufficiently rapid kinetics to enable library equilibration within a reasonable time frame; and (4) allow for conditions whereby library exchange may be stopped. Many types of processes have been reported as suitable for DCLs; one of the most commonly employed is thiol-disulfide exchange. This exchange chemistry is convenient because it occurs at a rate fast enough for screening at or near physiological pH (pH approximately 7.4–8.0), and in the context of nucleic acids has little interaction with the biological target. While useful during library selection, the thiol-disulfide exchange chemistry suffers from the fact that disulfide molecules are quickly cleaved by intracellular glutathione, which is present at low millimolar concentrations. Because of this, disulfide molecules identified from a screen must generally be resynthesized with a bioisostere for the disulfide moiety before they are suitable for bioassay. The exchange between substituted hydrazines and hydrazones also has merits in a DCL (Frei et al., 2017; Simpson, Pittelkow, Watson, & Sanders, 2010). This exchange occurs under mild aqueous conditions, but it occurs slowly at physiological pH. Simultaneous or orthogonal disulfide and acylhydrazone exchange has also been reported (Gromova, Ciszewski, & Miller, 2012).

2.3 Simple library generation and equilibration characterization

Verification of exchange within a library, and an understanding of the time required for the composition of the library to reach equilibrium, is necessary prior to conducting a DCL screen. This determination is best performed with a simplified version of the planned DCL. To accomplish this, a DCL is assembled under the proposed exchange conditions. In the case of the thiol/disulfide DCL described herein, the exchange occurs in a buffered aqueous solution at a pH of 7.4. Assessment of library components in some manner (for example, by analytical HPLC) immediately after combining all components of the library is conducted to provide a time zero control. Then the amount of those species must be reassessed periodically until it is determined that an equilibrium is established.

One method to determine the time until equilibrium has been reached is to compare the final two measurements that are taken of the library species. If these measurements are not statistically different from each other, the library must have reached equilibrium before either of those measurements. For instance, an entirely solution-phase DCL is often monitored by HPLC. The library should be simple enough to allow most of the species within the library to be resolved. Disulfide DCLs of varying complexity and concentration appear to reach equilibrium on the timescale of hours to days. To determine the exact equilibration time of a particular library, analytical HPLC could be performed on an aliquot of the library immediately after assembly, and then at time points every 12 or 24 h afterward. When two successive chromatograms are produced that are not significantly different, the library must have reached equilibrium before the first of those time points. A more in depth discussion of equilibration within a library of this type has been undertaken by other authors (Moore & Zimmerman, 2000).

The production of species more complex than the monomers within a library is a necessary component of a DCL. While the equilibration time can be determined as above, characterization of exchange products is necessary to confirm that the intended library species are being dynamically produced. Mass spectrometry has been especially useful for this purpose. Library species are present in a complex mixture and are often difficult to isolate. Sampling the library, and subjecting this to LC-MS or MALDI-MS allows for the simultaneous identification of at least hundreds of unique ion peaks. The mass spectra can be examined to determine if dimers or oligomers of the DCL monomers are present as expected (Gromova et al., 2012).

2.4 A complex thiol-disulfide library

The synthesis of monomeric species to form a DCL can be expeditiously performed through Solid-Phase Peptide Synthesis (SPPS). The protection and deprotection conditions discussed herein constitute a standard Fmoc/tBu strategy. Here, we discuss the synthesis and evaluation of an RBDCL successfully employed for the identification of peptide-based binders to specific RNA sequences. The methodology is transferrable to many different library configurations.

2.4.1 Solution-phase monomer synthesis

This library contains three amino acid positions, and a capping heterocycle. Exactly one amino acid position is always cysteine. The next position contains one of five amino acids, and the last position contains one of five different amino acids. Amino acids and heterocyclic caps were chosen with three criteria in mind. First, molecular weights of the monomers needed to have the least amount of overlap possible, so as to simplify mass spectral analysis. Second, library constituents needed to yield the greatest chemical diversity possible within the constraints of the overall design. Finally, although some positively charged residues were included, their use was limited to prevent the formation of very highly charged species in the DCL. Synthesis of the library monomers was accomplished by splitting the synthesis six ways: first, into one group of monomers that would remain tethered to resin beads during screening, and second, each of these groups was split into three portions, each of which had the cysteine installed in one of the three possible positions. Installation of further diversity was accomplished using standard split-pool methods within each of the six larger pools. After synthesis was completed, the product was a mixture of an equal molar amount of every species that constitute one of the possible permutations from Fig. 1. The protocol below describes how to produce one of three portions of the solution-phase monomers needed for the complete DCL. This portion includes cysteine is in the first position, as defined in the C to N direction.

A 20 mL polymer chromatography column was filled with 500 mg of Wang Resin (100–200 mesh size, 1 mmol resin sites per gram of resin). To this, 10 eq. (1620 mg, 5 mmol) of 1,1′-carbonyldiimidazole was added in 12 mL of DMF and agitated by rotation for 12 h. The column was evacuated of the solution-phase via vacuum, and the resin was washed three times with 15 mL of DCM. To the vessel, 10 eq. (421 μL, 5 mmol) of 1,3-diaminopropane was added in 12 mL of DMF, and again this was rotated for 12 h. The solution-phase was removed, and the resin was washed as above. Cysteine was then coupled to the resin by adding the following solution: 3 eq. (879 mg, 1.5 mmol) of Fmoc-Cys(Trt)-OH, 3 eq. (570 mg, 1.5 mmol) of HBTU, and 5 eq. (424 μL, 2.5 mmol) of DIPEA in 10 mL of DMF. After this solution was rotated for 1 h, the resin was washed as above. The Fmoc group was removed by adding 10 mL of 20% piperidine in DMF (v/v) to the resin and rotating this for 30 min. Washing was performed as above, and the resin was split equally into 5 smaller columns with a volume of 12 mL each (2 mL bed volume). To each of the 5 columns were added 3 eq. (114 mg, 0.3 mmol) of HBTU and 5 eq. (85 μL, 0.5 mmol) of diisopropylethylamine (DIPEA). To each of the columns, only one of the following reagents was also added: 3 eq. (141 mg, 0.3 mmol) of Fmoc-Lys(Boc)-OH, 3 eq. (186 mg, 0.3 mmol) of Fmoc-His(Trt)-OH, 3 eq. (170 mg, 0.3 mmol) of Fmoc-Ser(Trt)-OH, 3 eq. (116 mg, 0.3 mmol) of Fmoc-Phe-OH, or 3 eq. (99 mg, 0.3 mmol) of Fmoc-Ala-OH•$2H_2O$. After these reagents were added, 5 mL of DMF was added to each, and they were agitated by rotation for 30 min. After washing as above, an Fmoc deprotection step was

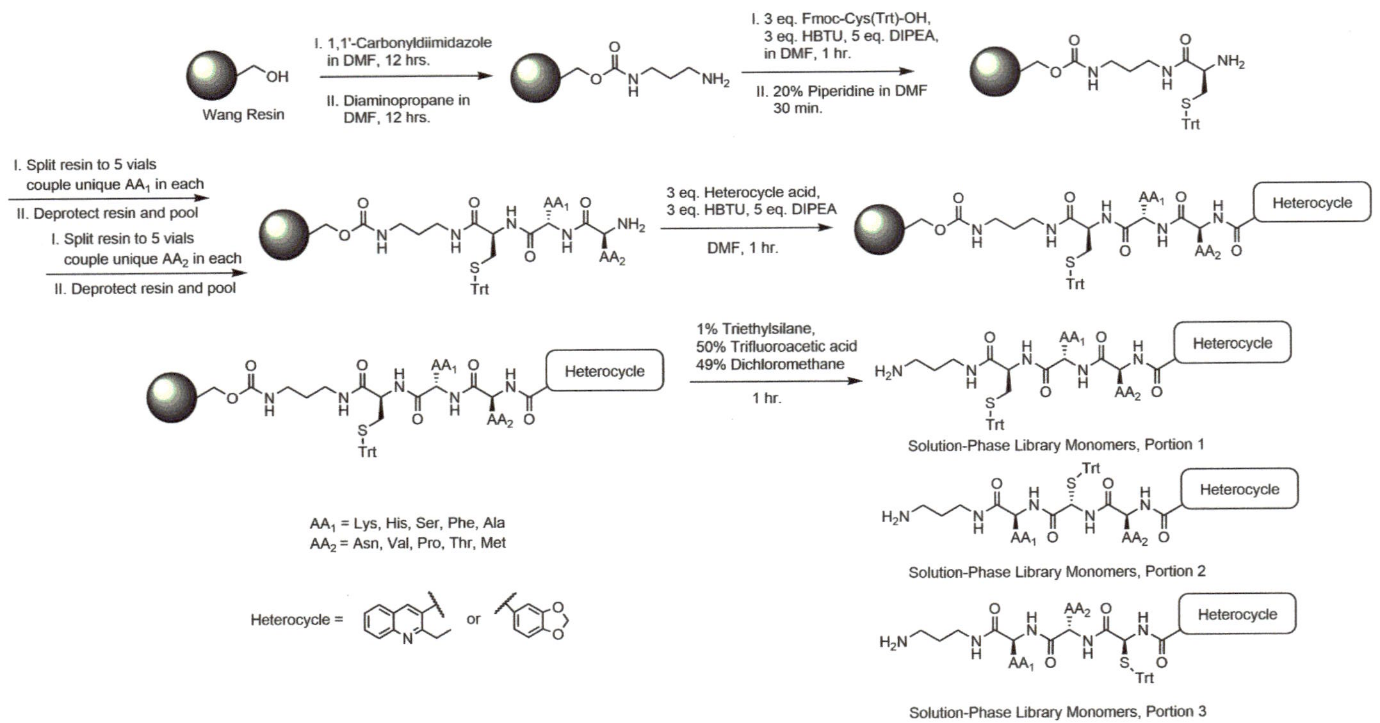

Fig. 1 Synthetic scheme describing the synthesis of one third of the solution-phase monomers that are necessary for the initial RNA target screening.

performed as above. The resin was then pooled into one vessel, mixed, and then split back into five vessels. To each vial, HBTU and DIPEA were added as above. One of the following amino acid derivatives was added to each vial: 3 eq. (178 mg, 0.3 mmol) of Fmoc-Asn(Trt)-OH, 3 eq. (102 mg, 0.3 mmol) of Fmoc-Val-OH, 3 eq. (101 mg, 0.3 mmol) of Fmoc-Pro-OH, 3 eq. (175 mg, 0.3 mmol) of Fmoc-Thr(Trt)-OH, or 3 eq. (99 mg, 0.3 mmol) of Fmoc-Met-OH. The solvent, DMF, was added as above, and the vials were rotated as above. A wash step was performed, followed by an Fmoc deprotection step in each vial. The resin was then pooled and split into two vials. To each vial, HBTU and DIPEA were added as above. Three equivalents (357 mg, 2.25 mmol) of 3-carboxy-2-ethyl-3-quinolinium chloride was added to one vial, and 3 eq. (452 mg, 2.25 mmol) of piperonylic acid chloride was added to the other. The solvent was added, and the vials were rotated as above.

After washing this portion of the library, it was combined with the other two portions that were intended to produce solution-phase monomers. The other two portions were synthesized in a similar fashion, but cysteine was incorporated at the second or third position. Following one more wash step, the resin was side chain deprotected and the monomers were cleaved from the resin by the addition of 10 mL of a 1% triethylsilane, 50% trifluoroacetic acid, 49% DCM solution for 1 h (caution: addition of these reagents to the resin causes an exotherm that may be sufficient to cause the cap on the synthesis column to loosen unless the system is vented several times in the first few minutes of reaction). The solution was separated from the resin, and compounds were then precipitated from the solution by the addition of cold (−20 °C) diethyl ether. The precipitated compounds were pelleted by centrifugation, and the ether was decanted. Fresh cold ether was added, the pellet was resuspended by vortexing, and the ether removed as before. This process was performed a total of five times to purify the compounds. After decanting the ether for the final time, the pellet was dissolved in ddH_2O and lyophilized to yield an off-white powder (McNaughton, 2007).

2.4.2 Resin-bound monomer synthesis

The resin-bound monomer synthesis diverges from the solution-phase monomer synthesis above in that a photo-cleavable linker (Fmoc-Anp-OH) is attached to the resin before the synthetic steps above begin, and the diaminopropane addition is replaced with Fmoc-Ahx-OH. Also, the cysteine that is coupled to the resin-bound monomers is in the form of Fmoc-Cys(S*t*Bu)-OH. The other synthetic steps are entirely the same (Fig. 2).

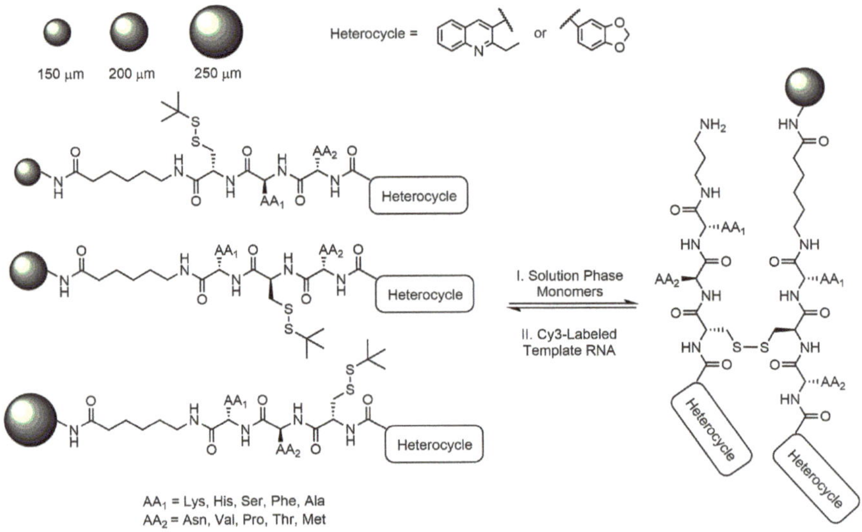

Fig. 2 An illustration of the structure of the resin-bound species, and the generation of dimeric species via disulfide exchange.

To aid in later deconvolution, three aliquots of resin were used to synthesize these monomers, each with a defined diameter of the beads. Bead diameter was used as a marker for cysteine position. Each resin aliquot had a dry mass of 1 g. The resin aliquots had the following properties: resin A, 140–170 μm, 0.45 mmol/g, 0.86 nmol/bead; resin B, 200–250 μm, 0.24 mmol/g, 1.50 nmol/bead; resin C, 280–320 μm, 0.23 mmol/g, 3.5 nmol/bead. The loading capacities of resin B and resin C were reduced so that beads from all resin aliquots had the same nmol/bead loading capacity. This was done by adding methoxyacetic acid to the first coupling reaction. To resin A was added: 3 eq. (583 mg, 1.35 mmol) of Fmoc-Anp-OH, 3 eq. (513 mg, 1.35 mmol) of HBTU, and 5 eq. (391 μL, 2.25 mmol) of DIPEA in 10 mL ofDMF. To resin B was added: 1.71 eq. (332 mg, 0.77 mmol) of Fmoc-Anp-OH, 1.29 eq. (45 μL, 0.58 mmol) of methoxyacetic acid, 3 eq. (513 mg, 1.35 mmol) of HBTU, and 5 eq. (391 μL, 2.25 mmol) of DIPEA in 10 mL of DMF. To resin C was added: 0.72 eq. (138 mg, 0.32 mmol) of Fmoc-Anp-OH, 2.28 eq. (79 μL, 1.03 mmol) of methoxyacetic acid, 3 eq. (513 mg, 1.35 mmol) of HBTU, and 5 eq. (391 μL, 2.25 mmol) of DIPEA in 10 mL of DMF. All three of the resin vials were agitated for 4 h. All of the resin vials were then washed and Fmoc-deprotected as in the solution-phase monomer synthesis. The next reaction was performed in the same manner for all resin aliquots. Each aliquot was reacted with

3 eq. (477 mg, 1.35 mmol) of Fmoc-Ahx-OH, 3 eq. (513 mg, 1.35 mmol) of HBTU, and 5 eq. (391 μL, 2.25 mmol) of DIPEA in 10 mL of DMF. All resin aliquots were agitated for 4 h, then washed and Fmoc-deprotected.

2.5 Screening conditions for RBDCC library

Five solid phase reaction vessels, with a volume of 1.5 mL each, were each assembled with 450 of the resin-bound monomer beads. This should have statistically provided 150 beads of each size, and 387 nmol in total. The first vessel was filled with a 1 mL solution containing 1 μM of Cy3-HIV-1 FSS RNA in phosphate buffer (10 mM phosphate buffer, 250 mM NaCl, pH 7.5). The remaining vessels were filled with a solution containing everything above, as well as 30 μM of the heterogeneous solution-phase monomers (based on the average molecular weight).

The first vessel, with no solution-phase monomers, was rotated for 3 h, and then the resin was separated from the solution via vacuum filtration. The resin was washed with buffer three times and placed under a fluorescence microscope using a filter cube intended for Cy3. The lack of any beads that appeared to be markedly more fluorescent than any others was interpreted as a lack of affinity between the RNA and resin-bound monomers. The remaining vessels were rotated for 72 h. While working with each vial successively, the number of washing steps and the time for each wash was increased. The exposure time of the fluorescent microscope images was also decreased. The increased stringency of the washes and the microscopy allowed for only the highest affinity molecules to be identified in the screen. The last vial, that was analyzed with the highest stringency (four washes 90 s each, exposure time 50 ms), identified three beads that were highly fluorescent. Each bead was placed individually in a microcentrifuge tube, with 100 μL of acetonitrile:methanol (4:1), and irradiated by a 4 W, 365 nm hand-held lamp made by UVP (Upland, CA) overnight. The supernatant was then analyzed by MALDI-MS. The size of each bead defined the cysteine position. Together with the mass spectrometry peaks, this was used to determine the identity of the monomer present on the surface of each bead.

2.6 Identification and characterization of library hits

After the high affinity monomers have been determined by the initial RBDCC screening, the full hit compounds that bound the labeled RNA must be determined. These hit compounds must then be synthesized and characterized.

2.6.1 Secondary screen experimental

In the screening approach described in Section 2.5, stoichiometry is biased to ensure that "dimers" are primarily formed on resin rather than in solution. This means that the mass spectral analysis of library results typically yields information about the monomers selected, but may not allow for identification of selected dimer. To complete analysis, each of the hit monomers must be combined with itself and every other hit monomer to determine which dimer has the highest affinity for the target RNA. To do this, each hit monomer must be synthesized on a larger scale using the methods described above, without the split-pool steps. Then, a "secondary screen" is assembled in a 24-well plate (Fig. 3). Every combination of a single solution phase monomer with a single resin-bound monomer is assembled in the wells. The molarity of all species is the same as that in the initial screen above. Wells that contain a dimer with a higher affinity for the labeled RNA, on the surface of the beads, will display more intense fluorescence on the beads. The high affinity dimers identified by this screening should be the same dimers that were binding the labeled RNA to the resin beads in the initial screen. Assuming the RBDCC selection has been efficient, typically no more than three or four monomers need to be synthesized, and thus this step is not onerous.

2.6.2 Synthesis of lead compounds

The synthesis of species in the library and secondary screen is only intended to produce enough of each compound to perform screening. Characterization and biological testing require greater amounts of these species, so they

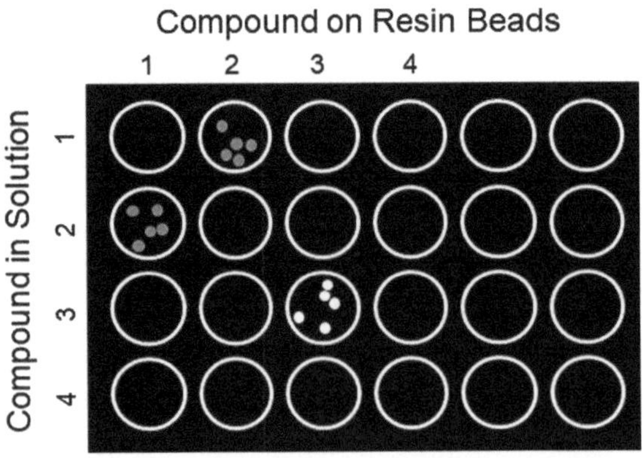

Fig. 3 A cartoon of what a secondary screen may look like using fluorescence microscopy. The white circles represent the walls of the wells in a 24-well plate, and the gray or white dots are fluorescent beads. In this example, compound 2-1 appears to have affinity for the RNA, but 3-3 is the highest affinity dimer.

must be produced at a larger scale. As previously mentioned, moieties that are present in library species solely for the purpose of screening (such as the labile disulfide) may also be altered at this juncture. For the library we have described, we have observed that an olefin or triazole moiety is capable of replacing the disulfide, while still retaining high affinity for the target. The olefin can be installed by adding amino acids with terminal olefin sidechains into the SPPS scheme where cysteine was previously (Palde, Ofori, Gareiss, Lerea, & Miller, 2010). Olefin metathesis of two monomers can then produces the dimeric species. The triazole moiety can be incorporated with amino acids bearing alkyne and azide side chains. A 1,3-dipolar cycloaddition, often referred to under the category of "click chemistry," can then be employed to produce the dimer (Hilimire et al., 2017). It has been useful to perform the dimerization reactions with one species still linked to resin beads, so that a heterodimer can be synthesized specifically.

Methylation of the amide nitrogen atoms within the peptide backbone of "hit" compounds has also been examined. Methylation provided these compounds with greater affinity and specificity than their non-methylated parent compounds. It was hypothesized that this is due to the methylation constraining the molecules to a subset of the conformers available to the parent compound. Peptide *N*-methylation is also recognized as a useful method to enhance bioavailability and biostability (Chatterjee, Rechenmacher, & Kessler, 2013).

2.6.3 Characterization of lead compound affinity and specificity

Once interesting new compounds have been found via a library, and they have been synthesized at a larger scale, the true affinity and specificity of the compounds must be determined. A diverse array of techniques for this type of characterization can be found. Surface Plasmon Resonance (SPR) analysis and Isothermal Titration Calorimetry (ITC) are techniques that have proven reliable for determining affinity constants with high accuracy. Other techniques that may be less precise, such as an RNA filter binding assay, have also been valuable to us to confirm relative affinity of several compounds for a given target (Gareiss et al., 2008). Determination of the affinity between a compound and its target is relatively straightforward with the techniques above, but determining the specificity of a compound is a more complex task. Initially proposed by Yitzhak Tor and colleagues, tRNA can serve as a useful initial test of specificity for a given RNA (Luedtke et al., 2003). a competition affinity assay can be performed where the target RNA is allowed to interact with the compound of interest and an excess of tRNA. If there is little or no change in the affinity for the target RNA under these conditions, the compound has specificity for the target relative to tRNA. This does not indicate that the compound has no appreciable affinity for all other RNAs (let alone all other biological molecules), but it is a simple first-pass assessment of compound specificity.

2.7 Increasing structural diversity within a library

All of the dimeric molecules formed in the library presented above have a similar shape. These structures were intended to mimic bis-intercalating peptide antibiotics echinomycin and similar molecules with known affinity for nucleic acid sequences (Olsen, Ramasamy, Bhat, Low, & Waring, 1986; Van Dyke & Dervan, 1984). Each has two somewhat linear monomers attached by a short chain of atoms that link them. An extension of this work (McAnany, Reichert, & Miller, 2016) included a subset of library monomers from the library described above, as well as two dithiols. This set of experiments was intended to test the ability of more complex structures (monomer-dithiol-monomer trimers) to form, while also probing the ability of the RNA recognition process to tolerate variations in linker geometry.

In these experiments, a small library consisting of three monomers was expanded with one of two unique dithiol species (D1 or D2, Fig. 4). The formation of trimeric monomer-dithiol-monomer species was observed by MALDI-MS, and the time until the library reached equilibrium was determined by HPLC by analogy to methods described above. One monomer used in the library was that previously shown to form a dimer able to selectively bind the HIV-1 FSS RNA. We hypothesized that this library would test the structural requirements for linker geometry in FSS RNA binding. Expected outcomes were that the library screen could yield a new, higher-affinity species incorporating a dithiol bridge, or, alternatively, the original dimer compound could be re-generated if a shorter linker was "preferred" by the RNA. Following equilibration of the DCL, we observed that the original dimer compound was amplified to the exclusion of bridged structures. These experiments demonstrated that the library can incorporate linker dithiols, and it can still amplify a dimer if that is the optimal structure. These results suggest that many linker lengths and functionalized linkers could be incorporated as fragments into a DCL and quickly screened for their ability to change the affinity or specificity of a hit compound.

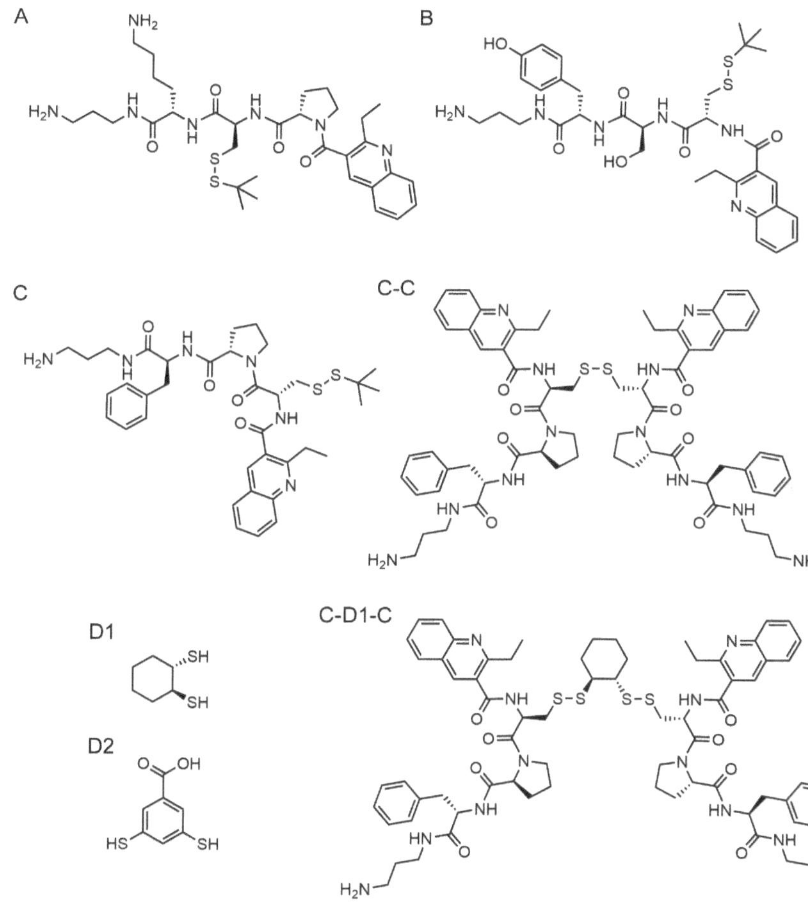

Fig. 4 Design of a simple library to test variation in linker geometry. Compounds A, B, and C are the monomeric components of the library. Monomers are able to come together to form disulfide dimers such as C-C, or they can react with bridging dithiols D1 and D2 to form species such as C-D1-C.

2.8 Incorporation of orthogonal exchange chemistries into a library

The bifunctional dithiols above proved to be a viable approach to increasing the complexity of library species beyond dimers. A disadvantage to this method, however, is that the bifunctional species have the ability to form polymeric species in solution, and they also do so in close proximity on a solid support. Another avenue that has been tested to expand library diversity without complications arising from oligomer formation is the inclusion of multiple exchange chemistries within a single library.

As an example of orthogonal exchange chemistries, hydrazide/hydrazone exchange and thiol/disulfide exchange were performed within one RBDCC. Bifunctional species bearing one hydrazone and one disulfide each were synthesized on resin beads. Solution phase monomers bearing one thiol, or one hydrazide were all added to the same pot with the resin beads. These experiments were performed at a pH of 7.4 in ammonium acetate buffer. To facilitate both forms of chemical exchange at this pH, thiopropanol and aniline were added to catalyze the disulfide and hydrazone exchange reactions respectively. Relative to any monomer species, thiopropanol was present in an equal molar amount and aniline was present at 37.5 eq. After 2 weeks of exchange, three of the potential six trimeric species were observed within the library by MALDI-MS (Gromova et al., 2012).

The results of these experiments demonstrated successful simultaneous use of two exchange chemistries, enabling the formation of trimeric species in a DCL. However, it is concerning that not all of the predicted trimeric species were observed. This could be due to the slower exchange rate of hydrazides with hydrazones at a near neutral pH, or the thermodynamic minimum of that library may simply not have included appreciable amounts of those species due to library self-selection. The latter possibility seems unlikely given the similarity in all of the monomeric species within the library. If the former possibility is correct, better catalysis of the hydrazone exchange reaction may avoid this stumbling block (Larsen, Pittelkow, Karmakar, & Kool, 2015).

3. Conclusions

DCLs, and particularly RBDCLs, have proven to be an effective method for rapidly identifying peptide-based compounds able to bind target RNA sequences with reasonable (micromolar or better) affinities and selectivities. These compounds are useful starting points for the synthesis of structures suitable for bioassay. Ongoing work in the field includes expansion of structural diversity, as well as use of fragments that more closely resemble drug-like molecules. We can anticipate that this strategy will continue to find utility in the discovery of novel RNA-binding molecules, thereby contributing to our understanding of sequence selectively binding RNA.

Acknowledgment

The authors would like to thank the NIH for the T32AI049815 grant.

References

Chatterjee, J., Rechenmacher, F., & Kessler, H. (2013). N-methylation of peptides and proteins: An important element for modulating biological functions. *Angewandte Chemie (International Ed. in English)*, *52*(1), 254–269. https://doi.org/10.1002/anie.201205674.

Childs-Disney, J. L., Parkesh, R., Nakamori, M., Thornton, C. A., & Disney, M. D. (2012). Rational design of bioactive, modularly assembled aminoglycosides targeting the RNA that causes myotonic dystrophy type 1. *ACS Chemical Biology*, 7(12), 1984–1993. https://doi.org/10.1021/cb3001606.

Corbett, P. T., Leclaire, J., Vial, L., West, K. R., Wietor, J.-L., Sanders, J. K. M., et al. (2006). Dynamic combinatorial chemistry. *Chemical Reviews*, *106*(9), 3652–3711. https://doi.org/10.1021/cr020452p.

Erlanson, D. A., McDowell, R. S., & O'Brien, T. (2004). Fragment-based drug discovery. *Journal of Medicinal Chemistry*, *47*(14), 3463–3482. https://doi.org/10.1021/jm040031v.

Frei, P., Pang, L., Silbermann, M., Eriş, D., Mühlethaler, T., Schwardt, O., et al. (2017). Target-directed dynamic combinatorial chemistry: A study on potentials and pitfalls as exemplified on a bacterial target. *Chemistry—A European Journal*, *23*(48), 11570–11577. https://doi.org/10.1002/chem.201701601.

Gareiss, P. C., Sobczak, K., McNaughton, B. R., Palde, P. B., Thornton, C. A., & Miller, B. L. (2008). Dynamic combinatorial selection of molecules capable of inhibiting the (CUG) Repeat RNA – MBNL1 interaction in vitro: Discovery of lead compounds targeting myotonic dystrophy (DM1). *Journal of the American Chemical Society*, *130*(48), 16254–16261. https://doi.org/10.1021/ja804398y.

Gromova, A. V., Ciszewski, J. M., & Miller, B. L. (2012). Ternary resin-bound dynamic combinatorial chemistry. *Chemical Communications (Cambridge)*, *48*(15), 2131–2133. https://doi.org/10.1039/c2cc17192a.

Hilimire, T. A., Chamberlain, J. M., Anokhina, V., Bennett, R. P., Swart, O., Myers, J. R., et al. (2017). HIV-1 frameshift RNA-targeted triazoles inhibit propagation of replication-competent and multi-drug-resistant HIV in human cells. *ACS Chemical Biology*, *12*(6), 1674–1682. https://doi.org/10.1021/acschembio.7b00052.

Keserű, G. M., Erlanson, D. A., Ferenczy, G. G., Hann, M. M., Murray, C. W., & Pickett, S. D. (2016). Design principles for fragment libraries: Maximizing the value of learnings from pharma fragment-based drug discovery (FBDD) programs for use in academia. *Journal of Medicinal Chemistry*, *59*(18), 8189–8206. https://doi.org/10.1021/acs.jmedchem.6b00197.

Larsen, D., Pittelkow, M., Karmakar, S., & Kool, E. T. (2015). New organocatalyst scaffolds with high activity in promoting hydrazone and oxime formation at neutral pH. *Organic Letters*, *17*(2), 274–277. https://doi.org/10.1021/ol503372j.

Li, J., Nowak, P., & Otto, S. (2013). Dynamic combinatorial libraries: From exploring molecular Recognition to systems chemistry. *Journal of the American Chemical Society*, *135*(25), 9222–9239. https://doi.org/10.1021/ja402586c.

Luedtke, N. W., Liu, Q., & Tor, Y. (2003). RNA – ligand interactions: Affinity and specificity of aminoglycoside dimers and acridine conjugates to the HIV-1 Rev response element. *Biochemistry*, *42*(39), 11391–11403. https://doi.org/10.1021/bi034766y.

McAnany, J. D., Reichert, J. P., & Miller, B. L. (2016). Probing the geometric constraints of RNA binding via dynamic covalent chemistry. *Bioorganic & Medicinal Chemistry*, *24*(17), 3940–3946. https://doi.org/10.1016/j.bmc.2016.02.029.

McNaughton, B. R. (2007). The development of methods. In *Static and dynamic combinatorial chemistry and their application to the discovery of molecules that bind the HIV-1 frameshift-inducing RNA stemloop*. Rochester, New York: University of Rochester. Doctor of Philosophy.

McNaughton, B. R., & Miller, B. L. (2006). Resin-bound dynamic combinatorial chemistry. *Organic Letters*, *8*(9), 1803–1806. https://doi.org/10.1021/ol060330+.

Moore, J. S., & Zimmerman, N. W. (2000). "Masterpiece" copolymer sequences by targeted equilibrium-shifting. *Organic Letters*, *2*(7), 915–918. https://doi.org/10.1021/ol0055723.

Morgan, B. S., Forte, J. E., & Hargrove, A. E. (2018). Insights into the development of chemical probes for RNA. *Nucleic Acids Research*, *46*(16), 8025–8037. https://doi.org/10.1093/nar/gky718 %J Nucleic Acids Research.

Ofori, L. O., Hilimire, T. A., Bennett, R. P., Brown, N. W., Smith, H. C., & Miller, B. L. (2014). High-affinity recognition of HIV-1 frameshift-stimulating RNA alters frameshifting in vitro and interferes with HIV-1 infectivity. *Journal of Medicinal Chemistry*, *57*(3), 723–732. https://doi.org/10.1021/jm401438g.

Olsen, R. K., Ramasamy, K., Bhat, K. L., Low, C. M. L., & Waring, M. J. (1986). Synthesis and DNA-binding studies of [lac2,lac6]TANDEM, an analog of des-N-tetramethyltriostin A (TANDEM) having L-lactic acid substituted for each L-alanine residue. *Journal of the American Chemical Society*, *108*(19), 6032–6036. https://doi.org/10.1021/ja00279a060.

Palde, P. B., Ofori, L. O., Gareiss, P. C., Lerea, J., & Miller, B. L. (2010). Strategies for recognition of stem – loop RNA structures by synthetic ligands: Application to the HIV-1 frameshift stimulatory sequence. *Journal of Medicinal Chemistry*, *53*(16), 6018–6027. https://doi.org/10.1021/jm100231t.

Shortridge, M. D., Wille, P. T., Jones, A. N., Davidson, A., Bogdanovic, J., Arts, E., et al. (2018). An ultra-high affinity ligand of HIV-1 TAR reveals the RNA structure recognized by P-TEFb. *Nucleic Acids Research*, *47*, 1523–1531. https://doi.org/10.1093/nar/gky1197.

Simpson, M. G., Pittelkow, M., Watson, S. P., & Sanders, J. K. M. (2010). Dynamic combinatorial chemistry with hydrazones: Cholate-based building blocks and libraries. *Organic & Biomolecular Chemistry*, *8*(5), 1173–1180. https://doi.org/10.1039/B917145B.

Telenti, A., Martinez, R., Munoz, M., Bleiber, G., Greub, G., Sanglard, D., et al. (2002). Analysis of natural variants of the human immunodeficiency virus type 1 gag-pol frameshift stem-loop structure. *Journal of Virology*, *76*(15), 7868–7873. https://doi.org/10.1128/JVI.76.15.7868-7873.2002.

Van Dyke, M. M., & Dervan, P. B. (1984). Echinomycin binding sites on DNA. *Science*, *225*(4667), 1122–1127. https://doi.org/10.1126/science.6089341 %J Science.

CHAPTER FIVE

A click chemistry assay to identify natural product ligands for pre-microRNAs

Amanda L. Garner[a,b,*], Daniel A. Lorenz[b], Erin E. Gallagher[a]
[a]Department of Medicinal Chemistry, College of Pharmacy, University of Michigan, Ann Arbor, MI, United States
[b]Program in Chemical Biology, University of Michigan, Ann Arbor, MI, United States
*Corresponding author: e-mail address: algarner@umich.edu

Contents

Abstract

Despite the great diversity of structure and function and relevance to human health, RNA remains an underexploited area of drug discovery. A major bottleneck toward this goal has been the identification of probes and drug leads that are specific for select RNAs and methods that will facilitate such discovery efforts. Our laboratory has recently developed an innovative approach for assaying RNA-small molecule interactions,

Methods in Enzymology, Volume 623
ISSN 0076-6879
https://doi.org/10.1016/bs.mie.2019.04.020

catalytic enzyme-linked click chemistry assay or cat-ELCCA, which is a functional assay that takes advantage of the power of catalytic signal amplification combined with the selectivity and bioorthogonality of click chemistry. Importantly, through application of this platform assay technology to the challenging problem of identifying selective inhibitors of pre-microRNA maturation, we identified natural products as a potential source of such compounds. Herein we describe this methodology in addition to the downstream pipeline toward the discovery of natural product ligands for pre-microRNAs. Through cat-ELCCA, our goal is to discover novel ligands to facilitate our investigation of RNA recognition by small molecules.

1. Introduction

RNA-targeted probe and drug discovery is an exciting, yet challenging area of medicinal chemistry (Connelly, Moon, & Schneekloth, 2016; Garner, 2018b; Thomas & Hergenrother, 2008). A major bottleneck toward promoting this field is the discovery of new chemical space for targeting RNAs. Key examples from the discovery of ribocil, a synthetic ligand for the FMN riboswitch (Howe et al., 2015), and RNA-binding SMN-C and NVS-SM1 SMN2 splicing modulators (Naryshkin et al., 2014; Palacino et al., 2015; Sivaramakrishnan et al., 2017), have provided recent evidence for the potential druggability of RNA. Yet, these molecules were identified from phenotypic screening approaches, and the direct targeting of RNA with small molecules remains difficult, particularly as most efforts have centered solely on rational design strategies using a limited number of potentially promiscuous chemotypes, such as Hoechst dyes and aminoglycosides (Lorenz & Garner, 2018).

Some of the earliest examples of RNA-binding molecules have come from natural products (NPs), and aminoglycoside, tetracycline and macrolide antibiotics function by binding to ribosomal RNA (rRNA) within the bacterial ribosome (Wilson, 2014). Historically, NPs have been invaluable for developing front-line drugs against cancer, infectious diseases, and parasites, and as tools for furthering our understanding of cellular biology (Newman & Cragg, 2012). In fact, a survey of all 246 anti-tumor agents approved by the FDA between the late 1940s and 2014 showed that >65% were either NPs or NP analogues, indicating the overwhelming success of NPs in the human pharmacopeia (Newman & Cragg, 2012). Despite this fact, the role of NPs in probe and drug discovery has diminished in recent years, in part due to the technical challenges of screening NP extract

(NPE) libraries using traditional targeted assay formats, such as fluorescence polarization (FP) and fluorescence resonance energy transfer (FRET) assays (Harvey, Edrada-Ebel, & Quinn, 2015).

To facilitate the discovery of new chemical entities for targeting RNAs, we have developed high-throughput screening (HTS) technology termed catalytic enzyme-linked click chemistry assay, or cat-ELCCA, which is a robust screening platform that does not suffer from compound interference (Garner, 2018a; Lorenz & Garner, 2016, 2018; Lorenz, Song, & Garner, 2015; Lorenz, Vander Roest, Larsen, & Garner, 2018). Key advantages of cat-ELCCA include its increased sensitivity due to catalytic signal amplification, negligible compound interference in comparison to traditional fluorescence-based assays due to added washing steps, and HTS applicability with Z' factors >0.6 using automated liquid handling (Garner, 2018a; Garner & Janda, 2010, 2011; Lorenz & Garner, 2016, 2018; Lorenz, Kaur, et al., 2018; Lorenz et al., 2015; Lorenz, Vander Roest, et al., 2018; Song, Menon, Mitchell, Johnson, & Garner, 2017; Zhang, Chung, & Oldenburg, 1999). As proof of concept for RNA-targeted probe discovery, we applied cat-ELCCA toward the goal of identifying selective and RNA-binding inhibitors of Dicer-mediated microRNA (miRNA or miR) maturation (Fig. 1) (Garner, 2018a; Lorenz & Garner, 2016, 2018; Lorenz et al., 2015; Lorenz, Vander Roest, et al., 2018). Using this approach, we have recently completed a HTS of commercial small molecules and NPEs to identify compounds and areas of chemical space capable of selectively inhibiting the processing of oncogenic pre-miR-21 (Lorenz, Vander Roest, et al., 2018). From this screening campaign, we excitingly found that NPs may be a potential source of such molecules, indicating that more highly functionalized molecular scaffolds, as represented by NPs, may be necessary for selectively disrupting pre-miRNA-Dicer interactions and recognizing simple hairpin structures like pre-miRNAs. Herein we describe our cat-ELCCA methodology and the subsequent downstream assays that can be used to identify natural product ligands for targeting pre-miRNAs.

2. cat-ELCCA for dicer-mediated pre-miRNA maturation

In order to be active, miRNAs undergo two maturation steps, one in the nucleus (pri-miRNA to pre-miRNA) and one in the cytoplasm (pre-miRNA to mature miRNA), mediated by the RNase III enzymes Drosha and Dicer, respectively (Fig. 1) (Ha & Kim, 2014). Thus, as a biological basis for the targeting of miRNAs, we hypothesized that compounds able to

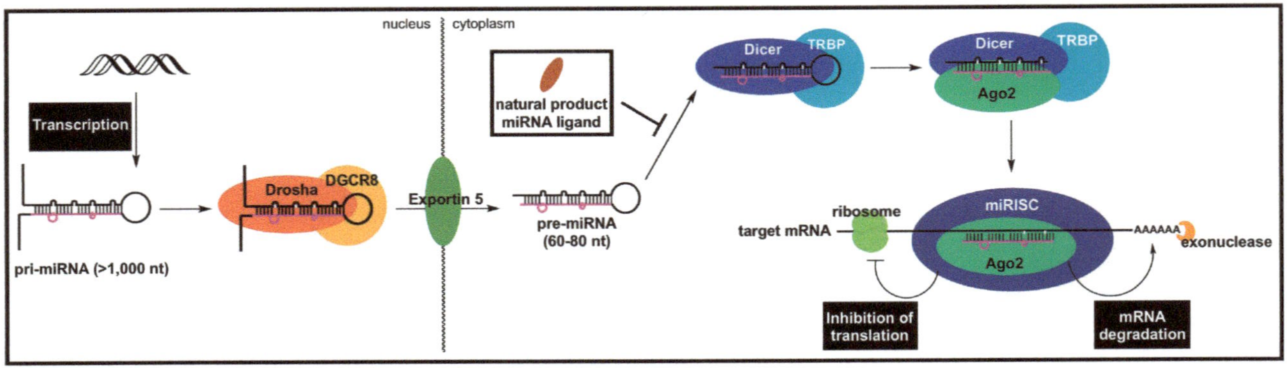

Fig. 1 miRNA biogenesis and activity, and the impact of targeting miRNAs with a natural product inhibitor of Dicer-mediated maturation. Ago2 = argonaute 2. DGCR8 = DiGeorge syndrome chromosomal region 8. miRISC = miRNA-loaded RNA-induced silencing complex. TRBP = TAR RNA binding protein.

productively bind to a pre-miRNA would inhibit miRNA maturation, and thereby, activity (Fig. 1). The targeting of pre-miRNAs was chosen due to their favorable size for assay development, screening and chemical synthesis, in addition to cytoplasmic localization for intracellular targeting with a small molecule.

The scheme for cat-ELCCA for Dicer-mediated pre-miRNA maturation is shown in Fig. 2 (Garner, 2018a; Lorenz et al., 2015; Lorenz & Garner, 2016, 2018). In brief, a chemically synthesized, *trans*-cyclooctene (TCO)-labeled pre-miRNA substrate is first immobilized into the wells of a 384-well streptavidin-coated microtiter plate. In the presence of Dicer, the terminal loop is cleaved, and following reaction with a methyltetrazine (mTet) modified-horseradish peroxidase (HRP), no signal is observed. In the presence of an inhibitor, such as an active NPE or purified pre-miRNA ligand (as shown), the loop is retained, resulting in catalytic signal amplification from the covalently click chemistry-linked HRP.

An overview of the work flow toward the goal of identifying selective natural product ligands for a pre-miRNA is shown in Fig. 3. Following HTS of NPEs, hit validation is first performed by re-testing positive NPEs in cat-ELCCA (step 1). Reproducible hit extracts are subsequently analyzed for selectivity by examining inhibitory activity against Dicer-mediated processing of a second pre-miRNA substrate using cat-ELCCA (step 2). The activity of validated hits showing selectivity for the pre-miRNA-of-interest are then tested in a gel-based secondary assay of Dicer activity to eliminate false hits due to the assay format (step 3). Iterative testing of further fractionated samples (step 4) in cat-ELCCA and the gel-based assay is performed until a purified, active natural product is identified (step 5). IC_{50} values in cat-ELCCA are then measured against the pre-miRNA target-of-interest, in addition to other substrates to assess selectivity (step 6).

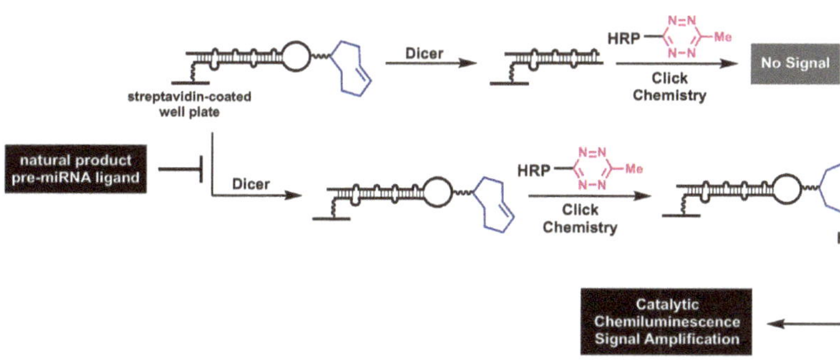

Fig. 2 cat-ELCCA for Dicer-mediated pre-miRNA maturation.

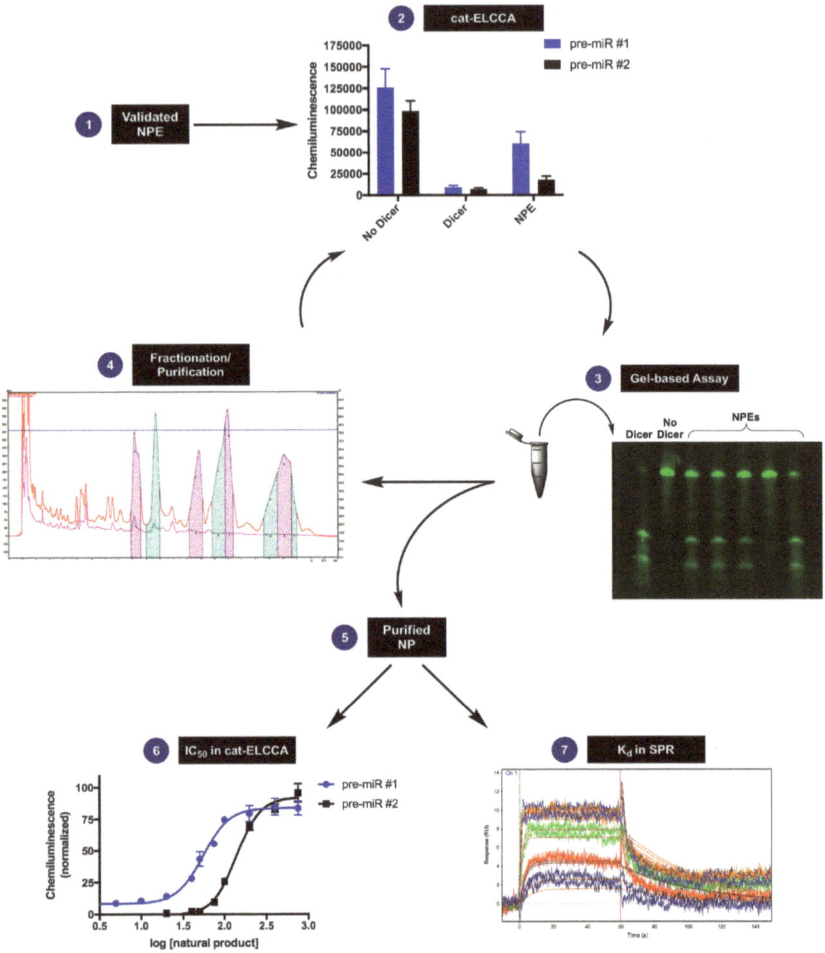

Fig. 3 Overview of the pipeline for the discovery of natural product ligands for pre-miRNAs.

Direct binding to the pre-miRNA is finally measured via surface plasmon resonance (SPR) spectroscopy to determine binding kinetics and affinity (step 7).

3. Preparation of assay reagents

3.1 Equipment

- Orbital shaker
- Refrigerated centrifuge

- Nanodrop spectrophotometer
- Centrifugal filter (10,000 MW cut-off)
- Mass spectrometer

3.2 Buffers and reagents

- Chemically synthesized pre-miRNA substrate (Dharmacon)
- Dimethyl sulfoxide (DMSO)
- (TCO)-PEG4-NHS (Click Chemistry Tools, A137)
- HRP (Pierce 31491)
- mTet-NHS (Click Chemistry Tools, 1128)
- Phosphate buffer (100 mM, pH 8.0)
- Sodium acetate buffer (3.0 M, pH 5.2)
- Cold ethanol
- Phosphate-buffer saline (PBS) (100 mM phosphate, 150 mM NaCl, pH 7.0)
- Bradford assay (Pierce)

3.3 Procedure

3.3.1 Preparation of pre-miRNA substrate

1. Design a pre-miRNA substrate (Fig. 4) to contain the following modifications: (1) An 18-atom polyethylene glycol (PEG)-based biotin linker at the 5′ phosphate terminus. Of note, modification is made at the 5′ end to maintain the 3′ 2-nucleotide overhang that is recognized by Dicer (Ha & Kim, 2014). (2) An aminoallyl uridine in place of a uridine or other nucleotide in the terminal loop that is cleaved by Dicer. Prior to use, the RNA must be deprotected, desalted and HPLC purified, which can be done by the manufacturer.
2. Dissolve the pre-miRNA substrate in phosphate buffer to a concentration of 1.0 mM.
3. Prepare a stock solution of (TCO)-PEG4-NHS in DMSO to a concentration of 10 mM.
4. Transfer 5.0 μL of pre-miRNA substrate solution and 5.0 μL of TCO solution to a 1.5 mL microcentrifuge tube. Shake the reaction mixture at 25 °C for 1 h.
5. Transfer the tube to an ice bucket, and precipitate the RNA by addition of 0.11 × volume sodium acetate (1.1 μL) and 4 × volume cold ethanol (40 μL). Centrifuge at 20,000 × *g* for 40 min at 4 °C.

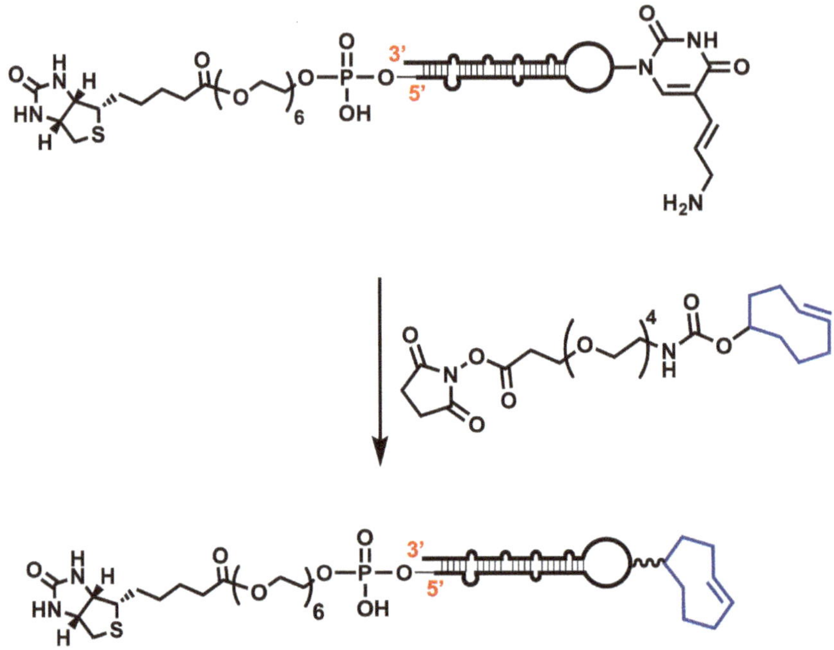

Fig. 4 Synthesis of TCO-labeled pre-miRNA substrate for cat-ELCCA.

6. Resuspend the RNA pellet in phosphate buffer at a concentration of 1.0 mM based on the amount of RNA originally added to the tube. RNA concentration can be measured using a Nanodrop spectrophotometer. Store at −80 °C for long-term use.

3.3.2 Preparation of labeled HRP

1. Prepare a stock solution of mTet-NHS in DMSO to a concentration of 100 mM.
2. Add 2.5 mg HRP to a microcentrifuge tube and dissolve in 186 μL of PBS. To the same tube, transfer 14 μL of mTet solution (Fig. 5). Shake the reaction mixture at 25 °C for 3 h.
3. Transfer the reaction mixture to a microcentrifuge concentrator (0.5 mL), concentrate to 50–100 μL and exchange the labeled HRP into PBS (5 × 400 μL) to remove unreacted mTet-NHS and DMSO.
4. Determine protein concentration via the Bradford assay using BSA standards.
5. Store at 4 °C for long-term use.

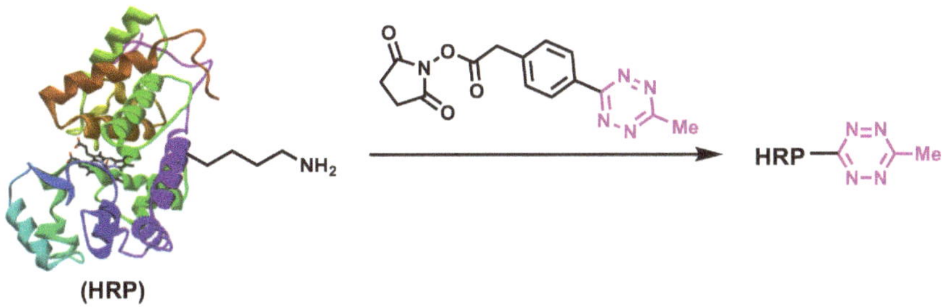

Fig. 5 Synthesis of mTet-labeled HRP for cat-ELCCA.

4. High-throughput screening of natural product extract libraries

Note: This protocol can also be performed using a multi-channel pipette.

4.1 Equipment

- Black, standard-capacity streptavidin-coated 384-well plates (Pierce 15,407)
- Plate washer (Biotek 405 ELX or equivalent)
- Plate centrifuge
- Plate tape (Pierce)
- Multidrop Combi or comparable reagent dispenser
- Refrigerator or cold room
- Sciclone liquid handler with V&P pintool (or analogous liquid handling set-up)
- Chemiluminescence plate reader (PHERAstar or equivalent)

4.2 Buffers and reagents

- Pre-miRNA substrate from Section 3.3.1
- Phosphate buffer (100 mM, pH 7.0)
- Dicer buffer (20 mM Tris (pH 7.5), 12 mM NaCl, 2.5 mM $MgCl_2$, 1 mM fresh DTT, 4.5% DMSO)
- NPE library (75 μg/mL) or purified NPs
- DMSO
- Human Dicer (108 μg/mL) (Lorenz & Garner, 2016; MacRae, Ma, Zhou, Robinson, & Doudna, 2008)
- mTet-HRP from Section 3.3.2

- Wash buffer (2 mM imidazole, 260 mM NaCl, 0.5 mM EDTA, 0.1% Tween-20, pH 7.0)
- SuperSignal West Pico (Pierce)

4.3 Procedure

4.3.1 Immobilization of pre-miRNA substrate

1. Wash the wells of the streptavidin plate with 3 × 50 μL of phosphate buffer using a plate washer.
2. Transfer 5 μL of pre-miRNA substrate (500 nM final) to the wells using a reagent dispenser.
3. Centrifuge the plates at 223 × *g* for 1 min.
4. Seal the plate with plate tape and incubate overnight at 4 °C.
5. Wash the wells with 3 × 50 μL of phosphate buffer using a plate washer.

4.3.2 Dicer-mediated pre-miRNA maturation

1. Transfer 5 μL of Dicer buffer to each of the wells using a reagent dispenser and centrifuge the plate at 223 × *g* for 1 min.
2. Add the NPEs to each well (50 nL; 75 μg/mL final) using a liquid handler. Of note, the same volume of DMSO should be added to the control wells. A representative plate map for the assay can be found in Fig. 6.

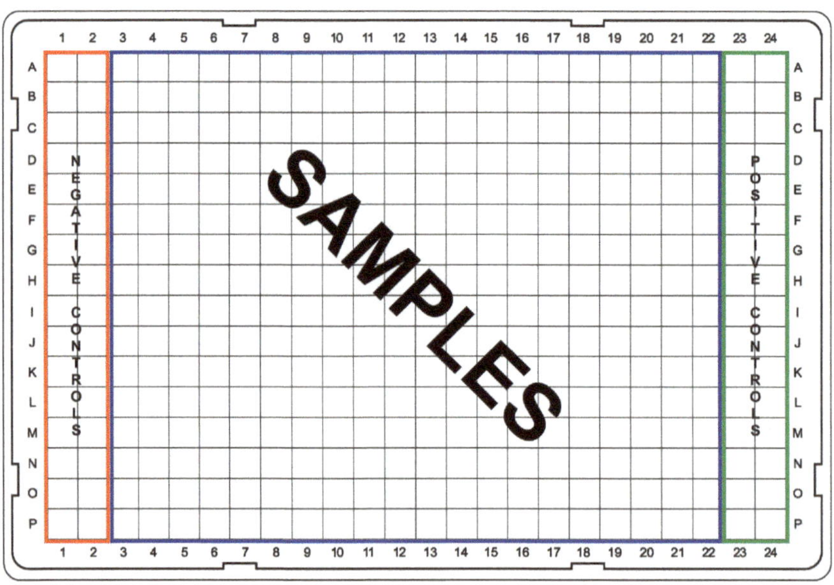

Fig. 6 Representative plate map for cat-ELCCA.

3. Incubate the plates at 25 °C for 15 min.
4. Transfer 5 μL of Dicer buffer containing 217 μg/nL Dicer (108 μg/mL Dicer, 5% glycerol, 0.01% Triton X-100 final; excess with respect to the concentration of immobilized pre-miRNA) using a reagent dispenser. An equivalent volume of Dicer buffer without enzyme should be added to the positive control wells (Fig. 6).
5. Centrifuge the plates at 223 × *g* for 1 min.
6. Seal the plate with plate tape and incubate for 5 h at 37 °C.
7. Wash the wells with 3 × 50 μL of phosphate buffer using a plate washer.

4.3.3 Click chemistry with mTet-HRP

1. Transfer 10 μL of mTet-HRP (750 nM final) to the wells using a reagent dispenser.
2. Centrifuge the plates at 223 × *g* for 1 min.
3. Seal the plate with plate tape and incubate for 2 h at 25 °C.
4. Wash the wells with 3 × 50 μL of wash buffer and 3 × 50 μL of phosphate buffer using a plate washer.

4.3.4 Chemiluminescence detection

1. Transfer 25 μL of SuperSignal West Pico solution, prepared according to the manufacturer's instructions, to the wells using a reagent dispenser.
2. Incubate for 5 min at 25 °C.
3. Measure chemiluminescence signal using a PHERAstar plate reader using the LUM plus module on autogain set to one of the positive control wells.

4.3.5 NPE hit validation and selectivity assessment

1. Following HTS, all NPEs showing ≥10% inhibition by plate based on the negative controls should be retested in triplicate for activity validation using cat-ELCCA as described above. NPEs showing inhibition at ≥3 standard deviations (SD) by plate from the negative controls can be considered confirmed hits.
2. Preliminary selectivity can be assessed by measuring inhibitory activity in cat-ELCCA using a second pre-miRNA substrate (prepared as in Section 3.3.1) and repeating procedures 4.3.1–4.3.5.
3. Dose-dependence of inhibition can be measured by performing cat-ELCCA across varying concentrations of NPEs or NPs to calculate IC_{50} values.

5. Gel-based assay of dicer activity

5.1 Equipment

- SDS PAGE equipment
- Fluorescence gel imager (Protein Simple FluorChem M or equivalent)

5.2 Buffers and reagents

- Pre-miRNA substrates from Section 3.3.1 (can be labeled or unlabeled)
- Human Dicer (108 μg/mL) (Lorenz & Garner, 2016; MacRae et al., 2008)
- Dicer buffer (20 mM Tris (pH 7.5), 12 mM NaCl, 2.5 mM $MgCl_2$, 1 mM fresh DTT, 40 U/mL RNaseOUT RNase inhibitor (Life Technologies))
- NPEs (75 μg/mL) or purified NPs
- DMSO
- 12.5% TBE-urea gels (Life Technologies)
- SYBR Gold nucleic acid stain (Life Technologies)
- miRNA marker (New England Biolabs)

5.3 Procedure

1. Prepare a solution of pre-miRNA (0.5 μL, 50 μM stock) in Dicer buffer (8.5 μL) in a microcentrifuge tube.
2. Add the NPE/NP or DMSO (0.5 μL) and incubate at 25 °C for 15 min.
3. Add Dicer in Dicer buffer (0.5 μL) and incubate at 37 °C for 30 min.
4. Analyze the Dicer digestion reaction (10-μL total volume) by running the sample on a 12.5% TBE-Urea gel. Load a miRNA marker for visualization of 17, 21 and 25 residue small RNAs as a control. Electrophorese to obtain separation of the unreacted pre-miRNA and mature miRNA strands.
5. Stain the gel using SYBR Gold as according to the manufacturer's instructions.
6. Image on a fluorescence gel imager. If the NPE/NP is able to inhibit Dicer maturation, some retention of the pre-miRNA band will be observed as shown in step 3 of Fig. 3. If Dicer is still able to process the pre-miRNA, smaller miRNA bands will be observed.

6. SPR assay for pre-miRNAs

6.1 Equipment

- SPR (SensiQ Pioneer or equivalent)
- QDAT software

6.2 Buffers and reagents

- BioCap SPR chips
- Pre-miRNA substrate from Section 3.3.1
- Purified natural products
- Running buffer (50 mM Tris (pH 7.6), 150 mM NaCl, 5% glycerol, 0.05% Tween-20, 1 mM magnesium chloride, 5% DMSO)
- Conditioning buffer (1 M NaCl, 50 mM NaOH)
- 0.5% SDS solution
- High salt running buffer (50 mM Tris (pH 7.6), 850 mM NaCl, 5% glycerol, 0.05% Tween-20, 1 mM magnesium chloride, 5% DMSO)
- Regeneration buffer (50 mM Tris (pH 7), 260 mM NaCl, 0.5 mM EDTA, and 0.1% Tween-20)

6.3 Procedure

1. Dock a BioCap chip into the SPR.
2. Prime the instrument 5 × with running buffer.
3. Condition the chip with two 30 μL injections of conditioning buffer, one 30 μL injection of 0.5% SDS, a third 30 μL injection of conditioning buffer, and one 30 μL injection of high salt running buffer.
4. Prepare the NPs to be tested by dissolving in 800 μL running buffer. Serially dilute the stock solution to make working stock solutions at 5 different concentrations (400 μL total volume). Typical starting concentrations for analysis are 3–50 μM.
5. Immobilize the pre-miRNA on channel 1 with a 100 μL injection of 750 nM RNA in high salt running buffer at a flow rate of 10 μL/min to a final RU of about 2600 to saturate the chip. Channels 2 and 3 should be used as identical reference channels.
6. Test the compounds in random order by injecting 50 μL of each compound at 50 μL/min with a dissociation time of 60 s.

7. Every four injections, inject 50 μL of running buffer at a flow rate of 50 μL/min and a dissociation time of 60 s as a blank run.
8. After each compound or blank injection, regenerate the chip with a 10 μL injection of regeneration buffer at a flow rate of 10 μL/min.
9. Analyze the data using the manufacturer provided QDAT software.

7. Outlook

This chapter describes the use of cat-ELCCA, a biochemical assay that uses click chemistry-mediated catalytic signal amplification, for the discovery of new chemical space for targeting miRNAs. Although we have described its application toward natural products discovery, the pipeline outlined could also be used for the identification of small molecule pre-miRNA ligands from commercial or industrial compound collections. As the discovery of direct-binding, functional inhibitors of RNA biology is becoming a high priority in RNA-targeted drug discovery campaigns, assays like cat-ELCCA will play an important role in these early stage efforts. Further, as we have also demonstrated the applicability of cat-ELCCA to RNA-protein interactions (Lorenz, Kaur, et al., 2018), another important area of nucleic acid biology with therapeutic potential, we foresee that many important tools for interrogating this new frontier in drug discovery will arise from this platform assay technology.

Acknowledgments

This work was supported by a pilot grant from the University of Michigan Center for the Discovery of New Medicines and the NIH (R01 GM118329 to A.L.G. and T32 GM007767 to E.E.G.).

References

Connelly, C. M., Moon, M. H., & Schneekloth, J. S., Jr. (2016). The emerging role of RNA as a therapeutic target for small molecules. *Cell Chemical Biology*, *23*, 1077–1090.

Garner, A. L. (2018a). cat-ELCCA: Catalyzing drug discovery through click chemistry. *Chemical Communications*, *54*, 6531–6539.

Garner, A. L. (2018b). *RNA Therapeutics; Topics in Medicinal Chemistry. vol. 27*. Springer.

Garner, A. L., & Janda, K. D. (2010). cat-ELCCA: A robust method to monitor the fatty acid acyltransferase activity of ghrelin O-acyltransferase (GOAT). *Angewandte Chemie, International Edition*, *49*, 9630–9634.

Garner, A. L., & Janda, K. D. (2011). A small molecule antagonist of ghrelin O-acyltransferase (GOAT). *Chemical Communications*, *47*, 7512–7514.

Ha, M., & Kim, V. N. (2014). Regulation of microRNA biogenesis. *Nature Reviews. Molecular Cell Biology*, *15*, 509–524.

Harvey, A. L., Edrada-Ebel, R., & Quinn, R. J. (2015). The re-emergence of natural products for drug discovery in the genomics era. *Nature Reviews. Drug Discovery*, *14*, 111–129.

Howe, J. A., Wang, H., Fischmann, T. O., Balibar, C. J., Xiao, L., Galgoci, A. M., et al. (2015). Selective small-molecule inhibition of an RNA structural element. *Nature*, *526*, 672–677.

Lorenz, D. A., & Garner, A. L. (2016). A click chemistry-based microRNA maturation assay optimized for high-throughput screening. *Chemical Communications*, *52*, 8267–8270.

Lorenz, D. A., & Garner, A. L. (2018). Approaches for the discovery of small molecule ligands targeting microRNAs. *Topics in Medicinal Chemistry*, *27*, 79–110.

Lorenz, D. A., Kaur, T., Kerk, S. A., Gallagher, E. E., Sandoval, J., & Garner, A. L. (2018). Expansion of cat-ELCCA for the discovery of small molecule inhibitors of the pre-let-7-Lin28 RNA-protein interaction. *ACS Medicinal Chemistry Letters*, *9*, 517–521.

Lorenz, D. A., Song, J. M., & Garner, A. L. (2015). High-throughput platform assay technology for the discovery of pre-microRNA-selective small molecule probes. *Bioconjugate Chemistry*, *26*, 19–23.

Lorenz, D. A., Vander Roest, S., Larsen, M. J., & Garner, A. L. (2018). Development and implementation of an HTS-compatible assay for the discovery of selective small-molecule ligands for pre-microRNAs. *SLAS Discovery: Advancing Life Sciences R & D*, *23*, 47–54.

MacRae, I. J., Ma, E., Zhou, M., Robinson, C. V., & Doudna, J. A. (2008). In vitro reconstitution of the human RISC-loading complex. *Proceedings of the National Academy of Sciences of the United States of America*, *105*, 512–517.

Naryshkin, N. A., Weetall, M., Dakka, A., Narasimhan, J., Zhao, X., Feng, Z., et al. (2014). Motor neuron disease. SMN2 splicing modifiers improve motor function and longevity in mice with spinal muscular atrophy. *Science*, *345*, 688–693.

Newman, D. J., & Cragg, G. M. (2012). Natural products as sources of new drugs over the 30 years from 1981 to 2010. *Journal of Natural Products*, *75*, 311–335.

Palacino, J., Swalley, S. E., Song, C., Cheung, A. K., Shu, L., Zhang, X., et al. (2015). SMN2 splice modulators enhance U1-pre-mRNA association and rescue SMA mice. *Nature Chemical Biology*, *11*, 511–517.

Sivaramakrishnan, M., McCarthy, K. D., Campagne, S., Huber, S., Meier, S., Augustin, A., et al. (2017). Binding to SMN2 pre-mRNA-protein complex elicits specificity for small molecule splicing modifiers. *Nature Communications*, *8*, 1476.

Song, J. M., Menon, A., Mitchell, D. C., Johnson, O. T., & Garner, A. L. (2017). High-throughput chemical probing of full-length protein-protein interactions. *ACS Combinatorial Science*, *19*, 763–769.

Thomas, J. R., & Hergenrother, P. J. (2008). Targeting RNA with small molecules. *Chemical Reviews*, *108*, 1171–1224.

Wilson, D. N. (2014). Ribosome-targeting antibiotics and mechanisms of bacterial resistance. *Nature Reviews. Microbiology*, *12*, 35–48.

Zhang, J.-H., Chung, T. D. Y., & Oldenburg, K. R. (1999). A simple statistical parameter for use in evaluation and validation of high throughput screening assays. *Journal of Biomolecular Screening*, *4*, 67–73.

CHAPTER SIX

Differentiation and classification of RNA motifs using small molecule-based pattern recognition

Giacomo Padroni, Christopher S. Eubanks, Amanda E. Hargrove*
Department of Chemistry, Duke University, Durham, NC, United States
*Corresponding author: e-mail address: amanda.hargrove@duke.edu

Contents

Methods in Enzymology, Volume 623
ISSN 0076-6879
https://doi.org/10.1016/bs.mie.2019.05.022

Abstract

Understanding how to design small molecules that target coding and non-coding RNA has the potential to exponentially increase the number of therapeutically-relevant druggable targets, which are currently mostly proteins. However, there is limited information on the principles at the basis of RNA recognition. In this chapter, we describe a pattern-based technique that can be used for the simultaneous elucidation of RNA motifs and small molecule features for RNA selective recognition, termed Pattern Recognition of RNA by Small Molecules (PRRSM). We provide protocols for the computational design and synthetic preparation of an RNA training set as well as how to perform the assay in plate reader format. Furthermore, we provide details on how to perform and interpret the statistical analysis and indicate possible future extensions of the technique. By combining insights into characteristics of the small molecules and of the RNA that leads to differentiation, PRRSM promises to accelerate the elucidation of the determinants at the basis of RNA recognition.

1. Introduction

1.1 RNA as a drug target

RNA molecules are key players in a myriad of biological processes ranging from transfer of genetic information to regulation of gene expression (Cech & Steitz, 2014; Morris & Mattick, 2014). More recently, non-protein coding (nc) RNA transcripts, which constitute ~70% of transcribed RNA in eukaryotic cells, have emerged as interesting targets for disease diagnosis and therapy as dis-regulation of these RNAs has been implicated in various types of cancers and neurodegenerative diseases (Cech & Steitz, 2014; Connelly, Moon, & Schneekloth, 2016; Jarroux, Morillon, & Pinskaya, 2017). The identification of RNA motifs correlated to these disease states has the potential to exponentially increase the number of druggable targets in cells (Warner, Hajdin, & Weeks, 2018). Small and large RNA transcripts alike exert their function in both normal and disease states when folded in active conformations that are stabilized by smaller structural motifs, which represent plausible therapeutic targets.

Targeting these structures with small molecules is particularly appealing for the tunability of the physicochemical properties and increased cellular uptake relative to oligonucleotides used in antisense strategies (Donlic & Hargrove, 2018; Sztuba-Solinska, Chavez-Calvillo, & Cline, 2019; Warner et al., 2018). However, the only FDA-approved small molecule drugs targeting RNA act on the bacterial ribosome. This is in part due to a dearth of information from both the small molecule design perspective as well as principles regarding RNA recognition; it is challenging to design small molecules that specifically target RNA when there is a lack of structural information for therapeutically plausible RNA targets.

1.2 Techniques for RNA structural determination and small molecule interactions

Current structure determination techniques include chemical and enzymatic probing (Ziehler & Engelke, 2000), which combines computational and experimental methods such as selective 2′-hydroxyl acetylation by primer extension (SHAPE) (Wilkinson, Merino, & Weeks, 2006), dimethylsulfate (DMS) (Tijerina, Mohr, & Russell, 2007), and light activated structural examination of RNA (LASER) (Ackermann & Famulok, 2013; Feng et al., 2018; Tius & Kawakami, 1995) for 2D structures; and nuclear magnetic resonance (NMR) (Salmon, Yang, & Al-Hashimi, 2014), X-ray diffraction (Cate & Doudna, 2000), small-angle X-ray scattering (Chen & Pollack, 2016) and cryo electron microscopy (Razi, Ortega, & Britton, 2016) for 3D structures. Indeed, the structural information derived from these techniques has provided evidence that the topologies and the dynamics of RNA motifs are associated with their molecular recognition properties including small molecule interactions.

Comprehensive analysis of the binding signatures (i.e., equilibrium dissociation constants, thermodynamic and kinetic parameters) of small molecules interacting with a series of RNA targets can be obtained combining a variety of biophysical methods, including NMR (Patwardhan et al., 2017), isothermal titration calorimetry (ITC) (Salim & Feig, 2009), surface plasmon resonance (SPR) (Hendrix, Priestley, Joyce, & Wong, 1997), Förster resonance energy transfer (FRET) (Xie, Dix, & Tor, 2009), ultraviolet (UV) melting spectroscopy (McPike, Sullivan, Goodisman, & Dabrowiak, 2002), and indicator displacement assays (IDA) (Patwardhan, Cai, Newson, & Hargrove, 2019; Zhang, Umemoto, & Nakatani, 2010). Despite being extremely insightful, the combination of RNA structural determination and small molecule binding techniques can be time- and material-consuming and often limited to singular RNA structures.

In order to overcome these limitations and accelerate both the discovery of RNA druggable motifs and RNA-targeting molecular scaffolds, high-throughput techniques capable of simultaneously screening the features of small molecules and RNA structures have started to emerge (Velagapudi & Disney, 2014).

In this method, we describe the experimental details of pattern recognition of RNA by small molecules (PRRSM), a succinct method that allows simultaneous elucidation of the structural features of a variety of RNA motifs and the governing principles for their recognition by a set of small molecules.

1.3 Pattern recognition background

Pattern recognition techniques are formidable methods for the analysis of a variety of interactions in complex chemical mixtures (Fig. 1) (Folmer-Andersen, Kitamura, & Anslyn, 2006; Kitamura, Shabbir, & Anslyn, 2009; Umali & Anslyn, 2010). Similar to the olfactory and gustatory systems,

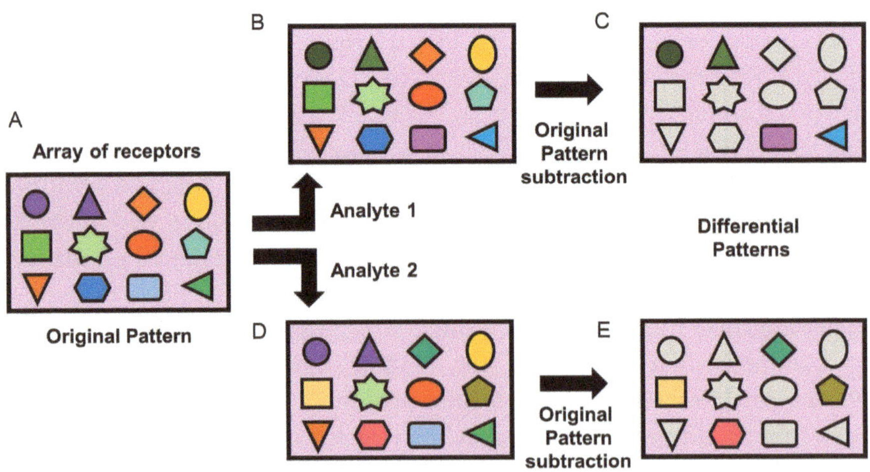

Fig. 1 Example of pattern-based assay. Analytes of the original array of receptors (A) are examined for differential interactions (B and D) to produce a differential pattern (C and E) after subtraction of the background. These patterns can be used to classify unknown analytes.

pattern-based sensing can differentiate among a large variety of similar "stimuli" such as nitrated explosives (Hughes, Glenn, Patrick, Ellington, & Anslyn, 2008), tannins in wine (Umali et al., 2011), ions in water (Palacios, Nishiyabu, Marquez, & Anzenbacher, 2007) and normal, cancerous and metastatic cells (Bajaj et al., 2009). One of the main advantages of this approach is that receptor-based sensing does not require strong and selective binding for a single analyte, but a moderate differential binding for a series of analytes, dramatically reducing the amount of structural information and synthetic effort required for studying receptor-analyte interactions. Sensor systems such as displacement of indicators, colorimetric indicators, or fluorescent labels are required to measure the interactions between analytes and receptors (Umali & Anslyn, 2010). Because of the multitude of data generated from the simultaneous analysis of a variety of analytes and receptors, pattern recognition techniques rely on multivariate statistical analysis such as Principal Component Analysis (PCA) and Linear Discriminant Analysis (LDA) that allows for patterns to become observable by reducing the data dimensionality. Advantageously, the unbiased classification of PCA can also be used for understanding the principles of the analytes-receptor interaction by examining the loading factors of each principal component. On the other hand, LDA is a supervised statistical analysis that achieves maximum differentiation between pre-defined groups and provides the classification power of the assay.

1.4 PRRSM overview

PRRSM is a pattern-based method that differentiates RNA motifs (analytes) by exploiting a series of small molecule receptors (Eubanks, Forte, Kapral, & Hargrove, 2017; Eubanks & Hargrove, 2017, 2019) (Fig. 2). PRRSM differentiation is achieved by measuring the variation of fluorescence intensity induced by the interaction between fluorescently labeled RNA motifs and a small molecule library. The clustering of PRRSM can be used to derive features and properties of both small molecules and RNA motifs that underlie selective recognition. Furthermore, the differentiation of RNA structure by small molecules has the potential to identify targetable motifs in structurally unknown RNA constructs.

In this method, we describe the details of the design and experimental procedures of PRRSM. Section 2 describes how to design a training set of fluorescently labeled RNA motifs using a computational approach, the choice of the small molecule receptors and some considerations for

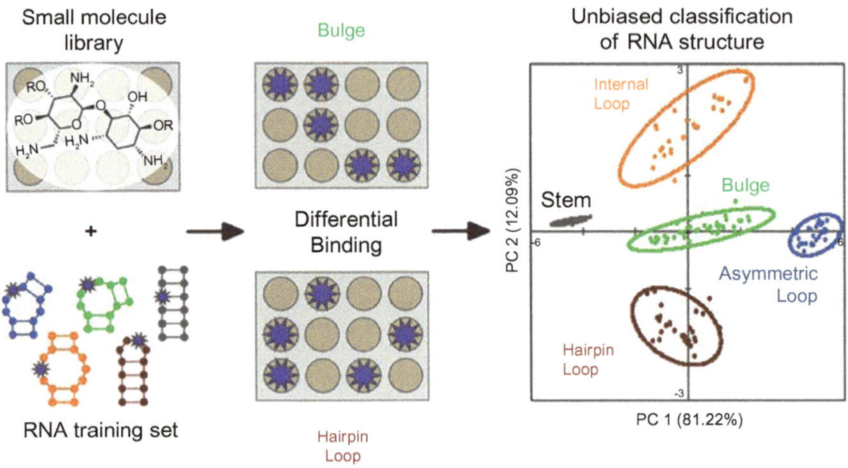

Fig. 2 Overview of PRRSM. The differential binding of a small molecule library (receptors) is used to classify an RNA training set (16 fluorescence labeled motifs). *Reprinted with permission from Eubanks, C. S., Forte, J. E., Kapral, G. J., & Hargrove, A. E. (2017). Small molecule-based pattern recognition to classify RNA structure.* Journal of the American Chemical Society, 139*(1), 409–416. Copyright 2017 American Chemical Society.*

achieving the optimal conditions for the assay. Sections 3 and 4 describe the synthetic procedures for the preparation of the fluorescent probe benzofuranyl uridine (BFU) as phosphoramidite monomer and its incorporation in RNA structures using an automated solid phase oligonucleotide synthesizer. Section 3 also describes how to synthetically modify the aminoglycoside receptors used here to increase their structural diversity. Section 5 describes how to conduct the assay in a plate reader format. Section 6 describes how to perform PCA and leave-one-out cross validation (LOOCV) and how to correctly use the data to achieve the optimal differentiation and interpretation. Section 7 briefly describes how PRRSM can complement other structural prediction techniques for determining targetable RNA motifs in unknown RNA constructs.

2. Assay design and experimental considerations

2.1 Computational sequence design of the training set

The first step of PRRSM is the design of a training set of sequences containing the RNA motifs of interest. The following considerations were obtained from our proof of concept study that focused on bulges (BLG), hairpin loops (HPL), internal loops (IL), asymmetric internal loops (AIL) and stems (STM) (Eubanks et al., 2017). Because of their intrinsically dynamic nature, RNA structures are normally characterized by multiple energetically stable conformations and, therefore, a rational design is required for minimizing the risk of multiple secondary structures present in solution that could lead to data misinterpretation. In PRRSM, the RNA cassettes flanking the variable motifs are kept constant in order to prevent a fluorescence variation due to changes in the RNA sequence or off-target small molecule binding (Fig. 3). The cassettes should contain sufficient GC content to ensure a melting temperature of at least 60 °C and minimize the possibility of partial unfolding events under experimental conditions. Computationally-predicted melting temperatures obtained from web server softwares such as Mfold (Zuker, 2003) can be used for this purpose. Furthermore, at least one GC base pair should be introduced at the terminal positions of the structures of interest in order to minimize "slipped" folding. The size and sequences of the RNA motifs (variable portions) can be computationally designed to mimic naturally occurring RNA secondary

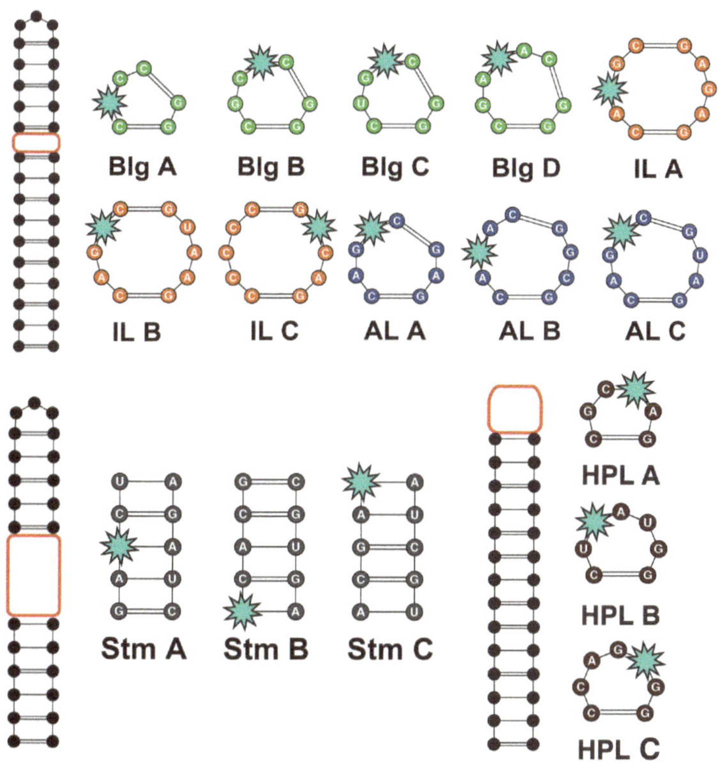

Fig. 3 Schematic representation of the RNA training set design. *Adapted from Eubanks, C. S., & Hargrove, A. E. (2017). Sensing the impact of environment on small molecule differentiation of RNA sequences.* Chemical Communications, *53(100), 13363–13366 with permission of the Royal Chemical Society.*

structures. For example, the size of the motifs described here was chosen according to their natural abundance by using RNA Secondary Structure and Statistical Analysis database (RNA STRAND) (Andronescu, Bereg, Hoos, & Condon, 2008). The sequences can mimic naturally occurring motifs or be obtained by randomization. In all the cases, the expected folding should be computationally confirmed. Lastly, the synthetic accessibility of the structures should be considered. The current protocol for PRRSM is restricted to short oligonucleotides (~40 nts) that are near the upper limit for RNA solid phase synthesis, which is required for selective incorporation of the fluorescent label. Further restrictions might arise from the poor synthetic tractability of certain sequences on the synthesizer.

2.2 Choice and localization of the fluorescence sensor

In PRRSM, the sensor that measures RNA-small molecule interactions is a fluorescent nucleotide inserted into selected positions of the RNA motifs. The ideal solvatochromic fluorophore for this purpose should have minimal impact on the overall RNA topologies and hydrogen bonding network, have good quantum yield, and be sensitive to environmental changes. Despite the higher steric hindrance, the uridine analog BFU was deemed superior to the more common 2-aminopurine (2-AP) because BFU does not mismatch with other nucleotides (2-AP can pair with U and C), and it has enhanced quantum yield and sensitivity to environmental changes such as solvation and base stacking (Tanpure & Srivatsan, 2011). The fluorescence sensor should be placed at the most flexible position of the RNA motif to allow the minimal perturbations induced by small molecules binding to cause the greatest fluorescence variation and minimize possible structural interference arising from the fluorophore moiety (Fig. 3). A computational approach can be used to identify this position. For example, structures can be subjected to MonteCarlo simulations using the FARFAR algorithm available in the Rosie web server to generate 1000 possible conformations (Moretti, Lyskov, Das, Meiler, & Gray, 2018). For each motif, 20 representative conformations can be clustered according to root-mean-squared deviation (RMSD) relative to the lowest energy structure to generate a structural ensemble (Cheng, Chou, & Das, 2015). The most flexible residues can be identified by visually comparing the frequency of nucleotides flipped out the stack of the helix in the structural ensembles. Alternatively, if a large number of conformations are analyzed, a pseudo-dihedral angle defined by the following four points can be calculated: (a) the center of mass of the 3′ base of the potentially flipped base in analysis; (b) center of mass of the sugar of the 3′ base; (c) the center of mass of the sugar of the base in analysis; (d) the center of mass of the base in analysis (Hart, Nyström, Öhman, & Nilsson, 2005). These values can be calculated for each base of the RNA motifs by using the cpptraj module included in the AMBER package (Roe & Cheatham, 2013). More flexible positions can be identified by plotting the frequency of distribution of this value of the nucleobases in analysis.

When choosing the substitution, priority should be given to flexible uridine residues to minimize the influence of hydrogen bonding networks and mismatched conformations.

2.3 Small molecule library choice and preparation

Small molecule aminoglycosides were identified as ideal starting receptors to develop PRRSM. First, this class of compounds has relatively high affinity for a variety of RNA constructs. The lack of selectivity for specific RNA motifs due to electrostatic interactions is expected to minimally impact pattern recognition experiments, where the only requirement is differential binding affinity, normally in the range from nanomolar to micromolar equilibrium dissociation constants (Chittapragada, Roberts, & Ham, 2009). Furthermore, aminoglycosides are commercially available, cheap, and relatively easy to functionalize in the event increased chemical diversity is required for the assay.

The same considerations can potentially be extended to new classes of small molecules with moderate binding affinity for the RNA structures of interest.

2.4 Considerations on conditions, buffer and concentrations

Despite the low volume of sample required for PRRSM, optimization of RNA and small molecule concentrations should be performed to avoid wasting precious material. RNA concentration at around 200 nM should ensure the optimal signal-to-noise ratio in the fluorescence measurement for most modern plate readers. The choice of the buffer also influences the intensity of the fluorescence signal. In our hands, a phosphate buffer (25 mM NaCl, 4 mM $MgCl_2$, 0.1 mM EDTA, pH 7.3, 25 °C, Buffer A) resulted in enhanced fluorescence outputs compared to buffers such as cacodylic, HEPES and Tris. However, no major differences were observed between phosphate and Tris in PRRSM structural differentiation. Variations to these conditions might increase or decrease clustering in PRRSM (Eubanks & Hargrove, 2017). For example, the assay tolerates pH ranges between 6 and 8 as minimal differences in predictive power are observed. However, more acidic conditions (pH 5) ablate the differentiation, presumably due to conformational and electrostatic perturbations of the aminoglycoside ligands. Variations of ionic strength also affect PRRSM. As expected, high sodium concentration (140 mM) reduces the predictive power presumably due to the weaker affinity of aminoglycosides for RNA under these conditions. Surprisingly, both high concentration and removal of Mg^{2+} (10 mM) do not affect motif classification but reduce the differentiation of individual sequences. Increasing the temperature to physiological conditions (37 °C) slightly reduces the predictive power relative to the standard conditions, presumably due to reduced binding affinity of aminoglycosides for RNA structures. However, physiological temperature in combination with a molecular crowder (PEG-12000) known to destabilize RNA structures to mimic in vivo conditions slightly reduces motif classification but enhances individual sequence separation. Altogether, these considerations indicate that buffer and experimental conditions can be adjusted for optimal motif classification or enhanced individual sequences differentiation to provide insight into the nature of the molecular recognition.

3. Synthetic preparation of BFU and guanidinylated aminoglycoside

3.1 Equipment for organic synthesis

- Rotary evaporator
- Vacuum manifold (Schlenk line or similar) connected to inert gas cylinder (N_2)
- Microwave reactor (Biotage® Initiator+ or similar)
- Ventilated fume hood
- Lyophilizer
- NMR spectrometer (400 or 500 MHz)
- HPLC (fitted with C18 column such as Phenomenex Luna 5 μm C18 100A 150 × 4.6 mm or similar)
- Mass spectrometer (Electrospray Ionization by direct infusion or coupled with a liquid chromatography apparatus)
- Standard organic chemistry glassware
- Water purification system (ELGA PURELAB Flex Veolia Water Technologies or similar)
- Dry solvents apparatus (Pure Solv Innovative Technology or similar)
- Thin layer chromatography (TLC)
- 230–400 mesh silica gel (Silicycle or similar)
- UV lamp (UVP-UVGL-25, 4 W, 254/365 nm or similar)

– Standard PPE such as protective glasses, nitrile gloves and flame-retardant lab coat (*Note*: Any specific additional PPE and SOPs required are indicated in the procedure).

3.2 Reagents

All the reagents and solvents were used as purchased without further purification unless otherwise stated. Reagents and solvents were purchased from Sigma-Aldrich, Glen Research, Chem Impex, Fisher, Acros and Oakwood Chemicals. Deuterated solvent for NMR characterization were purchased from Cambridge Isotope Laboratories.

3.3 Preparation of guanidinylated aminoglycosides

Increasing the chemical diversity of the receptors is often a good strategy to enhance the differential binding to RNA motifs. For example, the amino group of aminoglycosides can be further functionalized by introducing guanidine moieties, which were previously demonstrated to enhance RNA recognition (Luedtke, Baker, Goodman, & Tor, 2000; Luedtke, Carmichael, & Tor, 2003). A representative synthetic scheme and procedure for the aminoglycoside kanamycin (**1**) is shown below (Fig. 4). The same procedure might be applied to similar small molecules.

Preparation of **2**

a. Dissolve kanamycin (0.041 mmol, 1 equiv.) in deionized water (0.35 mL) and 1,4-dioxane (1.71 mL, 0.02 M) into a 5 mL round bottom flask.

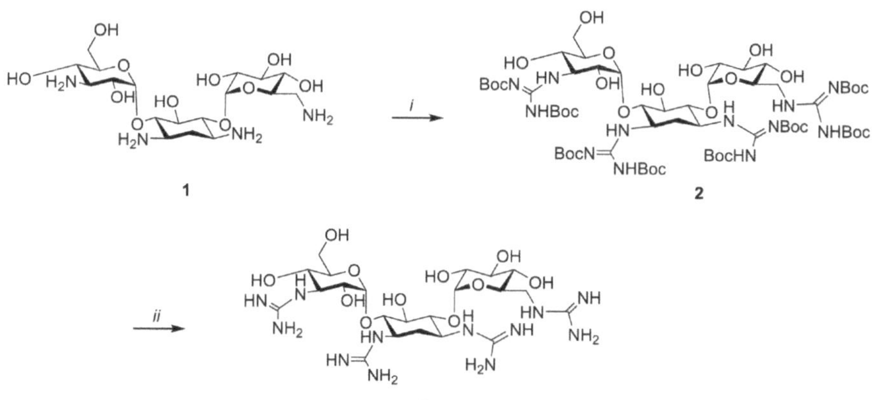

Fig. 4 Synthesis of guanidinylated kanamycin (**3**) Reagents and conditions: (i) *N,N′*-di-Boc-N″-triflylguanidine (8 equiv.), TEA (10 equiv.), water/dioxane, 3 d, rt. (ii) 1 M HCl aq., EtOAc, 4 h, rt.

b. While stirring, add *N,N′*-di-Boc-*N″*-triflylguanidine (8 equiv.).
c. After 5 min of stirring, slowly add triethylamine (10 equiv.) and stir for 3 days at room temperature. Check the progress by TLC (DCM/MeOH 90:10; Rf starting material ~0.05; Rf product ~0.5).
d. Extract with CH_2Cl2 (×3) and wash with brine (×3).
e. Dry the organic layer using anhydrous Na_2SO_4, filter, and evaporate under vacuum.
f. Purify the residue by silica column flash chromatography (90:10 DCM/MeOH) and dry in vacuo to yield the desired product as a white solid (yield 90%).
g. Confirm identity and purity by ^{1}H and ^{13}C NMR, and HRMS (Eubanks et al., 2017).

Preparation of **3**

a. Dissolve **2** (0.034 mmol, 1 equiv.) in ethyl acetate (0.86 mL, 0.03 M) and 1 M aqueous HCl (0.074 mL).
b. Stir the solution at room temperature for 4 h and check by TLC (DCM/MeOH 95:5; Rf starting material ~0.85; Rf product ~0.05).
c. Dilute the solution with toluene (3.2 mL) and concentrate in vacuo.
d. Resuspend **3** in water (4 mL) and lyophilize (yield 95%).
e. Confirm identity and purity by ^{1}H and ^{13}C NMR and HRMS (Eubanks et al., 2017).

3.4 Preparation of BFU phosphoramidite

The monomeric phosphoramidite of the fluorescence label BFU to be used for solid phase synthesis is prepared starting from the commercially available 5-iodouridine (**4**) in five steps (Fig. 5). Pd-catalyzed Suzuki-Miyaura coupling is used to connect the benzofuran moiety at the C5 position of uridine. Two-step one-pot sequential protection of the 3′-5′ hydroxyl groups and 2′ hydroxyl group, followed by selective deprotection of the 3′ and 5′ positions afford intermediate **7**. Protection of the 5′-position using DMT, followed by installation of the O-phosphoramidite group at the 3′ hydroxyl provide the final monomeric building block **9**.

Preparation of **5**

a. Dissolve 5-iodouridine (**4**, 0.70 mmol, 1 equiv.), 2-benzofuranylboronic acid (1.3 equiv.), and potassium hydroxide (2.0 equiv.) in 4.0 mL (0.175 M) of deionized water in a microwave vial (*Note*: choose appropriate microwave vial according to vendor indication).
b. Purge the solution with a nitrogen flow for 20 min.

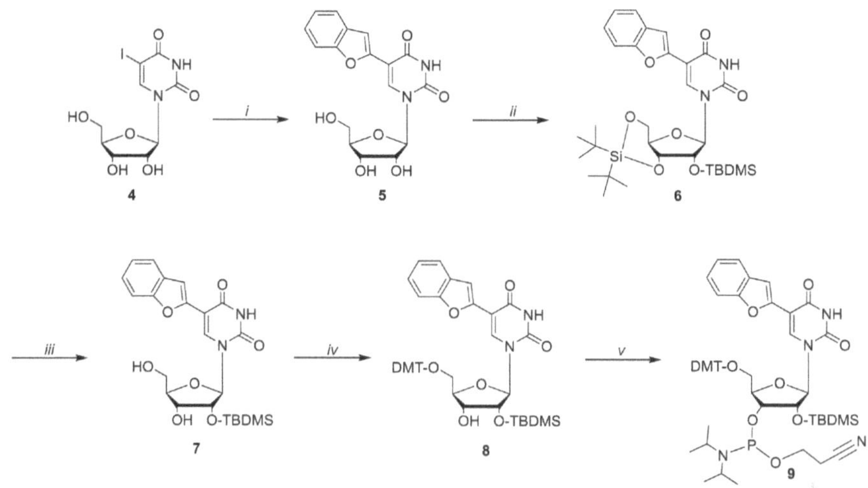

Fig. 5 Synthesis of the BFU phosphoramidite (**9**). Reagents and conditions: (i) BFBA (1.3 equiv.), KOH (2 equiv.), Na_2PdCl_4 (10 mol%), water, 1 h, 90 °C, MW. (ii) TBS-triflate (1.2 equiv.), dry DMF, 15 min, 0 °C; then imidazole (12 equiv.), 10 min, rt.; then TBDMS-Cl (4 equiv.), 3 h, 60 °C. (iii) HF-pyridine (~70% HF, 5 equiv.), pyridine (10 equiv.), dry DCM, 2 h, −5 °C to rt. (iv) DMT-Cl (5 equiv.), dry pyridine, 6 h, rt. (v) 2-cyanoethyl N,N,N′,N′-tetraisopropylphosphorodiamidite (3 equiv.), ETT (1.3 equiv.), dry DCM, 3 h, 65 °C, MW.

c. Add an aqueous solution of Na_2PdCl_4 (10 mol% relative to **4**) and purge with a nitrogen flow for further 5 min.
d. Seal and stir under microwave irradiation at 90 °C, high absorption, for 1 h.
e. Chill the vial on ice for 30 min and collect the white precipitate by vacuum filtration.
f. Rinse the solid with water (×2), and hexanes (×3).
g. Resuspend the solid in ~5 mL of deionized water and lyophilize to yield a white fluffy solid (yield 75%).
h. Confirm identity and purity by ^1H and ^{13}C NMR, HRMS and RP-HPLC (Eubanks et al., 2017).
i. A small impurity attributed to de-glycosilated product can be present. This can be removed by silica gel column chromatography (EtOAc/MeOH 95:5) or carried onto the next step and removed in the following purification.

Preparation of **6** (*Note*: this reaction must be performed under anhydrous conditions)

a. Dissolve compound **5** (0.88 mmol, 1 equiv.) in anhydrous *N,N*-dimethylformamide (4.4 mL, 0.2 M) in a 15 mL round bottom flask and stir at −5 °C under argon atmosphere.

b. Slowly add di-tert-butylsilyl-bis(trifluoromethansulfonate) ($(t\text{-}Bu)_2Si(OTf)_2$) (1.2 equiv.) drop-wise (*Note*: a rate of addition of 1drp/5s or slower is necessary for the reaction to proceed) and allow to react at room temperature for 15min.

c. Check starting material consumption by TLC (EtOAc:MeOH 9:1; Rf starting material ~0.65). If the starting material is still present, slowly add ($(t\text{-}Bu)_2Si(OTf)_2$) in 0.25 equiv. increments until completion.

d. Quench the solution by adding imidazole in portions (12.0 equiv.) and stir for 10min at room temperature.

e. Add *tert*-butyldimethylsilyl chloride (4.0 equiv., check Safety Data Sheet (SDS) for safety recommendations) and attach a reflux condenser to the reaction flask.

f. Heat the solution to 60°C for 3h and confirm the progress of the reaction by TLC (Hexanes/EtOAc 80:20; Rf starting material ~0.1; Rf product ~0.7).

g. Allow the reaction to cool to room temperature and chill the flask in an ice bath before adding water (10× dilution).

h. Collect the white precipitate by vacuum filtration.

i. Isolate the pure product **6** via silica column flash chromatography (gradient from 100% DCM to DCM/MeOH 98:2) as a white solid (yield 89%).

j. Confirm identity and purity by ^{1}H and ^{13}C NMR, HRMS and RP-HPLC (Eubanks et al., 2017).

Preparation of **7**

a. Dissolve **6** (0.28mmol, 1 equiv.) in freshly distilled anhydrous CH_2Cl_2 (1.4mL, 0.2M) in a polypropylene test tube and stir at −5°C.

b. Prepare a separate solution by diluting HF-pyridine (~70% HF, 5.0 equiv.) with anhydrous pyridine (10 equiv.) at 0°C in a polypropylene test tube (*IMPORTANT safety note*: HF is strongly corrosive and should be used in a ventilated fume hood. SDS should be consulted before use. Avoid the use of glass throughout the procedure until the reaction is quenched. Additional PPE such as neoprene gloves, apron and face-mask should be used when performing the reaction).

c. Add slowly the HF-pyridine solution to the CH_2Cl_2 solution and stir for 2h. Check the progress by TLC (Hexanes/EtOAc 85:15; Rf starting material ~0.6; Rf product=~0.35; TLC sample can be obtained by adding 1μL of reaction mixture to 100μL of saturated aqueous Na_2CO_3 solution. Product was extracted with 50μL of EtOAc and spotted on the TLC plate).

d. Dilute the mixture with CH_2Cl_2 (5 ×) and quench by slowly adding a saturated, aqueous Na_2CO_3 solution (5 ×).
e. Separate the organic layer and wash with sat. $NaHCO_3$ (× 2), and brine solution (×2).
f. Dry the organic layer with anhydrous Na_2SO_4, filter, and dry in vacuo.
g. Purify the crude product via silica column flash chromatography (DCM/MeOH 98:2) to yield **7** as a white solid (yield 81%).
h. Confirm identity and purity by ^{1}H and ^{13}C NMR, HRMS and RP-HPLC (Eubanks et al., 2017).

Preparation of **8** (*Note*: this reaction must be performed under anhydrous conditions)

a. Dissolve dried **7** (0.25 mmol, 1 equiv.) in anhydrous pyridine (2.5 mL, 0.1 M) while stirring at 0 °C under a stream of nitrogen.
b. Add 4,4′-dimethoxytrityl chloride (5.0 equiv.), allow the solution to warm to room temperature and stir for 6 h. Check reaction progress by TLC (Hexanes/EtOAC 65:35 + 3% TEA; Rf starting material ~0.4; Rf product ~0.6).
c. Pour the solution in water and extract with DCM (3 ×).
d. Wash the organic layer with saturated aqueous $NaHCO_3$ (× 2) and then brine (×2).
e. Dry the organic layer using anhydrous Na_2SO_4, filter, and dry in vacuo.
f. Purify the crude product via silica column flash chromatography (gradient from Hexanes/EtOAc 97:3 + 3% TEA to 50:50 + 3% TEA) to yield **8** as a white solid (yield 80%) (*Note*: silica gel required pretreatment with Hexanes:EtOAc 97:3 + 3% TEA by vigorous stirring for at least 10 min).
g. Confirm identity and purity by ^{1}H and ^{13}C NMR, HRMS and RP-HPLC (Eubanks et al., 2017).

Preparation of **9** (*Note*: this reaction must be performed under anhydrous conditions)

a. Add anhydrous **8** (0.193 mmol, 1 equiv.) and 5-ethylthio-1H-tetrazole (1.3 equiv.) to an oven-dried microwave vial purged with argon and leave under high-vacuum for 12 h.
b. Dissolve the solids using freshly distilled anhydrous CH_2Cl_2 (1.2 mL, 0.16 M) under argon atmosphere.
c. Add reagent 2-Cyanoethyl *N,N,N′,N′*-tetraisopropylphosphorodiamidi te (3 equiv.) dropwise via syringe.
d. Stir the solution under microwave irradiation at 65 °C, low absorption, for 3 h.
e. Check reaction progress by TLC (Hexanes/EtOAc 75:25 + 3% MeOH; run the TLC plate twice to provide enough separation between

diasteroisomers and starting material; Rf starting material ~0.65; Rf product 1–0.6; Rf product 2–0.7).

f. Load the crude directly to a silica column flash chromatography (gradient from 10% EtOAc (HPLC grade) in pentane (HPLC grade) + 3% TEA to 80% EtOAc (HPLC grade) in pentane (HPLC grade) + 3% TEA; *Note*: silica gel requires pretreatment with pentane:EtOAc 90:10 + 3% TEA by vigorous stirring for at least 10 min).

g. Collect both diastereomers and dry in vacuo.

h. Confirm the identity and purity by ^{1}H and ^{31}P NMR. An H-phosphonate impurity of 2–10% of the total yield can be present (^{31}P δ 8.01 and 7.57 ppm) and is not expected to interfere with the solid phase synthesis. If a larger amount of impurities is present, the crude can be re-purified by re-crystallization from DCM using cold hexanes or a short silica gel column purification using DCM/ACN/TEA 98:1:1) (Eubanks et al., 2017).

i. Due to instability, the compounds should be stored at −20 °C, in the dark, under inert atmosphere and readily used for solid phase synthesis.

4. Preparation of the RNA training set

4.1 Equipment

- Oligonucleotide synthesizer (MerMade 6 &12 or similar)
- Control Pore Glass columns (1000 Å from Glen Research or similar)
- Vacuum manifold
- Speed-vacuum evaporator or constant airflow streamer
- Polydivinylbenzene 4,4′ dimethoxytrityl affinity columns (GlenPack cartridges or similar)
- Centrifuge (Eppendorf 5810R or similar)
- Nanodrop spectrometer
- Gel electrophoresis apparatus
- Gel Imager (Biorad or similar)

4.2 Reagents

All the reagents are used without further purification unless otherwise stated. Reagents and solvents were purchased from Sigma-Aldrich and Glen Research. Ac-A-CE, Ac-C-CE, Ac-G-CE, U-CE (purchased from ChemGenes) and BFU phosphoramidite solutions should be used within a week from their preparation.

4.3 Automated solid-phase synthesis of the training set

The BFU-labeled RNA strands were synthesized using a MerMade 6 & 12 synthesizer. A detailed description of the use of this machine is beyond the scope of this chapter and can be found in the manual provided by the supplier. For this reason, only an overview of the process will be provided.

a. Dissolve the commercially purchased orthogonally protected Ac-A-CE, Ac-G-CE, Ac-C-CE, U-CE, and the synthesized BFU phosphoramidite in anhydrous acetonitrile at a final concentration of 67 mM.

b. Choose the appropriate columns containing the desired first 3′ nucleotide attached.

c. A script is prepared containing the following information: desired sequences; conditions for deprotection of the 4,4′ dimethoxytrityl group with 3% trichloroacetic acid in dichloromethane (typically two cycles, 20 s wait time, 12 s drain time, 2 s equalize time); conditions for coupling of the next nucleotide using a tetrazole catalyst 0.25 M ETT in acetonitrile (typically two cycles, 180 s wait time, 12 s drain time, 2 s equalize time); conditions of capping of the unreacted 5′ hydroxyl groups with a mixture of acetic anhydride (Cap A) and *N*-methylimidazole (Cap *B*) dissolved in tetrahydrofuran/pyridine (typically two cycles, 14 s wait time, 12 s drain time, 2 s equalize time); conditions for the oxidation of the phosphodiester bond with 0.1 M iodine in pyridine/water mixture (typically one cycle, 40 s wait time, 14 s drain time, 2 s equalize time); the DMT-on option.

d. Trityl log can be used to check coupling efficiency (normally between 85% and 100%), if available.

4.4 Cleavage, deprotection and purification of RNA

a. After completion of the synthesis, insert the RNA columns on the vacuum manifold, and elute the RNA by adding 333 μL of a 1:1 solution of 30% aq. ammonium hydroxide and 30% aq. methylamine and let the solution to drain under gravity for 7.5 min.

b. Drain the remaining solution under vacuo and repeat step (a) twice.

c. Incubate the collected solution at room temperature for 2 h to remove the acetyl protecting group from the nucleobases.

d. Remove the solvent using a speed-vac or by flushing the solution with a gentle stream of air (*Note*: exhaustive drying should be achieved to obtain optimal yield).

e. To deprotect the 2′ hydroxyl group, dissolve the obtained crystals in 115 μL dimethylsulfoxide, 60 μL triethylamine, and 75 μL of triethylamine:hydrogen fluoride (30%) and heat for 2.5 h at 65 °C (*IMPORTANT safety note*: HF is strongly corrosive and should be used in a ventilated fume hood. Avoid the use of glass throughout the procedure until the reaction is quenched. Additional PPE such as neoprene gloves, apron and face-mask should be used when performing the reaction).

f. After cooling to room temperature, add 1.75 mL of quenching buffer (Glen Research) to the solutions.

g. Insert a 3–5 μm polydivinylbenzene 4,4′ dimethoxytrityl affinity column into the appropriate inserts of the vacuum manifold.

h. Precondition the column by sequentially adding 0.5 mL acetonitrile and 1 mL of a 2 M solution of triethylammonium acetate (TEAA) in water. Drain the liquid using the vacuum after each addition.

i. Add the RNA solutions to the column and wash with 1 mL of a 1:9 acetonitrile:2 M triethylammonium acetate solution, 1 mL of RNase free water, 1 mL of 2% TFA solution in water (×2), 1 mL RNase free water (×2) and discard the eluted solutions.

j. In a clean and sterile vial, elute the RNA from the column by adding 1 mL of 1 M ammonium bicarbonate:30% acetonitrile.

k. Perform ethanol precipitation by adding 10% volume of sodium acetate (3 M pH 5.2) and 3 volume of ice-cold absolute ethanol.

l. Leave at −20 °C for 16 h and centrifuge for 2 h at 4 °C at 4000 rpm.

m. Remove the supernatant after centrifugation.

n. Repeat the ethanol addition and centrifugation steps (k–m) (×2).

o. Lyophilize the residue to remove any remaining ethanol.

p. Dissolve the RNA sequences in phosphate buffer (10 mM NaH_2PO_4, 25 mM NaCl, 4 mM $MgCl_2$, 0.5 mM EDTA, pH 7.3), and analyze the concentration using a Nanodrop spectrophotometer.

q. Determine the purity of the RNA with a pre-cast 15% polyacrylamide gel electrophoresis (PAGE) run with 1 × Tris, borate, and EDTA (TBE) buffer at 180 V for ~2 h.

r. Stain the gel using diamond dye nucleic acid for 20 min and image the gel using a UV gel imager.

5. Assay protocol

5.1 Equipment

- Calibrated Analytical balance
- Nanodrop spectrophotometer
- 96-well plates (Corning 96 well cell culture plate—V bottom or similar)

- 384-well plates (Corning 384 well, low volume, round black bottom or similar)
- Pipette set (Eppendorf research plus or similar)
- Multichannel pipette (Eppendorf 10–100 or similar)
- Electronic dispenser (Eppendorf Repeater E3 or similar)
- Sterilized pipette tips
- Plate-reader (Molecular Devices SpectraMax or BMG CLARIOstar or similar)
- Shaker (Corning PC-101; New Brunswick Scientific Excella E2 Platform or similar)
- Centrifuge compatible with well plates (Eppendorf 5810R or similar)

5.2 Buffers and reagents

- Buffer A: 25 mM NaCl, 4 mM $MgCl_2$, 0.1 mM EDTA, pH 7.3; or variants of this buffer
- Molecular biology grade DMSO
- RNase free water (nuclease-free)
- Small molecules aminoglycosides stock solution (2 mM in 10% DMSO in water)
- RNA training set stock solution in Buffer A.

5.3 Procedure

a. Weigh the commercially purchased and synthesized small molecules (amikacin, apramycin, 2-deoxystreptamine, dihydrostreptomycin, kanamycin, neamine, neomycin, streptomycin, sisomicin, guanidino-paromomycin, and guanidino-kanamycin) and dissolve them in 10% DMSO in RNase-free water for a final concentration of 2 mM. Vortex, spin-down and visually inspect the mixture to ensure complete solubilization.

b. Perform two serial dilutions for each small molecule in two separated 96-well plates (plate A and plate B). Table 1 indicates the amount of buffer and stock solutions required in each well. Each solution should be mixed thoroughly to ensure homogeneity.

c. In a 384-well plate, add small molecule (10 μL) with a 10 μL multichannel pipette by filling the 16 wells in a single column by alternating solutions from the 96-well plate B and A.

d. Dilute the stock solutions of each RNA to 400 nM concentration using Buffer A.

Table 1 Serial dilutions of aminoglycosides performed in two 96-well plates (plate A and B).

Equiv. (plate A)	Vol (μL)	Buffer vol (μL)	SM conc. (μM)	Equiv. (plate B)	Vol (μL)	Buffer vol (μL)	SM conc. (μM)
16	0.96	299.04	6.4	20	1.2	298.8	8
10	187.50	112.50	4	14	210.00	90.00	5.6
7	210.00	90.00	2.8	8	171.43	128.57	3.2
5	214.29	85.71	2	6	225.00	75.00	2.4
3	180.00	120.00	1.2	4	200.00	100.00	1.6
2	200.00	100.00	0.8	2.5	187.50	112.50	1
1	150.00	150.00	0.4	1.5	180.00	120.00	0.6
0	0.00	300.00	0	0.5	100.00	200.00	0.2

The volumes listed in the first row indicate the amount of the 2 mM stock solution to be diluted with buffer. The volumes in the following rows indicate the amount of the solution from the well above to be diluted with buffer to obtain the desired small molecule (SM) concentration.

e. Add 10 μL of the RNA solutions to each well for a final concentration of 200 nM. A repeater pipette should be used for accelerating this procedure and to reduce possible photobleaching of the fluorophore.
f. Cover the plate and shake at 100 rpm for 5 min to ensure homogeneous solutions.
g. Centrifuge the plate at 3000 rpm for 1 min to remove any air bubble.
h. Incubate the plate in the dark for 15 min to reach the binding equilibrium.
i. Insert the plate into the plate reader and measure the fluorescence intensity of each well by exciting the samples at 322 nm and reading emission at 455 nm. 50 flashes/read (*Note*: for measurement at 37 °C, an optically clear seal should be used to minimize evaporation and reduce error. Excitation and emission wavelengths are specific for BFU and should be adjusted for different fluorophores).
j. Export the raw fluorescence data in an excel format.
k. Repeat the experiment in triplicate.

6. Statistical methods for data analysis

6.1 Overview of principal component analysis and leave-one-out cross validation

Large amounts of data often require a differential statistical method to ease the interpretation. PCA is a non-biased statistical method that reduces the dimensionality of the dataset by utilizing its inherent variance (Jolliffe & Cadima, 2016; Wenderski, Stratton, Bauer, Kopp, & Tan, 2015). In PCA, a covariance matrix is created from a multivariate dataset, from which eigenvalues and eigenvectors (non-zero vectors that change by the factor eigenvalue) are determined. The dataset is transposed into new principal component space, which is dependent on the variance of the dataset and therefore might produce clustering. The principal components are ranked according to the percentage of variance explained, and normally 95% of the variance is required for a complete analysis of the dataset.

PCA does not inherently provide the predictive power of the analysis. For this reason, another statistical approach is required to assess the significance of the clustering generated from PCA and possible overtraining, and therefore to determine if the model can be used in predictive assays. Leave-one-out-cross validation (LOOCV) is a validation method that partitions the entire dataset between a training set and a smaller validation subset (Kim, 2009). A predicting model is generated from the training set and the validation set is analyzed and compared to its determination in the original entire dataset by using the RNA motifs as the basis of the prediction. The process is normally iterated by using different partitions and the averaged prediction is used to validate the predictive power of the experimental dataset. A predictive power of 100% indicates that LOOCV recapitulates the PCA clusters as if the data were not removed, while lower percentages showing deviations from the PCA plot and the validation method. A low predictive power might indicate potential underfitting (i.e., not enough training data to generate a good predictive model) or overfitting (i.e., the predictive model cannot successfully generalize new observations because the training set is too large) and adjustment of the dimension of the training data might be required (Stewart, Ivy, & Anslyn, 2014).

6.2 Equipment

- PC work station (Dell Work station equipped with Intel (R) Xeon (R) CPU E5-2680 v2 @ 2.80GHz 2 processors or similar)
- Excel Microsoft office
- XLSTAT package
- RStudio package

6.3 Principal component analysis (PCA) in PRRSM

PRRSM requires the entirety of the raw fluorescence data from the aminoglycoside titrations experiments to achieve structural classification.

This dataset can be organized in an excel sheet in which the fluorescence of each aminoglycoside (column) is associated with the relative RNA structure (row). PCA can be performed using XLSTAT software as reported below.

a. In excel, open the XLSTAT menu bar and select "*Analyzing data—Principal Component Analysis*"

b. In the windows, insert the "*observations/variable table*" by selecting the cells containing the raw fluorescence.

c. Select "*Variable labels*" and insert the cell containing the labels (the labels should indicate motifs or individual sequences identifiers).

d. Select the PCA type and in the "outputs" panel the desired values necessary for the analysis.

e. Press "Ok" to perform PCA.

f. Use Scree plot to assess the percentage of variance explained by each principal component and choose which type of graphs (2D or 3D) will be a better representation of the variance.

g. Use the biplots to assess the clustering.

h. Insert confidence ellipses by replotting the "factor scores" of the selected principal components table using a scatter plot (Visualize data-Scatter Plot) selecting the options "confidence ellipses."

6.4 Leave-one-out-cross validation (LOOCV)

This analysis can be performed in RStudio using the following procedure:

a. Import the PCA dataset as .csv file.

b. Load the caret and klaR libraries using the commands.

```
"library(caret)"
"library(klaR)"
```

c. Partition the dataset with a partition fraction of 0.875 using the command (*Note:* this is an eightfold cross validation, but any value between 0 and 1 can be used).

```
trainIndex <- createDataPartition(`Nameof the dataset here_1`$
Name, p=0.875, list=FALSE)
```

d. Name the partitions of the "*trainindex*". The training set and validation set are renamed to "*data_train*" and "*data_test*", respectively, using the following command:

```
data_train <- `name of the imported file`[ trainIndex,]
data_test <- ` name of the imported file`[-trainIndex,]
```

e. Run a Naïve Bayes algorithm of the training set to create a predictive model (named "model") using the following command:

```
model <- NaiveBayes(Name~., data=data_train)
```

f. Make predictions using the validation set (data_test) using the following command:

```
predictions <- predict(model, data_test)
```

g. Summarize the results by exporting a confusion matrix and a predictive percentage table (Tables 2–4) for each motif using the following command:

```
confusionMatrix(predictions$class, data_test$Name).
```

6.5 Data processing

As discussed in Section 2.4, the predictive power of PRRSM is strongly dependent on the assay conditions such as temperature, ionic strength, pH and buffers. Furthermore, data processing might also affect the predictive power and hide subtle differences within RNA motifs/sequences. After performing PCA with the entirety of the dataset, the loading plots relative to each small molecule should be analyzed and compared for each motif. In case some aminoglycosides do not show differential binding (i.e., loading factor

Table 2 Example of confusion matrix and statistics reference obtained from PRRSM using 10 mM NaH_2PO_4, 25 mM NaCl, 4 mM $MgCl_2$, 0.5 mM EDTA, 8 mM PEG 12000, pH 7.4 at 37 °C.

Prediction	AIL	Bulge	HPL	IL	Stem
AIL	6	0	0	0	0
Bulge	0	8	0	0	0
HPL	0	0	6	0	0
IL	0	0	0	6	0
Stem	0	0	0	0	6

Table 3 Example of overall statistics obtained from PRRSM using 10 mM NaH_2PO_4, 25 mM NaCl, 4 mM $MgCl_2$, 0.5 mM EDTA, 8 mM PEG 12000, pH 7.4 at 37 °C.

Accuracy	1
95% CI	(0.8911, 1)
No Information Rate	0.25
P-Value [Acc > NIR]	<2.2e-16
Kappa	1
Mcnemar's Test P-Value	NA

Table 4 Example of statistics by class obtained from PRRSM using 10 mM NaH_2PO_4, 25 mM NaCl, 4 mM $MgCl_2$, 0.5 mM EDTA, 8 mM PEG 12000, pH 7.4 at 37 °C.

	Class: AIL	Class: Bulge	Class: HPL	Class: IL	Class: Stem
Sensitivity	1.0000	1.0000	1.0000	1.0000	1.0000
Specificity	1.0000	1.0000	1.0000	1.0000	1.0000
Pos Pred Value	1.0000	1.0000	1.0000	1.0000	1.0000
Neg Pred Value	1.0000	1.0000	1.0000	1.0000	1.0000
Prevalence	0.1875	0.2500	0.1875	0.1875	0.1875
Detection Rate	0.1875	0.2500	0.1875	0.1875	0.1875
Detection Prevalence	0.1875	0.2500	0.1875	0.1875	0.1875
Balanced Accuracy	1.0000	1.0000	1.0000	1.0000	1.0000

close to 0), their fluorescence data can be removed to increase the clustering of the PCA. Removal of entire motifs might also increase the overall classification of PRRSM. For example, RNA stem sequences are normally poorly differentiated in PRRSM due to lower fluorescence intensities of BFU and reduced binding affinity of aminoglycosides, and their removal from the dataset enhances the predictive power of the remaining sequences (Fig. 6). Furthermore, the amount of data points should be adjusted in each case for achieving optimal differentiation as a limited number of observations (i.e., number of receptors and/or limited data points from the serial dilutions) might not generate a clear classification.

6.6 Interpretation of PRRSM

Information on selective RNA recognition can be derived by complementing PRRSM with the computational analysis of small molecules and RNA structures. For example, correlation between small molecules chemoinformatic properties and their PCA loading factor might be used to discern trend in molecular properties (such as hydrogen bond donor/acceptor, aromaticity, number of rotatable bonds and other physicochemical parameters) responsible for selective recognition of RNA sequences and motifs (Fig. 7). RNA motifs features exploited by small molecules for their differentiation (such as size, shape, and flexibility) can be identified via structural analysis of RNA motifs using computational tools, including structure prediction programs and molecular dynamics simulations. Furthermore, differences in the PCA plot obtained by varying experimental conditions

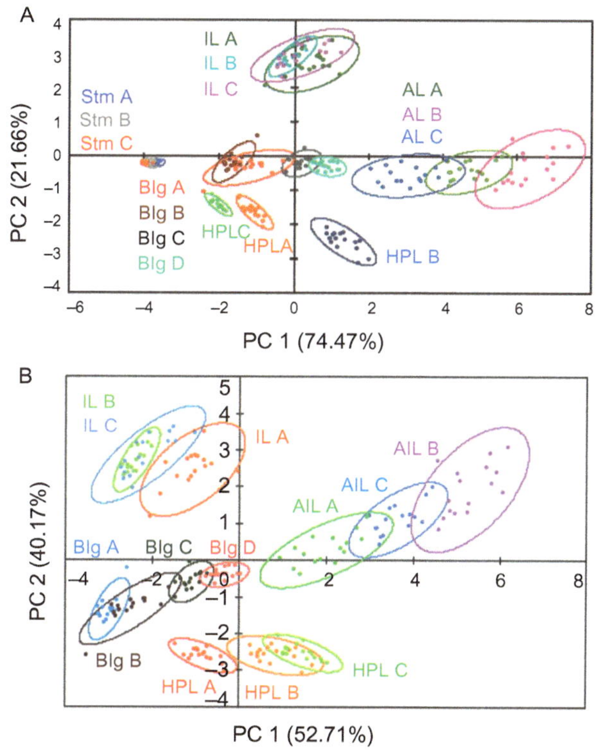

Fig. 6 (A) PCA plot of the entire training set obtained in 10 mM NaH_2PO_4, 25 mM NaCl, 4 mM $MgCl_2$, 0.5 mM EDTA, 8 mM PEG 12000, pH 7.4 at 37 °C. (B) PCA plot after removal of the stems under the same experimental conditions. *Adapted from Eubanks, C. S., & Hargrove, A. E. (2017). Sensing the impact of environment on small molecule differentiation of RNA sequences.* Chemical Communications, 53*(100), 13363–13366 with permission of the Royal Chemical Society.*

(i.e., ionic strength, buffer, pH, temperature and presence of a molecular crowder) can be used to gain important information on the determinants at the basis of the interaction. For example, in our case study involving aminoglycoside-RNA interactions (Eubanks & Hargrove, 2017), PRRSM demonstrated how electrostatic interactions are responsible for the binding affinity (i.e., decreased predictive power at high ionic concentration) but might compromise binding selectivity when their contribution to the binding is increased (i.e., decreased predictive power at lower pH), though other contributions from RNA structural rearrangement cannot be ruled out. Additionally, PRRSM suggested that the dynamic nature of the RNA motifs might be exploited for distinguishing sequences containing the same

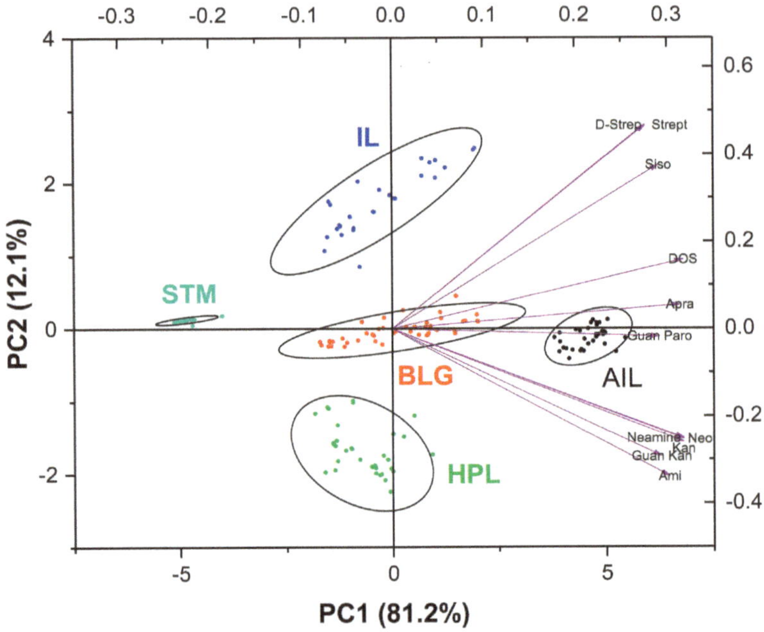

Fig. 7 PCA plot of the training set including the loading factors (purple arrows). From this plot is deduced that while all the aminoglycosides contribute PC1, 2-DOS, apramycin and guanidinylated paranomycin have none or limited contribution to differentiation along PC2.

motif. In particular, the reduced separation in PCA observed for experimental conditions that affect conformational flexibility of the RNA (i.e., high magnesium concentration) and the enhanced predictive power achieved using conditions that mimic a cellular environment (i.e., 37 °C, PEG 1200) in which higher conformational freedom is expected, suggest that increased conformational dynamics might favor RNA selective recognition.

Altogether, PRRSM can be used to gain information on both RNA and small molecule properties that confer selective RNA recognition.

7. PRRSM as potential structure and function prediction tool

Along with providing important information on selective RNA recognition by small molecules, the differential binding of PRRSM has the potential to be complementary to other prediction tools for identifying motifs of structurally unknown RNA analytes. In proof of concept studies involving structurally known RNA constructs, PRRSM correctly classified both the apical loop and the bulge regions of the HIV1-TAR construct despite the presence of both motifs (Eubanks et al., 2017). Furthermore, PRRSM was sensitive to structural rearrangement of RNA constructs such as the PreQ1 and fluoride riboswitch and correctly identified motifs and essential residues for RNA folding (Eubanks et al., 2019). We envisage that these classification features have the potential to be extended to unknown RNA constructs. In an ideal experiment, the fluorescent label BFU is placed at selected positions of the RNA sequence that are predicted to be part of structural motifs or considered important for structural rearrangement in case a folding event is hypothesized. The PCA or preferably LDA plots defined by an RNA training set containing known structural motifs is then input with the fluorescence variation induced by small molecules binding at each position of the unknown structure. The localization of each input will be used to determine if the modified position fall into specific secondary structures and/or is involved in structural rearrangements. A further potential extension would be using PRRSM to classify unknown structural rearrangement induced by base modifications and/or protein binding and tertiary motifs. Ultimately, we envision that by exploiting the relationship between structure and function of RNA, PRRSM could be utilized for investigating and classifying the function of unknown RNA structures after generation of a dedicated "functional" training set.

8. Summary and conclusion

Targeting RNA (both coding and non-coding) is an attractive strategy for increasing the number of druggable targets in vivo. Small molecules are particularly attractive for their tunable physicochemical properties and increased cellular uptake relative to oligonucleotides. However, there is currently a gap in knowledge of the principles governing RNA recognition by small molecules, and in particular the descriptors of small molecules that enable selective binding, as well as which RNA structures may be most targetable. Our strategy (PRRSM) is a high-throughput method to determine the binding properties of both small molecules and RNA constructs with the advantage that strong and selective small molecule binders are not required. Furthermore, differential small molecule recognition can be exploited to interrogate the conformational properties, including conformational switching, of biologically relevant RNA constructs. This critical insight into discernible structural patterns of common RNA topologies will provide the basis for the rational design of selective RNA ligands to target therapeutically important ncRNA fragments.

Acknowledgments

We would like to thank all members of the Hargrove lab for providing valuable feedback on the chapter. The authors acknowledge financial support for this work from Duke University and the US National Institute of General Medical Sciences (NIGMS) Maximizing Investigator's Research Award (MIRA) Grant/Award Number: R35GM124785.

References

Ackermann, D., & Famulok, M. (2013). Pseudo-complementary PNA actuators as reversible switches in dynamic DNA nanotechnology. *Nucleic Acids Research*, *41*, 4729–4739.

Andronescu, M., Bereg, V., Hoos, H. H., & Condon, A. (2008). RNA STRAND: The RNA secondary structure and statistical analysis database. *BMC Bioinformatics*, *9*(1), 340.

Bajaj, A., Miranda, O. R., Kim, I.-B., Phillips, R. L., Jerry, D. J., Bunz, U. H. F., et al. (2009). Detection and differentiation of normal, cancerous, and metastatic cells using nanoparticle-polymer sensor arrays. *Proceedings of the National Academy of Sciences of the United States of America*, *106*(27), 10912–10916.

Cate, J. H., & Doudna, J. A. (2000). [12] Solving large RNA structures by X-ray crystallography. In *Methods in Enzymology*: *Vol. 317*. *RNA–ligand interactions, part A*. (pp. 169–180). Elsevier.

Cech, T. R., & Steitz, J. A. (2014). The noncoding RNA revolution—Trashing old rules to forge new ones. *Cell*, *157*(1), 77–94.

Chen, Y., & Pollack, L. (2016). SAXS studies of RNA: Structures, dynamics, and interactions with partners. *Wiley Interdisciplinary Reviews: RNA*, 7(4), 512–526.

Cheng, C. Y., Chou, F.-C., & Das, R. (2015). Chapter two—Modeling complex RNA tertiary folds with Rosetta. In S.-J. Chen & D. H. Burke-Aguero (Eds.), *Computational methods for understanding riboswitches* (pp. 35–64). Elsevier.

Chittapragada, M., Roberts, S., & Ham, Y. W. (2009). Aminoglycosides: Molecular insights on the recognition of RNA and aminoglycoside mimics. *Perspectives in Medicinal Chemistry*, *3*, 21–37.

Connelly, C. M., Moon, M. H., & Schneekloth, J. S., Jr. (2016). The emerging role of RNA as a therapeutic target for small molecules. *Cell Chemical Biology*, *23*(9), 1077–1090.

Donlic, A., & Hargrove, A. E. (2018). Targeting RNA in mammalian systems with small molecules. *Wiley Interdisciplinary Reviews: RNA*, *9*(4), e1477.

Eubanks, C. S., Forte, J. E., Kapral, G. J., & Hargrove, A. E. (2017). Small molecule-based pattern recognition to classify RNA structure. *Journal of the American Chemical Society*, *139*(1), 409–416.

Eubanks, C. S., & Hargrove, A. E. (2017). Sensing the impact of environment on small molecule differentiation of RNA sequences. *Chemical Communications*, *53*(100), 13363–13366.

Eubanks, C. S., & Hargrove, A. E. (2019). RNA structural differentiation: Opportunities with pattern recognition. *Biochemistry*, *58*(4), 199–213.

Eubanks, C. S., Zhao, B., Patwardhan, N. N., Thompson, R. D., Zhang, Q., & Hargrove, A. E. (2019). Visualizing RNA conformational changes via pattern recognition of RNA by small molecules. *Journal of the American Chemical Society*, *141*(14), 5692–5698.

Feng, C., Chan, D., Joseph, J., Muuronen, M., Coldren, W. H., Dai, N., et al. (2018). Light-activated chemical probing of nucleobase solvent accessibility inside cells. *Nature Chemical Biology*, *14*, 276.

Folmer-Andersen, J. F., Kitamura, M., & Anslyn, E. V. (2006). Pattern-based discrimination of enantiomeric and structurally similar amino acids: An optical mimic of the mammalian taste response. *Journal of the American Chemical Society*, *128*(17), 5652–5653.

Hart, K., Nyström, B., Öhman, M., & Nilsson, L. (2005). Molecular dynamics simulations and free energy calculations of base flipping in dsRNA. *RNA*, *11*(5), 609–618.
Hendrix, M., Priestley, E. S., Joyce, G. F., & Wong, C.-H. (1997). Direct observation of aminoglycoside – RNA interactions by surface plasmon resonance. *Journal of the American Chemical Society*, *119*(16), 3641–3648.
Hughes, A. D., Glenn, I. C., Patrick, A. D., Ellington, A., & Anslyn, E. V. (2008). A pattern recognition based fluorescence quenching assay for the detection and identification of nitrated explosive analytes. *Chemistry—A European Journal*, *14*(6), 1822–1827.
Jarroux, J., Morillon, A., & Pinskaya, M. (2017). History, discovery, and classification of lncRNAs. In M. Rao (Ed.), *Long non coding RNA biology. Advances in experimental medicine and biology, Vol. 1008*, Singapore: Springer.
Jolliffe, I. T., & Cadima, J. (2016). Principal component analysis: A review and recent developments. *Philosophical Transactions of the Royal Society A: Mathematical, Physical and Engineering Sciences*, *374*(2065), 20150202.
Kim, J.-H. (2009). Estimating classification error rate: Repeated cross-validation, repeated hold-out and bootstrap. *Computational Statistics & Data Analysis*, *53*(11), 3735–3745.
Kitamura, M., Shabbir, S. H., & Anslyn, E. V. (2009). Guidelines for pattern recognition using differential receptors and indicator displacement assays. *The Journal of Organic Chemistry*, *74*(12), 4479–4489.
Luedtke, N. W., Baker, T. J., Goodman, M., & Tor, Y. (2000). Guanidinoglycosides: A novel family of RNA ligands. *Journal of the American Chemical Society*, *122*(48), 12035–12036.
Luedtke, N. W., Carmichael, P., & Tor, Y. (2003). Cellular uptake of aminoglycosides, guanidinoglycosides, and poly-arginine. *Journal of the American Chemical Society*, *125*(41), 12374–12375.
McPike, M. P., Sullivan, J. M., Goodisman, J., & Dabrowiak, J. C. (2002). Footprinting, circular dichroism and UV melting studies on neomycin B binding to the packaging region of human immunodeficiency virus type-1 RNA. *Nucleic Acids Research*, *30*(13), 2825–2831.
Moretti, R., Lyskov, S., Das, R., Meiler, J., & Gray, J. J. (2018). Web-accessible molecular modeling with Rosetta: The Rosetta online server that includes everyone (ROSIE). *Protein Science*, *27*(1), 259–268.
Morris, K. V., & Mattick, J. S. (2014). The rise of regulatory RNA. *Nature Reviews Genetics*, *15*, 423.
Palacios, M. A., Nishiyabu, R., Marquez, M., & Anzenbacher, P. (2007). Supramolecular chemistry approach to the design of a high-resolution sensor array for multianion detection in water. *Journal of the American Chemical Society*, *129*(24), 7538–7544.
Patwardhan, N. N., Cai, Z., Newson, C. N., & Hargrove, A. E. (2019). Fluorescent peptide displacement as a general assay for screening small molecule libraries against RNA. *Organic & Biomolecular Chemistry*, *17*(7), 1778–1786.
Patwardhan, N. N., Ganser, L. R., Kapral, G. J., Eubanks, C. S., Lee, J., Sathyamoorthy, B., et al. (2017). Amiloride as a new RNA-binding scaffold with activity against HIV-1 TAR. *Medicinal Chemistry Communications*, *8*(5), 1022–1036.
Razi, A., Ortega, J., & Britton, R. A. (2016). The impact of recent improvements in cryo-electron microscopy technology on the understanding of bacterial ribosome assembly. *Nucleic Acids Research*, *45*(3), 1027–1040.
Roe, D. R., & Cheatham, T. E. (2013). PTRAJ and CPPTRAJ: Software for processing and analysis of molecular dynamics trajectory data. *Journal of Chemical Theory and Computation*, *9*(7), 3084–3095.
Salim, N. N., & Feig, A. L. (2009). Isothermal titration calorimetry of RNA. *Methods*, *47*(3), 198–205.

Salmon, L., Yang, S., & Al-Hashimi, H. M. (2014). Advances in the determination of nucleic acid conformational ensembles. *Annual Review of Physical Chemistry*, *65*(1), 293–316.

Stewart, S., Ivy, M. A., & Anslyn, E. V. (2014). The use of principal component analysis and discriminant analysis in differential sensing routines. *Chemical Society Reviews*, *43*(1), 70–84.

Sztuba-Solinska, J., Chavez-Calvillo, G., & Cline, S. E. (2019). Unveiling the druggable RNA targets and small molecule therapeutics. *Bioorganic & Medicinal Chemistry*, *27*(10), 2149–2165.

Tanpure, A. A., & Srivatsan, S. G. (2011). A microenvironment-sensitive fluorescent pyrimidine ribonucleoside analogue: Synthesis, enzymatic incorporation, and fluorescence detection of a DNA abasic site. *Chemistry—A European Journal*, *17*(45), 12820–12827.

Tijerina, P., Mohr, S., & Russell, R. (2007). DMS footprinting of structured RNAs and RNA–protein complexes. *Nature Protocols*, *2*, 2608.

Tius, M. A., & Kawakami, J. K. (1995). The reaction of XeF2 with trialkylvinylstannanes: Scope and some mechanistic observations. *Tetrahedron*, *51*(14), 3997–4010.

Umali, A. P., & Anslyn, E. V. (2010). A general approach to differential sensing using synthetic molecular receptors. *Current Opinion in Chemical Biology*, *14*(6), 685–692.

Umali, A. P., LeBoeuf, S. E., Newberry, R. W., Kim, S., Tran, L., Rome, W. A., et al. (2011). Discrimination of flavonoids and red wine varietals by arrays of differential peptidic sensors. *Chemical Science*, *2*(3), 439–445.

Velagapudi, S. P., & Disney, M. D. (2014). Two-dimensional combinatorial screening enables the bottom-up design of a microRNA-10b inhibitor. *Chemical Communications*, *50*(23), 3027–3029.

Warner, K. D., Hajdin, C. E., & Weeks, K. M. (2018). Principles for targeting RNA with drug-like small molecules. *Nature Reviews Drug Discovery*, *17*, 547–558.

Wenderski, T. A., Stratton, C. F., Bauer, R. A., Kopp, F., & Tan, D. S. (2015). Principal component analysis as a tool for library design: A case study investigating natural products, brand-name drugs, natural product-like libraries, and drug-like libraries. In J. Hempel, C. Williams, & C. Hong (Eds.), *Chemical Biology. Methods in Molecular Biology, Vol. 1263*, New York, NY: Humana Press.

Wilkinson, K. A., Merino, E. J., & Weeks, K. M. (2006). Selective 2′-hydroxyl acylation analyzed by primer extension (SHAPE): Quantitative RNA structure analysis at single nucleotide resolution. *Nature Protocols*, *1*, 1610.

Xie, Y., Dix, A. V., & Tor, Y. (2009). FRET enabled real time detection of RNA-small molecule binding. *Journal of the American Chemical Society*, *131*(48), 17605–17614.

Zhang, J., Umemoto, S., & Nakatani, K. (2010). Fluorescent indicator displacement assay for ligand – RNA interactions. *Journal of the American Chemical Society*, *132*(11), 3660–3661.

Ziehler, W. A., & Engelke, D. R. (2000). Probing RNA structure with chemical reagents and enzymes. *Current Protocols in Nucleic Acid Chemistry*. 6.1.1-6.1.21.

Zuker, M. (2003). Mfold web server for nucleic acid folding and hybridization prediction. *Nucleic Acids Research*, *31*(13), 3406–3415.

CHAPTER SEVEN

Ligand-observed NMR techniques to probe RNA-small molecule interactions

David R. Calabrese[†], **Colleen M. Connelly**[†], **John S. Schneekloth Jr.***
Chemical Biology Laboratory, National Cancer Institute, Frederick, MD, United States
*Corresponding author: e-mail address: schneeklothjs@mail.nih.gov

Contents

Abstract

A growing understanding of the structure and function of RNA has revealed it as a key regulator of gene expression and disease. A multitude of noncoding functions apart from the central roles of RNA in coding for and facilitating protein biogenesis has stimulated research into RNA as a pharmacological target. Despite many exciting advances, RNA remains an understudied target for small molecules, and techniques to investigate RNA-binding molecules are still emerging. A key stumbling block in this area has been validation of RNA-small molecule interactions. Our laboratory has recently used multiple ligand-observed NMR techniques in this regard, including CPMG and WaterLOGSY. This work describes methods to use these techniques in the context of studying RNA-ligand interactions.

[†] These authors contributed equally to this work.

Methods in Enzymology, Volume 623
ISSN 0076-6879
https://doi.org/10.1016/bs.mie.2019.05.030

1. Introduction

Recent years have seen increasing research in RNA as a target for small molecule drug discovery (Connelly, Moon, & Schneekloth, 2016; Donlic & Hargrove, 2018; Morgan, Forte, & Hargrove, 2018; Warner, Hajdin, & Weeks, 2018). This interest has been motivated by multiple concurrent observations. The ENCODE project has revealed that while >80% of the human genome is transcribed into RNA, just 3% of these transcripts code for proteins (Hangauer, Vaughn, & McManus, 2013). Additionally, the discovery that many RNAs exhibit noncoding regulatory functions has suggested new opportunities for controlling biological pathways (Cech & Steitz, 2014). Advances in experimentally informed RNA structure prediction such as SHAPE (Deigan, Li, Mathews, & Weeks, 2009) and DMS-seq (Bevilacqua, Ritchey, Su, & Assmann, 2016) have revealed that many RNAs exhibit higher order structures *in vivo*, and that these structures can correlate with functions such as translation efficiency (Mustoe, Busan, et al., 2018; Mustoe, Corley, Laederach, & Weeks, 2018). Disease states can also be caused by aberrant RNA regulation (Cooper, Wan, & Dreyfuss, 2009; Esteller, 2011; Scotti & Swanson, 2016). Collectively, these discoveries provide a justification for attempts to develop RNA-targeted small molecule probes or therapeutics.

One challenge that has limited the development of RNA-binding small molecules is the difficulty in characterizing RNA-ligand interactions. Many highly useful approaches are available to quantify affinity constants or obtain other biophysical measurements (Blakeley et al., 2012). However, these biophysical methods can often be highly system dependent, and it is time intensive to determine which technique will work for an RNA of interest. Even when techniques are successful, experimental setup or protocols can be lengthy, and analysis can be restricted to a small number of compounds in a reasonable period of time. Thus, techniques that rapidly provide information about ligands of interest have substantial utility.

Ligand-observed NMR experiments provide a range of information about the RNA-binding properties of small molecule ligands (Bertini, McGreevy, & Parigi, 2012). In general, these techniques rely on observation of the chemical shifts of protons in the small molecule ligand and their altered behavior in the presence/absence of a biomolecule of interest. As such, the techniques are relatively sensitive and are "label-free": no modification of the RNA or small molecule is required to generate binding data.

However, a limitation that comes with these approaches is that they often do not generate information about the binding site of the compound on the RNA. None the less, these techniques have proven to be highly useful in the area of nucleic acid-ligand interactions. One notable use is in the validation of hits from high throughput screening campaigns, where ligand-observed NMR serves as an orthogonal method to rapidly demonstrate direct binding of a small molecule to the RNA or DNA of interest (Abulwerdi et al., 2019; Calabrese et al., 2018; Connelly et al., 2019; Garavis et al., 2014).

This chapter describes advantages and applications of ligand-observed NMR techniques to the field of RNA-targeted ligand discovery. More specifically, both Carr–Purcell–Meiboom–Gill (CPMG) (Hoffmann, Sobstyl, & Badali, 2009) and Water-Ligand Observed via Gradient SpectroscopY (WaterLOGSY) (Dalvit, Fogliatto, Stewart, Veronesi, & Stockman, 2001; Dalvit et al., 2000) techniques are discussed. These and other ligand-observed NMR experiments such as Saturation Transfer Difference (STD) have proven to be highly useful in drug discovery (Abulwerdi et al., 2019; Bertini et al., 2012; Wagstaff, Taylor, & Howard, 2013; Walpole, Monaco, Nepravishta, & Angulo, 2019), where they have been widely adopted as methods to screen for new ligands of protein targets (fragment-based drug discovery) and validate hits from screens as binding to proteins of interest (ruling out false positives). Such approaches are well documented for protein targets, and the related experimental theory is also extensively described (Aguirre, Cala, & Krimm, 2015; Becker, Bhattiprolu, Gubensak, & Zangger, 2018; Bertini et al., 2012; Li & Kang, 2017). Therefore, this work describes the utility of ligand observed NMR experiments in the specific case of RNA-ligand interactions, and the reader is directed to references above for more comprehensive discussions on the development, theory, and other uses of ligand-observed NMR techniques.

1.1 Carr–Purcell–Meiboom–Gill (CPMG)

CPMG NMR is a straightforward and reliable approach for validating binding of a ligand to RNA, and has been used more broadly in drug discovery to observe protein/ligand binding (Bertini et al., 2012; Hoffmann et al., 2009). CPMG is a relaxation-edited, ligand-observed ^{1}H NMR technique that employs differences in relaxation properties between the nuclei in macromolecules and nuclei in small molecules to probe binding. More specifically, CPMG utilizes a pulse sequence that exploits the differences in transverse relaxation times (T_2), which are directly dependent on the molecular rotational correlation time (τ_c) and thus on the molecular weight of a molecule or complex. Small molecules relax slowly (long T_2, second scale) exhibiting sharp, well defined peaks. In contrast, higher molecular weight species, such as proteins or nucleic acids, relax quickly (short T_2, millisecond scale) and tend to show broad NMR signals. Importantly, when small molecules are transiently bound to macromolecules, they exhibit changes in relaxation times that can be easily observed using CPMG pulse sequences.

The CPMG pulse measures T_2 relaxation time by following the signal decay as a function of the relaxation delay. A time delay of 100–400 ms is used before acquisition to filter out the signal of more rapidly relaxing molecules, thus selectively removing the signals from those ligands bound to fast relaxing macromolecules from the spectrum. Shorter CPMG delay times are used for high affinity ligands, since their relaxation is more strongly influenced upon binding. Longer relaxation delays can be used to study weakly binding ligands. Comparison of the resulting NMR spectra in the absence and presence of the macromolecule can be used to validate binding, as observed by altered peaks with decreased signal intensities and/or broadened line-widths (Fig. 1). Importantly, spectra are normalized to internal, nonbinding control molecules to avoid misinterpretation. CPMG experiments work well when the ligand is in 20-fold excess of the RNA. This ratio is important because larger ligand/RNA ratios may mask binding due to the large signal from the free ligand. With smaller ratios, effects can become difficult to visualize as peaks may become completely attenuated and not observable. In addition to validating binding, CPMG can also be used to determine dissociation constants by monitoring the linewidth or signal intensities as a function of the binding partner concentrations (Becker et al., 2018; Moschen et al., 2016). It can also be used in competition experiments with known, weaker binders.

CPMG provides an easy to interpret NMR spectrum, allowing for straightforward determination of binding of low molecular weight compounds. CPMG has similar sample and affinity requirements to STD and WaterLOGSY (as discussed below) and can therefore be used at the same stage of drug discovery or ligand validation and is often used in parallel with these techniques. CPMG requires low mole fractions of the RNA target, therefore only small quantities of RNA are needed (typically ~0.5–1.5 nmol/sample). In addition, CPMG does not require isotope labeling of either the ligand or RNA, making it an accessible and cost-effective NMR technique. Because CPMG monitors only the ^{1}H NMR signals from the ligand it is highly sensitive, and the entire data acquisition can be completed in under 10 min. CPMG is applicable to low-molecular weight compounds that bind to proteins/nucleic acids with on–off rates in the micro- to millisecond time scale, corresponding to affinities in the micromolar to millimolar range, and can in some cases be used to determine dissociation constants as well as binding (Becker et al., 2018; Moschen et al., 2016).

1.2 WaterLOGSY

WaterLOGSY is a ligand-observed NMR technique that can be used in target-directed drug screening or ligand validation to assess the binding of molecules to RNA (Becker et al., 2018; Cressina, Chen, Abell, Leeper, & Smith, 2011; Dalvit et al., 2001; Davis, Dunican, & Strobel, 2011). WaterLOGSY effects are achieved by applying indirect saturation of the RNA by selective saturation of bulk water protons. Magnetization is first transferred from the water to the RNA and then to the compounds interacting with the RNA. The magnetization transfer from the water to the RNA occurs through two pathways. First, transfer of magnetization can occur from water to labile RNA protons (i.e., exchangeable protons on heteroatoms in the RNA) proximal to the ligand binding site. The high number of exchangeable protons and buried water molecules in RNA contributes to the sensitivity of this method. Second, spin diffusion can occur between water and remote labile protons on heteroatoms in the RNA. Spin diffusion is facilitated by the residence times of water in binding pockets (few ns to hundreds of μs) being longer than the time to observe intermolecular water-macromolecule Nuclear Overhauser Effects (NOEs) (~300 ps), and shorter than the chemical shift time scale (ms). As a result, NOEs between water and binding/non-binding molecules can be distinguished through this method. Binding molecules interact with proton spins of inverted water causing negative cross-relaxation rates, where non-binding molecules have positive cross-relaxation rates. Thus, signals of non-binding molecules are oppositely phased and weaker than signals from binding ligands. In each sample, an internal, nonbinding control is included for comparison. For convenience, our laboratory has used *N*-Methyl-L-Valine for this purpose due to characteristic signals of the methyl groups in the aliphatic region and good solubility in a variety of buffers (Calabrese, Chen, et al., 2018; Calabrese, Zlotkowski, et al., 2018; Connelly et al., 2019; Zlotkowski et al., 2017). In practice, any suitable control may be used (including the solvent peak).

There are several different WaterLOGSY pulse sequences that have been developed (Fisher, 2015; Gossert, Henry, Blommers, Jahnke, & Fernandez, 2009). Generally, all of these sequences contain selective water excitation at the beginning. Many contain excitation sculpting pulses to induce water suppression in the final spectra as well. Variations in methods involve optimizing the mixing times and the recovery delay or adding a 180° inversion pulse in the middle of the mixing time to avoid the recovery of the equilibrium magnetization of the macromolecule and ligand protons.

WaterLOGSY has been widely used for detecting macromolecule-ligand interactions for RNA, DNA, and proteins because it is efficient and easy to interpret. As described above, in WaterLOGSY experiments peaks from molecules that bind to the RNA exhibit phasing opposite to the nonbinding control, providing a straightforward readout of the binding interaction. While WaterLOGSY spectra are reported in different ways, in this article we display ligand peaks not engaged in binding interactions phased down (negative), and binding interactions phased up (positive). WaterLOGSY is RNA inexpensive, requiring 1.5 nmoles of RNA or less for a sample with 100 μM compound, while running at a short time of 20–30 min. Unlike other NMR methods like STD, WaterLOGSY is not negatively impacted by local proton deficiencies or molecular motion that hinders intermolecular ^{1}H–^{1}H dipole interaction networks (Dalvit et al., 2000). Like CPMG, WaterLOGSY experiments do not require labeling of the RNA or small molecule. In comparison to small RNAs, larger RNA structures have longer rotational correlation times which causes them to be more efficient at magnetization transfer, leading to stronger signals. Thus, WaterLOGSY can be used with large, unlabeled RNAs. Further, multiple ligands can be screened simultaneously, allowing for a higher throughput if desired. Finally, there are very few false positives or negatives in WaterLOGSY experiments. As a result of these benefits, WaterLOGSY has been used as a screening method for fragments or weaker ligands and readily identifies ligands with K_Ds in the μM to mM range (Dalvit et al., 2000). Similar values are often seen in hits from larger scale high throughput screens as well, making the technique further useful for hit validation.

One final benefit of WaterLOGSY is that it can identify small molecule aggregation under the experimental conditions. Low molecular weight, hydrophobic compounds such as those often found in screening libraries have been documented to occasionally aggregate in aqueous buffers (Irwin et al., 2015). These aggregates can be relatively small and soluble (not visible to the naked eye) or can manifest as obviously visible precipitates or cloudiness in the assay buffer. Aggregates can cause false positives in enzymatic assays as well as biophysical assays, and while aggregation does not preclude the ability to specifically bind to the target of interest, it does make accurate measurements challenging. Because these aggregates are effectively high molecular weight colloids, the behavior of aggregates in WaterLOGSY experiments are comparable to small molecules bound to macromolecules. More specifically, peaks in compounds that aggregate exhibit phasing opposite to the nonbinding control, even in the absence of RNA. For this reason, it is both useful and imperative to perform WaterLOGSY in the absence and presence of the RNA of interest. Such an experimental design will both rule out false positive aggregators as well as confirm genuine binding to the RNA.

2. Materials

1. RNaseZap Decontamination Reagent (Invitrogen) or equivalent
2. 1% Nitric acid solution
3. 6M Sodium hydroxide solution
4. Unlabeled RNA of interest (IDT DNA, Dharmacon, or prepare by *in vitro* transcription)
5. Vivaspin 500 3kDa MWCO microcentrifuge filter (GE Healthcare) or equivalent
6. UltraPure Nuclease-free distilled water (Invitrogen)
7. Buffer prepared with UltraPure Nuclease-free distilled water (e.g., phosphate, Tris-d_{11} or HEPES-d_{18} buffers)
8. 1.5mL microcentrifuge tubes (Eppendorf)
9. Microcentrifuge
10. Thermo Scientific NanoDrop 2000 Spectrophotometer
11. Heat block or thermalcycler
12. Ligand of interest
13. *N*-Methyl-L-valine (Millipore Sigma)
14. Dimethyl sulfoxide-d_6 "100%" (DMSO-d_6; Cambridge Isotope Laboratories, Inc.)
15. Deuterium oxide "100%" (D_2O; Cambridge Isotope Laboratories, Inc.)
16. Shigemi tubes (5mm, D_2O) or standard 5mm NMR tubes
17. MicroCal ThermoVac apparatus
18. Bruker AVANCE III 500MHz spectrometer fitted with a TCI Cryoprobe with Z-gradient or equivalent
19. Data processing and analysis program (e.g., Mnova, topspin, CARA, etc.)

3. Methods

3.1 Sample preparation

3.1.1 General considerations

To avoid RNase contamination, all buffers should be prepared with Ultra-Pure RNase-free Water and all surfaces and equipment decontaminated with RNaseZap (Ambion) prior to RNA handling. All pipet tips and microcentrifuge tubes should be sterilized prior to use. Deprotected and HPLC purified oligonucleotides can be purchased from Dharmacon (ThermoFisher) or IDT DNA. To avoid contamination of samples, NMR tubes should be soaked with 1% HNO_3 overnight before/after use, rinsed with 6 M NaOH, washed with ethanol (1 ×) followed by and Milli-Q H_2O (3 ×), and dried.

3.1.2 Buffer selection

When choosing a buffer composition, it should be noted that many buffer components will contain protons themselves that will be observed in the ^1H NMR spectrum. The buffer ^1H signals will likely be more intense due to higher concentrations than the ligand, making it harder to observe the ligand signals and convoluting the data and interpretation. Therefore, a buffer that contains the fewest NMR active elements as possible is preferred. This can be accomplished by using phosphate-based buffers, since they contain no protons, or by using deuterated forms of the buffer components such as Tris-d_{11} or HEPES-d_{18}. Deuterated buffer components can often be purchased from Sigma or Cambridge Isotope Laboratories. If the ability to observe exchangeable protons is important, do not use deuterated solvents or solvents with deuterated components.

3.1.3 Buffer exchanging the RNA

Care should be taken to buffer exchange the RNA, for example, with a microcentrifuge filter to remove residual ethanol from RNA synthesis and purification that will lead to additional ^1H signals. Buffer exchange columns often have glycerol on the membrane when packaged, which will give distinct ^1H peaks that may mask the ligand signals. Before loading the RNA sample onto column, remove glycerol by thoroughly washing the column with Ultrapure RNase-free H_2O.

3.1.4 Choosing an internal, non-binding control

For the ligand observed NMR techniques discussed here, spectra are referenced to an internal, nonbinding control molecule to avoid misinterpretation. In CPMG, the internal control is used to normalize the signal intensity of the spectra in the presence or absence of RNA. For WaterLOGSY, the internal control is used to assess the phasing of ligand peaks to determine binding, nonbinding, and aggregation. In the experiments described here, the internal control *N*-Methyl-L-valine is used for its distinct peaks (e.g., two doublets at 0.90–0.95 ppm) that often do not overlap with the druglike molecules screened. *N*-Methyl-L-valine also demonstrates good solubility in the range of buffers used for these experiments. For WaterLOGSY, signals corresponding to added solvents (e.g., DMSO) can also be used to assess phasing of ligand peaks.

3.1.5 Ratio of ligand to RNA

For successful WaterLOGSY or CPMG NMR experiments, the ligand needs to be in excess of the RNA. In practice, an RNA:ligand molar ratio on the order of 1:20–1:100 is appropriate depending on the system and on the expected binding affinity. Typical ligand concentrations range from 100 to 500 μM, allowing for robust detection of the ligand signals in NMR experiments. Because similar ratios of RNA to ligand can be used for both CPMG and WaterLOGSY, the same samples can often be used for both experiments. For CPMG in particular, the choice of ligand–RNA ratio is important, as careful tuning of the ligand concentration can have a substantial impact on the quality of CPMG data (Williams & Daviter, 2013).

3.1.6 Experimental procedure

1. Resuspend the lyophilized RNA in UltraPure distilled H_2O to a concentration of 100 μM.
2. Thoroughly wash a 3 kDa MWCO microcentrifuge filter with UltraPure RNase-free H_2O (3×) and centrifuge to remove residual glycerol. Buffer exchange the RNA into the buffer of choice (e.g., 50 mM Tris-d_{11}, pH 7.5, 100 mM KCl, 1 mM $MgCl_2$ in UltraPure distilled H_2O) using centrifugal filtration according to the manufacturer's protocol. Centrifugation should be performed at 4 °C.
3. Anneal the RNA according the optimal procedure for the RNA of interest (e.g., heat to 95 °C for 3 min, followed by slowly cooling to room temperature for 30 min).

4. Prepare the ligand at a concentration of 10 mM in 100% DMSO-d_6 and the internal control *N*-Methyl-L-valine at a concentration of 10 mM in the NMR buffer.
5. For both CPMG and WaterLOGSY, data is interpreted by comparison of the resulting NMR spectra in the absence and presence of the RNA. Therefore, two samples should be prepared containing ligand only or ligand in the presence of RNA under the identical sample conditions (including ligand concentration, buffer, pH, etc.). Order of addition can impact ligand solubility and should be considered. Sample components should be combined in the following order in a microcentrifuge tube: DMSO-d_6, ligand, D_2O, buffer, negative control, followed by RNA. Between each addition, pipet mix or vortex and check the solution for precipitation. Precipitation can lead to poor shimming, broader line width, baseline distortion, and poor water suppression.
6. Prepare the ligand only sample as follows with the indicated final concentrations:
   a. 5% DMSO-d_6, ligand (100–300 μM), 10% D_2O, *N*-methyl-L-valine (100–300 μM, equal concentration to ligand) and bring to volume with buffer. Note: 5% DMSO-d_6 is the final concentration of the sample.
   b. A total volume of 300 μL (Shigemi tube) or 500 μL (standard NMR tube) should be used. Shigemi tubes allow for a lower volume and decreased quantity of ligand and RNA but are more susceptible to problems during acquisition due to bubbles within the solution.
   c. If necessary for compound solubility, 10% DMSO-d_6 can be used. In this case, no added D_2O is necessary, as the NMR can be locked onto the DMSO-d_6 during acquisition. In this situation, use a DMSO-matched Shigemi tube. Care should be taken as higher DMSO concentrations can impact the structure of the RNA (Lee, Vogt, McBrairty, & Al-Hashimi, 2013).
7. Prepare the ligand and RNA sample as follows with the indicated final concentrations:
   a. 5% DMSO-d_6, ligand (100–300 μM), 10% D_2O, *N*-methyl-L-valine (100–300 μM, equal concentration to ligand), RNA (at a 1:20–1:100 ratio to ligand), and bring to volume with buffer. Note: 5% DMSO-d_6 is the final concentration of the sample.

  b. As with the ligand only sample, a total volume of 300 μL (Shigemi tube) or 500 μL (standard NMR tube) should be used.
  c. As with the ligand only sample, if necessary for compound solubility, 10% DMSO-d_6 can be used. In this case, no added D_2O is necessary, as the NMR can be locked onto the DMSO-d_6 during acquisition. In this situation, use a DMSO-matched Shigemi tube. Care should be taken as higher DMSO concentrations can impact the structure of the RNA.
8. Incubate the samples on ice for 10 min.
9. If Shigemi tubes are being used, degas the samples at 20 °C for 15 min in a MicroCal ThermoVac apparatus. Any air bubbles, visible or non-visible "nanobubbles," can cause poor shimming, broader line-widths, baseline distortion, decreased NOE transfer, and most importantly poor water suppression.
10. Transfer the samples to NMR tubes, being careful to avoid the addition of bubbles.

3.2 Excitation sculpting

Prior to running a CPMG or WaterLOGSY experiment, a reference 1D-^{1}H spectrum is collected for each sample using the "zgesgp" excitation sculpting water suppression pulse sequence. The NMR procedures described below are written for Bruker instruments using the Topspin software package, however, these experiments can be carried out on any modern NMR spectrometer.

3.2.1 Experimental procedure

1. Set the temperature of the probe to 298 K and equilibrate.
2. Create a new experiment ("new"). Select the "zgesgp" experiment pulse sequence to run an excitation sculpting NMR. Click "getprosol" to set the probe head and solvent-dependent parameters. Select either "H_2O + D_2O_salt" for samples with 10% D_2O or "HDMSO" for samples with 10% d_6-DMSO in the solvent dropdown list.
3. Tune the spectrometer and match ("atma") then lock the sample ("lock"; select the solvent).
4. Shim the sample ("topshim gui"). Change the dropdown menu from "optimization" to "solvent suppression." Change the dropdown menu labeled "tune" to ("Z-X-Y-XZ-YZ-Z"). If using a Shigemi NMR tube, enter "shigemi" into the box labeled "parameters" and click the

adjacent check box. Click start. After the shimming is finished the final B_0 should be below 1 Hz. If not, the shim should be rerun. Click exit when B_0 is optimized.

   a. If the shimming in the final spectra is poor, run manual shimming. Click the "shim" tab in the BSMS control suite. To increase the line height in the lock display window, change the values of the Z, Z^2, Z^3, Z^4, Z^5, Y, X, XZ, and YZ shims. Change these shims values until there is no further increase. If the line height reaches the top of the window, click on the "Gain" tab and lower that value until the line height to the middle of the window. When optimized, change the gain value for the line height to be one third away from the top of the lock display window.

5. Determine the 90-degree pulse ("pulsecal").

6. View the FID shape for optimization ("gs"). Minimization of the FID size leads to better water suppression. Optimization of the water suppression is very important to achieve interpretable spectra. Click the "pulse" tab and change "P16" to 2 s. Next click the "offset" tab to change the offset frequency. Alter the frequency at the rate of 5 Hz at a time until a minimal FID size is achieved. Generally, the higher the DMSO content in the NMR sample, the larger the offset change is needed. Once the FID has been minimized click the button "save all."

7. Click the "acqupars" tab. Here number of scans, dummy scans, and sweep width can be altered. At ligand concentrations of 100 μM, 64 scans and 4 dummy scans should be sufficient. The number of scans will need to be adjusted based on the concentration of ligand. Sweep width can be changed depending on the ligand chemical shifts.

8. Run the pulse sequence ("zg") to acquire data. The total experiment time is 3.5 min.

9. When the pulse sequence is finished, the spectra can be viewed ("efp"). Examine the spectra for shimming of the solvent peak, the water suppression, and for impurities. For both CPMG and WaterLOGSY experiments, this spectrum will be used as a reference. If there are any problems, they can be addressed at this time, and then repeat the experiment.

3.3 CPMG

3.3.1 Experimental procedure

Prior to CPMG, a reference 1D-^{1}H spectrum is collected for each sample using the "zgesgp" excitation sculpting water suppression pulse sequence as described above. After a satisfactory excitation sculpting experiment has been run on the sample, a CPMG experiment can be performed. The experiment discussed here utilizes a CPMG pulse sequence with presaturation water suppression (Hoffmann et al., 2009). Note that the parameters provided here are starting points and can be systematically altered to provide optimal spectra for each sample.

1. Create a new experiment ("new"). The name of the experiment should be the same, but the "expno" (experiment number) should be increased by one from the previous experiment. Click "getprosol" to set the probe head and solvent-dependent parameters. Change the pulse sequence to CPMG with Presaturation Water Suppression ("cpmgpr1d").
2. Calibrate the length of the 90-degree pulse ("pulsecal").
3. View the FID shape for optimization ("gs"). Minimization of the FID size leads for better water suppression. Click the "pulse" tab and change "P16" to 2 s. Next click the "offset" tab to change the offset frequency. Start with the final frequency from the excitation sculpting experiment. Alter the frequency at a rate of 5 Hz at a time until a minimal FID size is achieved. Generally, the higher the DMSO content in the NMR sample, the larger the offset change needed. Once the FID has been minimized click the button "save all."
4. Click the "Acqupars" tab to adjust the following pulse parameters:
   a. L4—enter values between 100 and 600 to get the respective CPMG filter length. Choice of L4 depends on the approximate K_D of the ligand, and suggested values include:
      i. 100–200 for nM range K_D ligands
      ii. 200–400 for μM range K_D ligands
      iii. 400–600 for mM range K_D ligands
   b. D1—relaxation delay must be longer than 5 s in order to negate T1 effects, ideally 10–15 s
   c. D20 = 0.001 s
   d. Number of scans = 64 (can be increased when using lower concentration of ligand)
5. Run the pulse sequence ("zg") to acquire data. The total experiment time is around 4 min per sample. Experiment progress can be monitored in real time by entering "tr" into the command line to transfer the current scan to your memory, enter "efp" to see the spectrum.

6. When the pulse sequence is finished, the spectra can be viewed ("efp"). Examine the spectra for shimming of the solvent peak, the water suppression, and for impurities. If there are any problems, they can be addressed at this time, and then repeat the experiment.
7. Run the same pulse of both samples, with and without RNA.

3.3.2 Data analysis and interpretation

1. Load the CPMG spectra of the ligand without RNA and ligand with RNA into mNOVA (or similar) software.
2. On each individual spectrum, set the reference peak. This can be either DMSO, the non-binding control, or an alternative internal standard.
3. Superimpose the two spectra in order to compare the signal intensities and line-widths.
4. Examine the peaks from the non-binding control: The peaks should have the same chemical shifts in both spectra. Adjust the intensity of the non-binding control peaks to be equal in the two spectra.
5. Examine the peaks corresponding to the ligand:
   a. If these peaks are identical or near identical in both spectra, this indicated that this molecule does not bind to the RNA.
   b. If these peaks in the presence of RNA are significantly attenuated or broadened (>20%), this indicates that the ligand binds to the RNA. Occasionally, chemical shift perturbations may also be observed for the ligand signals, providing further evidence of the ligand binding to the RNA. Example data is illustrated in Fig. 1.

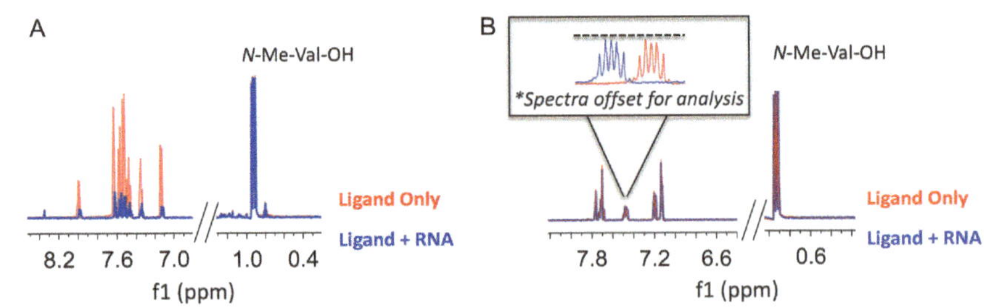

Fig. 1 CPMG NMR Data. (A) Validated small molecule-RNA interaction. Ligand peaks in the aromatic region are attenuated in the presence of RNA, while *N*-Methyl-L-valine (*N*-Me-Val-OH) peaks remain unaffected. (B) Non-binding compound. Peak attenuation is not observed in the presence of RNA. Inset illustrates offset spectra for purposes of visualization (chemical shift perturbation or peak attenuation are not observed).

3.4 WaterLOGSY

3.4.1 Experimental procedure

After an excitation sculpting experiment has been run on the sample and the spectrum is satisfactory, a WaterLOGSY can be run. Note that the parameters provided here are starting points and can be systematically altered to provide optimal spectra for each sample.

1. Create a new experiment ("new"). The name of the experiment should be the same, but the "expno" (experiment number) should be increased by one from the previous experiment. Click "getprosol" to set the probe head and solvent-dependent parameters. Change the pulse sequence to "ephogsygpno."
2. Determine the 90-degree pulse ("pulsecal").
3. View the FID shape for optimization ("gs"). Minimization of the FID size leads to better water suppression. Click the "pulse" tab and change "P16" to 2s. Next click the "offset" tab to change the offset frequency. Start with the final frequency for the excitation sculpting experiment. Alter the frequency at the rate of 5 Hz at a time until a minimal FID size is achieved. Generally, the higher the DMSO content in the NMR sample, the larger the offset change is needed. Once the FID has been minimized click the button "save all."
4. Click the "acqupars" tab. Here number of scans, dummy scans, and sweep width can be altered. At ligand concentrations of 100 μM, 512 scans and 32 dummy scans should be sufficient. Sweep width can be changed depending on the ligand chemical shifts.
5. Run the pulse sequence ("zg") to acquire data. The total experiment time is around 30 min.
6. When the pulse sequence is finished, the spectrum can be viewed ("efp"). Examine the spectrum for shimming of the solvent peak, the water suppression, and for impurities. If there are any problems, they can be addressed at this time, and then repeat the experiment. Do not phase the spectrum. This step is critical as changing the phasing of the spectrum at this point can cause incorrect interpretation of the data.

3.4.2 Data analysis and interpretation

Data can be processed using mNOVA (or similar) software.

1. Load the spectra of the excitation sculpting of the ligand without RNA, the WaterLOSGY of the ligand without RNA, and the WaterLOSGY of the ligand with RNA.

2. On each individual spectrum, set the reference peak. This can be either DMSO, the non-binding control, or an alternative internal standard.
3. Stack the three spectra.
4. Examine the peaks from the non-binding control. These peaks should have the same chemical shift in all three spectra. The peak should be positively phased in the excitation sculpting and negatively phased in both WaterLOGSY spectra.
5. Examine the peaks from the ligand. The peaks should be positively phased in the excitation sculpting experiment.
   a. If these peaks are positively phased in the WaterLOGSY spectra without RNA, this indicates that the ligand aggregates at these concentrations, and the WaterLOGSY spectrum with RNA can be ignored. Exchangeable protons on the ligands (e.g., N—H or O—H) will phase positively in all WaterLOGSY spectra, regardless of aggregation, so it is important to make these observations with non-exchangeable protons (e.g., C—H). Importantly, positive phasing of exchangeable protons does not indicate aggregation.
   b. If ligand peaks are negatively phased in both WaterLOGSY spectra, this indicates that this molecule does not bind to the RNA.
   c. If these peaks are negatively phased in the WaterLOGSY spectrum without RNA and positively phased in the WaterLOGSY spectrum with RNA, this indicates that the ligand binds to the RNA. The ligand peaks in the excitation sculpting and the WaterLOGSY with RNA spectra can be compared to observe chemical shift perturbations and peak broadening for further evidence of the ligand binding to the RNA. Example data can be seen in Fig. 2.

4. Outlook

This chapter describes the use of ligand-observed NMR to study RNA-binding small molecules. As efforts toward targeting RNA with small molecules increase, new approaches will continue to be developed to validate these compounds. Ligand observed NMR approaches that have enjoyed success in protein-targeted drug discovery will continue to find expanded use with RNA targets as well. Furthermore, these and related label-free techniques to study small molecule-RNA interactions will continue to provide rapid, high value information to validate RNA-binding properties of small molecules. Thus, ligand-observed NMR techniques provide an important tool for chemists studying RNA as a small molecule drug target.

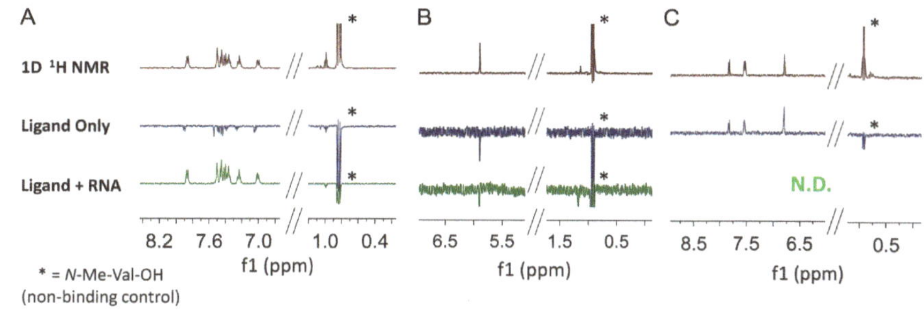

Fig. 2 WaterLOGSY NMR data. Illustrated are a standard, excitation sculpted ^{1}H NMR spectrum (top), WaterLOGSY in the absence of RNA (middle), and WaterLOGSY in the presence of RNA (bottom). Internal, non-binding control peaks (*N*-Methyl-L-valine) are illustrated with an asterisk (*). N.D. = not determined. (A) Example of an RNA-binding small molecule. Note that ligand peaks change phase in the presence of RNA in comparison to non-binding control. (B) Example of a non-binding compound. No phase change is observed in the presence of RNA. (C) Example of an aggregating small molecule. Note that ligand peaks are phased opposite of non-binding control even in the absence of RNA.

Acknowledgments

This work was supported by the Center for Cancer Research of the National Cancer Institute, NIH (1 ZIA BC011585 04). We thank Joseph J. Barchi for helpful discussions.

References

Abulwerdi, F. A., Xu, W., Ageeli, A. A., Yonkunas, M. J., Arun, G., Nam, H., et al. (2019). Selective small-molecule targeting of a triple helix encoded by the long noncoding RNA, MALAT1. *ACS Chemical Biology*, *14*(2), 223–235. https://doi.org/10.1021/acschembio.8b00807.

Aguirre, C., Cala, O., & Krimm, I. (2015). Overview of probing protein-ligand interactions using NMR. *Current Protocols in Protein Science*, *81*. https://doi.org/10.1002/0471140864.ps1718s81. 17.18.11-24.

Becker, W., Bhattiprolu, K. C., Gubensak, N., & Zangger, K. (2018). Investigating protein-ligand interactions by solution nuclear magnetic resonance spectroscopy. *Chemphyschem*, *19*(8), 895–906. https://doi.org/10.1002/cphc.201701253.

Bertini, I., McGreevy, K. S., & Parigi, G. (2012). *NMR of biomolecules: Towards mechanistic systems biology*. Weinheim, Germany: Wiley-VCH.

Bevilacqua, P. C., Ritchey, L. E., Su, Z., & Assmann, S. M. (2016). Genome-wide analysis of RNA secondary structure. *Annual Review of Genetics*, *50*, 235–266. https://doi.org/10.1146/annurev-genet-120215-035034.

Blakeley, B. D., DePorter, S. M., Mohan, U., Burai, R., Tolbert, B. S., & McNaughton, B. R. (2012). Methods for identifying and characterizing interactions involving RNA. *Tetrahedron*, *68*(43), 8837–8855. https://doi.org/10.1016/j.tet.2012.07.001.

Calabrese, D. R., Chen, X., Leon, E. C., Gaikwad, S. M., Phyo, Z., Hewitt, W. M., et al. (2018). Chemical and structural studies provide a mechanistic basis for recognition of the MYC G-quadruplex. *Nature Communications*, *9*(1), 4229. https://doi.org/10.1038/s41467-018-06315-w.

Calabrese, D. R., Zlotkowski, K., Alden, S., Hewitt, W. M., Connelly, C. M., Wilson, R. M., et al. (2018). Characterization of clinically used oral antiseptics as quadruplex-binding ligands. *Nucleic Acids Research*, *46*(6), 2722–2732. https://doi.org/10.1093/nar/gky084.

Cech, T. R., & Steitz, J. A. (2014). The noncoding RNA revolution-trashing old rules to forge new ones. *Cell*, *157*(1), 77–94. https://doi.org/10.1016/j.cell.2014.03.008.

Connelly, C. M., Moon, M. H., & Schneekloth, J. S., Jr. (2016). The emerging role of RNA as a therapeutic target for small molecules. *Cell Chemical Biology*, *23*(9), 1077–1090. https://doi.org/10.1016/j.chembiol.2016.05.021.

Connelly, C. M., Numata, T., Boer, R. E., Moon, M. H., Sinniah, R. S., Barchi, J. J., et al. (2019). Synthetic ligands for PreQ1 riboswitches provide structural and mechanistic insights into targeting RNA tertiary structure. *Nature Communications*, *10*(1), 1501. https://doi.org/10.1038/s41467-019-09493-3.

Cooper, T. A., Wan, L., & Dreyfuss, G. (2009). RNA and disease. *Cell*, *136*(4), 777–793. https://doi.org/10.1016/j.cell.2009.02.011.

Cressina, E., Chen, L. H., Abell, C., Leeper, F. J., & Smith, A. G. (2011). Fragment screening against the thiamine pyrophosphate riboswitch thiM. *Chemical Science*, *2*(1), 157–165. https://doi.org/10.1039/c0sc00406e.

Dalvit, C., Fogliatto, G., Stewart, A., Veronesi, M., & Stockman, B. (2001). WaterLOGSY as a method for primary NMR screening: Practical aspects and range of applicability. *Journal of Biomolecular NMR*, *21*(4), 349–359. https://doi.org/10.1023/A:1013302231549.

Dalvit, C., Pevarello, P., Tato, M., Veronesi, M., Vulpetti, A., & Sundstrom, M. (2000). Identification of compounds with binding affinity to proteins via magnetization transfer from bulk water. *Journal of Biomolecular NMR*, *18*(1), 65–68.

Davis, J. H., Dunican, B. F., & Strobel, S. A. (2011). glmS riboswitch binding to the glucosamine-6-phosphate alpha-anomer shifts the pKa toward neutrality. *Biochemistry*, *50*(33), 7236–7242. https://doi.org/10.1021/bi200471c.

Deigan, K. E., Li, T. W., Mathews, D. H., & Weeks, K. M. (2009). Accurate SHAPE-directed RNA structure prediction. *Proc. Nat. Acad. Sci.* *106*, 97–102.

Donlic, A., & Hargrove, A. E. (2018). Targeting RNA in mammalian systems with small molecules. *Wiley Interdisciplinary Reviews. RNA*, *9*(4), e1477. https://doi.org/10.1002/wrna.1477.

Esteller, M. (2011). Non-coding RNAs in human disease. *Nature Reviews. Genetics*, *12*(12), 861–874. https://doi.org/10.1038/nrg3074.

Fisher, J. (2015). *Modern NMR techniques for synthetic chemistry*. Boca Raton: CRC Press/Taylor & Francis Group.

Garavis, M., Lopez-Mendez, B., Somoza, A., Oyarzabal, J., Dalvit, C., Villasante, A., et al. (2014). Discovery of selective ligands for telomeric RNA G-quadruplexes (TERRA) through 19F-NMR based fragment screening. *ACS Chemical Biology*, *9*(7), 1559–1566. https://doi.org/10.1021/cb500100z.

Gossert, A. D., Henry, C., Blommers, M. J., Jahnke, W., & Fernandez, C. (2009). Time efficient detection of protein-ligand interactions with the polarization optimized PO-WaterLOGSY NMR experiment. *Journal of Biomolecular NMR*, *43*(4), 211–217. https://doi.org/10.1007/s10858-009-9303-5.

Hangauer, M. J., Vaughn, I. W., & McManus, M. T. (2013). Pervasive transcription of the human genome produces thousands of previously unidentified long intergenic noncoding RNAs. *PLoS Genetics*, *9*(6), e1003569. https://doi.org/10.1371/journal.pgen.1003569.

Hoffmann, M. M., Sobstyl, H. S., & Badali, V. A. (2009). T2 relaxation measurement with solvent suppression and implications to solvent suppression in general. *Magnetic Resonance in Chemistry*, *47*(7), 593–600. https://doi.org/10.1002/mrc.2434.

Irwin, J. J., Duan, D., Torosyan, H., Doak, A. K., Ziebart, K. T., Sterling, T., et al. (2015). An aggregation advisor for ligand discovery. *Journal of Medicinal Chemistry*, *58*(17), 7076–7087. https://doi.org/10.1021/acs.jmedchem.5b01105.

Lee, J., Vogt, C. E., McBrairty, M., & Al-Hashimi, H. M. (2013). Influence of dimethylsulfoxide on RNA structure and ligand binding. *Analytical Chemistry*, *85*(20), 9692–9698. https://doi.org/10.1021/ac402038t.

Li, Y., & Kang, C. (2017). Solution NMR spectroscopy in target-based drug discovery. *Molecules*, *22*(9), 1399. https://doi.org/10.3390/molecules22091399.

Morgan, B. S., Forte, J. E., & Hargrove, A. E. (2018). Insights into the development of chemical probes for RNA. *Nucleic Acids Research*, *46*(16), 8025–8037. https://doi.org/10.1093/nar/gky718.

Moschen, T., Grutsch, S., Juen, M. A., Wunderlich, C. H., Kreutz, C., & Tollinger, M. (2016). Measurement of ligand-target residence times by (1)H relaxation dispersion NMR spectroscopy. *Journal of Medicinal Chemistry*, *59*(23), 10788–10793. https://doi.org/10.1021/acs.jmedchem.6b01110.

Mustoe, A. M., Busan, S., Rice, G. M., Hajdin, C. E., Peterson, B. K., Ruda, V. M., et al. (2018). Pervasive regulatory functions of mRNA structure revealed by high-resolution SHAPE probing. *Cell*, *173*(1). https://doi.org/10.1016/j.cell.2018.02.034. 181–195.e118.

Mustoe, A. M., Corley, M., Laederach, A., & Weeks, K. M. (2018). Messenger RNA structure regulates translation initiation: A mechanism exploited from bacteria to humans. *Biochemistry*, *57*(26), 3537–3539. https://doi.org/10.1021/acs.biochem.8b00395.

Scotti, M. M., & Swanson, M. S. (2016). RNA mis-splicing in disease. *Nature Reviews. Genetics*, *17*, 19–32. https://doi.org/10.1038/nrg.2015.3.

Wagstaff, J. L., Taylor, S. L., & Howard, M. J. (2013). Recent developments and applications of saturation transfer difference nuclear magnetic resonance (STD NMR) spectroscopy. *Molecular BioSystems*, *9*(4), 571–577. https://doi.org/10.1039/c2mb25395j.

Walpole, S., Monaco, S., Nepravishta, R., & Angulo, J. (2019). STD NMR as a technique for ligand screening and structural studies. *Methods in Enzymology*, *615*, 423–451. https://doi.org/10.1016/bs.mie.2018.08.018.

Warner, K. D., Hajdin, C. E., & Weeks, K. M. (2018). Principles for targeting RNA with drug-like small molecules. *Nature Reviews. Drug Discovery*, *17*(8), 547–558. https://doi.org/10.1038/nrd.2018.93.

Williams, M. A., & Daviter, T. (2013). *Protein-ligand interactions: Methods and applications* (2nd ed.). New York: Humana Press; Springer.

Zlotkowski, K., Hewitt, W. M., Sinniah, R. S., Tropea, J. E., Needle, D., Lountos, G. T., et al. (2017). A small-molecule microarray approach for the identification of E2 enzyme inhibitors in ubiquitin-like conjugation pathways. *SLAS Discovery: Advancing Life Sciences R & D*, *22*, 760–766. 2472555216683937. https://doi.org/10.1177/2472555216683937.

CHAPTER EIGHT

Synthetic bPNAs as allosteric triggers of hammerhead ribozyme catalysis

Yufeng Liang, Jie Mao, Dennis Bong*

Department of Chemistry and Biochemistry, The Ohio State University, Columbus, OH, United States
*Corresponding author: e-mail address: bong.6@osu.edu

Contents

Abstract

The biochemistry and structural biology of the hammerhead ribozyme (HHR) have been well elucidated. The secondary and tertiary structural elements that enable sugar–phosphate bond scission to be catalyzed by this RNA are clearly understood. We have taken advantage of this knowledge base to test the extent to which synthetic molecules, may be used to trigger structure in secondary structure and tertiary interactions and thereby control HHR catalysis. These molecules belong to a family of molecules we generally call "bPNAs" based on our work on bifacial peptide nucleic acid (bPNA). This family of molecules displays the "bifacial" heterocycle melamine, which acts as a base-triple upon capturing two equivalents of thymine or uracil. Loosely structured internal

Methods in Enzymology, Volume 623
ISSN 0076-6879
https://doi.org/10.1016/bs.mie.2019.04.028

oligouridylate bulges of 4–20 nucleotides can be restructured as triplex hybrid stems upon binding bPNAs. As such, a duplex stem element can be replaced with a bPNA triplex hybrid stem; similarly, a tertiary loop–stem interaction can be replaced with a loop–bPNA–stem complex. The ability to control RNA structure–function facilitates elucidation of these critical aspects of RNA recognition. In this chapter, we discuss how bPNAs are prepared and applied to study structure–function turn on in the hammerhead ribozyme system.

1. Introduction

This chapter describes the application of bPNAs for triggering RNA structure–function in the hammerhead ribozyme, by virtue of the melamine base triple. Melamine hydrogen bonding has been studied extensively in organic solvents (ten Cate, Huskens, Crego-Calama, & Reinhoudt, 2004; Whitesides et al., 1995) and the solid state (Ranganathan, Pedireddi, & Rao, 1999). The energetic signatures of assembly (Ma & Bong, 2011a) are identical to those of nucleic acid duplexes (Kool, 2001; SantaLucia & Hicks, 2004; Yakovchuk, Protozanova, & Frank-Kamenetskii, 2006). Accordingly, melamine recognition of native bases (Arambula, Ramisetty, Baranger, & Zimmerman, 2009; Ma & Bong, 2011b, 2011c; Zhou & Bong, 2013) in a range of contexts has been observed (Ariga & Kunitake, 1998; Kawasaki, Tokuhiro, Kimizuka, & Kunitake, 2001; Ma & Bong, 2013), presaging the possibility of triplex formation with melamine and oligo T/U domains. Unlike triplex formation with native nucleic acids (Duca, Vekhoff, Oussedik, Halby, & Arimondo, 2008; Felsenfeld, Davies, & Rich, 1957; Pilch, Levenson, & Shafer, 1990; Soto, Loo, & Marky, 2002) involving both Watson–Crick (WC) and Hoogsteen base pairing at high salt (Rougee et al., 1992; Xodo, Manzini, & Quadrifoglio, 1990) and peptide nucleic acids (Nielsen, 1999) which invade native duplex structures, melamine triplex hybridization unites two noninteracting oligo-T/U domains. Our group found that α-PNAs (Mittapalli et al., 2007) displaying melamine at alternate residue sidechains form *triplex* hybrids with oligo T/U DNA/RNA; we called these peptides bifacial peptide nucleic acid (bPNA) (Piao, Xia, & Bong, 2013; Zeng, Pratumyot, Piao, & Bong, 2012). These bPNAs bind T-rich DNA to form thermally stable ($T_m \sim 58\,°C$) triplex domains that can block enzymatic processing of DNA and RNA (Piao et al., 2013; Xia, Piao, Fredrick, & Bong, 2013). The triplex hybrids can functionally replace stems in RNAs and DNAs, serving as triggers of molecular recognition and catalysis (Piao, Xia, Mao, & Bong, 2015; Xia, Piao, & Bong, 2014). The bPNA family of melamine-bearing molecules has expanded beyond peptides to include peptoids (Mao & Bong, 2015), polyacrylates (Zhou, Xia, & Bong, 2015) and synthetic small molecules (Mao, DeSantis, & Bong, 2017). Herein, we focus on the use of bPNA peptides and small molecules (Fig. 1) to trigger folding and function in the hammerhead ribozyme RNA (HHR) scaffold to facilitate a deeper understanding of RNA recognition and catalysis.

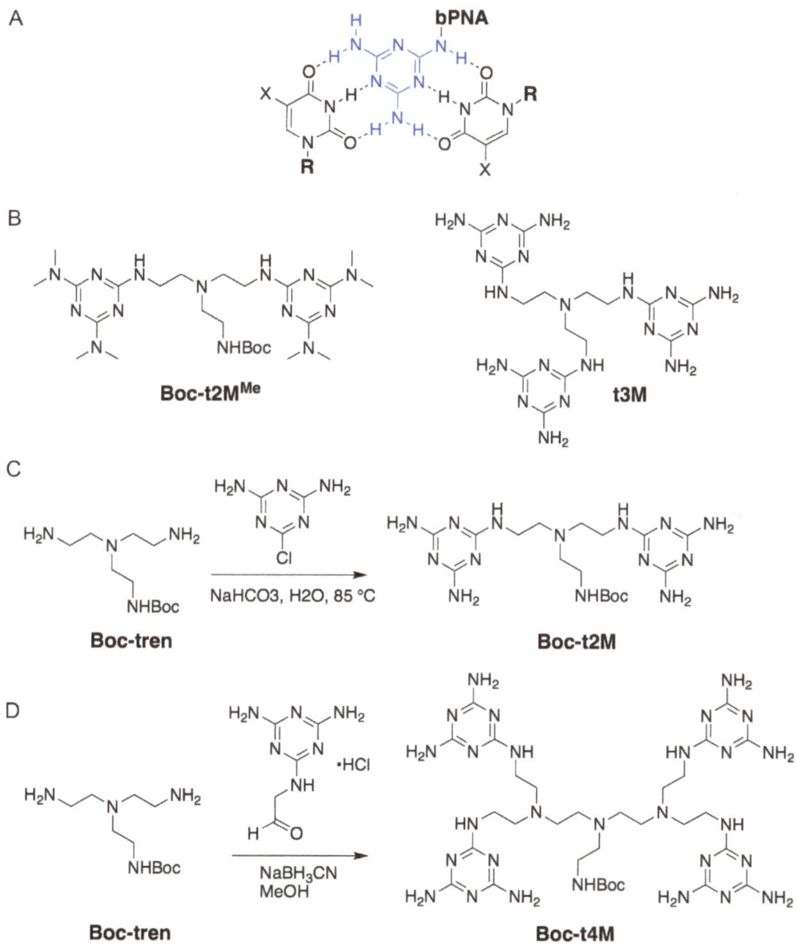

Fig. 1 (A) Base-triple formation of T (X = CH_3, R = DNA) or U (X = H, R = RNA) and melamine attached to bPNA. (B) Methylated Boc-t2M and t3M structures. Synthetic approaches to the oligoethyleneimine (tren) bPNA scaffold from mono-Boc tren using (C) nucleophilic aromatic substitution to form Boc-t2M and (D) reductive alkylation to form Boc-t4M.

The mini hammerhead ribozyme system (Robertson & Scott, 2007; Uhlenbeck, 1987) features a 19 nt catalytic core supported by three structural stems. The function of the HHR is to cleave itself, and this catalytic function is dependent on well-folded secondary structure in the stems and a tertiary contact between the loop of stem 2 and stem 1. It is possible to use bPNAs to trigger both secondary and tertiary structure in the HHR scaffold, replacing these native structures with bPNA triplex hybrids. Below we provide detailed protocols for the synthesis of bPNAs and design of HHRs under the control of bPNA triggers.

2. Synthetic protocols

Preparation of bPNAs is relatively straightforward but requires close attention. The following synthetic procedures have been developed to avoid chromatography where possible, but final purification of peptides requires HPLC. Oligoethyleneimine scaffolds (tren-derived) may be purified without column chromatography.

2.1 Materials and instrumentation for chemical synthesis

All chemicals were used without further purification from commercial sources, unless otherwise noted. Solid-phase peptide synthesis was carried out under nitrogen on an AAPTEC 96 automated peptide synthesizer equipped with a 40-well reaction block. MALDI-Mass spectra were acquired on Bruker Microflex MALDI-TOF instrument using RP mode. Peptide samples were dissolved in ACN/H_2O 1:99 (v/v) with 0.1% TFA while α-cyano-4-hydroxycinnamic acid was saturated in ACN/H_2O 1:1 (v/v) with 0.1% TFA used as the matrix. Peptide and matrix were mixed in 1:1 ratio and then spotted on MSP 96 target ground steel (Bruker Daltonics) for mass determination. Electrospray mass spectroscopy was acquired on a Bruker MicroTOF equipped with an electrospray ionization source under positive mode. Mass spectrometry instruments were provided by a grant from the Ohio BioProducts Innovation Center. NMR spectra were acquired on a Bruker Avance DPX 400 instrument.

2.2 Synthesis of bPNA amino acid derivatives

2.2.1 Boc-K^M-OH

This lysine derivative is prepared from Boc-Lys-OH, and must be converted to the Fmoc derivative prior to automated solid-phase peptide synthesis. It is not possible to start from the Fmoc derivative due to basic conditions required for installation of the melamine ring. However, the derivatization from Boc-Lys-OH is high yielding (95%), straightforward, and does not require chromatography. Commercially available Boc-Lys-OH (10 g, 40.6 mmol) was suspended in 120 mL H_2O with NaOH (2 eq, 3.25 g, 81.2 mmol) and 2,4-diamino-6-chlorotriazine (1.25 eq, 7.4 g, 50.7 mmol) and heated to 85 °C for 5 h. Thin layer chromatography (5,1 dichloromethane-methanol) confirmed full consumption of starting material. The reaction was cooled to room temperature and filtered to remove unreacted chlorotriazine. The aqueous solution was cooled in an ice bath and the pH was adjusted to 5 using 1 N HCl, resulting in a white precipitate, which was collected via filtration and washed with cold water then dried under vacuum. This material is the derivative Boc-Lys(M)-OH, or Boc-K^M-OH, and it should be analytically pure at this point.

2.2.2 *Fmoc-K^M-OH*

The Boc-K^M-OH (10 g, 28 mmol) deprotected by dissolution in neat TFA (80 mL) and 1 mL of water was added as scavenger; alkylated product was never observed, and addition of water may be unnecessary. The mixture was stirred at room temperature until complete as judged by electrospray mass spectroscopy and TLC (~1 h). The reaction was concentrated to an oil under reduced pressure, but still contains significant quantities of TFA, which is difficult to fully remove. This residue was resuspended in 200 mL water and the solution was neutralized to pH 7 using $NaHCO_3$. The solution was rendered basic with another 1.5 eq of $NaHCO_3$ and cooled in an ice bath. A solution of Fmoc-OSU (14 g, 42 mmol) in 200 mL dioxane was also ice-bath cooled and added portionwise to the lysine derivative. The resulting mixture was stirred at 0 °C for 1 h and allowed to warm to room temperature overnight. The reaction was worked up via removal of dioxane under reduced pressure, addition of water (150 mL), and extraction of the aqueous layer with ethyl acetate (2 × 100 mL). The aqueous layer was acidified to pH 1 with 1 N HCl and extracted again with EtOAc (3 × 120 mL). The combined organic layers were concentrated under vacuum. The resulting residue was purified CH_2Cl_2 to yield Fmoc-K^M-OH as a white solid (8 g, 60%). Alternatively, a lower yielding purification (44%) may be performed by trituration of the crude residue with ethanol (~200 mL).

2.3 Solid-phase synthesis of bPNAs

Peptide synthesis was performed on an AAPPTEC Apex 396 Peptide Synthesizer using Rink Resin LS (100–200 mesh) equipped with a 40-well synthesis block. The melamine base does not require protection from SPPS or cleavage conditions and generally can be handled like any other Fmoc-amino acid derivative. To minimize aggregation on resin due to the melamine rings a low-loading resin (0.2 mmol/g) is recommended. Standard solid phase conditions may be used. The following solutions should be prepared:

A. 0.25 M solution of Fmoc amino acid with 0.25 *M* HOBt in NMP. (HOBt is commercially available as the hydrate and this is acceptable)

B. 20% Piperidine in NMP (by volume)

C. Diisopropylcarbodiimide (DIC), neat

Approximately 100–150 mg of resin is added to a reaction well, which accommodates a recommended maximum ~2 mL liquid volume when resin is swollen. Prior to synthesis, resin is swollen in DMF for 45 min then drained under N_2 pressure and deprotected with piperidine solution B. The swelling, deprotection, washing, coupling, shrinking, and cleavage protocols are as follows.

2.3.1 Swelling

1 × 45 min. Two mL of DMF is delivered to the resin in the reaction chamber and agitated for 45 min at 475 rpm, then drained under N_2 pressure for 2 min.

2.3.2 Deprotection cycle

2 × 5 min. Two mL of piperidine solution is delivered via liquid handler to reaction chamber and reaction block is agitated at 475 rpm for 5 min, then drained under N_2 pressure (2 min). This is repeated. A wash cycle follows.

2.3.3 Wash cycle

3 × 2 min. Two mL DMF is delivered to the reaction chamber and reaction block is agitated at 475 rpm for 2 min, then drained under N_2 pressure (2 min).

2.3.4 Coupling cycle

1 × 45–60 min. Two mL (0.5 mmol) of amino acid/HOBt solution (A) is added to washed resin, followed by 75 μL (0.48 mmol) DIC solution (C). Reaction block is agitated at 475 rpm for 45–60 min. Standard amino acid derivatives (e.g., Fmoc-Glu(OtBu)-OH) are coupled for 45 min while Fmoc-K^M-OH is coupled for 60 min. Coupling may be extended to 90 min for longer peptides. After coupling is complete, reaction is drained under N_2 pressure (2 min).

2.3.5 Methanol wash

3 × 5 min. Two mL methanol is added to washed resin and agitated for 5 min at 475 rpm, then drained for 2 min. This is repeated two additional times, then drained under N_2 pressure for 15 min to dry the shrunken resin.

Synthesis procedures are typically the following sequence:

1. Swell
2. Deprotection
3. Wash
4. Coupling

Repeat

5. Final wash
6. Methanol wash

Note that there is no wash after coupling as this is not necessary. After the methanol wash, the peptide resin should then be dry, but clumped together and easy to transfer by spatula into a preweighed 20 mL glass scintillation vial. This resin is then washed three times with diethyl ether, removing ether each time with a glass pipette. The resin should be further shrunken at this time. The vial is covered with a clean kimwipe which is secured over the mouth of the vial with an elastic band, and then evacuated in a lyophilization bell to dry the resin completely. The vial is then weighed to obtain the weight of the peptide-resin. It is important to obtain an accurate weight in order to perform the peptide cleavage from resin. The following cleavage protocol is followed:

2.3.6 Peptide cleavage from resin

For every 100 mg of peptide resin, 1 mL of cleavage cocktail is used. For bPNAs containing only K^M and glutamic acid or lysine, there is no need for a complex scavenger mix. Peptides were cleaved for 1 h at RT using a 95:5 volume ratio of TFA to water. The cleavage reaction including resin was filtered through cotton into a 50-mL conical tube and the peptides were precipitated from TFA by addition of 30–50 mL chilled ether then centrifuged to yield a white pellet. This pellet was resuspended in chilled ether and repelleted three times (3 × 20 mL) to remove as much TFA as possible.

2.3.7 HPLC purification

Crude peptides were dissolved in solvent A and purified by HPLC on a semiprep (8 mL/min) C_{18} reversed phase column using a two solvent gradient. Solvent A = 0.1% TFA, 1% acetonitrile, 99% water; solvent B = 90% acetonitrile, 10% water, 0.07% TFA. bPNA peptides typically elute around

40% B, depending on sequence. Typical UV detector settings for peptides are at 230 nm while melamine itself can be observed as a distinct peak at 238 nm under acidic conditions. Most purifications are set to observe 238 nm, with a final check for purity at 230 nm. Purified peptide fractions were concentrated under vacuum to remove acetonitrile then lyophilized. Though peptides are purified on a standard gradient optimized for concentrated elution and efficiency, peptide purity is evaluated on an analytical column with an optimized isocratic gradient and confirmed by MALDI-TOF using α-cyano-4-hydroxycinnamic acid matrix.

2.4 Oligoethyleneimine bPNAs by nucleophilic aromatic substitution

There are generally two methods employed in the preparation of oligoethyleneimine scaffolds displaying melamine: (1) nucleophilic aromatic substitution (NAS) of chloride for amine, or (2) reductive alkylation of amine using melamine acetaldehyde (Fig. 1). The first method is more straightforward and can proceed using only commercially available materials, but affords less control over substitution. The second involves more steps but can be highly controlled. We discuss both approaches below.

2.4.1 t3M from NAS

This molecule can bind to a tetrauridylate bulge but only uses 2 of 3 rings in the binding. This binding event can serve as a folding trigger for RNA. If the presence of a third nonparticipating melamine ring is acceptable, then it is recommended to prepare this molecule, as it is the easiest to synthesize. Commercially available tris(2-aminoethyl)amine (0.75 mL, 5 mmol) was dissolved in 30 mL H_2O, followed by addition of 2,4-diamino-6-chlorotriazine (4.36 g, 30 mmol) and $NaHCO_3$ (1.5 g, 18 mmol). Diaminochlorotriazine may be purchased but the synthesis from cyanuric chloride and ammonium hydroxide in water is straightforward. The reaction was heated to 85 °C and left to stir overnight. After cooling to room temperature, the solution was filtered, the solid was washed with water twice, and dried under vacuum (1.63 g, 73%). The crude solid was further purified by prep HPLC using gradient of 0% to 15% acetonitrile over 40 min.

2.4.2 Boc-tren

In order to prepare a tren scaffold with only two melamines (t2M), a protected derivative, mono-Boc tren, is required. Though this starting material is only available in modest yield, the preparation is convenient.

Tris(2-aminoethyl)amine (6 mL, 40 mmol) was dissolved in 50 mL DCM and cooled in an ice bath. A solution of Boc anhydride (1.09 g, 5 mmol) in 20 mL DCM was added dropwise, followed by dropwise addition of triethylamine (0.7 mL, 5 mmol) in 20 mL DCM. The reaction was warmed up to room temperature slowly and stirred overnight. After removal of the solvent under reduced pressure, the crude oil was resuspended in 50 mL H_2O and extracted with DCM (50 mL × 5). The organic phase was dried over Na_2SO_4 and solvent removed under reduced pressure. Chromatographic purification of the crude product (DCM:MeOH = 5:1, 2% concentrated ammonium hydroxide) yielded 700 mg (57%) of a light yellow oil.

2.4.3 t2M from NAS

6-chloro-2,4-diamino-1,3,5-triazine(550 mg, 3.78 mmol) and $NaHCO_3$ (500 mg, 5.95 mmol) were added into a 40 mL H_2O solution of mono-Boc protected tren (220 mg, 0.89 mmol). The slurry was heated to 85 °C and reacted overnight. The reaction was then cooled to room temperature and filtered. The solid was washed twice with water and dried under vacuum. The crude product was then purified by silica gel chromatography using DCM:MeOH:conc NH_4OH = 80:20:2 to yield Boc-t2M as a white solid (90 mg, 22%).

Boc t2M was dissolved in trifluoroacetic acid (TFA) and reacted at room temperature for 10 min. The TFA was evaporated under N_2 flow, the residue was dissolved in H_2O, and solution lyophilized to obtain the TFA salt of t2M.

2.4.4 t2M^{Me} from NAS

An appropriate control molecule to evaluate selective binding of melamine to T/U domains is the *N*-methylated melamine derivative. These molecules have the same overall electrostatics and similar shape, but with hydrogen bond donors replaced with methyl groups. Though this control changes steric interactions somewhat, this is unavoidable. The tetramethylated chloro triazine can be prepared in a manner similar to the diamino chlorotriazine by reaction of cyanuric chloride with dimethylamine instead of ammonium hydroxide. Thus, 6-chloro-N^2,N^2,N^4,N^4-tetramethyl-2,4-diamino-1,3,5-triazine (80 mg, 0.4 mmol) and $NaHCO_3$ (84 mg, 1 mmol) were added into 10 mL 1,4-dioxane:H_2O = 1:1 solution of mono Boc-protected tren (40 mg, 0.162 mmol). The slurry was heated to 85 °C and reacted overnight. The reaction was then cooled to RT and extracted with DCM twice. The organic layer was then dried and condensed to obtain a crude mixture which was purified by silica gel chromatography using 2% MeOH in DCM with 2% concentrated ammonia in MeOH. The product was obtained as a white solid (45 mg, 48.4%). Boc $t2M^{Me}$ was dissolved in TFA and reacted at room temperature for 10 min. The TFA was evaporated under N_2 flow, the residue was dissolved in H_2O and lyophilized to obtain the TFA salt of $t2M^{Me}$.

2.5 Oligoethyleneimine bPNAs from reductive alkylation

Notes. The reductive alkylation procedure to prepare oligoethyleneimine bPNAs is efficient but requires careful attention. Like all reductive alkylation/aminations, the reaction requires slightly acidic conditions, but acidity <3 will result in significant quantities of aldehyde reduction. The aldehyde used is melamine acetaldehyde, and the corresponding alcohol is difficult to remove completely from the oligoethyleneimine bPNA product, and thus its production should be minimized. We have found that sequential addition of substoichiometric aliquots of melamine acetaldehyde to the amine substrate and reaction at RT or 50 °C for 30 min generates the imine, which can then be completely reduced with $NaBH_3CN$ in 30 min. Thus, it is recommended to aliquot reagents in this alternating sequence, allowing imine formation at elevated temperature, then cooling to RT for the addition of reductant. Though laborious, this procedure allows for complete conversion to the per-alkylated scaffold within several hours, with minimum alcohol production, and the product may be purified by trituration. Additionally, given the sensitivity of the reaction to acid, care must be taken to completely remove excess acid from starting materials. The hydrochloride salt of the melamine acetaldehyde is preferred, but this starting material should be completely free of excess HCl used in the production of the free aldehyde from protected acetal. These reactions may be followed by analytical RP-HPLC (Fig. 2).

2.5.1 Boc-t2M

Dissolve Boc-ethylenediamine (0.15 g, 0.94 mmol) in 20 mL of methanol in a 50-mL round bottom flask. Prepare four aliquots of melamine acetaldehyde hydrochloride (0.11 g, 0.54 mmol per aliquot) and $NaBH_3CN$ (33.9 mg, 0.54 mmol per aliquot). Add two aliquots of melamine acetaldehyde hydrochloride to the methanol solution and stir for 30 min at room temperature to allow imine formation to proceed. The hydrochloride salt is preferred here to generate the slightly acidic conditions favorable for reductive alkylation. Add aliquots of aldehyde and borohydride sequentially

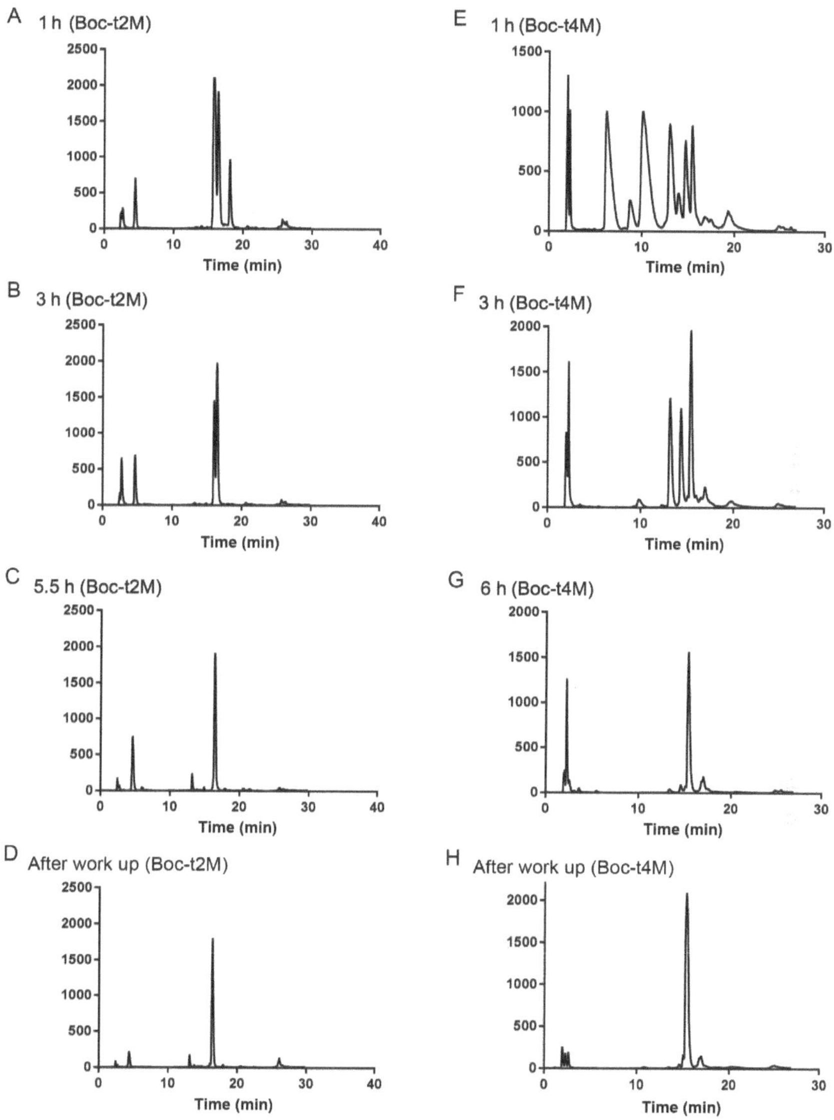

Fig. 2 Analytical C_{18}RP-HPLC traces of reaction progress for reductive alkylation with melamine acetaldehyde of (A–D) Boc-t2M from Boc-ethylenediamine and (E–H) Boc-t4M from Boc-tren. t2M gradient: 0%B (5 min); 0–20%B (5 min); 20%B (10 min); 20–100%B (5 min); 5 min B-wash. t4M gradient: 10%B (5 min); 10–30%B (10 min) 30%B (5 min); 30–100%B (2 min); 5 min B-wash. Early products observed and identified by ESI-MS in (A) are: monoalkylated (15.9 min), dialkylated (16.5 min), and monoalkylated + imine (18.2 min). Early products in Boc-t4M synthesis identified in (B) are: monoalkylated (6.2 min), dialkylated (8.8, 10.3 min), trialkylated (13.1, 14.1 min), tetraalkylated (14.7 min), and trialkylated + imine (15.5 min).

every 30 min over 3–4 h. A major side product of the reaction is the alcohol from reduction of melamine acetaldehyde, which can be difficult to separate from the desired product. If the pH drops to <3, then significant aldehyde reduction is observed. It was empirically determined that addition of aldehyde and reductant in aliquots over time minimizes formation of this side product as compared to addition of larger portions at the start of the reaction. A typical treatment is shown below:

Reaction start:

(1) 2 Aliquots (1.15 eq) melamine acetaldehyde with Boc-ethylenediamine, stir 30 min
(2) 1 Aliquot $NaBH_3CN$ added, stirred 30 min
(3) 1 Aliquot melamine acetaldehyde added, react 30 min
(4) 1 Aliquot $NaBH_3CN$ added, react 30 min
(5) 1 Aliquot melamine acetaldehyde and 1 aliquot $NaCNBH_3$ added, react 30 min
(6) Final aliquot of $NaCNBH_3$ added

Monitor the reaction by analytical HPLC to follow the progression of the alkylation (Fig. 2). The imine intermediate may be detected by HPLC/MS (ESI). To work up the reaction, reduce the amount of methanol to ~4 mL and transfer to two 2-mL-centrifuge tubes. Centrifuge to obtain the solid and discard the methanol solution. The residue will be sticky, but can be triturated and sonicated in acetone three times, using 3 mL acetone for each wash. Each time after trituration, centrifuge to obtain the solid and discard the acetone. Combine the solid into one centrifuge tube and repeat the trituration procedure with ~0.5 mL of neutral water to wash away inorganic salts. A final trituration with 0.5 mL of ethanol yields a solid that can be air-dried to yield 210 mg white powder (0.45 mmol, yield = 48%).

2.5.2 Boc-t4M

The procedure to produce Boc-t4M is similar to Boc-t2M, using portionwise addition of melamine acetaldehyde and $NaBH_3CN$ to an amine scaffold to minimize production of the reduced acetaldehyde side product. To prepare Boc-t4M, dissolve Boc-tren (0.2 g, 0.81 mmol) in 40 mL methanol. Prepare five aliquots of melamine acetaldehyde (0.166 g, 1 eq, 0.81 mmol per aliquot) and $NaBH_3CN$ (51 mg, 1 eq, 0.81 mmol). Reactants are aliquoted into the reaction every 30 min until addition is complete. The reaction rate slows considerably as it progresses, resulting in small quantities of trialkylated material in addition to the desired t4M. This may be separated, or further reacted by addition of more aldehyde and borohydride with extended reaction times.

Work up the reaction as described for t2M, by concentration, sedimentation and washing with acetone and water. Isolate the final product and wash with 5 mL of ethanol, then air-dry to obtain 480 mg of white powder (0.56 mmol, yield = 69%). The reaction can be followed by RP-HPLC (Fig. 2), though HPLC is not needed for purification.

3. Design and preparation of U-site HHR constructs

3.1 Materials, instrumentation, and general notes for transcription

DNA templates were purchased from integrated DNA technologies (IDT) purified by gel prior to use in transcription. SYBR® gold was purchased from Thermo Fisher Scientific. DNA and RNA stock solutions were prepared in deionized water and concentrations were determined by measuring solution absorbance at 260 nm by Thermo Fisher Nanodrop 2000. The promoter sequence in each DNA template is underlined and does not appear in the final RNA transcripts. Buffers were prepared in house with the following composition: 1 × PBS (137 m*M* NaCl, 2.7 m*M* KCl and 10 m*M* PO_4^{3-}, pH 7.4), Tris–Cl (50 m*M* tris(hydroxymethyl)aminomethane, pH 7.6, adjusted with HCl). RNA transcription buffer (10 ×) was 1 *M* HEPES-KOH pH 7.5, 100 m*M* $MgCl_2$, 20 m*M* Spermidine-HCl, 400 m*M* DTT.

3.1.1 General protocol for transcription

RNAs were prepared via runoff transcription using wild-type T7 RNA polymerase (Milligan & Uhlenbeck, 1989). A general starting point for transcriptions used the following conditions: 1 × transcription buffer with an additional 10 m*M* DTT, 10 m*M* $MgCl_2$, 20 m*M* rNTP, 7.5% glycerol, 350 n*M* DNA template, and 2 μL of T7 polymerase stock per 100 μL of transcription reaction solution. These conditions require optimization for new template sequences. The transcriptions were run for 2 h at 37 °C, then quenched with 1.5 eq of EDTA and an equal volume of 2 × TBE/urea loading buffer. Samples were heated at 95 °C for 5 min, cooled to room temperature, and loaded onto acrylamide (5% bisacrylamide) denaturing (8 *M* urea) gel (8.3 cm × 7.3 cm × 1.5 mm). The RNA product bands were visualized on the gel by UV shadowing and cut from the gel. Desired RNA can be isolated from the cut band by standard "crush and soak" (gel is forced through a syringe and RNA extracted into water by overnight agitation at RT) or electroelution, which generally gives higher recovery. Extracted or electroeluted RNA was precipitated by addition of 10% volume of 5 *M*

NH_4OAc pH 5.2 and 2.5 volume equivalents of 200 proof ethanol, and incubation at −20 °C for at least 2 h. RNA was centrifuged down at high speed for 10 min at 4 °C and the pellet was washed with 500 μL of ice-cold 70% ethanol in water and centrifuged again at high speed for 10 min at 4 °C. RNA pellets were then dried in a speed-vac at room temperature for 20 min and suspended in water and stored at −20 °C. After confirming RNA purity by denaturing gel, this pelleted material can be used to prepare stock solutions.

3.1.2 General design considerations and challenges in transcription

Transcription targets containing extended oligo-U domains can be challenging as these domains are considered "slippery" for the polymerase and can also be read as termination signals (Martin & Tinoco, 1980; von Hippel, 1998; Yarnell & Roberts, 1999), thus leading to less pure transcription reactions with truncations and closely eluting products. This problem is exacerbated when the starting point for transcription is the U-domain. Yield and purity of transcription are greatly improved by the addition of several non-U nucleotides (preferably a GC rich sequence) prior to the U-site. While it is difficult to predict which transcripts will be the most problematic, it is generally true that shorter, internal oligo-U sites give better results than longer sites near start of transcription. If it is unavoidable to have the U-site near the start of the transcript, then addition of a few nts prior can be very helpful. In addition, a broad search of conditions for optimization of transcription can dramatically improve the yield. In some cases, the use of T7 mutants (Lyon & Gopalan, 2018) can improve purity and allow useful quantities of U-domain transcripts to be obtained when it is not possible using wild-type T7, though reaction rates, and sometimes overall yield, are decreased. The transcripts described below were amenable to our standard transcription protocol using wild-type T7, despite extended oligo-U domains.

3.2 Design of modified HHR transcripts with U-sites for allosteric binding

Installation of an allosteric site that is sensitive to triggering by bPNA requires replacement of a structural element with a oligo-U site that is typically less structured. Upon hybridization with a bPNA molecule, the duplex stem is replaced with a triplex stem. This concept appears to be largely backbone independent. In the minimal HHR scaffold, there are three structural stems that support a catalytic site of ~19 nts. The sequence identity of the stems is not critical, though there needs to be structure. Full activity of the HHR requires an additional tertiary interaction between loop 2 and stem 1 that creates a more closed HHR structure, and further twists the substrate strand sugar–phosphate backbone into a reactive conformation for cleavage at the active site. This tertiary interaction is a Hoogsteen base pair between U and A (Robertson & Scott, 2007). HHR activity is greatly reduced by stem replacement with oligo-U loops and fully restored by bPNA peptide triplex hybridization. Further, the tertiary interaction between loop and stem may also be ablated by replacement of the loop and stem sequence with oligo-U sequences and restored by binding of oligoethyleneimine bPNAs (t2M, t4M) that bridge the loop and stem domains to restore the tertiary contact. Below, we describe our design methodology for these two systems.

3.3 Design and use of stem/loop replacement HHR sequences

Replacement of stem 2 alone with an oligo-U loop does indeed decrease HHR self-cleavage in the minimal HHR, but significant activity remains at 10 mM Mg^{2+}. Replacement of two stems with oligo-U yields U-(2,3) HHR, which is not detectably functional in self-cleavage at high magnesium ion concentration (Fig. 3). This U-mutant was produced by transcription. The DNA template is shown below, with the promoter region underlined and the regions corresponding to stem 2 and 3 regions shown in bold.

3.3.1 Double stem replacement HHR: The U-(2,3) ribozyme sequence

This type I HHR mutant was designed to retain the sequence of the catalytic site while replacing duplex stems 2 and 3 with oligo-U loops (Fig. 3). We replaced all three stems (1,2,3) that support the catalytic domain, but this system did not refold into a catalytically active form by triplex hybridization with bPNA. We speculate that the replacement of all three stems with oligo-U domains may result in misfolding upon triplex hybridization. It may yet be possible to design a triple stem-replaced construct that could fold into a catalytically active form. However, a DNA template for the U-(2,3) HHR can designed as indicated below. Two different binding sites were prepared, of different sizes. Stem-loop 2 was replaced with 16 nt loop, comprised of a U6-GNRA-U6 domain. It was anticipated that the U6-domains would form a triplex stem with bPNA, capped with a stable GNRA (GAGA) loop. Similarly, stem-loop 3 was replaced with a larger 24 nt loop, U10-GAGA-U10. Stem 1, formed from the 5′ terminus and the domain following the substrate sequence, was left intact. A autonomously folding tRNA-Lys module was fused to the 3′ end of stem 1, so that the HHR cleavage reaction would

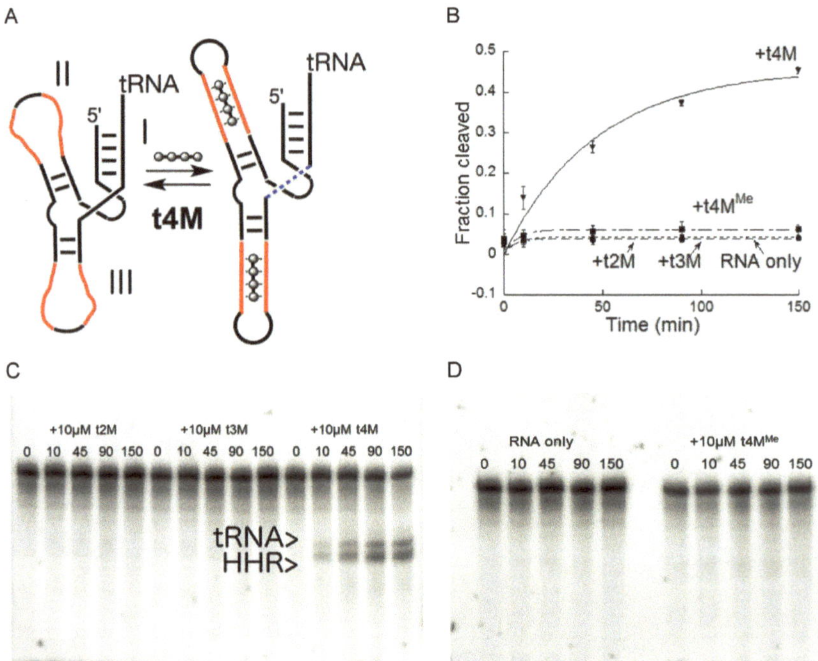

Fig. 3 (A) Illustration of U-(2,3) HHR cleavage triggered by stem refolding. The U-domains (*red*) bind to t4M, inducing cleavage of the scissile sugar phosphate bond (*dashed blue line*). (B) Results of gel quantification, in triplicate. (C) Representative gel analysis of U-(2,3) HHR cleavage reactions at 10 m*M* $MgCl_2$. Top band indicates intact U-(2,3) ribozyme, bottom two bands indicate the cleavage products. (D) Negligible cleavage is observed under the same conditions without bPNAs or with methylated t4M.

produce a stable, structured RNA leaving group that could be easily detected and stained on gel. The DNA template for T7 polymerase transcription, and the transcription product is shown below, with the stem replacement domains shown in bold and T7 promoter underlined.

DNA template: 5′-TCA CTG TAA AGA GGT GTT GGT TCT CTT AAT CTT TAA CTT AAA AGG TTA ATG CTA AGT TAG CTT TAC AGT GCG AC**A AAA AAA AAA TCT CAA AAA AAA AA**G TTT CG**A AAA AAT CTC AAA AAA** CTC ATC AGG CAC TGC CTA TAG TGA GTC GTA TTA ATT TC-3′.

RNA transcript: 5′-GG CAG UGC CUG AUG AGU **UUU UUG AGA UUU UUU** CGA AAC **UUU UUU UUU UGA GAU UUU UUU UUU** GUC GCA CUG UAA AGC UAA CUU AGC AUU AAC CUU UUA AGU UAA AGA UUA AGA GAA CCA ACA CCU CUU UAC AGU GA-3′.

3.3.2 Tertiary contacts and stem replacement: S-U4n and S-U4n-2bp HHR

Stem restructuring has effects at both ends of the stem. In an HHR stem 2 replacement experiment, loss of structure destabilizes the catalytic site at the base of the stem, but also results in loss of a critical tertiary interaction between the loop of stem 2 and stem 1. This tertiary contact is a Hoogsteen AU base pair formed between an A in the loop atop stem 2 and a U in stem 1. This contact is restored upon triplex hybridization with oligoethyleneimine bPNA (t4M), a notion that was tested in the two constructs, S-U4n HHR and S-U4n-2bp HHR. As mentioned previously, a single stem 2 replacement results in significant retention of cleavage activity at 10 m*M* Mg^{2+} concentration; however, minimal cleavage is seen at 0.1 m*M* Mg^{2+}. The construct S-U4n HHR is a stem replacement mutant of HHR in which the four base pair stem is replaced with two U4 domains, while retaining the native GUGA loop (Fig. 4). This stem replacement is a much smaller perturbation than in.

U-(2,3) HHR, and is designed to accommodate t4M binding. To stabilize the loop, a GC base pair is retained at the base of the loop as a clamp structure. Notably, addition of t4M can restore a significant degree of bond scission at 0.1 m*M* Mg^{2+}, while $t4M^{Me}$, which has base-triple formation blocked by methylation, has no effect on catalysis. This indicates that t4M base recognition of the U-site rescues HHR function. The S-U4n-2bp HHR was designed to test the extent to which this restructuring affects the known tertiary contact between the GUGA loop of stem 2 and stem 1 by using the identical stem 2 design, but simply increasing the length of stem 1 by 2 base pairs. This 2 bp insert extended the distance between the loop and the uracil in stem 1 that was known to a Hoogsteen base pair with the adenine in the loop of stem 2, and proved to be sufficient to completely shut down HHR cleavage. The ability to tune HHR reactivity through this subtle adjustment remote from the active site attests to the amenability of the HHR scaffold to engineering. The DNA templates and transcript sequences for the S-U4n and S-U4n-2bp HHRs are shown below, with the genetically encoded bPNA binding site shown in bold, tertiary loop-stem Hoogsteen contact shown in double underline, tRNA-Lys sequence italicized and T7 promoter region underlined.

S-U4n HHR. DNA template: 5′-GCC GGG GGT GGG ATT TGA ACC CAC GTA AGG CGG ATC TGC AGT CCG CTG CCT AGC CCC TAG ACT ACC CCG GCT GGT AGA CTG TGA CCC AGT CTC CTG GGT TTC G**AA AAC TCA CGA AAA** CTC ATC AGA CAG TCT ATA GTG AGT CGT ATT A-3′

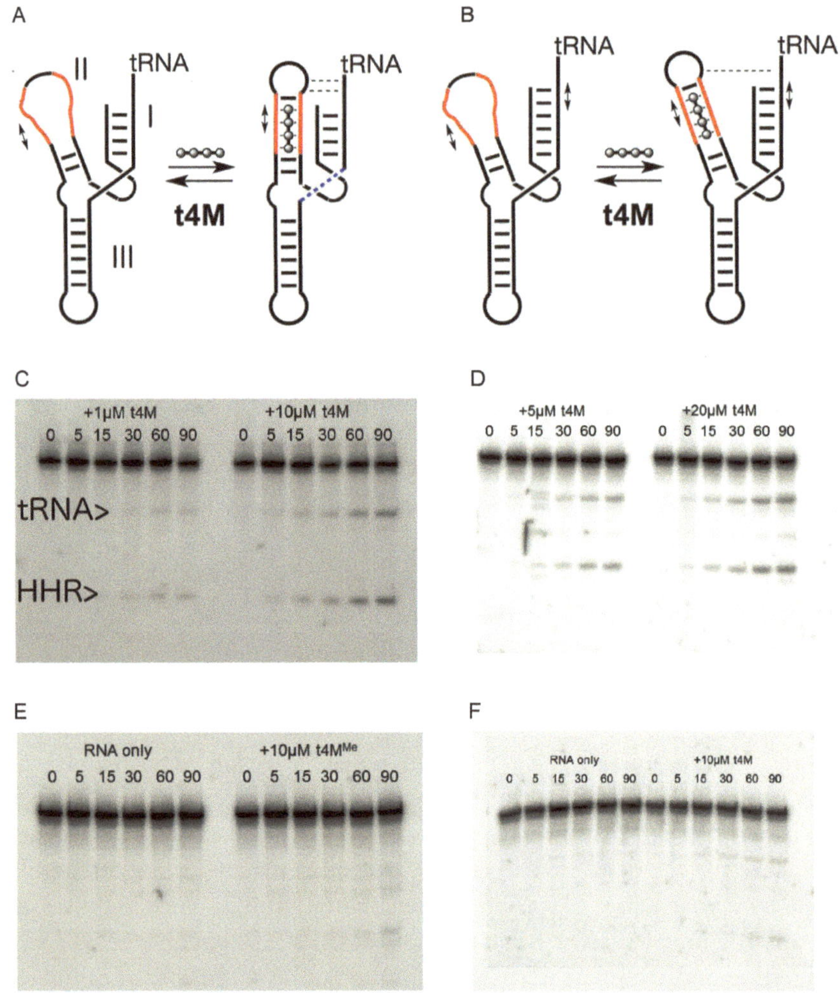

Fig. 4 S-U4n (A) and S-U4-2bp (B) HHR cleavage reactions triggered by tren bPNAs at 0.1 m*M* Mg^{2+}, analyzed by gel. Top band indicates intact S-U4n ribozyme, middle band indicates the tRNA cleavage product, bottom band indicates the ribozyme cleavage product. (C and D) S-U4n ribozyme rescued by t4M, under 0.1 m*M* Mg^{2+} conditions at increasing t4M concentration. Much weaker cleavage is seen with (E) methylated t4M control molecule and (F) the S-U4n-2bp ribozyme.

S-U4n HHR. RNA transcript: 5′-G ACU GUC UGA UGA G**UU UUC GUG** A^{L4}**GU UUU** CGA AAC CCA GGA GAC UGG GUC ACA G|UC U$^{1.7}$AC-*CAG CCG GGG UAG UCU AGG GGC UAG GCA GCG GAC UGC AGA UCC GCC UUA CGU GGG UUC AAA UCC CAC CCC CGG C-3′*

S-U4n-2bp. DNA template: 5′- GCC GGG GGT GGG ATT TGA ACC CAC GTA AGG CGG ATC TGC AGT CCG CTG CCT AGC CCC TAG ACT ACC CCG GCT GGT AGA GCC TGT GAC CCA GTC TCC TGG GTT TCG **AAA ACT CAC GAA AA**C TCA TCA GAC AGG CTC TAT AGT GAG TCG TAT TA-3′

S-U4n-2bp. RNA transcript: 5′-GA**G C**CU GUC UGA UGA G**UU UUC GUG** A^{L4}**GU UUU** CGA AAC CCA GGA GAC UGG GUC ACA G|**GC** UC $^{1.7}$U AC-*CAG CCG GGG UAG UCU AGG GGC UAG GCA GCG GAC UGC AGA UCC GCC UUA CGU GGG UUC AAA UCC CAC CCC CGG C*-3′.

3.4 Design and use of U-loop replacement binary HHRs

Transcription of fully functional HHRs is challenging because the transcript will immediately cleave in transcription buffer, which contains 10 m*M* Mg^{2+} that favors bond scission. Stem replacement HHRs discussed above may be transcribed as a single RNA strand because the loss of secondary structure serves to suppress self-cleavage during the reaction and permit isolation. For loop replacement studies, the duplex stems supporting the catalytic site are intact, and thus a full-length HHR will completely self-cleave prior to isolation. For this reason, loop replacement studies must be carried out on binary HHRs, wherein one strand is an enzyme strand, and the other is a substrate strand. These binary HHRs exhibit significant rate dependence on the sequence identity of the stem 2 loop. However, though a GUGA loop is found in the wild-type HHR, with the adenine base of the loop interacting with uracil base in stem 1, a GUUU loop also functions well in cleavage. This suggests that the HHR loop-stem interaction may be quite plastic, and may accept alternative noncanonical interactions such as a GU wobble pair in addition to the AU Hoogsteen pair. To study the turn on of tertiary interactions with t4M, we transcribed the enzyme strand for a binary HHR in which the GUGA loop of stem 2 was replaced with UUUU. To provide a baseline of reactivity, the HHR enzyme strand with the native GUGA loop was also transcribed. It is convenient to follow this reaction by fluorescently tagged RNA substrates, which were purchased from IDT that are annealed with the enzyme strand (Fig. 5), and unreactive until triggered with Mg^{2+} ions and t4M bPNA. This avoids potential problems of refolding the enzyme and substrate strands together with a large tRNA fusion on the substrate strand. The DNA templates and transcript sequences are shown below, with the promoter sequence underlined and the critical

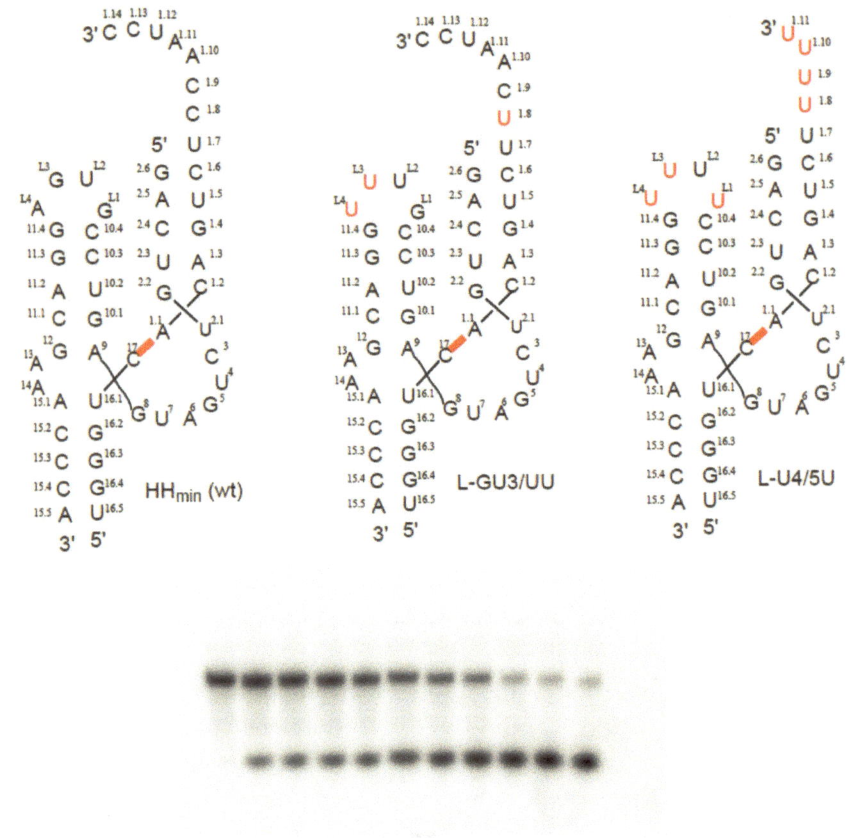

Fig. 5 (*Top*) Secondary structure representation of the binary HHRs. Scissile bond shown in *red* between C17 and A1.1. Modified bases in the loop and stem I shown in *red*. (*Below*) Representative gel image for L-U4/5U binary ribozyme cleavage. Top band is the intact Cy3 labeled substrate strand RNA, bottom band indicates the cleavage product, with 10 μ*M* t4M, under 0.1 m*M* Mg^{2+} conditions. Aliquots (*left to right*) were taken at 0, 15, 30, and 45 s, and 1, 2, 3, 5, 15, 30, and 60 min.

loop shown in bold. The relevant nucleotide modifications in enzyme and substrate strands are shown in double underline.

- *WT HHR enzyme strand* DNA template: 5′-TGG GTT TCG TCC TCA CGG ACT CAT CAG ACA GTC TAT AGT GAG TCG TAT TA-3′
- *WT* HHR enzyme strand: 5′-G ACU GUC UGA UGA GUC C**GU GA**G GAC GAA ACC CA-3′.
- *L-U4* HHR enzyme strand DNA template: 5′-TGG GTT TCG TCC AAA AGG ACT CAT CAG ACA GTC TAT AGT GAG TCG TAT TA-3′

- *L-U4* HHR enzyme strand: 5′-G ACU GUC UGA UGA GUC CUU UUG GAC GAA ACC CA-3′
- *Substrate strands* (Scissile bond: *C-A*.)
  WT-HHR: 5′ Cy3-UGGGUCA*C–A*GUCUCCAAUCC-3′
  2U: 5′ Cy3-UGGGUCA*C–A*GUCUUCAAUCC-3′
  5U: 5′ Cy3-UGGGUCA*C–A*GUCUUUUU-3′

4. Triggering HHR catalysis with bPNAs

4.1 General reaction protocols

4.1.1 Single-stranded HHRs

A stock solution of ribozyme (625 nM) and tren derivatives (12.5 μ*M*) in 1.25 × Tris–Cl buffer (pH 7.6) was annealed at 75 °C. An aqueous $MgCl_2$ solution was added to the annealed solution to initiate the reaction at 37 °C. The final concentrations were: 500 n*M* ribozyme, 10 μ*M* tren derivatives and 10 m*M* Mg^{2+} in 1 × Tris–Cl buffer (pH 7.6). Aliquots were taken at 0, 10, 45, 90, 150 min, quenched with urea-EDTA, and frozen immediately on dry ice. Samples were analyzed on 12% denaturing acrylamide gel and stained with SYBR® gold. This strategy has been demonstrated with both peptide bPNAs (Xia et al., 2014) and oligoethyleneimine bPNAs (Mao et al., 2017). We focus our protocol description on the oligoethyleneimine scaffold, but the procedures for use with peptide are identical.

4.1.2 Binary HHRs

Samples for single turnover experiments were prepared by making a stock solution with concentrations of the following species: 5 μ*M* of enzyme strand, 400 n*M* of the Cy3 labeled substrate strand, and 10 μ*M* tren derivatives in 1.25 × Tris–Cl buffer (pH 7.6). The mixture was heated at 75 °C for 2 min and equilibrated at 27 °C for 10 min. $MgCl_2$ was then added to initiate the reaction at 27 °C. The final concentrations were: 4 μ*M* enzyme strand, 320 n*M* Cy3 labeled substrate strand, 10 μ*M* tren derivatives and 0.1 m*M* Mg^{2+} in 1 × Tris–Cl buffer (pH 7.6). Aliquots were taken at 0, 15, 30, 45 s and 1, 2, 3, 5, 15, 30, 60 min, quenched with urea-EDTA, and frozen immediately on dry ice. Samples were analyzed on 20% denaturing acrylamide gel.

4.2 Experimental design and optimizing conditions

4.2.1 Cleavage reactions of single-stranded HHRs

The single-stranded HHR U-mutants such as U-(2,3), S-U4n and S-U4n-2bp HHRs can be transcribed under normal conditions and do not self-cleave appreciably, despite 10 mM $MgCl_2$ in the reaction. However, their reactivity remains sensitive to Mg^{2+} concentration, and the combination of both t4M and Mg^{2+} is required to trigger RNA cleavage. The presence of the tRNA fusions in these constructs yields two RNAs of comparable size upon cleavage, the HHR itself and the tRNA. These two RNAs may be easily observed by gel (Fig. 4). Addition of the methylated t4M molecule generally does not cause HHR cleavage. This control is essential to rule out the effects of the oligoamine backbone itself; molecules of this type can affect RNA function, but typically at much higher concentrations (Pingoud et al., 1984; Quigley, Teeter, & Rich, 1978). Reactions should be run in triplicate and gels quantified using standard software (Imagequant) to obtain HHR cleavage rates (Fig. 4). Reactivity should be very reproducible for each construct, resulting in relatively tight error bars. If the reaction rate does not allow monitoring by reasonable data time point sampling, the rate may be tuned by decreasing or increasing Mg^{2+} concentration to decrease or increase reaction rate, respectively. Decreasing Mg^{2+} to 0.1 mM enabled convenient monitoring of HHR cleavage in the sU4n HHR as a function of t4M concentration. Typically, cleavage reactions were monitored at a single t4M concentration over time. In addition, rates were measured as a function of increasing t4M concentrations. In this way, it can be clearly established that the t4M molecules can directly influence HHR cleavage.

4.2.2 Cleavage reactions of binary HHRs

Binary HHRs were annealed together prior to use in cleavage reactions as described above. Interestingly, it is possible to mix different substrate strands with each HHR enzyme strand, yielding cleavage rates reflective of their dependence on t4M as well as substrate–enzyme matching. While both loop modified L-GU3 and LU4 HHRs do not effectively cleave the native substrate, L-GU3 can cleave a substrate strand that features U's at both position 1.7 (native) and 1.8 in stem 1, which suggests that the L1 guanine may be forming a tertiary contact with the U 1.8. The labeled substrate strand provides a convenient way to quantitatively follow the cleavage reaction, signaling unreacted and cleaved substrate with the same dye (Fig. 5).

5. Summary

The synthetic bPNAs described herein can bridge native and artificial architectures, uniting nucleic acid biotechnology (Burnett & Rossi, 2012) with new materials (Roh, Ruiz, Peng, Lee, & Luo, 2011). This protocol provides a general guide on the synthesis of bPNA peptides and oligoethyleneimine derivatives that have been used to trigger cleavage in engineered hammerhead ribozymes. This strategy enables additional control over RNA structure; this facilitates elucidation of critical aspects of RNA recognition that contribute to RNA functions, including catalysis.

Acknowledgments

This work was supported in part by funding from the National Institutes of Health (GM111995-01A1 to D.B.), NASA (GRT00044810 to D.B.), and NSF (DMR 1802432 to D.B).

References

Arambula, J. F., Ramisetty, S. R., Baranger, A. M., & Zimmerman, S. C. (2009). A simple ligand that selectively targets CUG trinucleotide repeats and inhibits MBNL protein binding. *Proceedings of the National Academy of Sciences of the United States of America*, *106*, 16068–16073. https://doi.org/10.1073/pnas.0901824106.

Ariga, K., & Kunitake, T. (1998). Molecular recognition at air-water and related interfaces: Complementary hydrogen bonding and multisite interaction. *Accounts of Chemical Research*, *31*, 371–378. Retrieved from http://pubs3.acs.org/acs/journals/doilookup?in_doi=10.1021/ar970014i.

Burnett, J. C., & Rossi, J. J. (2012). RNA-based therapeutics: Current progress and future prospects. *Chemistry & Biology*, *19*, 60–71. https://doi.org/10.1016/j.chembiol.2011.12.008.

Duca, M., Vekhoff, P., Oussedik, K., Halby, L., & Arimondo, P. B. (2008). The triple helix: 50 years later, the outcome. *Nucleic Acids Research*, *36*(16), 5123–5138. https://doi.org/10.1093/nar/gkn493.

Felsenfeld, G., Davies, D. R., & Rich, A. (1957). Formation of a three-stranded polynucleotide molecule. *Journal of the American Chemical Society*, *79*, 2023–2024. https://doi.org/10.1021/ja01565a074.

Kawasaki, T., Tokuhiro, M., Kimizuka, N., & Kunitake, T. (2001). Hierarchical self-assembly of chiral complementary hydrogen-bond networks in water: Reconstitution of supramolecular membranes. *Journal of the American Chemical Society*, *123*, 6792–6800. Retrieved from http://www.ncbi.nlm.nih.gov/entrez/query.fcgi?cmd=Retrieve&db=PubMed&dopt=Citation&list_uids=11448183.

Kool, E. T. (2001). Hydrogen bonding, base stacking, and steric effects in DNA replication. *Annual Review of Biophysics and Biomolecular Structure*, *30*(1), 1–22. https://doi.org/10.1146/annurev.biophys.30.1.1.

Lyon, S., & Gopalan, V. (2018). A T7 RNA polymerase mutant enhances the yield of 5'-thienoguanosine-initiated RNAs. *Chembiochem: A European Journal of Chemical Biology*, *19*(2), 142–146. https://doi.org/10.1002/cbic.201700538.

Ma, M., & Bong, D. (2011a). Determinants of cyanuric acid and melamine assembly in water. *Langmuir: The ACS Journal of Surfaces and Colloids*, *27*, 8841–8853. https://doi.org/10.1021/la201415d.

Ma, M., & Bong, D. (2011b). Directed peptide assembly at the lipid-water interface cooperatively enhances membrane binding and activity. *Langmuir: The ACS Journal of Surfaces and Colloids*, *27*, 1480–1486. https://doi.org/10.1021/la104405r.

Ma, M., & Bong, D. (2011c). Protein assembly directed by synthetic molecular recognition motifs. *Organic & Biomolecular Chemistry*, *9*, 7296–7299. https://doi.org/10.1039/c1ob05998j.

Ma, M., & Bong, D. (2013). Controlled fusion of synthetic lipid membrane vesicles. *Accounts of Chemical Research, 46*, 2988–2997.

Mao, J., & Bong, D. (2015). Synthesis of DNA-binding peptoids. *Synlett: Accounts and Rapid Communications in Synthetic Organic Chemistry, 26*(11), 1581–1585. https://doi.org/10.1055/s-0034-1380698.

Mao, J., DeSantis, C., & Bong, D. (2017). Small molecule recognition triggers secondary and tertiary interactions in DNA folding and hammerhead ribozyme catalysis. *Journal of the American Chemical Society, 139*(29), 9815–9818. https://doi.org/10.1021/jacs.7b05448.

Martin, F. H., & Tinoco, I., Jr. (1980). DNA-RNA hybrid duplexes containing oligo(dA:rU) sequences are exceptionally unstable and may facilitate termination of transcription. *Nucleic Acids Research, 8*(10), 2295–2299. Retrieved from https://www.ncbi.nlm.nih.gov/pubmed/6159577.

Milligan, J. F., & Uhlenbeck, O. C. (1989). [5] Synthesis of small RNAs using T7 RNA polymerase. In *Vol. 180. Methods in Enzymology* (pp. 51–62). Academic Press. https://doi.org/10.1016/0076-6879(89)80091-6.

Mittapalli, G. K., Reddy, K. R., Xiong, H., Munoz, O., Han, B., De Riccardis, F., et al. (2007). Mapping the landscape of potentially primordial informational oligomers: Oligodipeptides and oligodipeptoids tagged with triazines as recognition elements. *Angewandte Chemie, International Edition, 46*, 2470–2477. https://doi.org/10.1002/anie.200603207.

Nielsen, P. E. (1999). Peptide nucleic acid. A molecule with two identities. *Accounts of Chemical Research, 32*(7), 624–630. https://doi.org/10.1021/ar980010t.

Piao, X., Xia, X., & Bong, D. (2013). Bifacial peptide nucleic acid directs cooperative folding and assembly of binary, ternary, and quaternary DNA complexes. *Biochemistry, 52*, 6313–6323. https://doi.org/10.1021/bi4008963.

Piao, X., Xia, X., Mao, J., & Bong, D. (2015). Peptide ligation and RNA cleavage via an abiotic template interface. *Journal of the American Chemical Society, 137*(11), 3751–3754. https://doi.org/10.1021/jacs.5b00236.

Pilch, D. S., Levenson, C., & Shafer, R. H. (1990). Structural analysis of the (dA)10.2(dT)10 triple helix. *Proceedings of the National Academy of Sciences of the United States of America*, Vol. 87, 1942–1946. Retrieved from http://www.ncbi.nlm.nih.gov/entrez/query.fcgi?cmd=Retrieve&db=PubMed&dopt=Citation&list_uids=2308955.

Pingoud, A., Urbanke, C., Alves, J., Ehbrecht, H. J., Zabeau, M., & Gualerzi, C. (1984). Effect of polyamines and basic proteins on cleavage of DNA by restriction endonucleases. *Biochemistry, 23*(24), 5697–5703. Retrieved from https://www.ncbi.nlm.nih.gov/pubmed/6098296.

Quigley, G. J., Teeter, M. M., & Rich, A. (1978). Structural analysis of spermine and magnesium ion binding to yeast phenylalanine transfer RNA. *Proceedings of the National Academy of Sciences of the United States of America, 75*(1), 64–68. Retrieved from https://www.ncbi.nlm.nih.gov/pubmed/343112.

Ranganathan, A., Pedireddi, V. R., & Rao, C. N. R. (1999). Hydrothermal synthesis of organic channel structures: 1:1 hydrogen-bonded adducts of melamine with cyanuric and trithiocyanuric acids. *Journal of the American Chemical Society, 121*, 1752–1753. Retrieved from http://pubs3.acs.org/acs/journals/doilookup?in_doi=10.1021/ja983928o.

Robertson, M. P., & Scott, W. G. (2007). The structural basis of ribozyme-catalyzed RNA assembly. *Science, 315*(5818), 1549–1553. https://doi.org/10.1126/science.1136231.

Roh, Y. H., Ruiz, R. C. H., Peng, S., Lee, J. B., & Luo, D. (2011). Engineering DNA-based functional materials. *Chemical Society Reviews, 40*(12), 5730–5744. https://doi.org/10.1039/c1cs15162b.

Rougee, M., Faucon, B., Mergny, J. L., Barcelo, F., Giovannangeli, C., Garestier, T., et al. (1992). Kinetics and thermodynamics of triple-helix formation: Effects of ionic strength and mismatches. *Biochemistry*, *31*, 9269–9278. Retrieved from http://www.ncbi.nlm.nih.gov/entrez/query.fcgi?cmd=Retrieve&db=PubMed&dopt=Citation&list_uids=1390713.

SantaLucia, J., Jr., & Hicks, D. (2004). The thermodynamics of DNA structural motifs. *Annual Review of Biophysics and Biomolecular Structure*, *33*, 415–440. https://doi.org/10.1146/annurev.biophys.32.110601.141800.

Soto, A. M., Loo, J., & Marky, L. A. (2002). Energetic contributions for the formation of TAT/TAT, TAT/CGC(+), and CGC(+)/CGC(+) base triplet stacks. *Journal of the American Chemical Society*, *124*, 14355–14363. Retrieved from http://www.ncbi.nlm.nih.gov/entrez/query.fcgi?cmd=Retrieve&db=PubMed&dopt=Citation&list_uids=12452709.

ten Cate, M. G. J., Huskens, J., Crego-Calama, M., & Reinhoudt, D. N. (2004). Thermodynamic stability of hydrogen-bonded nanostructures: A calorimetric study. *Chemistry*, *10*(15), 3632–3639. https://doi.org/10.1002/chem.200400085.

Uhlenbeck, O. C. (1987). A small catalytic oligoribonucleotide. *Nature*, *328*(6131), 596–600. https://doi.org/10.1038/328596a0.

von Hippel, P. H. (1998). An integrated model of the transcription complex in elongation, termination, and editing. *Science*, *281*(5377), 660–665. Retrieved from https://www.ncbi.nlm.nih.gov/pubmed/9685251.

Whitesides, G. M., Simanek, E. E., Mathias, J. P., Seto, C. T., Chin, D., Mammen, M., et al. (1995). Noncovalent synthesis: Using physical-organic chemistry to make aggregates. *Accounts of Chemical Research*, *28*, 37–44. Retrieved from http://pubs3.acs.org/acs/journals/doilookup?in_doi=10.1021/ar00049a006.

Xia, X., Piao, X., & Bong, D. (2014). Bifacial peptide nucleic acid as an allosteric switch for aptamer and ribozyme function. *Journal of the American Chemical Society*, *136*(20), 7265–7268. https://doi.org/10.1021/ja5032584.

Xia, X., Piao, X., Fredrick, K., & Bong, D. (2013). Bifacial PNA complexation inhibits enzymatic access to DNA and RNA. *Chembiochem: A European Journal of Chemical Biology*, *15*, 31–36.

Xodo, L. E., Manzini, G., & Quadrifoglio, F. (1990). Spectroscopic and calorimetric investigation on the DNA triplex formed by d(CTCTTCTTTCTTTTCTTTCTTCTC) and d(GAGAAGAAAGA) at acidic pH. *Nucleic Acids Research*, *18*, 3557–3564. https://doi.org/10.1093/nar/18.12.3557.

Yakovchuk, P., Protozanova, E., & Frank-Kamenetskii, M. D. (2006). Base-stacking and base-pairing contributions into thermal stability of the DNA double helix. *Nucleic Acids Research*, *34*, 564–574. https://doi.org/10.1093/nar/gkj454.

Yarnell, W. S., & Roberts, J. W. (1999). Mechanism of intrinsic transcription termination and antitermination. *Science*, *284*(5414), 611–615. Retrieved from https://www.ncbi.nlm.nih.gov/pubmed/10213678.

Zeng, Y., Pratumyot, Y., Piao, X., & Bong, D. (2012). Discrete assembly of synthetic peptide-DNA triplex structures from polyvalent melamine-thymine bifacial recognition. *Journal of the American Chemical Society*, *134*(2), 832–835. https://doi.org/10.1021/ja2099326.

Zhou, Z., & Bong, D. (2013). Small-molecule/polymer recognition triggers aqueous-phase assembly and encapsulation. *Langmuir: The ACS Journal of Surfaces and Colloids*, *29*, 144–150. https://doi.org/10.1021/la304457y.

Zhou, Z., Xia, X., & Bong, D. (2015). Synthetic polymer hybridization with DNA and RNA directs nanoparticle loading, silencing delivery, and aptamer function. *Journal of the American Chemical Society*, *137*(28), 8920–8923. https://doi.org/10.1021/jacs.5b05481.

CHAPTER NINE

Using Rosetta for RNA homology modeling

Andrew M. Watkins[a], Ramya Rangan[b], Rhiju Das[a,b,*]
[a]Department of Biochemistry, Stanford University School of Medicine, Stanford, CA, United States
[b]Biophysics Program, Stanford University, Stanford, CA, United States
*Corresponding author: e-mail address: rhiju@stanford.edu

Contents

Abstract

The three-dimensional structures of RNA molecules provide rich and often critical information for understanding their functions, including how they recognize small molecule and protein partners. Computational modeling of RNA 3D structure is becoming increasingly accurate, particularly with the availability of growing numbers of template structures already solved experimentally and the development of sequence alignment and 3D modeling tools to take advantage of this database. For several recent "RNA puzzle" blind modeling challenges, we have successfully identified useful template structures and achieved accurate structure predictions through homology modeling tools developed in the Rosetta software suite. We describe our semi-automated methodology here and walk through two illustrative examples: an adenine riboswitch aptamer, modeled from a template guanine riboswitch structure, and a SAM I/IV riboswitch aptamer, modeled from a template SAM I riboswitch structure.

Methods in Enzymology, Volume 623
ISSN 0076-6879
https://doi.org/10.1016/bs.mie.2019.05.026

1. Introduction

RNA plays a host of critical functional roles in cells, from translation regulation to catalysis (Gesteland, Cech, & Atkins, 2005). To achieve these functions, many noncoding RNAs (ncRNAs) take on complex 3D folds, with secondary structure helical elements positioned by structured junctions and tertiary contacts. As experimental characterization of these structures can be challenging and time-consuming, there is strong interest in developing computational strategies to predict these structures using physical modeling or knowledge-based fragment and motif sampling (Das, Karanicolas, & Baker, 2010; Ditzler, Otyepka, Sponer, & Walter, 2010; Laing & Schlick, 2011; Sim, Minary, & Levitt, 2012; Xu & Chen, 2018). To make better use of available information, these computational strategies can be augmented with homology modeling, where a portion of the modeled coordinates is built based on a previously solved homologous structure (Piatkowski et al., 2016; Rother, Rother, Boniecki, Puton, & Bujnicki, 2011).

As the number of experimentally characterized RNA structures grows, the potential for accurate homology modeling for new molecules increases as well. The Protein Data Bank now has over 1500 non-redundant RNA structures at high or medium resolution (4.0 Å resolution or better) (Leontis & Zirbel, 2012), and these deposited structures represent 87 different Rfam families and 21 different Rfam clans (Kalvari et al., 2018). These structures provide a wealth of diverse potential templates for new modeling challenges. Furthermore, the availability of large databases of RNA sequences across species has generated more opportunities to discover homologous sequences. Alignment software can account for covariation in RNA secondary structure, providing more accurate alignments than those based exclusively on sequence (Kalvari et al., 2018; Nawrocki & Eddy, 2013). Expert inspection and biochemical analysis routinely reveal homology between distantly related ncRNA classes even prior to structural characterization. Recent examples include the lariat capping ribozyme, whose catalytic core is homologous to the group I self-splicing intron (Einvik, Nielsen, Westhof, Michel, & Johansen, 1998); the group II self-splicing introns, homologous to the eukaryotic spliceosome (Toor, Keating, Taylor, & Pyle, 2008); extended "sub-motifs" shared between adenosylcobalamin (AdoCbl) and flavin mononucleotide (FMN) riboswitches (Barrick & Breaker, 2007; Jaeger, Verzemnieks, & Geary, 2009); homology between glutamine and "downstream peptide" riboswitches (Ames & Breaker, 2014); and shared binding sites for S-adenosyl methionine (SAM) across distinct riboswitch classes (Mirihana Arachchilage, Sherlock, Weinberg, & Breaker, 2018; Weinberg et al., 2008).

We have found that riboswitches—genetic control elements that respond to the presence of small molecules (Serganov & Patel, 2012)—are particularly amenable to template identification and then computational homolog modeling of structure. Often, multiple classes of riboswitches bind the same or similar ligand and possess identifiable sequence homology to each other and to solved riboswitch structures. Riboswitches are also excellent use cases for homology modeling because their functional state requires a stably-folded ligand binding site, entailing intricate 3D folds in ligand-recognition domains—folds that are unlikely to be recovered by *de novo* RNA modeling. If ligand binding sites from previously solved riboswitches can be identified for a new riboswitch, borrowing that structural information can greatly improve modeling accuracy.

Recent blind modeling challenges in the RNA Puzzles trials (Miao et al., 2017) support the view that RNA homology modeling can be accurate and biologically useful. We have used identifiable but at times distant homologies to previously solved structures to achieve accurate blind models of numerous riboswitch and other ncRNA targets, including the GIR1 lariat-capping ribozyme, the adenosylcobalamin riboswitch, the glutamine riboswitch, the Zika xrRNA, and the SAM I/IV riboswitch (RNA Puzzles 5, 6, 8, 14, 18; Fig. 1). In several cases, we were further able to correctly predict ligand-binding sites, based on clustering of conserved residues in these 3D folds (Miao et al., 2017). While addressing these blind challenges, we have developed a framework for homology modeling of RNA structures with Rosetta computational tools. Depending on the target and modeling sub-problem, either a fragment assembly algorithm (Fragment Assembly of RNA with Full-Atom Refinement, or FARFAR) (Cheng, Chou, & Das, 2015; Das et al., 2010) or a high resolution fragment-free algorithm called stepwise Monte Carlo (SWM) (Watkins et al., 2018) are best suited to the modeling challenge. FARFAR modeling is the more well-developed approach and the method has been previously reviewed (Cheng, Chou, & Das, 2015), albeit not yet for homology modeling problems. SWM is a newer method that seeks higher resolution. In the case of Zika xrRNA, recent homology modeling with SWM correctly predicted all noncanonical base pairs of the RNA, a previously unmet challenge in RNA computational structural modeling (Fig. 1D) (Watkins et al., 2018).

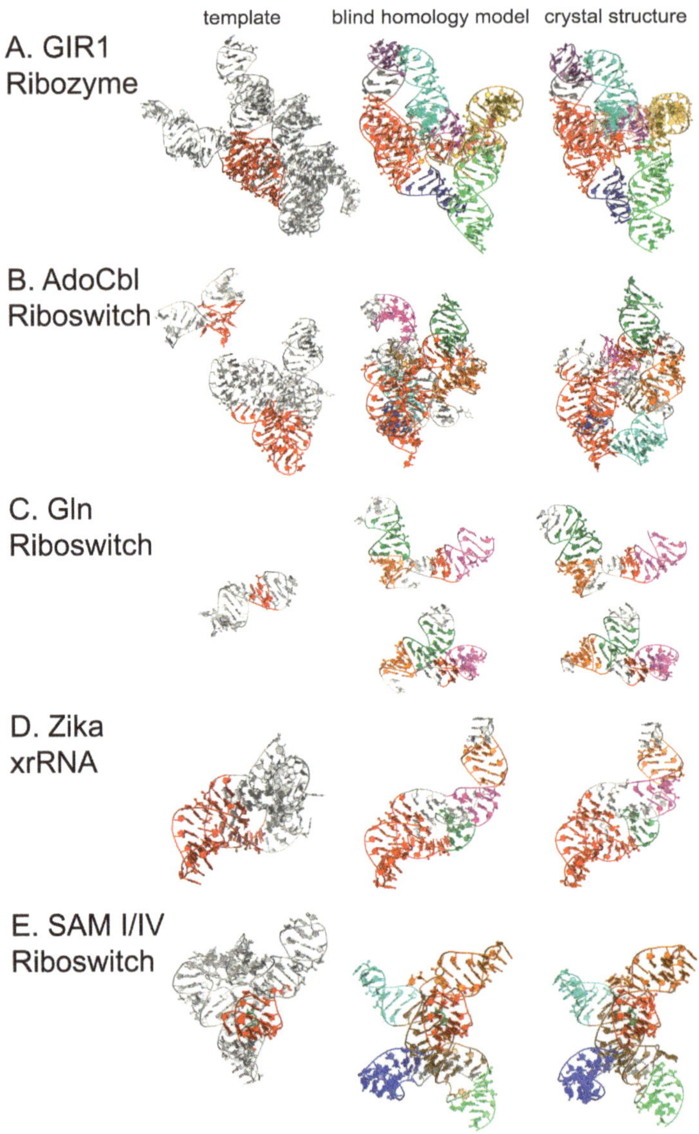

Fig. 1 Prior use of homology modeling in blind challenges using Rosetta algorithms. In RNA Puzzles 5, 6, 14, 18, and 8, previously solved structures provided substantial insight that guided modeling. Each subpanel displays the template structure (left; portion employed in red), the best submitted model (middle), and the eventual crystal structure (right). (A) The GIR1 ribozyme contained a helix arrangement found in the related Azoarcus group I intron. PDB codes: 3BO3 (Lipchock & Strobel, 2008) and 4P8Z (Meyer et al., 2014). (B) The AdoCbl riboswitch contained a T-loop architecture also found in earlier structures of FMN riboswitches, and a kink-turn motif that we obtained

In this review, we describe our approach for homology modeling, from obtaining an appropriate template to the computational strategies for completing missing coordinates, as outlined in Fig. 2. We then illustrate this approach by walking through worked examples for two challenges: first, an artificial challenge modeling an adenine riboswitch based on a well-known guanine riboswitch structure; second, an actual blind challenge (RNA Puzzle 8) where we modeled the SAM I/IV riboswitch. Before describing this Rosetta-focused methodology, we note that other groups have also developed excellent tools outside Rosetta for RNA homology modeling, and we advise the reader to try multiple tools and check for consensus (Piatkowski et al., 2016; Popenda et al., 2012; Xu & Chen, 2018). We also note that current homology modeling methods can be computationally expensive, especially if only a fraction of a target RNA has an identifiable template in a previously solved structure and the rest must be modeled *de novo*. Readers should be aware that computer clusters with ~100 CPUs are therefore required for the methods below. Last, we note that, at this time, computational strategies for RNA structural modeling must be augmented with human insight to identify and best use homology information; several specific examples of the value of expert inspection are described below.

2. Method

Identifying a homologous solved structure can substantially reduce the unknown portion of a structure that must be modeled through computationally expensive algorithms, allowing for the generation of models with higher accuracy with the same amount of computer time. For illustration, in the allied field of protein structural modeling, modeling based on

from the structure of the MBP-L30e-mRNA complex. PDB codes: 2YIE (Vicens, Mondragon, & Batey, 2011); 4GXY (Peselis & Serganov, 2012); and 1T0K (Chao & Williamson, 2004). (C) Both bound (top) and unbound (bottom) structures of the glutamine riboswitch contain a sarcin-ricin loop. PDB codes: 1Q9A (Correll, Beneken, Plantinga, Lubbers, & Chen, 2003); 5DDO (Ren et al., 2015); and 5DDP (Ren et al., 2015). (D) The Zika xrRNA features substantial homology to Murray Valley Encephalitis xrRNA, but the presence of a pseudoknot not found in the original crystal changes the overall fold architecture. PDB codes: 4PQV (Chapman et al., 2014) and 5TPY (Akiyama et al., 2016). (E) The SAM I/IV riboswitch, later discussed as a "worked example" here, was RNA Puzzle 8; in the original blind modeling, we used a template region around the SAM binding site. PDB codes: 2GIS (Montange & Batey, 2006) and 4OQU (Trausch et al., 2014).

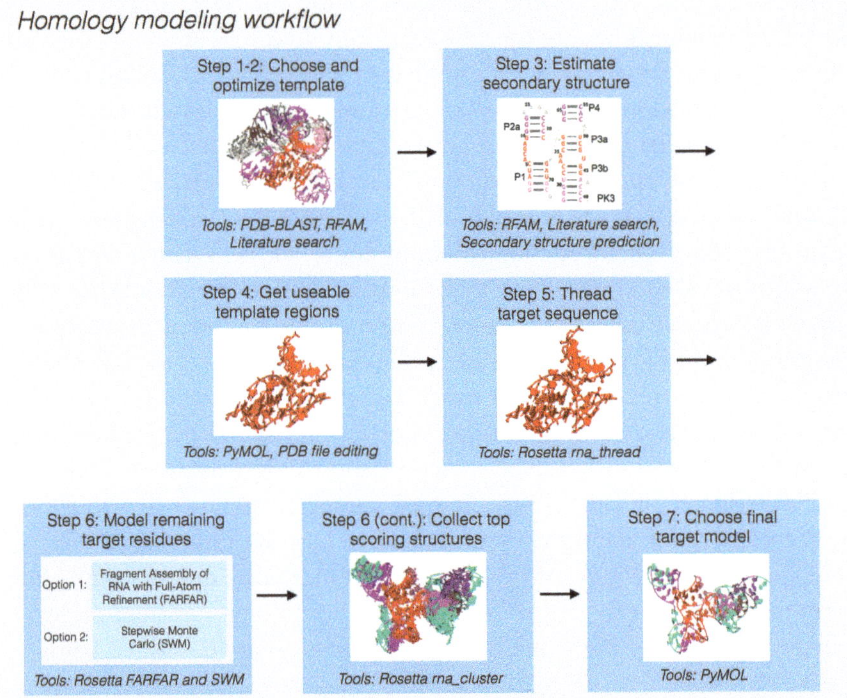

Fig. 2 Workflow for homology modeling with the Rosetta framework. Preparing homologous regions requires selecting an appropriate template structure and refining that structure within the Rosetta score function using ERRASER-Phenix. Here, the template is colored at helical regions (shades of purple and pink) and in an elaborately folded ligand-binding core (red) that serves to seed modeling runs. The secondary structure of the target sequence should be estimated. Useable template regions should be selected that lack experimental artifacts and are sufficiently similar to the modeled structure. The desired sequence is then threaded onto the template structure. Finally, remaining regions of the structure are modeled *de novo* in the context of this threaded structure using modeling with either FARFAR or SWM. New non-template coordinates are shown in teal.

templates has yielded substantially more accurate models than *de novo* modeling, even for proteins with limited similarity to existing structures (Moult, Fidelis, Kryshtafovych, Schwede, & Tramontano, 2018). However, after identifying a solved structure with substantial homology, modeling a new RNA molecule remains challenging. Judgments must be made as to which portions of the template should be used for the new structure, and significant portions of the structure may remain incomplete, requiring *de novo* modeling. Indeed, sequence discrepancies between the modeled structure and template in non-helical regions can critically impact the global fold. In this section, we provide guidelines to help address these challenges, describing a series of steps for homology modeling from selecting a template structure to retrieving final coordinates.

2.1 Selecting candidate template structures

Homology modeling begins with identifying a candidate template structure. Often the target molecule is already known to fall into a known ncRNA family, and a literature search can provide an expert sequence alignment for that family based on manual curation. This scenario is typically the case for new riboswitches, which are indeed often identified on the basis of such sequence alignments and careful, manual detective work leveraging expert knowledge of biochemistry and molecular biology (Weinberg et al., 2017). If the function of the target molecule is known, a simple PDB search (https://www.rcsb.org) for the name of the riboswitch class, e.g., "guanine riboswitch," will often uncover templates with previously solved structure.

If homologies of the target molecule to previously solved structures are not immediately apparent, an automated back-up method is to use the PDB's advanced search, which includes an option to BLAST a target sequence. If any structures result, the PDB will automatically identify an approximate sequence alignment. If a PDB-BLAST fails, the Rfam database can carry out a sophisticated search of an input sequence against its large archive of ncRNA families and return a sequence alignment. The Rfam alignment can also help identify additional members of the same family from the PDB.

Sometimes these homology searches result in multiple possible template structures for the target. Although multiple templates may be used to seed parallel modeling runs, it is still a good idea to choose one template, or at least a subset of those available to optimize use of computational time. The two most important parameters in making this choice are the resolution of the structure in question and the sequence similarity to the target. As a rule of thumb, as long as the resolution is sufficient to place atomic coordinates, it should be a secondary concern; sequence similarity is more important in choosing between possible templates.

Finally, we note that template structures do not have to be large or complex to substantially improve or accelerate modeling. Many RNA subsequences form structural *modules* which form extremely similar folds, varying little with their structural context (Miao & Westhof, 2017; Westhof, Masquida, & Jossinet, 2011). As a result, small sub-structures of a handful of nucleotides, which would never appear in any homology search, may still be of use. By far the most valuable such sub-structures to identify are tertiary contacts, because they can anchor the relative orientation of local structures that are distal in the RNA sequence and secondary structure. Two common tertiary contact modules are the intercalated T-loop (Chan, Chetnani, & Mondragon, 2013) and the tetraloop/receptor interaction (Jaeger, Michel, & Westhof, 1994; Wu, Chai, Fraser, & Zimmerly, 2012). If sequences compatible with either of these interactions may be found, it is likely worthwhile to attempt some modeling runs with these interactions seeded in, using a T-loop containing PDB (such as a tRNA structure) or a tetraloop/receptor (such as the *Tetrahymena* ribozyme's P4–P6 domain) as a miniature template. Packages like RMdetect and FR3D can discover some simple RNA motifs automatically, including kink turns (Cruz & Westhof, 2011; Sarver, Zirbel, Stombaugh, Mokdad, & Leontis, 2008), although we typically still recommend inspecting sequence-conserved regions of the predicted secondary structure of target molecules for evidence of tertiary motifs like T-loops and tetraloop receptors, which are not well-captured by any currently available automated program.

2.2 Optimizing the templates

Due to the challenges associated with manually fitting atomic coordinates, especially with the medium resolution (2.5–3.5 Å diffraction resolution) often obtained for RNA structures, experimental structures almost always contain flaws and geometric errors that can impede accurate modeling. These errors are particularly pervasive for coordinates deposited before 2014, after which the PDB began its new deposition and validation system (Chou, Echols, Terwilliger, & Das, 2016; Chou, Sripakdeevong, Dibrov, Hermann, & Das, 2013; Keating & Pyle, 2012; Read et al., 2011; Wang et al., 2008). Sometimes these flaws are innocuous, but other times they can change functional conclusions, and it is difficult to identify what will happen *a priori*. While the PDB keeps records of MolProbity analyses to catalog clashes and geometrical outliers (Chen et al., 2010), rather than merely selecting the least flawed option, there are facilities to eliminate geometrical outliers while maintaining minimal deviations from the parent structure. These include ERRASER-Phenix (Chou et al., 2013, 2016), RNABC (Wang et al., 2008), and RCrane (Keating & Pyle, 2012). For the Rosetta homology modeling methods described herein, we recommend application of ERRASER-Phenix as at least a final step, because ERRASER can refine the structure into the same force field as used in subsequent Rosetta modeling.

2.3 Estimating target secondary structure

To choose relevant regions of the template structure and to complete 3D modeling of new regions of the target structure, the target's secondary structure provides important, often critical, constraints. For regions of the target molecule homologous to a template of known structure, the secondary structure can be inferred through homology. For other parts, various tools are available for predicting the secondary structure of an RNA from its sequence. The most powerful source of secondary structure information are expert analyses in the literature, which typically integrates all available biochemical, evolutionary, and prior structure information about the target. If experimental facilities are available, multidimensional chemical mapping experiments offer reliable RNA secondary structures (Tian & Das, 2016; Cheng, Kladwang, Yesselman, & Das, 2017); it is worth noting here that SHAPE and DMS-guided approaches are simpler but have been less reliable (Kladwang, VanLang, Cordero, & Das, 2011; Tian & Das, 2016; Miao et al., 2017, 2015). If an Rfam family or an expert-curated sequence alignment exists for the target sequence, the secondary structure constructed from a sequence alignment by tools like Infernal (Nawrocki & Eddy, 2013) will be the most reliable estimate for the target secondary structure. If an Rfam family or extensive sequence alignment is not available, tools for secondary structure prediction including RNAstructure (Reuter & Mathews, 2010), NUPACK (Dirks & Pierce, 2003), ViennaRNA (Lorenz et al., 2011), and CONTRAfold (Do, Woods, & Batzoglou, 2006) are appropriate although none reliably return information on pseudoknots. All these tools can be used to generate secondary structures for the sequence at hand, and the resulting predictions should be compared to find consensus helices.

2.4 Identifying useable portions of the template

Once a template structure is optimized and the target's secondary structure is predicted, sections of the template that will be used for further modeling should be identified. One possible template inaccuracy is the difference between crystallization and biologically relevant conditions. Regions of the template with crystal artifacts should be excluded in further modeling.

For instance, in our homology modeling of the Zika xrRNA (Fig. 1D) (Watkins et al., 2018), we identified a previously solved structure of the Murray Valley Encephalitis xrRNA, but it crystallized as a dimer, replacing an internal pseudoknot with an intermolecular contact, meaning that despite high sequence similarity, almost half of the template structure had to be discarded.

The sequence match to the homologous structure will be the primary determinant for which remaining regions of the template structure to keep. In addition, any sections in which the new sequence's assumed secondary structure disagrees with the template structure should not be included, as illustrated by the Zika xrRNA case described above. Any non-helical regions should be left out if replacing with the new sequence would break hydrogen bonding patterns. For example, if a GCAA tetraloop in the template is replaced with a UUCG tetraloop, it is better to model this tetraloop *de novo* or to find a template from another structure, for instance by searching the RNA 3D motif atlas (Petrov, Zirbel, & Leontis, 2013). With these exclusions, some regions of the structure may no longer be connected by conserved tertiary contacts or junctions; an example occurred in the adenosylcobalamin riboswitch where a kink turn and a complex T-loop/pseudoknot motif were drawn from two unconnected templates (separate red regions in Fig. 1B). Separated regions can be included as distinct rigid bodies in later modeling steps and allowed to reorient with respect to each other as intervening loops and helices are remodeled.

In deciding on useable portions of the template, it is helpful to additionally consider all available data on the sequence to be modeled. Biochemical data such as mutational analyses may point to divergence from the homologous structure. If a sequence alignment is available, covariance analysis can suggest new contacts or delineate regions in which the template is not applicable. Last, chemical mapping data, particularly new multidimensional experiments (Tian & Das, 2016), can be informative in detecting or confirming homologies of the target molecule to previously solved structures. As noted above for secondary structure, literature analyses of the RNA class, if available, remain the most valuable source for template identification.

2.5 Threading the target sequence onto template structure

Even for regions of the template that can be retained for the modeling task, the target sequence being modeled may deviate from the template structure's sequence. For example, an AU base pair in the template structure may be replaced with a GC base pair in the target molecule. As a first step, the target sequence must be threaded onto the template, such that the RNA backbone remains constant while the corresponding nucleotides change to the target sequence. If the sequence alignment is ambiguous, this step and further steps should be applied to each reasonable threading—or correspondence—of the modeled sequence to the template. To obtain the threaded structure, one can use the `rna_thread` Rosetta application (see Appendix for example command-line).

Depending on the sequence changes being made, small-scale optimization of the threaded structure may be necessary prior to further modeling. If all the changes are in helix sequences, then the local geometry is likely agnostic to threading. In particular, helix structure will be unaltered when the sequence changes between Watson Crick pairs. For helix sequence changes that convert Watson-Crick pairs to G-U wobble pairs and *vice versa*, the user may wish to leave those base pairs or the entire helix out of the template so that they are rebuilt in the next step.

2.6 Modeling the target structure

Once the threaded template is optimized, the task remaining is to model the regions of the structure that lack homology to the template; doing so will set the relative orientation of any templated sections and provide 3D coordinates for every atom in the target RNA. As noted in the Introduction, two algorithms are available in Rosetta for modeling RNA folds: a medium-resolution method (Fragment Assembly of RNA with Full-Atom Refinement, or FARFAR) (Das et al., 2010) and a method that can achieve higher resolution but at high computational expense (stepwise Monte Carlo, or SWM) (Watkins et al., 2018). Detailed command-lines for these two approaches will be presented when discussing the worked examples.

Choosing between these algorithms involves three considerations:

a. The size of the target complex, and of the region that requires new modeling.

If high resolution is desired, SWM is the method of choice, but involves significantly more computation per model than FARFAR. Large structures and extensive remodeling challenges are possible but computationally expensive with SWM. As a rule of thumb, SWM modeling scales with the number of nucleotides or fixed helices that need to be rebuilt. Each such element requires 50 cycles of SWM simulation (typically a few minutes) to solve, up to problems with around

20 elements (several hours) per model. For example, if a problem involves three helices of known structure, interconnected by three loops with a total of 10 nucleotides, 650 cycles will be required to complete SWM models, corresponding to about an hour of simulation time per model. Ideally, hundreds of models need to be created, so even with a computer cluster with 100 CPUs, such a SWM modeling problem will require overnight runs. Challenges of more than 30 residues are better approached using FARFAR, as the lower resolution achieved will be offset by the greater number of completed trajectories.

b. The number of helices with length changes between template and target.

FARFAR has an efficient method for accurately recapitulating realistic helix flexibility; SWM samples one residue at a time and has an inefficient method for adding base-paired residues. Therefore, if there are major helix length changes between template and target, FARFAR is currently recommended.

c. The presence of multiway junctions and tertiary contacts.

Multiway junctions or tertiary contacts often form the structural core of complex RNA folds and can feature intricate combinations of non-Watson-Crick base pairs whose high-resolution geometry is essential to an accurate global fold. Ideally, as many as possible of these interactions may be found in the template structure. Otherwise, these features should be modeled using SWM. If there are many such complex interactions in a fold, one may decompose the structure into independent SWM jobs, and subsequently assemble successful sub-problem solutions using FARFAR or SWM.

A combination of FARFAR and SWM can be helpful for achieving high accuracy models. Initial FARFAR modeling may provide hypotheses for regions of the global structure that should be proximal; these proximal helices and junction residues can be modeled in more detail in separate smaller SWM runs. Alternatively, if a multiway junction or tertiary contact is essential for achieving the correct global fold, it is feasible to first generate models for this region with SWM, and then seed these elements into a FARFAR modeling round as rigid bodies to generate more accurate global models.

During modeling runs, a choice needs to be made for whether to allow regions drawn from templates to be optimized away from their starting structures. Usually, it is best to keep these regions fixed so as to preserve the structural information from the starting template; however, we typically allow for minimization of nucleotides near the edges of these regions to allow for small changes that can propagate and "relax" newly built regions into more energetically favored configurations (see Appendix for example command-lines and explanations). In some cases, including the walk-through examples below, we also carry out pre-optimization of template segments through the `rna_minimize` command; this step automatically uses coordinate constraints to the structure's initial coordinates for its first round of minimization, ensuring that any large clashes are relieved without major global changes to the RNA structure (see Appendix for example command-lines).

2.7 Model selection

For both FARFAR and SWM algorithms, hundreds to thousands of independent trajectories that stochastically build the new regions are carried out. With either sampling strategy, the presence of numerous independently built models with low RMSD to the top-scoring model suggests sufficient sampling. A plot of Rosetta score versus the RMSD to the top scoring model can be used to quickly assess the conformational space accessed over all modeling runs. For any given Rosetta score cutoff (say, a cutoff for the top 10 models), observing multiple structures close in RMSD to the top structure indicates support for the best scoring model. In addition to score v. RMSD plots, inspection of the top 10 or 20 models in 3D molecular viewers like Pymol allows for visual assessment of whether any subset of these models have converged to the same global fold, a hallmark of convergence of the modeling and typically a good sign of accuracy (Cheng, Chou, Kladwang, et al., 2015; Kappel et al., 2018; Shortle, Simons, & Baker, 1998). The resulting collection of top-scoring models are then the candidates for the predicted structure. More quantitatively, the mutual RMSD between the top scoring models can help assess convergence to the native structure, with lower mutual RMSD indicating more accurate models (Kappel et al., 2018).

3. Worked examples

This section briefly describes two worked examples of Rosetta RNA homology modeling, following the steps outlined above and in Fig. 2. For both examples, we test both alternatives for the *de novo* modeling step (SWM vs. FARFAR) to illustrate the rationale and outcomes involved in that choice. Readers wishing to work through these examples may also be interested in command-lines and a publicly available repository of computer files; see Appendix.

3.1 The adenine riboswitch

Riboswitches fold to form selective, tight pockets to bind their target ligands. A classic example is the purine riboswitch (Fig. 3), where a Watson-Crick base pair between the ligand and nucleotide 74, which is always a pyrimidine, defines the binding selectivity of the riboswitch. As a first example of Rosetta homology modeling, we model the *V. vulnificus* adenine riboswitch structure, comparing modeling using the FARFAR and SWM approaches.

The first steps in the method are to select and optimize an appropriate template structure. For this test, we select the *B. subtilis* guanine riboswitch (PDB ID 1Y27) (Serganov et al., 2004) and we run the structure through the ERRASER-Phenix pipeline. In the third step, we estimate the secondary structure for the target adenine riboswitch, which has only small deviations from the template's secondary structure (Fig. 3). In the fourth step of the method, we identify homologous regions in the template and target to determine which regions of the template structure to use. Here, there are sequence changes in RNA helices, and while these can affect the ensemble of favored structures for each helix (Yesselman et al., 2018), those changes are typically overwhelmed by a well-conserved tertiary context, as we see here. We excise the residues whose mutations fall in loops or that break or form Watson-Crick pairs (as in P2), marked as mismatches in Fig. 3. In the fifth step, we thread the target sequence onto the remaining backbone.

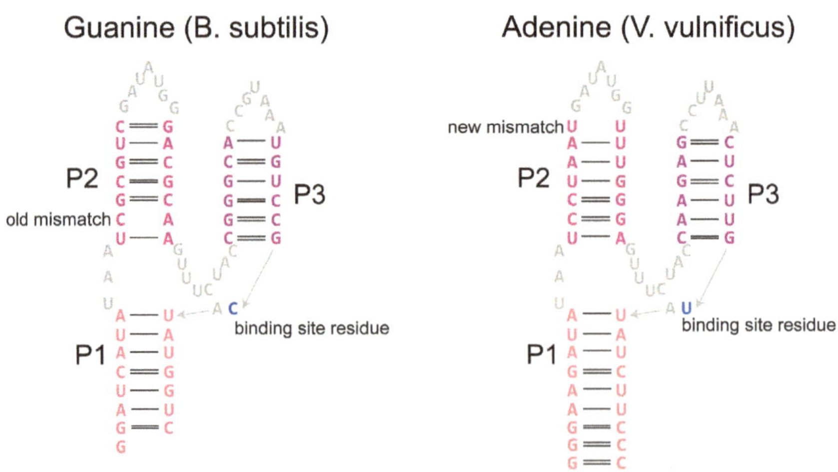

Fig. 3 The secondary structure and sequence of two purine riboswitches are highly similar, with sequence deviations outside of helices marked in blue.

Now we can move to the sixth step of the method, as outlined in Fig. 2: modeling the remaining residues in the target structure. We compare two simulation approaches for completing the modeling. We perform a simulation with SWM to build in the mutated residues, followed by a simulation with FARFAR just to extend the P1 helix. In the second approach, we perform a simulation where we model in both the mutated residues and the P1 helix extension with FARFAR. For the ligand in each simulation case, we place a new adenine in a perfect Watson-Crick orientation with U74, mapped from a C in the template guanine riboswitch structure. (While SWM can explicitly remodel ligand binding modes, this may not be necessary for an initial simulation or for a simple binding mode like a single Watson-Crick base pair.) For each type of simulation, we observed convergence by comparing the top 10 models (as ranked by Rosetta energy), and we chose the top model as our best structure (Fig. 4). We note that in

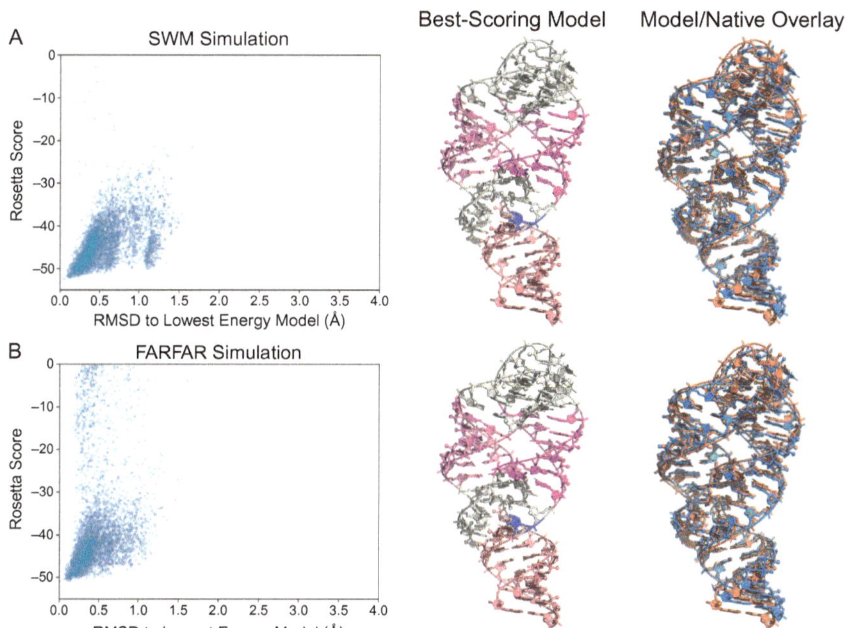

Fig. 4 On the left, plots comparing model score to the RMSD of the lowest energy model for SWM simulations (A) and FARFAR simulations (B) show convergence to the lowest energy structure. In the middle column are the lowest energy models from the SWM (top) and FARFAR (bottom) simulations. In the right column, the top scoring structures (salmon) are overlaid on the crystal structure of the target (blue, PDB ID: 4TZY) (Zhang & Ferre-D'Amare, 2014), which was not used in homology modeling.

both cases, the mutual RMSD between the top 10 structures is <0.5 Å, indicating strong convergence and suggesting that the lowest energy model will likely be close to the actual structure of this adenine riboswitch. In the seventh step, we therefore choose the lowest energy model from each run as the final model.

In this case, the experimental structure of the adenine riboswitch is known, and so the accuracy of the Rosetta homology models can be assessed. Compared to a crystal structure of this RNA (PDB ID: 4TZY) (Zhang & Ferre-D'Amare, 2014), the numerical values of the RMSD of the homology model to the experimental structure are around 1.4 Å (Fig. 4, Table 1). This RMSD is not quite atomic accuracy (which would be sub-Angstrom) and is greater than the deviation between crystal structures of the same molecule solved by independent laboratories (e.g., <0.5 Å RMSD between the adenine riboswitch structures 4TZY (Zhang & Ferre-D'Amare, 2014) and 1Y26 (Serganov et al., 2004)). Nevertheless, the modeling confirms that the guanine and adenine riboswitch aptamers can adopt the same fold around distinct purines, and the deviation is much smaller than the length scale of nucleotides (~6 Å from one nucleotide to the next). Indeed, not much flexibility had to be modeled; there was little chance with either SWM or FARFAR of the riboswitches adopting a fold with substantial deviations. Remodeling the binding site across similar binding modes is a task to which either SWM or FARFAR is comfortably amenable.

3.2 SAM I/IV riboswitch

The prior example was an illustration of what can be done for a ligand binding RNA with a well-conserved global architecture with small sequence changes in helices, loops, and the ligand binding site that contribute to selectivity. Modeling within a single family of folds, as above, is a very different scale of challenge from modeling one fold family based on another, which

Table 1 Accuracy achieved in 200 CPU-hour simulations, using either SWM for mutations and FARFAR for helix extension or using exclusively FARFAR, achieve high-resolution models of the adenine riboswitch. RMSD (in Å) is to the actual crystal structure of the adenine riboswitch (PDB ID: 4TZY) (Zhang & Ferre-D'Amare, 2014).

Simulation	RMSD (low-E)	RMSD (best of 5)	Best RMSD	# structures
SWM	1.485	1.477	1.404	5000
FARFAR	1.479	1.478	1.393	4960

can involve global "rewiring" of the strands connecting otherwise homologous core structural elements. We illustrate the case of cross-family modeling with the SAM I/IV riboswitch.

As the first step of homology modeling, we must identify a candidate template structure. The SAM I/IV riboswitch was proposed as an RNA-puzzle (Miao et al., 2017) at a time when multiple structures of the SAM I riboswitch had already been solved by crystallography, but no riboswitches of the SAM IV or SAM I/IV class had been solved (Schroeder, Daldrop, & Lilley, 2011). The most conserved region of the SAM I/IV riboswitch appeared highly homologous to the SAM I riboswitch, and the literature proposed homologies at the family level (Weinberg et al., 2008) (Fig. 5). Taking the next two steps, we optimize the structure for SAM I riboswitch (2YGH) (Schroeder et al., 2011) with ERRASER-Phenix, and we draw out the secondary structure of SAM I/IV (Fig. 5). Then, we isolate template regions homologous to the target structure. Specifically, the red region of Fig. 5, which contains the ligand binding site as well as three of the four helices comprising the core junctional architecture of the riboswitch, is highly

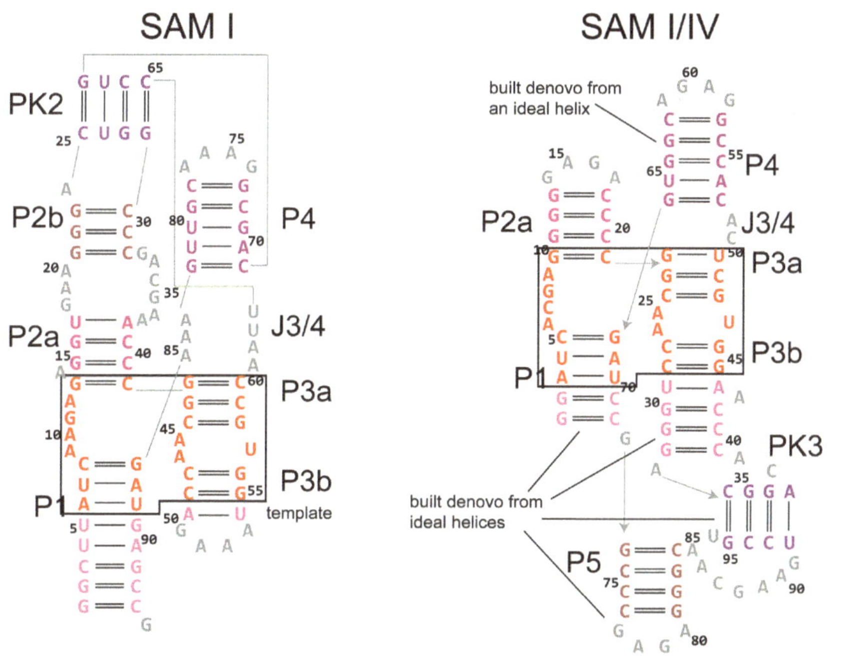

Fig. 5 The secondary structure and sequence of SAM I and SAM I/IV riboswitches are highly dissimilar, except for a well-conserved ligand binding site (red).

conserved. To model the SAM I/IV riboswitch "puzzle," we take this conserved binding site exactly, excising it from the SAM I riboswitch structure and threading on the target sequence, the fifth step of our homology modeling procedure (Fig. 2).

For the next step, we need to plan out how to model new regions of the target SAM I/V riboswitch *de novo* that are not shared with the SAM I template. The two families appear quite different outside the core (Fig. 5). Compared to the template SAM I riboswitch, in the SAM I/IV riboswitch there is a 16-nt deletion around P2, a 10-nt insertion in P3, the J3/4 loop truncated by 6 nt, and P1 is shortened by 6 nt but with an additional 4 bp helix P5 added (Fig. 5). We first summarize the regions that remain for *de novo* modeling after using the SAM I binding site homology, and we then outline the specific commands and Rosetta utilities that can be employed to carry out the modeling.

Compared to the SAM I structure, in the SAM I/IV riboswitch, P2a needs be truncated and then a GNRA tetraloop needs to be added to the remaining four Watson-Crick base pairs; this addition can be completed with FARFAR or SWM modeling. For a change in J3/4, the problem amounts to solving for the best orientation of two nucleotides (the CA loop). P4 is a 5 bp helix capped with a GNRA tetraloop and thus involves well-known RNA structures. That segment could be threaded from P4 in the template, but given its conventional structure, here we separately build it and let it be placed relative to the core as a fixed body in the next stage of *de novo* sampling of J3/4. These transformations occur on the "top" side of the template (Fig. 5) and are strictly independent of the other required modifications, described next.

On the "bottom" side of the template are the P3b and P5 modifications, which must be considered jointly and are the bulk of the challenge. P3b is extended by multiple base pairs and terminates in a loop implicated in a pseudoknot (PK3). At the same time, P5 is appended to the 3′ end of P1, terminating in a seven-residue loop, four bases in PK3, and a final nucleotide. Solving the joint fold of L3b, PK3, P5, and the 3′ loop is the major challenge.

We may solve these two challenges using either stepwise Monte Carlo or FARFAR. For the user, the requirements for running each method are broadly similar, including specification of a FASTA file (see Appendix) and a PDB file of input RNA coordinates representing the template region, renumbered to correspond to the SAM I/IV system (sequence in Appendix). Additionally, a FARFAR run requires the specification of an explicit secondary structure in "dot-bracket" notation. In FARFAR, the secondary structure may be used to impose energetic restraints to encourage base pair formation, or to guide the generation of fixed helical inputs. In contrast,

SWM currently requires the additional provision of fixed helical input structures. The secondary structure files and fixed helical inputs for use in FARFAR and SWM modeling are included in the Appendix.

With the sequence, secondary structure, and initial structure files on hand, we run simulations using two approaches, each with either FARFAR or SWM. In the first approach, we conduct a nearly naïve simulation, starting from the correct binding site but providing the rest of the structure simply as helices. In the second approach, we conduct strategic simulations which subdivided the problem into separate, more tractable modeling challenges—we first truncate P2a and complete the tetraloops capping P2a, P4, and P5. We then include the best-scoring models made for each of these sub-problems as rigid bodies in a subsequent simulation that tackles the P3b and P5 modifications. These strategic simulations reduce the scope of the problem from a problem that is very large for the SWM method (27 nucleotides to be built, seven input pieces of RNA) to something more manageable (15 nucleotides, 6 input pieces of RNA). We expected that the latter simulations would perform better because they can focus more computational power for the harder parts of the problem.

We used equal computational time for each case. The complexity of the naïve SWM challenge required simulations of five thousand Monte Carlo cycles. These simulations required on average 88,491 s, or almost 25 h, for each generated structure. Accordingly, we limited ourselves to 6000 CPU-hours to the other simulation conditions as well (Table 2). FARFAR completed an order of magnitude more total trajectories in the same amount of CPU time than SWM, for either style of problem specification. The top 10 structures from each simulation showed adequate convergence, with numerous independently modeled structures within 4 Å of the top scoring model, and with the top 10 structures having mutual RMSD <4 Å for the FARFAR simulations. As expected from this convergence level (Kappel et al., 2018), the models were also accurate compared to the actual

Table 2 6000 CPU-hour simulations with diverse starting assumptions achieve high-resolution models of the SAM I/IV riboswitch.

Simulation	RMSD (low-E)	RMSD (best of 5)	Best RMSD	# structures
SWM_naïve	10.914	3.137	3.137	222
FARFAR_naïve	5.050	4.344	3.495	12,966
SWM_strategic	4.077	3.932	2.931	691
FARFAR_strategic	5.379	4.195	3.106	17,755

experimental structure of the SAM I/IV riboswitch aptamer (PDB ID: 4L81) (Trausch et al., 2014), with the correct global fold of the RNA even in the newly modeled regions outside the ligand-binding core (Fig. 6). The best RMSD to native for the top five lowest scoring structures fell under 5 Å (Fig. 6). The RMSDs of the lowest energy models to the native structure

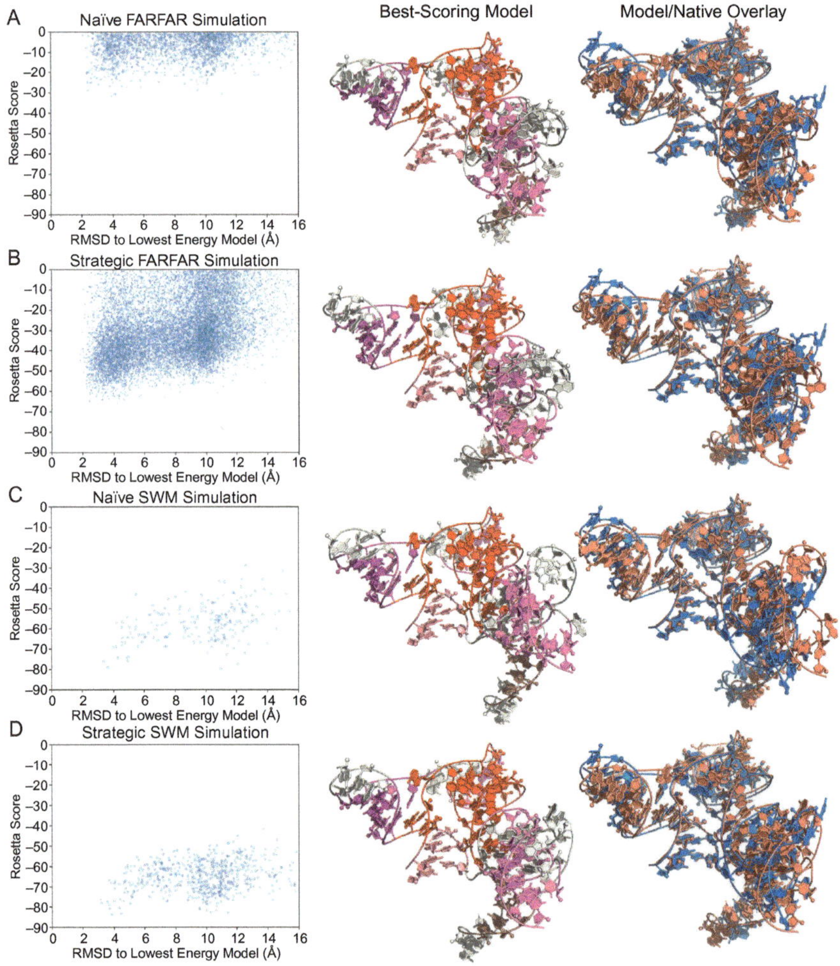

Fig. 6 Rosetta score in the default rna_res_level_energy4.wts scoring function, plotted against heavyatom RMSD to the lowest energy generated structure, comparing (A) SWM_naive, (B) FARFAR_naive, (C) SWM_strategic, and (D) FARFAR_strategic (blue). Top scoring structures for each simulation setting are depicted in the middle column with coloring corresponding to the secondary structure diagram (Fig. 4). In the right column, the top scoring structures (salmon) are overlaid on the crystal structure of the target (PDB ID: 4L81) (Trausch et al., 2014) which was not used in homology modeling (blue).

suggest that both SWM and FARFAR can achieve acceptable convergence and acceptable accuracy on this challenge, with FARFAR's ability to complete more trajectories making up for any shortcomings relative to the "high-resolution" SWM method.

4. Summary

Rosetta provides useful tools for RNA homology modeling, and their performance is well-suited to common challenges arising in current RNA structural biology. In the future, more of the steps discussed above for effective homology modeling will be automated. Very large or complex structures, for which homologous regions are limited or for which sequence/structure similarity to prior solved structures is hard to parse, currently remain inaccessible. With the acceleration of experimental methods such as cryo-EM and multidimensional chemical mapping, additional data may be employed to further accelerate these particularly challenging cases (Kappel et al., 2018; Tian & Das, 2016).

Appendix

Installation and setup

Documentation for installing Rosetta can be found here: https://www.rosettacommons.org/demos/latest/tutorials/install_build/install_build

After installing Rosetta, to install the python scripts distributed with Rosetta as `rna_tools`, execute the shell script `Rosetta/tools/rna_tools/INSTALL` (or add it to `.bashrc`).

In the next two Appendix sections, we will include command lines used for homology modeling simulations of the adenine riboswitch and SAM I/IV riboswitch. The command lines were run with Rosetta 3.10.

For each simulation step, we have included input files, output files, and command lines in the Github repository here:

https://github.com/everyday847/rosetta_rna_homology_modeling_examples

Note that when running the command lines in the following sections, it might be necessary to append "`.macosclangrelease`" or "`.linuxgccrelease`" to each executable name. Typing the executable as written below (for instance, `rna_thread`) and tab completing will show the name of the executable on your machine upon installation.

Adenine riboswitch simulation files and command lines

- The template structure used is the guanine riboswitch PDB structure 1Y27 (Serganov et al., 2004). For simplicity, we remove the ligand, water molecules, and metal ions.
- Threading the adenine riboswitch sequence onto the guanine riboswitch template. Template structures are passed in with the `-s` flag, and `-seq` indicates the sequence to place onto the template:

```
rna_thread -s 1y27_start_culled.pdb -seq ggaagauauaauccuaauga
uaugguuugggaguuucuaccaagagccuuaaacucuugauuaucuuc
```

- Minimizing the threaded template to relieve any clashes. The `-score:weights` flag here specifies the current best performing RNA scoring function, `rna_res_level_energy4.wts`. The `-restore_talaris_behavior` flag ensures that the score function exactly reproduces the default settings from when the weights were first optimized.

```
rna_minimize -s threaded.pdb -score:weights
stepwise/rna/rna_res_level_energy4.wts -
restore_talaris_behavior
```

- Since the threading application by default changes residue numbering to chain A, numbered sequentially from 1, at this point we restore standard purine riboswitch numbering. In this case, since the target structure has already been deposited with a particular numbering scheme, we can further confirm that our numbering matches up.

```
renumber_pdb_in_place.py threaded_minimize.pdb X:14-81
```

- Mismatched residues, including part of the P1 stem, are deleted from `threaded_minimize.pdb` via text editor: specifically, residues X:14–15, X:26, X:31, X:39, X:44, X:62, X:74, and X:81 are removed.
- FASTA file which defines the sequence for the complete adenine riboswitch system to be modeled, which includes a longer P1 stem than is present in the template. Note here that this FASTA file sequencing numbering must be consistent with the threaded template structure numbering after the steps above.

```
>4tzy_target.pdb X:13-83
gggaagauauaauccuaaugauaugguuugggaguuucuaccaagagccuuaaacucuug
auuaucuuccc
```

- Running FARFAR simulation for building remaining residues, left out from the template. The flags used here are as follows. The `-s` flag specifies the starting structure, which here is the threaded template structure. The `-native` flag specifies the native structure, only used for scoring the models for later RMSD analysis. The -minimize_rna flag indicates that the modeling will include energy minimization steps. As before, the -score:weights flag specifies the current best performing RNA scoring function, and the -restore_talaris_behavior flag includes global scoring corrections necessary for that score function. With the -extra_minimize_res flag, we can allow residues in the template structure that are near *de novo* modeled residues to resample and minimize their energies. Finally, -use_legacy_job_distributor is necessary for cluster execution using the rosetta_submit.py script, as described in the Github repository.

```
rna_denovo -s threaded_minimize_culled.pdb -native
4tzy_target.pdb -fasta target.fasta -minimize_rna true -
nstruct 50 -score:weights
stepwise/rna/rna_res_level_energy4.wts -
restore_talaris_behavior -use_legacy_job_distributor -
extra_minimize_res X:25 X:27 X:30 X:32 X:38 X:40 X:42 X:45
X:61 X:63 X:73 X:75 -out:file:silent farna_rebuild.out
```

- Running SWM simulation. The flags here are analogous to those in the FARFAR simulation above, with the addition of the -cycles flag which indicates the number of stepwise Monte Carlo cycles to perform.

```
stepwise -s threaded_minimize_culled.pdb -native
4tzy_target.pdb -fasta target.fasta -cycles 50 -nstruct 5 -
score:weights stepwise/rna/rna_res_level_energy4.wts -
restore_talaris_behavior
-use_legacy_stepwise_job_distributor -extra_min_res X:25
X:27 X:30 X:32 X:38 X:40 X:42 X:45 X:61 X:63 X:73 X:75 -
out:file:silent swm_rebuild.out
```

Each of the resulting 500 models are extracted to individual PDB files using extract_lowscore_decoys.py swm_rebuild.out 500 and subsequently finished (the P1 helix is extended) by FARFAR, generating 10 models each:

```
rna_denovo -s swm_rebuild.out.1.pdb -native 4tzy_target.pdb
-fasta target.fasta -minimize_rna true -nstruct 10 -
score:weights stepwise/rna/rna_res_level_energy4.wts -
restore_talaris_behavior -use_legacy_job_distributor -
extra_minimize_res X:25 X:27 X:30 X:32 X:38 X:40 X:42 X:45
X:61 X:63 X:73 X:75 -out:file:silent
threaded_minimize_stepwise_finished_by_farfar.out
```

- Retrieving top models. The `extract_lowscore_decoys` utility can be used to retrieve the top ten scoring models from each simulation:

```
extract_lowscore_decoys.py
threaded_minimize_stepwise_finished_by_farfar.out 10
extract_lowscore_decoys.py farna_rebuild.out 10
```

SAM I/IV simulation files and command lines

Many of the flags used in the command lines below are explained in more detail above in the adenine riboswitch appendix.

- The template structure for this example is the SAM I riboswitch structure (PDB ID: 2YGH) (Schroeder et al., 2011). For both FARFAR and SWM simulations, we input as a PDB file the section of the template that is homologous to the target, threaded with the target sequence. Below is the FASTA file corresponding to the inputted homology structure. As described above, this PDB file must be renumbered; for this, we use the standard numbering for the SAM I/IV riboswitch family. Command lines for these steps are analogous to the adenine riboswitch example above.

```
>start.pdb A:1-10 A:21-28 A:45-50 A:67-71
ggaucacgagcggcaaccggugcugaucc
```

- FASTA file for complete SAM I/IV riboswitch system to be modeled:

```
>sam_I/IV A:1-96
ggaucacgagggggagaccccggcaaccugggacggacacccaaggugcucacaccggag
acgguggauccggcccgagagggcaacgaaguccgu
```

- PDB files for non-template helices. For modeling helices of SAM I/IV that are not taken from the template structure, we can include PDB files corresponding to idealized A-form helices. These PDBs may be generated automatically by passing the corresponding sequence to rna_helix.py and renumbering (example below). Below, we list FASTA files for all the non-template helices used as inputs for FARFAR and SWM modeling.

```
rna_helix.py -o HELIX2.pdb -seq ggg ccc
renumber_pdb_in_place.py HELIX2.pdb A:11-13 A:18-20
```

All FASTA files for the input helix PDB files:

```
>HELIX2.pdb A:11-13 A:18-20
gggccc
>HELIX4.pdb A:29 A:44
ua
```

```
>HELIX5.pdb A:30-32 A:40-42
gggccc
>HELIX6.pdb A:53-57 A:62-66
caccgcggug
>HELIX7.pdb A:73-76 A:81-84
gcccgggc
>HELIX8.pdb A:34-37 A:92-95
cggauccg
```

- Secondary structure file for the SAM I/IV riboswitch in dot-bracket notation:

```
(((((....((((....))))(((..((((((.[[[[..))).))).)))..
(((((....))))))))))).((((....)))).......]]]].
```

Note: FARFAR supports traditional "dot-bracket" notation as well as the use of square, curly, and angular brackets for first through third order pseudoknots, and matching letters for higher order pseudoknots.

- Running naïve FARFAR simulation. In addition to the flags used above in the adenine riboswitch example, we use the flag `-allow_complex_loop_graph` to allow accurate scoring for the pseudoknotted structure being modeled, and we include `-superimpose_over_all` to compute RMSDs by aligning over all residues in this extensive modeling challenge:

```
rna_denovo -s starting_puzzle8_chunk.pdb HELIX2.pdb
HELIX4.pdb HELIX5.pdb HELIX6.pdb HELIX7.pdb HELIX8.pdb
-native 4L81.pdb -fasta target.fasta -save_times
-allow_complex_loop_graph true -superimpose_over_all
-cycles 100000 -nstruct 50 -use_legacy_job_distributor true
-score:weights stepwise/rna/rna_res_level_energy4.wts
-restore_talaris_behavior -minimize_rna true
-out:file:silent farna_rebuild.out
```

- Running naïve SWM simulation. Here and in later command lines, `-motif_mode` ensures that nucleotides on the ends of input structures may be energy-minimized along with newly built residues during simulation:

```
stepwise -s starting_puzzle8_chunk.pdb HELIX2.pdb
HELIX4.pdb HELIX5.pdb HELIX6.pdb HELIX7.pdb HELIX8.pdb -
native 4L81.pdb -fasta target.fasta -save_times -motif_mode
-allow_complex_loop_graph true -superimpose_over_all
```

```
-cycles 5000 -nstruct 5
-use_legacy_stepwise_job_distributor true
-score:weights stepwise/rna/rna_res_level_energy4.wts
-restore_talaris_behavior -out:file:silent swm_rebuild.out
```

- Generating intermediate inputs for strategic simulations:

The template-P2 combination:

```
target_1.fasta:
>starting_puzzle8_chunk.pdb A:1-28 A:45-50 A:67-71
ggaucacgagggggagaccccggcaaccgugcugaucc
stepwise -s starting_puzzle8_chunk.pdb HELIX2.pdb -native
4L81.pdb -fasta target_1.fasta -save_times -motif_mode
-allow_complex_loop_graph true -superimpose_over_all
-cycles 200 -nstruct 5000 -
use_legacy_stepwise_job_distributor true
-score:weights stepwise/rna/rna_res_level_energy4.wts
-restore_talaris_behavior
-out:file:silent swm_rebuild1.out
```

Finishing the P4 tetraloop:

```
rna_denovo -s HELIX6.pdb -working_res A:53-66 -native 4L81.pdb
-fasta target.fasta -save_times -motif_mode
-allow_complex_loop_graph true -superimpose_over_all
-cycles 10000 -nstruct 5000 -use_legacy_job_distributor true
-score:weights stepwise/rna/rna_res_level_energy4.wts
-restore_talaris_behavior -minimize_rna true
-out:file:silent farna_rebuild2.out
```

Finishing the P5 tetraloop:

```
rna_denovo -s HELIX7.pdb -working_res A:73-84 -native 4L81.pdb
-fasta rna_puzzle_8.fasta -save_times -motif_mode
-allow_complex_loop_graph true -superimpose_over_all
-cycles 10000 -nstruct 5000 -use_legacy_job_distributor true
-score:weights stepwise/rna/rna_res_level_energy4.wts
-restore_talaris_behavior -minimize_rna true
-out:file:silent farna_rebuild3.out
```

Extracting the top-scoring model from each simulation to seed the final simulation:

```
extract_lowscore_decoys.py swm_rebuild1.out 1
extract_lowscore_decoys.py farna_rebuild2.out 1
extract_lowscore_decoys.py farna_rebuild3.out 1
```

- Running strategic FARFAR simulation:

```
rna_denovo -s farna_rebuild3.out.1.pdb farna_rebuild2.out.1.
pdb swm_rebuild1.out.1.pdb HELIX4.pdb HELIX5.pdb HELIX8.pdb
-native 4L81.pdb -fasta target.fasta
-secstruct_file secstruct -save_times
-allow_complex_loop_graph true -superimpose_over_all
-cycles 100000 -nstruct 5000 -use_legacy_job_distributor true
-score:weights stepwise/rna/rna_res_level_energy4.wts
-restore_talaris_behavior -minimize_rna true
-out:file:silent farna_rebuild_final.out
```

- Running strategic SWM simulation:

```
stepwise -s farna_rebuild3.out.1.pdb farna_rebuild2.out.1.pdb
swm_rebuild1.out.1.pdb HELIX4.pdb HELIX5.pdb HELIX8.pdb
-native 4L81.pdb -fasta target.fasta -save_times -motif_mode
-allow_complex_loop_graph true
-superimpose_over_all -cycles 2000 -nstruct 5000
-use_legacy_stepwise_job_distributor true -score:weights step-
wise/rna/rna_res_level_energy4.wts
-restore_talaris_behavior -out:file:silent swm_rebuild_final.
out
```

- Obtaining low-energy models. To choose a subset of these models, we can use the following commands:

```
extract_lowscore_decoys.py farna_rebuild.out 10
extract_lowscore_decoys.py swm_rebuild.out 10
extract_lowscore_decoys.py farna_rebuild_final.out 10
extract_lowscore_decoys.py swm_rebuild_final.out 10
```

References

Akiyama, B. M., Laurence, H. M., Massey, A. R., Costantino, D. A., Xie, X., Yang, Y., et al. (2016). Zika virus produces noncoding RNAs using a multi-pseudoknot structure that confounds a cellular exonuclease. *Science*, *354*(6316), 1148–1152. https://doi.org/10.1126/science.aah3963.

Ames, T. D., & Breaker, R. R. (2014). Bacterial aptamers that selectively bind glutamine. *RNA Biology*, *8*(1), 82–89. https://doi.org/10.4161/rna.8.1.13864.

Barrick, J. E., & Breaker, R. R. (2007). The distributions, mechanisms, and structures of metabolite-binding riboswitches. *Genome Biology*, *8*(11), R239. https://doi.org/10.1186/gb-2007-8-11-r239.

Chan, C. W., Chetnani, B., & Mondragon, A. (2013). Structure and function of the T-loop structural motif in noncoding RNAs. *Wiley Interdisciplinary Reviews RNA*, *4*(5), 507–522. https://doi.org/10.1002/wrna.1175.

Chao, J. A., & Williamson, J. R. (2004). Joint X-ray and NMR refinement of the yeast L30e-mRNA complex. *Structure*, *12*(7), 1165–1176. https://doi.org/10.1016/j.str.2004.04.023.

Chapman, E. G., Costantino, D. A., Rabe, J. L., Moon, S. L., Wilusz, J., Nix, J. C., et al. (2014). The structural basis of pathogenic subgenomic flavivirus RNA (sfRNA) production. *Science*, *344*(6181), 307–310. https://doi.org/10.1126/science.1250897.

Chen, V. B., Arendall, W. B., 3rd, Headd, J. J., Keedy, D. A., Immormino, R. M., Kapral, G. J., et al. (2010). MolProbity: All-atom structure validation for macromolecular crystallography. *Acta Crystallographica Section D, Biological Crystallography*, *66*(Pt 1), 12–21. https://doi.org/10.1107/S0907444909042073.

Cheng, C. Y., Chou, F. C., & Das, R. (2015). Modeling complex RNA tertiary folds with Rosetta. *Methods in Enzymology*, *553*, 35–64. https://doi.org/10.1016/bs.mie.2014.10.051.

Cheng, C. Y., Chou, F. C., Kladwang, W., Tian, S., Cordero, P., & Das, R. (2015). Consistent global structures of complex RNA states through multidimensional chemical mapping. *eLife*, *4*,. e07600https://doi.org/10.7554/eLife.07600.

Cheng, C. Y., Kladwang, W., Yesselman, J. D., & Das, R. (2017). RNA structure inference through chemical mapping after accidental or intentional mutations. *Proceedings of the National Academy of Sciences of the United States of America*, *114*(37), 9876–9881. https://doi.org/10.1073/pnas.1619897114.

Chou, F. C., Echols, N., Terwilliger, T. C., & Das, R. (2016). RNA structure refinement using the ERRASER-phenix pipeline. *Methods in Molecular Biology*, *1320*, 269–282. https://doi.org/10.1007/978-1-4939-2763-0_17.

Chou, F. C., Sripakdeevong, P., Dibrov, S. M., Hermann, T., & Das, R. (2013). Correcting pervasive errors in RNA crystallography through enumerative structure prediction. *Nature Methods*, *10*(1), 74–76. https://doi.org/10.1038/nmeth.2262.

Correll, C. C., Beneken, J., Plantinga, M. J., Lubbers, M., & Chen, Y. (2003). The common and the distinctive features of the bulged-G motif based on a 1.04 a resolution RNA structure. *Nucleic Acids Research*, *31*(23), 6806–6818. https://doi.org/10.1093/nar/gkg908.

Cruz, J. A., & Westhof, E. (2011). Sequence-based identification of 3D structural modules in RNA with RMDetect. *Nature Methods*, *8*(6), 513–521. https://doi.org/10.1038/nmeth.1603.

Das, R., Karanicolas, J., & Baker, D. (2010). Atomic accuracy in predicting and designing noncanonical RNA structure. *Nature Methods*, 7(4), 291–294. https://doi.org/10.1038/nmeth.1433.

Dirks, R. M., & Pierce, N. A. (2003). A partition function algorithm for nucleic acid secondary structure including pseudoknots. *Journal of Computational Chemistry*, *24*(13), 1664–1677. https://doi.org/10.1002/jcc.10296.

Ditzler, M. A., Otyepka, M., Sponer, J., & Walter, N. G. (2010). Molecular dynamics and quantum mechanics of RNA: Conformational and chemical change we can believe in. *Accounts of Chemical Research*, *43*(1), 40–47. https://doi.org/10.1021/ar900093g.

Do, C. B., Woods, D. A., & Batzoglou, S. (2006). CONTRAfold: RNA secondary structure prediction without physics-based models. *Bioinformatics*, *22*(14), e90–e98. https://doi.org/10.1093/bioinformatics/btl246.

Einvik, C., Nielsen, H., Westhof, E., Michel, F., & Johansen, S. (1998). Group I-like ribozymes with a novel core organization perform obligate sequential hydrolytic cleavages at two processing sites. *RNA*, *4*(5), 530–541.

Gesteland, R. F., Cech, T. R., & Atkins, J. F. (2005). *The RNA world*. New York: Cold Spring Harbor Press.

Jaeger, L., Michel, F., & Westhof, E. (1994). Involvement of a GNRA tetraloop in long-range RNA tertiary interactions. *Journal of Molecular Biology*, *236*(5), 1271–1276. https://doi.org/10.1016/0022-2836(94)90055-8.

Jaeger, L., Verzemnieks, E. J., & Geary, C. (2009). The UA_handle: A versatile submotif in stable RNA architectures. *Nucleic Acids Research*, *37*(1), 215–230. https://doi.org/10.1093/nar/gkn911.

Kalvari, I., Argasinska, J., Quinones-Olvera, N., Nawrocki, E. P., Rivas, E., Eddy, S. R., et al. (2018). Rfam 13.0: Shifting to a genome-centric resource for non-coding RNA families. *Nucleic Acids Research*, *46*(D1), D335–D342. https://doi.org/10.1093/nar/gkx1038.

Kappel, K., Liu, S., Larsen, K. P., Skiniotis, G., Puglisi, E. V., Puglisi, J. D., et al. (2018). De novo computational RNA modeling into cryo-EM maps of large ribonucleoprotein complexes. *Nature Methods*, *15*(11), 947–954. https://doi.org/10.1038/s41592-018-0172-2.

Keating, K. S., & Pyle, A. M. (2012). RCrane: Semi-automated RNA model building. *Acta Crystallographica. Section D, Biological Crystallography*, *68*(Pt 8), 985–995. https://doi.org/10.1107/S0907444912018549.

Kladwang, W., VanLang, C. C., Cordero, P., & Das, R. (2011). Understanding the errors of SHAPE-directed RNA structure modeling. *Biochemistry*, *50*(37), 8049–8056. https://doi.org/10.1021/bi200524n.

Laing, C., & Schlick, T. (2011). Computational approaches to RNA structure prediction, analysis, and design. *Current Opinion in Structural Biology*, *21*(3), 306–318. https://doi.org/10.1016/j.sbi.2011.03.015.

Leontis, N. B., & Zirbel, C. L. (2012). Nonredundant 3D structure datasets for RNA knowledge extraction and benchmarking. In *RNA 3D structure analysis and prediction* (pp. 281–298). .

Lipchock, S. V., & Strobel, S. A. (2008). A relaxed active site after exon ligation by the group I intron. *Proceedings of the National Academy of Sciences of the United States of America*, *105*(15), 5699–5704. https://doi.org/10.1073/pnas.0712016105.

Lorenz, R., Bernhart, S. H., Honer Zu Siederdissen, C., Tafer, H., Flamm, C., Stadler, P. F., et al. (2011). ViennaRNA Package 2.0. *Algorithms for Molecular Biology*, *6*, 26. https://doi.org/10.1186/1748-7188-6-26.

Meyer, M., Nielsen, H., Olieric, V., Roblin, P., Johansen, S. D., Westhof, E., et al. (2014). Speciation of a group I intron into a lariat capping ribozyme. *Proceedings of the National Academy of Sciences of the United States of America*, *111*(21), 7659–7664. https://doi.org/10.1073/pnas.1322248111.

Miao, Z., Adamiak, R. W., Antczak, M., Batey, R. T., Becka, A. J., Biesiada, M., et al. (2017). RNA-puzzles round III: 3D RNA structure prediction of five riboswitches and one ribozyme. *RNA*, *23*(5), 655–672. https://doi.org/10.1261/rna.060368.116.

Miao, Z., Adamiak, R. W., Blanchet, M. F., Boniecki, M., Bujnicki, J. M., Chen, S. J., et al. (2015). RNA-puzzles round II: assessment of RNA structure prediction programs applied to three large RNA structures. *RNA*, *21*(6), 1066–1084. https://doi.org/10.1261/rna.049502.114.

Miao, Z., & Westhof, E. (2017). RNA structure: Advances and assessment of 3D structure prediction. *Annual Review of Biophysics*, *46*, 483–503. https://doi.org/10.1146/annurev-biophys-070816-034125.

Mirihana Arachchilage, G., Sherlock, M. E., Weinberg, Z., & Breaker, R. R. (2018). SAM-VI RNAs selectively bind S-adenosylmethionine and exhibit similarities to SAM-III riboswitches. *RNA Biology*, *15*(3), 371–378. https://doi.org/10.1080/15476286.2017.1399232.

Montange, R. K., & Batey, R. T. (2006). Structure of the S-adenosylmethionine riboswitch regulatory mRNA element. *Nature*, *441*(7097), 1172–1175. https://doi.org/10.1038/nature04819.

Moult, J., Fidelis, K., Kryshtafovych, A., Schwede, T., & Tramontano, A. (2018). Critical assessment of methods of protein structure prediction (CASP)-round XII. *Proteins*, *86*(Suppl 1), 7–15. https://doi.org/10.1002/prot.25415.

Nawrocki, E. P., & Eddy, S. R. (2013). Infernal 1.1: 100-fold faster RNA homology searches. *Bioinformatics*, *29*(22), 2933–2935. https://doi.org/10.1093/bioinformatics/btt509.

Peselis, A., & Serganov, A. (2012). Structural insights into ligand binding and gene expression control by an adenosylcobalamin riboswitch. *Nature Structural & Molecular Biology*, *19*(11), 1182–1184. https://doi.org/10.1038/nsmb.2405.

Petrov, A. I., Zirbel, C. L., & Leontis, N. B. (2013). Automated classification of RNA 3D motifs and the RNA 3D motif atlas. *RNA*, *19*(10), 1327–1340. https://doi.org/10.1261/rna.039438.113.

Piatkowski, P., Kasprzak, J. M., Kumar, D., Magnus, M., Chojnowski, G., & Bujnicki, J. M. (2016). RNA 3D structure modeling by combination of template-based method ModeRNA, template-free folding with SimRNA, and refinement with QRNAS. *Methods in Molecular Biology*, *1490*, 217–235. https://doi.org/10.1007/978-1-4939-6433-8_14.

Popenda, M., Szachniuk, M., Antczak, M., Purzycka, K. J., Lukasiak, P., Bartol, N., et al. (2012). Automated 3D structure composition for large RNAs. *Nucleic Acids Research*, *40*(14), e112. https://doi.org/10.1093/nar/gks339.

Read, R. J., Adams, P. D., Arendall, W. B., 3rd, Brunger, A. T., Emsley, P., Joosten, R. P., et al. (2011). A new generation of crystallographic validation tools for the protein data bank. *Structure*, *19*(10), 1395–1412. https://doi.org/10.1016/j.str.2011.08.006.

Ren, A., Xue, Y., Peselis, A., Serganov, A., Al-Hashimi, H. M., & Patel, D. J. (2015). Structural and dynamic basis for low-affinity, high-selectivity binding of L-glutamine by the glutamine riboswitch. *Cell Reports*, *13*(9), 1800–1813. https://doi.org/10.1016/j.celrep.2015.10.062.

Reuter, J. S., & Mathews, D. H. (2010). RNAstructure: Software for RNA secondary structure prediction and analysis. *BMC Bioinformatics*, *11*, 129. https://doi.org/10.1186/1471-2105-11-129.

Rother, K., Rother, M., Boniecki, M., Puton, T., & Bujnicki, J. M. (2011). RNA and protein 3D structure modeling: Similarities and differences. *Journal of Molecular Modeling*, *17*(9), 2325–2336. https://doi.org/10.1007/s00894-010-0951-x.

Sarver, M., Zirbel, C. L., Stombaugh, J., Mokdad, A., & Leontis, N. B. (2008). FR3D: Finding local and composite recurrent structural motifs in RNA 3D structures. *Journal of Mathematical Biology*, *56*(1–2), 215–252. https://doi.org/10.1007/s00285-007-0110-x.

Schroeder, K. T., Daldrop, P., & Lilley, D. M. (2011). RNA tertiary interactions in a riboswitch stabilize the structure of a kink turn. *Structure*, *19*(9), 1233–1240. https://doi.org/10.1016/j.str.2011.07.003.

Serganov, A., & Patel, D. J. (2012). Molecular recognition and function of riboswitches. *Current Opinion in Structural Biology*, *22*(3), 279–286. https://doi.org/10.1016/j.sbi.2012.04.005.

Serganov, A., Yuan, Y. R., Pikovskaya, O., Polonskaia, A., Malinina, L., Phan, A. T., et al. (2004). Structural basis for discriminative regulation of gene expression by adenine- and guanine-sensing mRNAs. *Chemistry & Biology*, *11*(12), 1729–1741. https://doi.org/10.1016/j.chembiol.2004.11.018.

Shortle, D., Simons, K. T., & Baker, D. (1998). Clustering of low-energy conformations near the native structures of small proteins. *Proceedings of the National Academy of Sciences*, *95*(19), 11158–11162. https://doi.org/10.1073/pnas.95.19.11158.

Sim, A. Y., Minary, P., & Levitt, M. (2012). Modeling nucleic acids. *Current Opinion in Structural Biology*, *22*(3), 273–278. https://doi.org/10.1016/j.sbi.2012.03.012.

Tian, S., & Das, R. (2016). RNA structure through multidimensional chemical mapping. *Quarterly Reviews of Biophysics*, *49*, e7. https://doi.org/10.1017/S0033583516000020.

Toor, N., Keating, K. S., Taylor, S. D., & Pyle, A. M. (2008). Crystal structure of a self-spliced group II intron. *Science*, *320*(5872), 77–82. https://doi.org/10.1126/science.1153803.

Trausch, J. J., Xu, Z., Edwards, A. L., Reyes, F. E., Ross, P. E., Knight, R., et al. (2014). Structural basis for diversity in the SAM clan of riboswitches. *Proceedings of the National Academy of Sciences of the United States of America*, *111*(18), 6624–6629. https://doi.org/10.1073/pnas.1312918111.

Vicens, Q., Mondragon, E., & Batey, R. T. (2011). Molecular sensing by the aptamer domain of the FMN riboswitch: A general model for ligand binding by conformational selection. *Nucleic Acids Research*, *39*(19), 8586–8598. https://doi.org/10.1093/nar/gkr565.

Wang, X., Kapral, G., Murray, L., Richardson, D., Richardson, J., & Snoeyink, J. (2008). RNABC: Forward kinematics to reduce all-atom steric clashes in RNA backbone. *Journal of Mathematical Biology*, *56*(1–2), 253–278. https://doi.org/10.1007/s00285-007-0082-x.

Watkins, A. M., Geniesse, C., Kladwang, W., Zakrevsky, P., Jaeger, L., & Das, R. (2018). Blind prediction of noncanonical RNA structure at atomic accuracy. *Science Advances*, *4*(5), eaar5316. https://doi.org/10.1126/sciadv.aar5316.

Weinberg, Z., Lunse, C. E., Corbino, K. A., Ames, T. D., Nelson, J. W., Roth, A., et al. (2017). Detection of 224 candidate structured RNAs by comparative analysis of specific subsets of intergenic regions. *Nucleic Acids Research*, *45*(18), 10811–10823. https://doi.org/10.1093/nar/gkx699.

Weinberg, Z., Regulski, E. E., Hammond, M. C., Barrick, J. E., Yao, Z., Ruzzo, W. L., et al. (2008). The aptamer core of SAM-IV riboswitches mimics the ligand-binding site of SAM-I riboswitches. *RNA*, *14*(5), 822–828. https://doi.org/10.1261/rna.988608.

Westhof, E., Masquida, B., & Jossinet, F. (2011). Predicting and modeling RNA architecture. *Cold Spring Harbor Perspectives in Biology*, *3*(2). a003632. https://doi.org/10.1101/cshperspect.a003632.

Wu, L., Chai, D., Fraser, M. E., & Zimmerly, S. (2012). Structural variation and uniformity among tetraloop-receptor interactions and other loop-helix interactions in RNA crystal structures. *PLoS One*, 7(11). e49225https://doi.org/10.1371/journal.pone.0049225.

Xu, X., & Chen, S. J. (2018). Hierarchical assembly of RNA three-dimensional structures based on loop templates. *The Journal of Physical Chemistry. B*, *122*(21), 5327–5335. https://doi.org/10.1021/acs.jpcb.7b10102.

Yesselman, J. D., Denny, S. K., Bisaria, N., Herschlag, D., Greenleaf, W. J., & Das, R. (2018). RNA tertiary structure energetics predicted by an ensemble model of the RNA double helix. *BioRxiv*, 341107. https://doi.org/10.1101/341107.

Zhang, J., & Ferre-D'Amare, A. R. (2014). Dramatic improvement of crystals of large RNAs by cation replacement and dehydration. *Structure*, *22*(9), 1363–1371. https://doi.org/10.1016/j.str.2014.07.011.

CHAPTER TEN

Monitoring co-transcriptional folding of riboswitches through helicase unwinding

Christopher P. Jones[a], Subrata Panja[b], Sarah A. Woodson[b], Adrian R. Ferré-D'Amaré[a,*]

[a]Biochemistry and Biophysics Center, National Heart, Lung and Blood Institute, Bethesda, MD, United States
[b]T.C. Jenkins Department of Biophysics, Johns Hopkins University, Baltimore, MD, United States
*Corresponding author: e-mail address: adrian.ferre@nih.gov

Contents

Abstract

In the cell, RNAs fold and begin to function as they are being transcribed. In contrast, in the laboratory, RNAs are typically studied after transcription is completed. Co-transcriptional folding can regulate the function of riboswitches and ribozymes and dictate the order of ribonucleoprotein assembly. Methods to observe and investigate RNA folding and activity during transcription are therefore desirable, yet synchronizing RNA polymerases and incorporating labels at specific sites for biophysical studies can be challenging. A recent methodological advance has been to harness highly processive, engineered "super-helicases" to unwind hybrid RNA-DNA duplexes, thereby releasing the RNA 5′–3′. When combined with single-molecule fluorescence detection,

Methods in Enzymology, Volume 623
ISSN 0076-6879
https://doi.org/10.1016/bs.mie.2019.05.031

RNA folding and concomitant activity can be studied in vitro in a manner that mimics vectorial folding during transcription. Herein, we describe methods for designing and preparing fluorescently labeled RNA-DNA duplex substrates for sequential helicase-dependent RNA folding experiments.

1. Introduction

The role of co-transcriptional folding in RNA function has been appreciated for four decades (Boyle, Robillard, & Kim, 1980; Nussinov & Tinoco, 1981; Pan & Sosnick, 2006). A small domain of RNA structure can typically fold in less time than is needed for synthesis of the next domain, given typical transcription speeds of 20–100 nt/s. As a result, the 5′-end of an RNA molecule begins to fold while the rest of the RNA is yet to be synthesized. The synthesis of the 5′ portion of a transcript prior to the complete synthesis of the entire transcript can temporally regulate the transcript's folding and function. While co-transcriptional folding can regulate the activity of any RNA, since folding is important to the function of most transcripts, the best studied examples involve RNAs that regulate gene expression. For example, the misfolding or delayed folding of the 5′ end of the transcript can decrease the activity of self-cleaving ribozymes, which in turn alters mRNA turnover or processing (Heilman-Miller & Woodson, 2003; Lazinski & Taylor, 1995; Pan, Artsimovitch, Fang, Landick, & Sosnick, 1999). Conversely, RNA folding during transcription is essential for riboswitches that regulate transcription termination and elongation in response to ligand binding (Jones & Ferré-D'Amaré, 2017) because the decision to stop or read through a terminator must be made before the polymerase passes by the terminator stem-loop. Elegant biophysical studies have shown that such riboswitches are "kinetically controlled" by the rates of strand synthesis, ligand binding, and RNA folding (Frieda & Block, 2012; Wickiser, Cheah, Breaker, & Crothers, 2005; Wickiser, Winkler, Breaker, & Crothers, 2005; Zhang, Lau, & Ferré-D'Amaré, 2010). Apart from better understanding how these RNA-based regulators function co-transcriptionally, investigating their mechanisms could provide insight into how to target them with inhibitors or engineer them for other purposes.

While measuring simultaneously the rates of RNA folding, ligand binding, and RNA synthesis for riboswitch or ribozyme function is technically challenging, individual rates have been determined using various techniques

(Frieda & Block, 2012; Wickiser, Cheah, et al., 2005; Wickiser, Winkler, et al., 2005). Single-molecule Förster resonance energy transfer (FRET) can provide relative populations of individual states and transition rates between states under different experimental conditions, such as ion or ligand concentrations and temperature (Brenner, Scanlan, Nahas, Ha, & Silverman, 2010; Fiegland, Garst, Batey, & Nesbitt, 2012; Liberman et al., 2015; Wood, Ferré-D'Amaré, & Rueda, 2012). While the ideal experiment is to observe RNA molecules fold during synthesis by RNA polymerase at different polymerase rates, this is challenging due to the difficulty of introducing site-specific fluorescent labels during transcription. A common alternative is to study different lengths of transcripts representing different points of elongation to understand their activity at equilibrium. This approach is equivalent to measuring different time points of an elongating RNA at a very slow elongation rate (i.e., assuming an equilibrium is reached). However, it may miss transient yet functionally important RNA structures.

An alternative approach to visualize the sequential folding of RNA in real time has recently been developed using engineered helicases (Hua, Panja, Wang, Woodson, & Ha, 2018). Crosslinked into their active unwinding conformation, these processive enzymes unwind thousands of base pairs of double-stranded DNA at ~140 nt/s (Arslan, Khafizov, Thomas, Chemla, & Ha, 2015). The robust activity of these engineered helicases was exploited to observe the sequential folding of a ribozyme. In this assay, the helicase translocates 3′ to 5′ along the DNA strand of an RNA/DNA hybrid (Arslan et al., 2015; Hua et al., 2018). As the helicase unwinds the duplex, the RNA strand is released in the 5′–3′ direction, allowing for folding in a pseudo-co-transcriptional manner (Fig. 1A). This approach not only avoids the complex pausing behavior of cellular polymerases but also allows fluorescent labels to be attached to the RNA strand at any location in advance so that folding kinetics can be observed by single-molecule FRET. Alternative methods for incorporating a fluorophore involve starting transcription with an RNA fragment synthesized in advance or walking the polymerase along the template a few nucleotides at a time (Liu et al., 2015; Perez-Gonzalez, Lafontaine, & Penedo, 2016). In both cases the labeling sites are restricted to the 5′ region of the RNA, which does not have an opportunity to fold during the transcription assay.

The helicase RNA folding assay is controlled by adding the reagents to the slide at specific times. The fluorescently labeled RNA-DNA duplex is hybridized with a biotinylated DNA oligomer and immobilized on slides

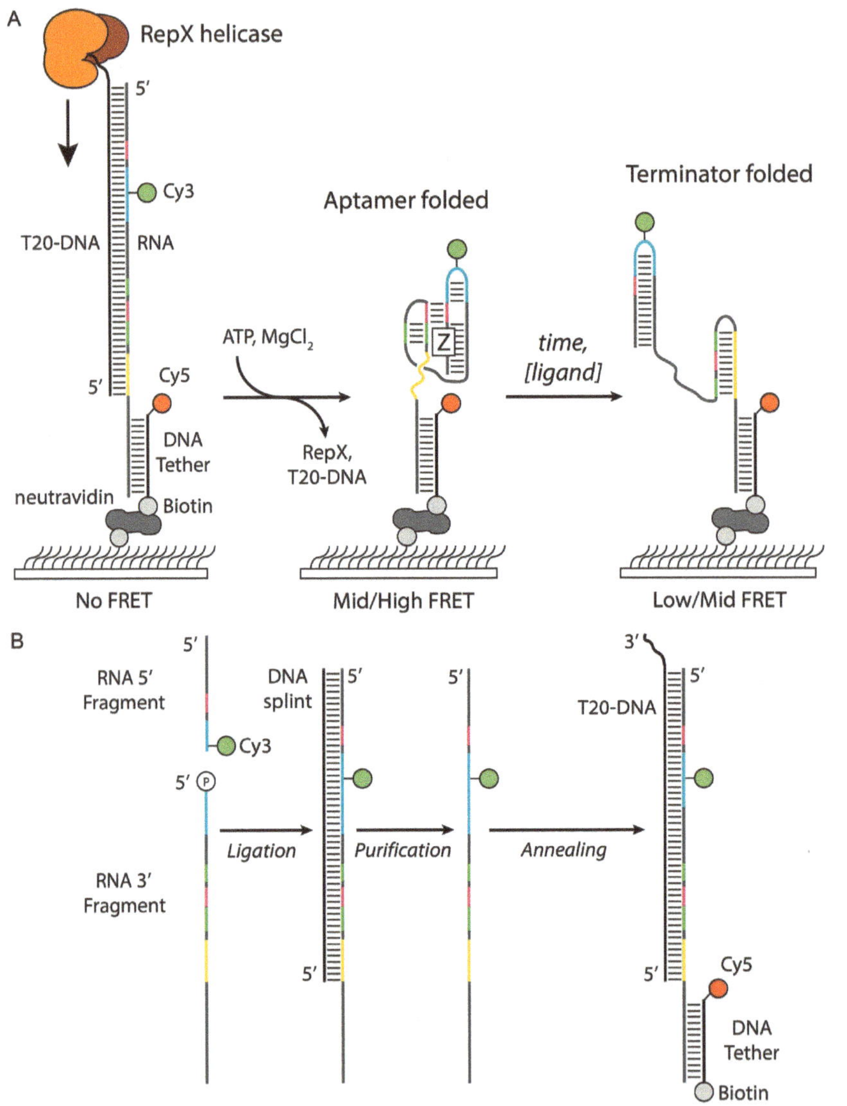

Fig. 1 Workflow of duplex preparation. (A) An example helicase unwinding event displaces the T20-DNA strand (black) to allow for the RNA strand (gray and colored) to sequentially fold in a 5′–3′ direction. After complete unwinding, only the Cy3-labeled RNA strand and Cy5-labeled biotinylated DNA strand remain. RNA folding should be observed as a decrease in the distance between the Cy3 and Cy5 dyes, increasing FRET efficiency. For riboswitches, folded states may contain mutually exclusive paired regions (pink and green) observed by changes in FRET. (B) The RNA 3′ fragment containing a 3′ region complementary to the biotinylated DNA is in vitro transcribed and purified. After phosphorylation at the 5′-end and purification, the 3′ fragment is ligated with T4 RNA

doped with neutravidin (Fig. 1A). The helicase is pre-loaded onto a single-stranded 3′ dT20 extension of the complementary DNA (T20-DNA). After unbound reagents are washed away, the unwinding reaction is started by adding ATP to the slide chamber. Other reagents, such as Mg^{2+} or riboswitch ligands, can be added to the slide at the same time. The folding kinetics of individual RNA molecules can be deduced from the changes in fluorescence intensity under a variety of conditions, including unwinding velocity. For example, the passing helicase enhances the fluorescence emission of Cy3, reporting the progress of unwinding. The folded state of the RNA can be inferred from the efficiency of FRET between a donor and acceptor pair of fluorophores attached to the RNA (Roy, Hohng, & Ha, 2008). Thus, ribozyme cleavage may be followed as the RNA folds and releases itself from the surface, and ribozyme or riboswitch folding may be observed as the RNA refolds in the presence of its ligand. In principle, this assay could be used to study any process for which directional RNA folding is necessary.

Herein we describe protocols for producing fluorescently labeled RNA-DNA hybrids for examining co-transcriptional folding or unwinding velocity using the helicase unwinding assay. We suggest methods for choosing and evaluating labeling sites, ligation and purification of fluorescently labeled oligonucleotides, and forming duplexes that serve as the substrates for unwinding experiments. A protocol for conducting helicase unwinding experiments is then described.

2. Design strategies for labeling sites

The goal is to capture changes to RNA structure by observing correlated changes in judiciously placed fluorescence dyes that do not perturb the system. This can be achieved using dye pairs (e.g., cyanine dyes Cy3 and Cy5) that undergo FRET (Roy et al., 2008) or by a single dye that fluctuates in intensity due to changes in its environment. In the latter case, interaction between the dye and helicase results in protein-induced fluorescence enhancement (PIFE), which is sensitive in the ~0–4 nm distance range

Ligase 2 to the chemically synthesized Cy3-labeled 5′ fragment via annealing to a DNA splint (black). After purification to remove the splint and unligated fragments, the ligated RNA is annealed to a T20-DNA (black) containing the dT20 helicase binding site at the 3′ end and a biotinylated, Cy5-labeled DNA tether for attachment to the streptavidin-coated slide surface.

(Hwang & Myong, 2014). Since the helicase unwinds in a single direction in these experiments (Fig. 1A), PIFE appears and disappears as the helicase passes by the Cy3 dye in the duplex.

2.1 Choice of labeling sites

As the RNA is extended in the hybrid duplex, placement of a Cy3 dye toward the 5′ end of the RNA and the Cy5 dye at the 3′ end will result in a low FRET efficiency for the initial duplex and increasing FRET efficiency as the RNA folds. Within the 5′ half of the RNA, the ideal labeling spot is a bulge or loop residue that is flexible and not highly conserved. To reduce the costs of synthesizing a fluorescently labeled oligonucleotide, the labeling position may be near the 3′ end of the RNA 5′ fragment, such as the penultimate 3′ nucleotide (e.g., U17 of an 18-nt oligonucleotide).

Riboswitches are commonly studied using inline probing, a chemical probing method that uses the natural instability of RNA to determine which nucleotides are protected from ligand binding (Regulski & Breaker, 2008). Flexible nucleotides within loops and bulges tend to be the most reactive, and paired or inflexible nucleotides tend to be less reactive. Inline probing is an alternative method to determine which nucleotides are ideal labeling sites. For riboswitches, this includes nucleotides that are reactive both in the presence and absence of ligand or across the range of conditions to be assayed. The choice of labeling site is greatly aided by a structural model of the RNA, to insure that the attached dye is unlikely to occlude a binding pocket or clash with nearby residues. The RNA's activity after fluorescent labeling should be evaluated by measuring the apparent dissociation constant of a ligand in single molecule experiments and comparing this value to a previously determined apparent dissociation constant for the unlabeled RNA. Alternatively, performing inline probing experiments with ^{32}P-labeled RNA with and without the fluorescent label can determine if the label perturbs binding and uses relatively little sample (Regulski & Breaker, 2008). If the RNA is a ribozyme, one should determine whether the attached fluorophores alter the ribozyme self-cleavage rate.

2.2 Strategies for obtaining RNAs for ligation

The preparation of substrates for helicase RNA folding experiments may require the ligation of two or more oligonucleotides, at least one of which contains a fluorescent label, and annealing of the labeled RNA strand to a cDNA and a tether DNA to create a hybrid duplex (Fig. 1B). Fluorescent labels may be introduced during or after chemical synthesis, as described elsewhere (Hermanson, 2018; Walter & Burke, 2000). Using a DNA splint, as previously described (Moore & Query, 2000), the fluorescently labeled RNA 5′ fragment is ligated to an in vitro transcribed RNA (the RNA 3′ fragment) (Fig. 1B). As the 3′ fragment must contain a 5′-monophosphate for the ligation to succeed, this RNA can be phosphorylated with PNK, phenol:chloroform extracted, and ethanol precipitated before the ligation (Fig. 1B). The DNA splint used to ligate the two RNAs can be the same DNA used later in annealing reactions (T20-DNA). The splint should anneal to as much of the RNA fragments as possible to maximize annealing while not being too close in size to the ligated RNA (i.e., 10 nt shorter or longer), which would result in band overlap during denaturing gel purification.

Unlabeled RNAs may be chemically synthesized or transcribed and purified following standard protocols (Ferré-D'Amaré & Doudna, 1996; Jones & Ferré-D'Amaré, 2014; Milligan & Uhlenbeck, 1989). An example template used for transcribing the ZTP riboswitch RNA 3′ fragment contains the T7 promoter sequence, hammerhead ribozyme (italicized), and the first 15 nt of the ZTP riboswitch sequence: 5′-TAA TAC GAC TCA CTA TA *ggg aga TAT AAC TGA TA ctg atg agt ccg tga gga cga aac ggt acc cgg tac cgt c* TAT CAG TTA TAT GAC. The underlined 11-nt sequence is complementary to the first 11 nt of the ZTP riboswitch sequence and, upon base pairing, positions the ribozyme cleavage site. After transcription, the RNA product will contain the ZTP riboswitch sequence, a 5′-UU linker, and a 3′ sequence (5′-CAU UAG CGG UAU UCC GGA AUU GCC GUA AUC GCG-3′) for annealing to a biotinylated DNA tether (5′-/Biotin/-CGC GAT TAC GGC AAT TCC GGA ATA CCG CTA ATG-3′) for attachment to the streptavidin slide surface. The template can be PCR-amplified with 2′-*O*-methyl modified primers as previously described (Kao, Zheng, & Rudisser, 1999), and the primer can be designed to create a template containing the 3′-biotin RNA sequence for attachment to the surface. For example, the reverse primer used for amplifying the ZTP riboswitch 3′ fragment for ligation, containing the 35-nt extension complementary to the DNA tether (underlined) and UU linker sequence (italicized), is 5′-mC mGCG ATT ACG GCA ATT CCG GAA TAC CGC TAA TG*A ACC GAC AAT CTA GGC TTG T*-3′, where the "m" indicates 2′-*O*-methylated residues. If the length of the RNA is similar in size to the 57-nt hammerhead RNA, the bands can overlap during separation via denaturing urea-PAGE, especially if the transcription yields are high. To improve separation, a stem-loop (e.g., 5′-CAT GTG GAA ACA CAT GAA or 5′-CAT GTG TGT GTG GAA ACA CAC ACA CAT GAA-3′) may be inserted prior to the 11-nt complementary sequence in the hammerhead ribozyme to increase its length without disrupting cleavage. Note that if a hammerhead ribozyme is not used during transcription, the 5′ triphosphorylated end must be dephosphorylated and re-phosphorylated (or converted into a monophosphate through the use of RNA 5′-pyrophosphohydrolase (Deana, Celesnik, & Belasco, 2008), available through NEB) prior to use and purified, and the ligation yields may be lower due to heterogeneity at the 5′ end of the RNA.

2.3 Design of complementary DNA and tether DNA

The 3′ end of the biotinylated DNA for tethering the complex to the slide may also be an ideal location for introducing a second dye (Cy5). This strategy (Fig. 1) is similar to placing a dye at the 3′ of the RNA without having to perform an additional labeling reaction or ligation, and it also allows for the RNA sequence to be changed relatively easily while the biotinylated, Cy5-labeled DNA tether can be kept constant. For example, a series of experiments with a riboswitch would compare the RNA with and without the terminator hairpin, or with mutations to change the terminator stability or ligand binding site. Using the design above, only the transcribed RNA 3′ fragment, reverse primer for amplifying the template to make that RNA, and unlabeled T20-DNA need to be synthesized to perform an additional series of experiments.

If folding brings the 5′ and 3′ ends of the RNA close together, a donor dye at the 5′ end and an acceptor dye on the tether DNA will be brought in close proximity in the folded state. The resulting FRET efficiency for the folded state may approach 1 (Hua, B., Jones, C. P., Ferré-D'Amaré, A. R., & Ha, T., unpublished observations; Hua et al., 2018). This could be a drawback if additional conformational changes occurring in the middle of the RNA are not observed because they do not disrupt pairing of the 5′ and 3′ ends.

For experiments exploiting multiple Cy3 PIFE peaks to measure helicase velocity, it is necessary to introduce more than one Cy3 dye into a single RNA prior to duplex annealing. This experiment takes advantage of the directional movement of the helicase across a Cy3 dye, which causes an increase in Cy3 fluorescence as the helicase approaches the dye during unwinding and a decrease in fluorescence as the helicase moves away (Hwang & Myong, 2014). The resulting peak can be used to indicate a location of the helicase on the template, and the observation of two peaks separated in time can be used to estimate the helicase unwinding rate under desired reaction conditions. Ligations involving more than two pieces of RNA can be performed simultaneously using the same protocols below although yields typically decrease as more fragments are introduced.

3. Preparation of biotinylated, fluorescently labeled RNA/DNA hybrid duplexes

3.1 Equipment

Bio-Rad C1000 Thermal Cycler or equivalent device
Table-top centrifuges for microcentrifuge and PCR tubes
Typhoon Trio Phosphorimager (GE Healthcare) or equivalent fluorescent gel imaging system
Slab gel electrophoresis apparatus, glass plates (16 cm wide, 25 cm long), 0.8 and 3 mm spacers, power supply
NanoDrop or equivalent UV–vis spectrophotometer
Sterile razor blades
Whatman Elutrap electroelution system
Amicon Ultra centrifugal filters, regenerated cellulose (3000 and 10,000 Da molecular weight cutoff), 0.5 mL (Millipore)
Amicon Ultrafree centrifugal filters, PVDF 0.22 μm, 0.5 mL (Millipore)
Total internal reflection fluorescence microscope, equipped with a 532-nm laser and a 633-nm laser, as described in Hua et al. (2014)

3.2 Reagents

T4 polynucleotide kinase (NEB), enzyme and buffers
T4 RNA Ligase 2 (NEB), enzyme and buffers
HEPES (Fisher), as a 0.5 M stock, adjusted to pH 7.4 with potassium hydroxide, sterile filtered
Potassium chloride (Fisher), as a 1.5 M stock, sterile filtered
RepX helicase (e.g., 5 μM stock), prepared as described previously (Arslan et al., 2015)
Biotinylated bovine serum albumin (BSA, Sigma)
Tween-20 (Sigma)
Neutravidin (Fisher)
Trolox (Sigma)
Glucose (Sigma)
Glucose oxidase (Sigma)
Catalase (Sigma)

Quartz slides coated with dichlorodimethylsilane-Tween-20 (Hua et al., 2014) or polyethylene glycol-Tween-20 (Ha et al., 2002)
Custom fluorescently labeled and biotinylated oligonucleotides (e.g., Trilink)
Unlabeled RNA oligonucleotides for ligation (~2.5 nmol or ~0.1 mg, Integrated DNA Technologies)
Phenol, buffered to pH 8 (Sigma)
Chloroform (Fisher)
Sodium acetate (Fisher), as a 3 M stock, pH 5.1
100% ethanol, ice cold
70% *v*/*v* ethanol in RNase-free water, ice cold
Urea (Fisher)
40% 19:1 Acrylamide:bis-acrylamide (Fisher)
10× TBE buffer (0.9 M Tris base, 0.9M Boric acid, 20 mM ethylenediaminetetraacetic acid, pH 8.0)
30% *w*/*v* ammonium persulfate (APS) (Fisher)
Tetramethylethylenediamine (TEMED) (Fisher)
SYBR Gold Nucleic Acid Gel Stain (Fisher), ethidium bromide, or equivalent nucleic acid stain
Ultrapure RNase-free water

4. Protocols

4.1 Ligation and purification of fluorescently labeled RNA

1. All oligonucleotides should be resuspended in RNase-free water (e.g., at a concentration of 200 μM and dilutions thereof) and checked for purity prior to ligation. For a two-piece ligation, this includes a RNA 5′ fragment, a RNA 3′ fragment, and a DNA splint complementary to both fragments (Fig. 1). Purity at each step is extremely important, as undesired products can introduce heterogeneity later during annealing steps. Work should be performed in the dark when using fluorescently-labeled RNAs, which should be kept on ice unless otherwise noted.
2. Although RNA ligations often succeed at 1:1:1 mixtures of the three components, optimization is sometimes required. For pilot experiments to optimize annealing with the DNA splint and ligation of the two RNA fragments, set up reactions consisting of 1:2:1, 1:1:1, 1:1:2, and 2:1:1 ratios of RNA 5′ fragment to RNA 3′ fragment to DNA splint. In addition, set up control 1:1:1 reactions in which one or two of each of the three reagents are excluded. If the ligation is

to be performed many times with different dye locations or mutated sequences, it may be beneficial to optimize the ligation first with unlabeled RNAs before proceeding with fluorescently labeled RNA. Omission of buffer and salt (50 mM HEPES, pH 7.4, and 150 mM KCl) diminished the efficiency of annealing (data not shown). While the ligation can also be performed with T4 DNA Ligase, inclusion of fluorescent labels near the ligation site greatly decreases ligation yields using this enzyme in our hands. For ligations involving more than two fragments of RNA, the ratio of T20-DNA:RNA with all RNA pieces in an equimolar ratio can be optimized first. A sample analytical scale protocol for a 1:1:1 mixture is given in the next step.

3. Mix the following components by pipetting and centrifuge briefly.
   1 μL of 10 μM RNA 5′ fragment (Cy3-labeled)
   1 μL of 10 μM RNA 3′ fragment
   1 μL of 10 μM DNA splint
   2 μL of 0.5 M HEPES-KOH, pH 7.4
   2 μL of 1.5 M KCl
   13 μL of RNase-free water
4. Anneal the mixture in a thermocycler by heating to 95 °C for 3 min and slowly cooling to 20 °C by decreasing the temperature 5 °C every 30 s. Save 2 μL of the annealed mixture for non-denaturing PAGE analysis. Lower annealing temperatures may be used for ligating shorter RNAs.
5. Add 5 μL of 10 × T4 RNA Ligase 2 reaction buffer (containing ATP and $MgCl_2$) and bring the volume up to 50 μL with RNase-free water. Add 1 μL of T4 RNA Ligase enzyme, mix gently by pipetting, and centrifuge briefly. Incubate for 1 h at 37 °C. Save an aliquot for denaturing PAGE analysis.
6. Check the success of the annealing step by running 2 μL aliquots on an 8% polyacrylamide 1 × TBE non-denaturing gel. Use a non-denaturing loading dye (25% glycerol and the refolding buffer with 0.001% *w/v* xylene cyanol, 0.001% *w/v* bromophenol blue). Image the gel with the fluorescence imager. The gel may also be stained with SYBR gold or ethidium bromide to identify unlabeled RNA fragments. If annealing is poor, the ligation is unlikely to succeed.
7. Check the success of the ligation reaction by running 2 μL aliquots on an 8% polyacrylamide 8 M Urea 1 × TBE denaturing gel. Use a denaturing loading dye (i.e., formamide with 0.001% *w/v* xylene cyanol, 0.001% *w/v* bromophenol blue) and heat samples at 95 °C for 5 min prior to loading. Image the gel with the fluorescence imager as before. Save successful ligations for purification of the ligated product.

8. Once an acceptable ligation yield has been achieved in the pilot reactions, the reactions may be scaled up 30 times and combined with successful small scale reactions for purification of the desired ligated product.
9. To purify the ligated RNA, prepare a preparative 8% polyacrylamide gel containing 8 M Urea and 1 × TBE (3 mm spacers). Omit bromophenol blue or xylene cyanol from the loading mix if the dyes co-migrate with the RNA. Heat samples at 95 °C for 5 min prior to loading to completely denature the ligation complexes. Perform all steps in the dark to avoid photobleaching fluorophores. At a 1.5 nmol ligation scale, the bands corresponding to unligated Cy3-labeled RNA and ligated Cy3-labeled RNA should be clearly visible by eye without UV shadowing (Fig. 2A). The gel should be run to maximize separation of the DNA splint and ligated fluorescently labeled RNA, allowing unreacted labeled RNA to run off the gel if necessary.

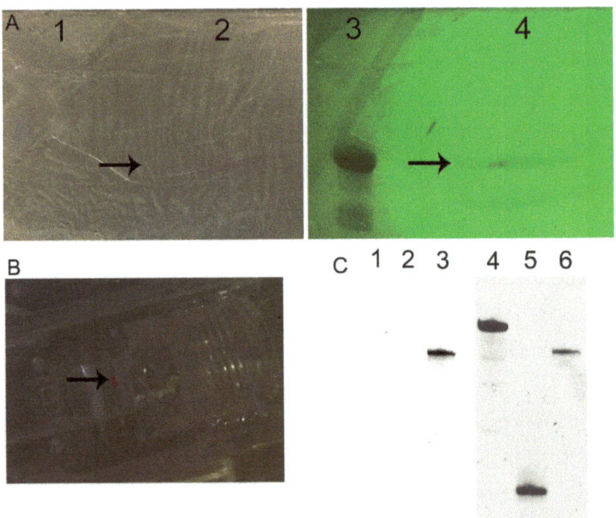

Fig. 2 Preparation of ligated Cy3-labeled RNA. (A) Example 1.5 nmol scale ligation purification denaturing PAGE gel viewed by eye (left) and by UV shadowing (right). Note the pink band for Cy3-labeled RNA (arrowed) running above the DNA splint. Lanes 1 and 3 contain a 128-nt marker, and lanes 2 and 4 contain the ligation mixture. (B) An Elutrap cell collection chamber contains labeled RNA (arrowed) after elution. (C) Post-ligation 8% denaturing PAGE imaged for fluorescence prior to staining (lanes 1–3) and after staining with ethidium bromide (lanes 4–6). As lanes 1 and 2 contain no fluorescent marker, the two gel images were aligned by eye. Lanes 1 and 4 contain a 150-nt marker, lanes 2 and 5 contain a 76-nt marker, and lanes 3 and 6 contain the 129-nt ligated Cy3-labeled RNA.

10. Cut out the ligated RNA product band (i.e., pink for Cy3) with a sterile razor blade. Elute the RNA from the gel slice overnight at 120 V in the dark in RNase-free 1 × TBE buffer using an Elutrap. Note that contamination of the Elutrap with RNase is disastrous. Wash the Elutrap instruments thoroughly with RNase-free water before and after use, store the Elutrap cells wrapped in Saran wrap when not in use, and use designated RNase-free tweezers for manipulating Elutrap cells and membranes. Never touch the inside surfaces of the Elutrap cells, breathe on the cells, or leave the cells in open air to collect dust for long. If contamination has occurred, extensive washing should be performed to remove RNase. The TBE running buffer used for elution should be RNase-free inside the elution chamber. An intensely colored dot of fluorescently labeled RNA should appear in the elution chamber when transfer is complete (Fig. 2B).
11. Concentrate the labeled RNA in a centrifugal concentrator, wash several times with RNase-free water, filter with a 0.22 μm centrifugal filter, and measure the absorbance at both 260 nm and the dye absorption wavelength. Calculate the concentration of the labeled species using both the dye and RNA extinction coefficients. Concentrations can be calculated using the Beer-Lambert Law ($A = \varepsilon c l$, where A is the absorbance, ε is the extinction coefficient at that wavelength, c is the concentration, and l is the cuvette path length). If the dye also absorbs at 260 nm, which is the case for Cy3, absorption of free dye can be measured to determine the correction value (b), which is the absorption at 260 nm (A_{260}) divided by the absorption at the fluorescence absorption wavelength (A_{fl}). The corrected RNA concentration at 260 nm can then be calculated by subtracting the contribution of the dye absorbance at 260 nm from the RNA absorbance at 260 nm ($A - bA_{fl} = \varepsilon c l$). For longer RNAs, this correction may be small and within error of determining the RNA concentration. If the RNA concentration calculated from the absorbance at 260 nm is significantly higher than the concentration calculated from the absorbance of the dye, contaminating DNA is likely present, or the dye labeling efficiency is poor. For RNAs containing more than one dye, the dye concentration should reflect the number of dye molecules per strand.
12. Assess the purity of the fluorescently labeled RNA by analyzing a small aliquot by denaturing PAGE. Image the gel for fluorescence and again after staining with a sensitive nucleic acid stain (SYBR Gold) to visualize both fluorescently labeled and unlabeled species

(Fig. 2C). The purity of the ligated RNA at this stage is extremely important for interpreting future pilot annealing reactions below. If more than one band is present, then the annealing mixture prepared below will be significantly more complicated.

13. Aliquot ligated RNA and store at −20 °C in the dark.

4.2 Duplex annealing for single molecule FRET

1. Single-molecule FRET experiments are typically conducted at final concentrations of ~60 pM labeled duplex, so the reaction scales below will produce enough sample for many experiments. The annealing reaction for single molecule FRET involves three oligonucleotides—the labeled RNA purified above, a biotinylated DNA tether for attachment to the FRET surface, and a T20-DNA (Fig. 1A). The latter is complementary to the riboswitch RNA and has 20 T residues added to the 3′ end for helicase binding prior to unwinding.
2. Anneal the complex by combining 1:0.9:10 ratio of RNA to biotinylated DNA to T20-DNA. Set up control reactions in which one or two reagents have been excluded, as well as controls containing only one reagent. To optimize annealing, it may necessary to titrate the ratio of T20-DNA. In the case of the ZTP riboswitch RNA, the primary source of heterogeneity is from the RNA, which dimerizes (Jones & Ferré-D'Amaré, 2015). Excess T20-DNA will be washed from the streptavidin-coated slide surface prior to initiating unwinding by adding ATP. An example recipe for annealing the complex is:
   1 μL of 2.5 μM ligated Cy3-labeled RNA
   0.9 μL of 2.5 μM Cy5-labeled biotinylated DNA
   1 μL of 25 μM T20-DNA
   2 μL of 0.5 M HEPES-KOH, pH 7.4
   2 μL of 1.5 M KCl
   13.1 μL of RNase-free water
3. Mix the reaction by pipetting and centrifuge briefly. In a thermocycler, heat the mixture at 95 °C for 3 min and slowly cool to 20 °C (i.e., decreasing temperature by 5 °C every 30 s).
4. Run an aliquot of the reaction and controls on an 8% polyacrylamide 1 × TBE non-denaturing gel to check for annealing (Fig. 3). Fluorescence imaging should be performed with both dye channels to check if the Cy3 and Cy5 signals overlap as expected. Annealing with T20-DNA should shift the mobility of the RNA upward. An additional

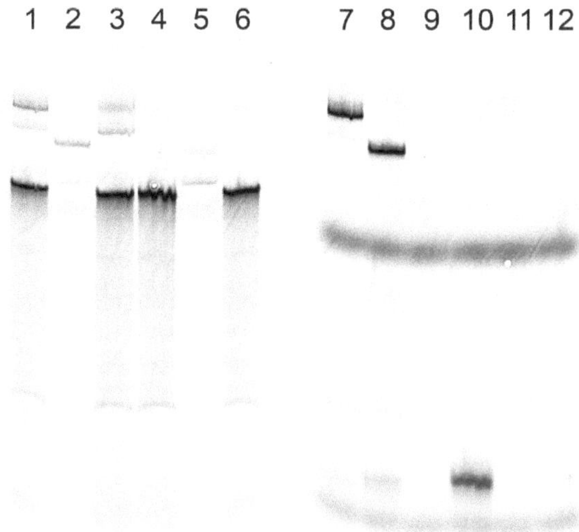

Fig. 3 Example duplex annealing gel. The same gel was imaged twice to detect Cy3 (lanes 1–6) and Cy5 (lanes 7–12). In this experiment, the RNA contains a Cy3 label, the biotinylated DNA contains a Cy5 label, and the T20-DNA also contains a Cy3 label, and the three were mixed in a 1:0.9:10 ratio. Lanes 1 and 7 contain all three oligonucleotides, lanes 2 and 8 contain RNA and biotinylated DNA, lanes 3 and 9 contain RNA and T20-DNA, lanes 4 and 10 contain biotinylated DNA and T20-DNA, lanes 5 and 11 contain RNA only, and lanes 6 and 12 contain T20-DNA only. Note that in this annealing reaction, a higher migrating species is present in the absence of biotinylated DNA, likely due to a greater than 1:1 interaction between the RNA and T20-DNA. Xylene cyanol dye in the nondenaturing gel loading buffer is also visible in lanes 7–12.

smaller upshift should be observed upon annealing with biotin-DNA. The sub-stoichiometric ratio of biotinylated DNA to RNA will result in two bands in the lane corresponding to the complexes with and without biotinylated DNA, one of which will not stick to the slide surface as it lacks a biotin. This band should be identified by its co-migration with the T20-DNA:RNA control annealing reaction lacking biotinylated DNA.

5. The control annealing mixture of RNA:biotin-DNA without T20-DNA should be saved for use as a control for FRET experiments described below as it represents the final equilibrium FRET state after unwinding (Fig. 1).
6. Store aliquots at $-20\,^{\circ}\mathrm{C}$ until further use.

4.3 Single molecule helicase unwinding

1. From above, complexes now available for single molecule measurements include the T20-DNA:RNA-Cy3:biotin-DNA ternary

complex and the RNA-Cy3:biotin-DNA control complex. Before conducting unwinding experiments, equilibrium measurements of the RNA-Cy3:biotin-DNA complex can be conducted to understand how the RNA folds in the absence of sequential folding.

2. A detailed protocol for the preparation of passivated quartz slides is available elsewhere (Hua et al., 2014). The slides used for these experiments have inlet and outlet holes at the ends, which can be attached to tubing and a syringe with epoxy (Roy et al., 2008) to allow for easy buffer washing and sample application. A single slide can contain multiple lanes made air-tight and separated from one another by double-sided tape. A ~1 cm piece of the wide end of a 200 μL pipette tip (cut with a serile razor blade) can be fixed to the slide with epoxy to create a small sample well of ~100 μL around the slide inlet hole. Similarly, a ~1 cm piece of the small end a 200 mL pipette tip can fixed to the outlet hole to create a port for syringe tubing. The buffer level in the sample well can be monitored as the buffer is drawn into the slide to avoid introducing air.
3. Wash the slide with 50 μL of 0.2 g/L biotinylated BSA in 10 mM Tris–HCl, pH 8.0, 50 mM NaCl (T50 buffer). Incubate for 5 min, and wash out unbound BSA with T50 buffer.
4. Wash the slide with 0.25 g/L neutravidin in T50 buffer. Incubate for 5 min, and wash out unbound neutravidin with T50 buffer.
5. Wash the slide with 60 pM fluorescently labeled complex freshly diluted in 50 mM HEPES-KOH, pH 7.4, 150 mM KCl, 10 mM $MgCl_2$ (RNA folding buffer), and incubate for 1 min. Wash out unbound nucleic acids with RNA folding buffer.
6. Check for the presence and density of the slide-bound complex by turning on lasers. Without imaging buffer, the dyes will photobleach rapidly. Turn off lasers, and move to a new region of the slide for subsequent data acquisition.
7. Wash the slide with 50 nM RepX in 10 mM Tris–HCl, pH 7.5, 0.3 M KCl, 10% v/v glycerol (RepX storage buffer). Incubate 1 min to allow RepX to bind. Wash out free RepX with RNA folding buffer.
8. Wash slide with RNA folding buffer containing 4 mM Trolox, 0.8% w/v glucose, 165 U/mL glucose oxidase, and 2170 U/mL catalase (imaging buffer). This buffer may be supplemented with suitable final reaction components (i.e., riboswitch ligand). Turn on lasers to check molecule density on the slide, laser brightness, and rate of dye photobleaching. Data may be collected at this point to observe the FRET

values for intact complex molecules prior to unwinding. Based on the designs above (Fig. 1), the FRET efficiencies of the hybrid duplex should be ~0.

9. To initiate unwinding, draw imaging buffer containing 1 mM ATP into the slide, and wait 1 min for unwinding to complete. Turn on lasers and collect data at various regions of the slide until a satisfactory number of molecules have been imaged.
10. Alternatively, for real-time imaging of duplex unwinding, begin recording movies a few seconds before drawing ATP into the slide. As only a single region of the slide will be available for data collection, it is essential to find an optimally dense region of the slide to maximize the number of molecules observed. Unwinding will proceed rapidly after ATP addition, and traces should complete within ~10–20 s for the ~100-bp RNA/DNA hybrid duplexes described herein. Data collection should continue until the majority of molecules have photobleached to discriminate single molecules from multiple overlapping molecules.
11. Traces are recorded and analyzed using custom code as previously described (Hua et al., 2018), with intensity corrections performed by plotting uncorrected FRET efficiency against uncorrected stoichiometry (Hellenkamp et al., 2018). A fraction of molecules may remain in a no-FRET state throughout the measurement due to either the absence of RepX on the duplex or the reannealing of T20-DNA immediately after unwinding.

5. Summary

These protocols describe how to produce fluorescently labeled substrates for performing helicase unwinding experiments, an approach to probe sequential RNA folding as it might occur during transcription. In this experiment, synthesis by RNA polymerase has been replaced with helicase-facilitated vectorial liberation from a hybrid duplex consisting of a fluorescently labeled RNA and unlabeled T20-DNA. An advantage of this approach is the ease with which substrates can be prepared, the ability to study ligands that might otherwise modulate RNA polymerase activity, and the recreation of transcription-like dynamics in a systematic manner. The DNA strand can be easily changed to incorporate base modifications that stabilize or destabilize the duplex, creating a potential avenue for future exploration.

Acknowledgments

We thank T.-J. Ha, B. Hua, P.J. Murray and R. Rosenthal for discussions. This work was supported in part by NSF MCB-1616081, K22HL139920-01 from the National Heart, Lung and Blood Institute (NHLBI) to C.P.J., and the intramural program of the NHLBI, NIH.

References

Arslan, S., Khafizov, R., Thomas, C. D., Chemla, Y. R., & Ha, T. (2015). Protein structure. Engineering of a superhelicase through conformational control. *Science*, *348*(6232), 344–347. https://doi.org/10.1126/science.aaa0445.

Boyle, J., Robillard, G. T., & Kim, S. H. (1980). Sequential folding of transfer RNA. A nuclear magnetic resonance study of successively longer tRNA fragments with a common 5′ end. *Journal of Molecular Biology*, *139*(4), 601–625.

Brenner, M. D., Scanlan, M. S., Nahas, M. K., Ha, T., & Silverman, S. K. (2010). Multivector fluorescence analysis of the xpt guanine riboswitch aptamer domain and the conformational role of guanine. *Biochemistry*, *49*(8), 1596–1605. https://doi.org/10.1021/bi9019912.

Deana, A., Celesnik, H., & Belasco, J. G. (2008). The bacterial enzyme RppH triggers messenger RNA degradation by 5′ pyrophosphate removal. *Nature*, *451*(7176), 355–358. https://doi.org/10.1038/nature06475.

Ferré-D'Amaré, A. R., & Doudna, J. A. (1996). Use of cis- and trans-ribozymes to remove 5′ and 3′ heterogeneities from milligrams of in vitro transcribed RNA. *Nucleic Acids Research*, *24*(5), 977–978.

Fiegland, L. R., Garst, A. D., Batey, R. T., & Nesbitt, D. J. (2012). Single-molecule studies of the lysine riboswitch reveal effector-dependent conformational dynamics of the aptamer domain. *Biochemistry*, *51*(45), 9223–9233. https://doi.org/10.1021/bi3007753.

Frieda, K. L., & Block, S. M. (2012). Direct observation of cotranscriptional folding in an adenine riboswitch. *Science*, *338*(6105), 397–400. https://doi.org/10.1126/science.1225722.

Ha, T., Rasnik, I., Cheng, W., Babcock, H. P., Gauss, G. H., Lohman, T. M., et al. (2002). Initiation and re-initiation of DNA unwinding by the *Escherichia coli* Rep helicase. *Nature*, *419*(6907), 638–641. https://doi.org/10.1038/nature01083.

Heilman-Miller, S. L., & Woodson, S. A. (2003). Perturbed folding kinetics of circularly permuted RNAs with altered topology. *Journal of Molecular Biology*, *328*(2), 385–394.

Hellenkamp, B., Schmid, S., Doroshenko, O., Opanasyuk, O., Kuhnemuth, R., Rezaei Adariani, S., et al. (2018). Precision and accuracy of single-molecule FRET measurements-a multi-laboratory benchmark study. *Nature Methods*, *15*(9), 669–676. https://doi.org/10.1038/s41592-018-0085-0.

Hermanson, G. T. (2018). *Bioconjugate techniques* (2nd ed.). New York: Academic Press.

Hua, B., Han, K. Y., Zhou, R., Kim, H., Shi, X., Abeysirigunawardena, S. C., et al. (2014). An improved surface passivation method for single-molecule studies. *Nature Methods*, *11*(12), 1233–1236. https://doi.org/10.1038/nmeth.3143.

Hua, B., Panja, S., Wang, Y., Woodson, S. A., & Ha, T. (2018). Mimicking co-transcriptional RNA folding using a superhelicase. *Journal of the American Chemical Society*, *140*(32), 10067–10070. https://doi.org/10.1021/jacs.8b03784.

Hwang, H., & Myong, S. (2014). Protein induced fluorescence enhancement (PIFE) for probing protein-nucleic acid interactions. *Chemical Society Reviews*, *43*(4), 1221–1229. https://doi.org/10.1039/c3cs60201j.

Jones, C. P., & Ferré-D'Amaré, A. R. (2014). Crystal structure of a c-di-AMP riboswitch reveals an internally pseudo-dimeric RNA. *The EMBO Journal*, *33*, 2692–2703. https://doi.org/10.15252/embj.201489209.

Jones, C. P., & Ferré-D'Amaré, A. R. (2015). Recognition of the bacterial alarmone ZMP through long-distance association of two RNA subdomains. *Nature Structural & Molecular Biology*, *22*(9), 679–685. https://doi.org/10.1038/nsmb.3073.

Jones, C. P., & Ferré-D'Amaré, A. R. (2017). Long-range interactions in riboswitch control of gene expression. *Annual Review of Biophysics, 46*, 455–481. https://doi.org/10.1146/annurev-biophys-070816-034042.
Kao, C., Zheng, M., & Rudisser, S. (1999). A simple and efficient method to reduce nontemplated nucleotide addition at the 3 terminus of RNAs transcribed by T7 RNA polymerase. *RNA, 5*(9), 1268–1272.
Lazinski, D. W., & Taylor, J. M. (1995). Intracellular cleavage and ligation of hepatitis delta virus genomic RNA: Regulation of ribozyme activity by cis-acting sequences and host factors. *Journal of Virology, 69*(2), 1190–1200.
Liberman, J. A., Suddala, K. C., Aytenfisu, A., Chan, D., Belashov, I. A., Salim, M., et al. (2015). Structural analysis of a class III preQ1 riboswitch reveals an aptamer distant from a ribosome-binding site regulated by fast dynamics. *Proceedings of the National Academy of Sciences of the United States of America, 112*(27), E3485–E3494. https://doi.org/10.1073/pnas.1503955112.
Liu, Y., Holmstrom, E., Zhang, J., Yu, P., Wang, J., Dyba, M. A., et al. (2015). Synthesis and applications of RNAs with position-selective labelling and mosaic composition. *Nature, 522*(7556), 368–372. https://doi.org/10.1038/nature14352.
Milligan, J. F., & Uhlenbeck, O. C. (1989). Synthesis of small RNAs using T7 RNA polymerase. *Methods in Enzymology, 180*, 51–62.
Moore, M. J., & Query, C. C. (2000). Joining of RNAs by splinted ligation. *Methods in Enzymology, 317*, 109–123.
Nussinov, R., & Tinoco, I., Jr. (1981). Sequential folding of a messenger RNA molecule. *Journal of Molecular Biology, 151*(3), 519–533.
Pan, T., Artsimovitch, I., Fang, X. W., Landick, R., & Sosnick, T. R. (1999). Folding of a large ribozyme during transcription and the effect of the elongation factor NusA. *Proceedings of the National Academy of Sciences of the United States of America, 96*(17), 9545–9550.
Pan, T., & Sosnick, T. (2006). RNA folding during transcription. *Annual Review of Biophysics and Biomolecular Structure, 35*, 161–175. https://doi.org/10.1146/annurev.biophys.35.040405.102053.
Perez-Gonzalez, C., Lafontaine, D. A., & Penedo, J. C. (2016). Fluorescence-based strategies to investigate the structure and dynamics of aptamer-ligand complexes. *Frontiers in Chemistry, 4*, 33. https://doi.org/10.3389/fchem.2016.00033.
Regulski, E. E., & Breaker, R. R. (2008). In-line probing analysis of riboswitches. *Methods in Molecular Biology, 419*, 53–67. https://doi.org/10.1007/978-1-59745-033-1_4.
Roy, R., Hohng, S., & Ha, T. (2008). A practical guide to single-molecule FRET. *Nature Methods, 5*(6), 507–516. https://doi.org/10.1038/nmeth.1208.
Walter, N. G., & Burke, J. M. (2000). Fluorescence assays to study structure, dynamics, and function of RNA and RNA-ligand complexes. *Methods in Enzymology, 317*, 409–440.
Wickiser, J. K., Cheah, M. T., Breaker, R. R., & Crothers, D. M. (2005). The kinetics of ligand binding by an adenine-sensing riboswitch. *Biochemistry, 44*(40), 13404–13414. https://doi.org/10.1021/bi051008u.
Wickiser, J. K., Winkler, W. C., Breaker, R. R., & Crothers, D. M. (2005). The speed of RNA transcription and metabolite binding kinetics operate an FMN riboswitch. *Molecular Cell, 18*(1), 49–60. https://doi.org/10.1016/j.molcel.2005.02.032.
Wood, S., Ferré-D'Amaré, A. R., & Rueda, D. (2012). Allosteric tertiary interactions preorganize the c-di-GMP riboswitch and accelerate ligand binding. *ACS Chemical Biology*, 7(5), 920–927. https://doi.org/10.1021/cb300014u.
Zhang, J., Lau, M. W., & Ferré-D'Amaré, A. R. (2010). Ribozymes and riboswitches: Modulation of RNA function by small molecules. *Biochemistry, 49*(43), 9123–9131. https://doi.org/10.1021/bi1012645.

CHAPTER ELEVEN

Strategies for understanding RNA recognition by X-ray crystallography and NMR methods

Aiai Sun, Kaiyi Huang, Luqian Zheng, Aiming Ren*
Life Sciences Institute, Zhejiang University, Hangzhou, Zhejiang, China
*Corresponding author: e-mail address: aimingren@zju.edu.cn

Contents

Abstract

One class of non-coding RNA molecules, termed riboswitches, are the regulatory elements with ability to control gene expression by sensing different cellular metabolites. Specific recognition of the corresponding ligand induced the conformation rearrangement of riboswitch, and thus turned on/off the down-stream gene expression. In this chapter, we will focus on two principal methods that are currently used to investigate the recognition mechanism of riboswitches, including X-ray crystallography and

Methods in Enzymology, Volume 623
ISSN 0076-6879
https://doi.org/10.1016/bs.mie.2019.05.024

NMR spectroscopy. First, we present that how to get the diffraction-quality crystals of riboswitches in specific functional states. Then we focus on how to use NMR spectroscopy as a tool to learn the dynamics of riboswitches. Detailed protocols are listed below.

1. Introduction

Many non-coding RNA molecules have been found with significant function in diverse biological process (Eddy, 2001). Riboswitches are conserved non-coding RNA elements located in 5′-untranslated regions of mRNA molecules (Breaker, 2011). These naturally occurring RNA elements usually consist of two domains: the highly conserved sensor domain for ligand recognition and the followed variable sequences, termed as expression platform (Nudler & Mironov, 2004). As illustrated in Fig. 1, riboswitches usually regulate the down-stream gene expression with different mechanisms at transcription or translation level (Serganov & Nudler, 2013).

Riboswitches have been found to sense almost 40 classes of diverse cellular metabolites including ions, purines and their derivatives, cofactors and their derivatives, amino acids and so on (Breaker, 2011; Greenlee et al., 2018). Reaching to a threshold level, these typical cellular metabolites would be recognized by the sensor domain of their corresponding riboswitches, which induce structural rearrangement of both the sensor domain and expression platform that will be able to regulate the down-stream gene expression (Fig. 1) (Serganov & Patel, 2012). The ligand specific recognition of a riboswitch is almost solo determined by the sensor domain. The sequences of the sensor domain in the same riboswitch class are highly conserved between different species. They usually adopt the common secondary and tertiary core fold. Each riboswitch class has their own unique conserved sequence and tertiary fold, even some of them recognize the same ligands (Batey, 2011; Serganov & Nudler, 2013).

To investigate the recognition mechanism of riboswitches, the ligand-induced conformational rearrangement between the sensor domain and the expression platform was needed to be captured, which involves at least two states of the sensor domain that binding with ligand (ligand-bound state) and without ligand (ligand-free state) (Bothe et al., 2011; Golden & Kundrot, 2003; Ke & Doudna, 2004; Serganov & Patel, 2012). The methods to solve or detect the riboswitch structures of ligand-free state and ligand-bound state and their conformational rearrangement dynamics are described below.

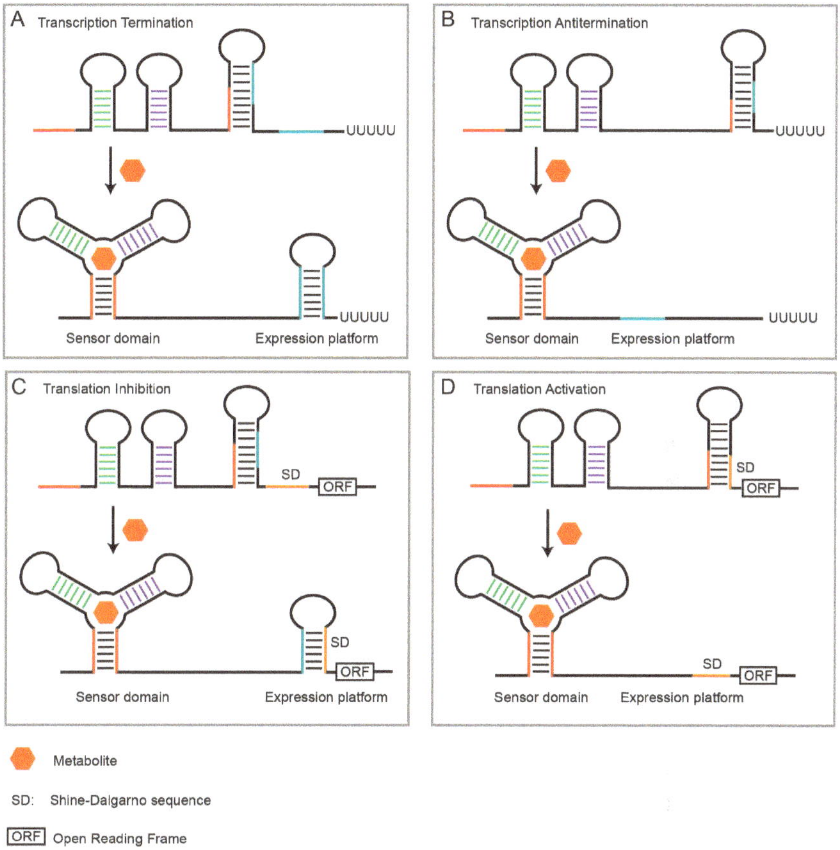

Fig. 1 Riboswitches regulate gene expression with diverse mechanisms. The typical cellular metabolites bind to their corresponding riboswitches with structural rearrangement in sensor domain accordingly that further induce the conformational change in the expression platform and control the down-stream gene expression *via* (A) Transcription Termination, (B) Transcription Antitermination, (C) Translation Inhibition, (D) Translation Activation.

2. Diverse methods used for analyzing RNA recognition pattern

Crystallography, cryo-electron microscopy (Cryo-EM) and nuclear magnetic resonance (NMR) are the principal methods currently used for structure determination. The length of most riboswitches varies from tens to hundreds of nucleotides (Breaker, 2011), which excludes cryo-EM to be applied for their structure determination currently. Most reported riboswitches structures in PDB data bank were solved by X-ray crystallography or NMR methods (Bothe et al., 2011; Golden & Kundrot, 2003). Both X-ray crystallography or NMR methods have their own advantages and disadvantages. We will summarize the accumulated practical experience of the past decades study in Section 4 in this chapter.

2.1 RNA crystallization

As to the overall workflow, the crystallization of riboswitches is essentially similar to proteins (Golden & Kundrot, 2003). Application of *in vitro* transcription (normally followed by a self-cleavage HDV or hammerhead ribozyme at 3′-end of the target RNA to ensure the homogeneity) or solid-phase synthesis, large amount of the target RNA could be generated (Pikovskaya, Serganov, Polonskaia, Serganov, & Patel, 2009). Following by different strategies of purification, highly pure target RNA could be prepared. Besides purity and homogeneity, some other aspects also need to be considered in RNA crystallization (Flores, Walshe, & Ataide, 2014).

First and foremost, the choice of the target RNA sequence is critical in crystallization (Pikovskaya et al., 2009). Riboswitches are usually identified from many different species, in which these riboswitches composed of not only the highly conserved sequence and motif, but also variable sequence and loop composition. It is recommended to start from three to five representative riboswitch sequences that contain most of the evolutionarily conserved residues, together with the shortest predicted helices and the fewest residue number of bulges and other irregularities. The variable loop regions are normally flexible and may interfere with crystal packing, thus it is better to replace the variable loop region with regular stable tetra-loop or some protein-bound loop that could be applied co-crystallization with related proteins. Details for these trials would not be discussed here and multiple successful cases are available for further reading (Dagenais, Girard, Bonneau, & Legault, 2017; Huang & Lilley, 2018; Liberman et al., 2015; Lu et al., 2008; Ren et al., 2015).

Second, the annealing operation of RNA sample is necessary before crystallization (Flores et al., 2014). The ligand selectivity of a riboswitch is critical for the gene-expression regulation activity, which depends on the tertiary fold of the riboswitch (Serganov & Nudler, 2013). However, in most cases, *in vitro* transcribed RNA is purified in denatured condition containing 8.0 M urea (Cheong, Hwang, Lee, Choi, & Cheong, 2004) that makes RNA molecules become disordered. The annealing procedure prior to crystallization is necessary to make these RNA molecules fold correctly and homogenously. The start annealing condition for regular RNA folding can be chosen with different temperature ranging from 50 to 90 °C for 1–5 min, and followed with fast cooling on ice (Pikovskaya et al., 2009). Cations, especially magnesium ion that would be able to facilitate the tertiary structure folding of RNA molecules and contribute to ligand recognition are usually needed in annealing buffer. The recommended concentration of magnesium ion to start is ranging from 1 mM to 10 mM, the optimal one of which that reinforce the homogeneity of RNA can be evaluated by the native PAGE assay. It should be kept in mind that high temperature and high concentration of magnesium ion could sometimes cause the degradation of some RNA molecules (Golden & Kundrot, 2003).

Third, solving the phase problem is sometimes still challenging for crystal structure determination (Golden & Kundrot, 2003). Similar to protein crystallization, two optional solutions are available for this challenge. One is introducing the heavy atom into the specific nucleic acid such as substitution of uridine residue with its derivative 5-bromouridine (Gilbert, Montange, Stoddard, & Batey, 2006). However, with 5-bromouridine replacement of uridine, sometimes good-quality crystals are not easy to obtain from its original crystallization conditions, because the introduction of bromine may interfere with the ligand specific recognition, the correct structural folding of the RNA molecules or the crystal packing. Thus, the positions to introduce 5-bromouridine replacements should be tested, which are usually located beyond recognition site. Crystals with successful introduction of 5-bromouridine at specific sites will diffract strong anomalous scattering signal, which is helpful for sequence tracing and structural model building. Another way is soaking crystals with heavy metal cations such as iridium and osmium (Zheng et al., 2017). It was reported that iridium has preference to bind with the major side of G·U wobble pair in a standard A-form helix (Golden & Kundrot, 2003). But iridium soaking may damage the crystal packing, which may also result in bad diffraction. Thus, different concentration of iridium with different soaking time are also needed to be tested in heavy metal cations soaking experiments.

2.2 NMR spectroscopy of RNA molecules

For decades, NMR spectroscopy has been a powerful tool used not only for structure determination but also for the dynamic study of some biological molecules (Bothe et al. 2011; Furtig, Richter, Wohnert, & Schwalbe, 2003).

The binding selectivity and affinity for ligands are two key parameters in discussion of riboswitches (Breaker, 2011). Riboswitches can recognize the corresponding metabolites normally with high selectivity and high affinity, but strongly discriminate against other related metabolites even in high concentrations (Serganov & Patel, 2012). The conformational changes in response to certain metabolites will provide the necessary information to learn the recognition mechanism of riboswitches (Serganov & Nudler, 2013).

The prerequisite for NMR spectroscopy is preparing milligram quantities of isotope-labeled (^{13}C, ^{15}N) RNA samples (Furtig et al., 2003). NMR spectroscopy does not require crystallization of samples, thus riboswitches with original sequences containing flexible loops are well suitable for NMR experiments (Bothe et al., 2011). Natural abundant carbon and nitrogen are ^{12}C isotope and ^{14}N isotope, both of which are not suitable for NMR measurement as ^{12}C is NMR inactive and ^{14}N possesses an electric quadrupole moment. Therefore, ^{12}C and ^{14}N have to be replaced with NMR-active ^{13}C and ^{15}N nuclei during sample preparation. To enhance the signal intensity of NMR, an RNA sample with millimolar concentration is recommended in NMR experiments (Cheong et al., 2004).

With ^{13}C, ^{15}N-labeled RNA sample in hand, the critical step in NMR spectroscopy is the resonance assignment of NMR-active atoms (^{1}H, ^{13}C, ^{15}N) in the molecule (Furtig, Richter, Bermel, & Schwalbe, 2004; Furtig et al., 2003; Ren et al., 2015). Every NMR-active atom shows a distinct chemical shift due to its certain chemical environment, which can be viewed as the only ID for further NMR data analysis. Compared to proteins composing of 20 amino acids with distinct side chains, RNAs consist of only four chemical structure-similar nucleotides. Moreover, most structured RNAs adopt A-form helix as their basic secondary structure element, which provides a similar chemical environment for nucleotides within the element. Less number of building units and similar chemical environment result in much more severe signal overlap in RNA spectra, which impedes the assignment of chemical shift for relatively large RNAs such as riboswitches (Furtig et al., 2003).

Whereas, in the non-canonical structural elements such as bulges, apical loops or internal loops, the observed chemical shifts are usually dispersed quite well due to the distinct local structure. The non-canonical structural elements in most cases play a more important role for ligand recognition of riboswitches (Breaker, 2011), and thus NMR is suitable for characterizing recognition mechanism of riboswitches (Ren et al., 2015). For structure determination of the large RNA or RNA-protein complex, multiple labeling techniques including site-specific labeling or segmental labeling should be introduced, which could drastically reduce the resonance overlap (Duss, Maris, von Schroetter, & Allain, 2010; Liu et al., 2015; Wunderlich et al., 2015).

NMR titration experiment is used to detect conformation rearrangement and binding interface during ligand binding (Lu et al., 2011). Unlabeled ligand that is dissolved in the NMR buffer is added into the isotope-labeled RNA sample with molecular ratio typically ranging from 0 to 10. At each ratio NMR 2D spectra of NH and/or CH group are acquired. The NMR signal of the nucleoside involved in ligand binding or conformational rearrangement will be perturbed obviously in spectra that gives out the related information with regard to ligand recognition (Ren et al., 2015).

2.3 Residual dipolar couplings measurement

Most structured RNA molecules contain at least two stems. These stem orientation varies during the conformational rearrangement of riboswitch RNA with or without ligand binding. Measuring residual dipolar couplings (RDCs), which therefore termed as RDC, is a promising NMR method for studying the dynamics of riboswitches (Getz, Sun, Casiano-Negroni, Zhang, & Al-Hashimi, 2007). *J* couplings, NOE contacts, residual dipolar couplings, and relaxation rates are NMR parameters acquired by different experiments with distinct representative principles (Furtig et al., 2003). With the given NMR-active atom modulated by dipole-dipole interaction from an adjacent nucleus with additional local magnetic field, residual dipolar couplings provide the long-range angular restraints (Getz et al., 2007).

Due to isotropic orientation distribution of a molecule in solution, the value of residual dipolar coupling averages to zero. With alignment media hampering the isotropic reorientation of RNA molecules, the RDC value becomes detectable (Bothe et al., 2011). Several alignment media are available for RNA study, among which the filamentous bacteriophage (Pf1 phage) is the most frequently used one. Composed of genomic DNA and coat protein, Pf1 phage is rich of negative charge and adopts a rod-like shape with about 20,000 Å in length and about 60 Å in diameter, and thus readily induces slight alignment of RNAs in solution (Getz et al., 2007).

The In-Phase/Anti-Phase (IPAP) approach (de Alba & Tjandra, 2004) is popular for the measurement of RDC data (Kern, Schanda, & Brutscher, 2008). With inphase and antiphase J coupling spectra recorded, the upfield and the downfield components of a doublet are separated by calculating sum and difference of these two spectra, which avoids introducing additional crowdedness to spectra.

3. Equipment and chemicals

3.1 Equipment

(1) PCR system
(2) Gel imaging systems
(3) Gel electrophoretic apparatus (GE Healthcare Life Sciences)
(4) Incubator and shaker
(5) Centrifuge
(6) Water bath kettle
(7) Thin layer chromatography plates, 20 × 20 cm and 37 × 45 cm
(8) Elutrap systems, each with four elutrap chambers, which are assembled with BT1 and BT2 membranes (GE Healthcare Life Sciences)
(9) Vacuum freeze-drying machine
(10) Nanodrop
(11) 24 or 96 well plates
(12) Centrifuge tubes
(13) Bruker Avance 600 spectrometer or other NMR spectrometer
(14) NMR tube
(15) Depth gauge

3.2 Chemicals

(1) 2 × PCR mix (TSINGKE Biological Technology)
(2) LB medium: 10 g tryptone, 5 g yeast extract, 10 g NaCl, which are dissolved in 1 L water
(3) Ampicillin
(4) PureLink™ Hipure Expi Plasmid Megaprep Kit/Cartridges (invitrogen by Thermo Fisher Scientific)
(5) Universal DNA purification kit (Tiangen Biotech Co., Ltd.)
(6) TIANprep Mini Plasmid Kit (Tiangen Biotech Co., Ltd.)
(7) Restriction enzymes and 10 × fast digest buffer (Thermo Fisher Scientific)
(8) Agarose
(9) EDTA (ethylenediamine tetraacetic acid disodium salt)
(10) 50 × TAE: 242 g Tris, 57.1 mL acetate, 100 mL 0.5 M EDTA pH 8.0, dissolved to 1 L

(11) 10× TBE: 108g Tris, 55g boric acid, 40mL 0.5M EDTA pH 8.0, dissolved to 1L
(12) 29:1 acrylamide: bis acrylamide
(13) T4 DNA Ligase and 10× Ligation Buffer (Takara Bio)
(14) DH5α Competent cells
(15) Diethyl Pyrocarbonate (DEPC).
(16) RNase ERASE
(17) Chloroform/isoamyl alcohol (24:1)
(18) Phenol
(19) Bacteriophage T7 RNA polymerase
(20) ATP, UTP, GTP, CTP
(21) Magnesium chloride
(22) Sodium acetate
(23) Sodium chloride
(24) Isopropanol
(25) Ethanol
(26) HEPES
(27) Dithiothreitol
(28) Spermidine
(29) JCSG Core I Suite, JCSG Core II Suite, JCSG Core III Suite, JCSG Core IV Suite, Index, Crystal Screen I/II, Natrix
(30) $^{13}C/^{15}N$ labeled nucleotide triphosphates
(31) Pf1 phage

4. Protocol

With a flow scheme depicted in Fig. 2, the protocol here mainly consists of three parts: RNA sample preparation, RNA crystallization and NMR spectroscopy of RNA dynamic study.

4.1 Preparation of RNA sample for crystallization or NMR spectroscopy

4.1.1 Preparation of the recombination plasmid

(1) DNA templates preparation. DNA templates containing the corresponding sequence of the target RNA with a self-cleavage hammerhead or HDV ribozyme at 3′-end are obtained by assembly PCR or synthesized complementary with a DNA synthesizer. For crystallization, the sequences in DNA templates are recommended to use the regular tetra-loops or accessory protein bound loops to replace the variable loop

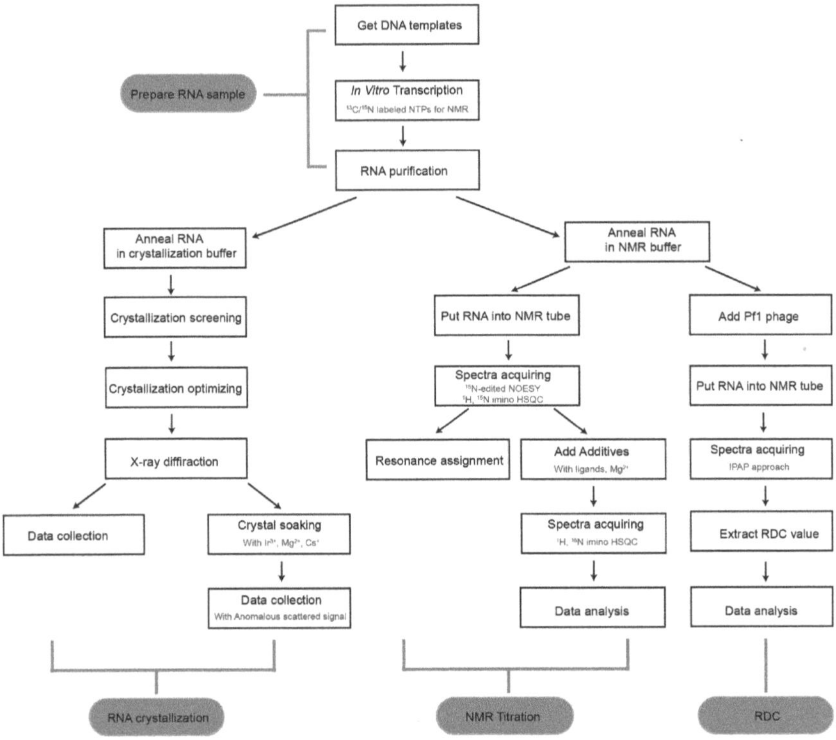

Fig. 2 The flow scheme of the methods mentioned in this chapter for studying riboswitch recognition mechanism.

regions. For NMR spectroscopy, DNA templates corresponding to the native riboswitch sequences are recommended.

(2) The recombinant plasmid preparation. Digest the PCR product and vector under the control of T7 promoter with the same restriction enzymes. In our case, restriction enzymes *Stu* I and *Hind* III are used for pUT7 vector and only *Hind* III is used for the PCR product. Then the digested PCR product and pUT7 vector are ligated with T4 ligase.

(3) Screening and selection of the correct constructs. Transform the ligated product into *E. coli* DH5α competent cell, select the grown single clone for further sequencing and save the right one for template preparation.

4.1.2 Preparation of DNA templates for transcription

(1) Transform recombinant plasmids with correct sequences into *E. coli* DH5α competent cell and inoculate the single clone into 3 mL LB medium supplemented with 100 μg/mL ampicillin followed by culturing at 37 °C for 4–6 h, then the medium volume is amplified to 1 L and continue culturing at 37 °C overnight.

(2) The overnight cultivated bacterial cells are harvested, and the plasmid DNA is purified with PureLink™ Hipure Expi Plasmid Megaprep Kit/Cartridges. The purified plasmid DNA is finally dissolved in 1.0–1.5 mL milli-Q water.

(3) Digest the recombinant plasmid containing the inserted DNA with restriction enzyme (in our case, *Hind* III restriction enzyme is used according to manufacturer procedure). To verify the complete digestion, gel electrophoresis with the concentration of 1% agarose is performed.

(4) Perform phenol/chloroform extraction. Add 150 μL chloroform/isoamyl alcohol mixture (24:1) to 600 μL digestion system, followed by adding 150 μL phenol, shaking gently in each step. Centrifuge the sample at 12,000 rpm for 5 min. Carefully collect the upper aqueous phase and repeat the above step only with chloroform/isoamyl alcohol mixture. The supernatant of last step is precipitated with isopropanol. The DNA pellet is washed by 80% ethanol. Perform the air drying for the DNA pellet and dissolve the dried DNA in DEPC-H_2O. Then the linearized DNA template can be used for *in vitro* transcription.

4.1.3 Preparation of RNA samples

(1) Before preparing the large amount of RNA samples, the transcription conditions need to be optimized in small scale to maximize the RNA yield. The tested conditions include concentrations of $MgCl_2$ and DNA template. A general transcription reaction mixture contains 20 μL mixture of 100 mM Na-HEPES buffer pH 7.9, 60 mM DTT, 2 mM spermidine, 4 mM of each ribonucleotide triphosphate, 60–180 ng/μL of DNA template, 15–30 mM $MgCl_2$, and 2 units of T7 RNA polymerase. The reaction mixture will be incubated at 37 °C for 2–3 h. Then additional $MgCl_2$ is added to a final concentration of 30 mM and the reaction mixture is incubated at 37 °C for another 30 min with the purpose to cleave the 3′-end ribozyme from target RNA more completely.

(2) Use Urea-PAGE gel to test the yield of in vitro transcription of different conditions in small scale and choose the one with the highest yield to perform the large scale transcription in a total volume of 20 mL at 37 °C for 4 h. Then adjust the concentration of $MgCl_2$ to 30 mM and incubate at 37 °C for another 30 min to cleave the 3′-end ribozyme.

For NMR spectroscopy, the riboswitches can be transcribed with $^{13}C/^{15}N$ labeled nucleotide triphosphates for introduction of isotope labels into the RNA samples.

(3) Freeze the transcription mixture quickly by liquid nitrogen; thaw the mixture and centrifuge at 12,000 rpm for 30 min at 4 °C to remove the precipitate of magnesium pyrophosphate. Transfer the supernatant into a new 50 mL tube carefully. Concentrate it with Amicon Ultra-15 Centrifugal Filter to 2–3 mL.

(4) Mix the RNA with an equal volume of 2 × urea loading buffer. Load the sample to a large polyacrylamide gel (PAGE) and run it for a proper time (normally varies from 12 to 20 h depending on the length of the target RNA) at 50 W. Cut the target RNA band from the PAGE gel, from which RNA is extracted by electro-elution with elutrap-system. Ethanol precipitation and the followed washing and drying are finally performed in order to get the dry and pure RNA sample.

(5) Dissolve the pure RNA sample in about 5 mL DEPC-H_2O. Filter the solution to remove the acrylamide particles. Precipitate filtered RNA with ethanol. Dry the pellet using the lyophilizer and then dissolve it in about 100 μL water. Test the concentration of RNA by NanoDrop™ 2000 spectrophotometer.

4.2 Crystallization of RNA sample

4.2.1 Crystallization screening

(1) Anneal the purified RNA with or without ligand at 65 °C for 5 min in buffer containing 50 mM HEPES pH 7.0, 50 mM NaCl and 5 mM $MgCl_2$, followed by cooling on ice for 0.5–1 h before setting up crystallization trials. The RNA concentration is usually 0.5 mM.

(2) Screen the crystallization conditions with different commercial kits including Index, Crystal Screen I/II (Hampton Research), Natrix/Natrix2 (Hampton Research), JCSG core I–IV suites (Qiagen). Crystallization of RNA sample is usually performed by the sitting-drop or hanging-drop vapor diffusion method at 16 °C after mixing sample with reservoir solution at a 1:1 ratio.

(3) Check the crystallization screening conditions to get the diffracted crystals and optimize the initial crystallization condition to improve the diffraction quality of the crystals if needed; if there's no crystals appear in crystallization screening, optimization of the RNA constructs will be necessary.

(4) Transfer crystals into the cryoprotectant solution before flash-freezing in liquid nitrogen. These crystals can be used for native X-ray diffraction data collection if the diffraction is good enough.

4.2.2 X-ray diffraction data collection and structure determination

(1) Native diffraction data collection of RNA crystals is done in a similar way as protein crystals. Owing to the versatility of RNA molecules, it's difficult to use molecular replacement (MR) method to solve the tertiary structure of new RNA molecules. Most of the time, single anomalous diffraction (SAD) method or multiple anomalous diffraction (MAD) method will be applied to solve RNA molecule structure. To collect the anomalous data of RNA molecules, heavy atoms with anomalous signal are introduced into the native crystals by soaking native crystals in the crystallization condition supplemented with different concentration of heavy atoms or by incorporation the heavy atom derivative nucleotides into RNA molecule in the step of RNA sample preparation.

(2) After the anomalous data collection, HKL2MAP (Pape & Schneider, 2004), PHENIX (Afonine et al., 2012) or other programs will be used to locate the heavy atom position in crystal and solve the phase problem. Then the initial structure model will be refined with PHENIX or REFMAC (Murshudov et al., 2011) programs.

(3) To determine the position of the binding divalent magnesium cations, crystals can be soaked with manganese cations and followed by the collection of anomalous signal of manganese to locate magnesium cations. For monovalent metal atoms, cesium cations can be applied in a similar way to locate the binding sites.

4.3 NMR spectroscopy of RNA sample

4.3.1 NMR sample preparation

(1) Choose suitable buffer condition. To improve signal intensity, the buffer pH under 7.0 with low salt concentration is strongly recommended, which is good for reducing the rate of $^1H/^1H$ exchange between RNA and solvent H_2O. In our case for L-glutamine riboswitch study, 50 mM potassium acetate at pH 6.8 is used. To determine the significance of magnesium ion, buffer with or without magnesium ion is adopted for the same spectrum recording.

(2) Exchange RNA sample into NMR buffer. The purified RNA samples are concentrated or dialysis into certain NMR buffer containing 5–10% D_2O for magnetic field locking.

(3) Anneal RNA sample. The annealing procedure of NMR sample is the same as that described in Section 4.2.1 with proper annealing temperature followed with fast cooling on ice.

(4) Clean NMR tube. For brand new tube, simple soaking in cleaning solutions and rinsing is enough. For used tube, metal-free sulfuric acid as a strong acid is recommended before rinsing. The cleaned tube is further dried by blowing a stream of filtered dry nitrogen gas against the bottom of the tube.

(5) Transfer sample into NMR tube. Carefully transfer RNA sample into NMR tube with pipette. Cap the tube and send it to spectrometer for data acquisition.

(6) Sample preparation for RDC. The stock Pf1 phage often needs to be customized so that it is prepared in the desired NMR buffer. The final concentration of Pf1 depends on the shape of RNA, and typically 10–25 mg/mL is suitable for riboswitches. Relatively small RDCs (within $\pm$40 Hz) are required for solution NMR studies, otherwise data analysis will be hampered by severe line broadening. An optimal pf1 concentration will lead to RDCs with maximum absolute value about 20–30 Hz.

4.3.2 NMR spectrum recording

(1) Injection of NMR tube. Adjust tube position in spinner according to the sample amount with sample depth gauge. Inject target NMR tube into the spectrometer under computer control.

(2) Tuning and matching probe. Start probe tune with one nuclei at once (first ^{15}N, then ^{13}C and last ^{1}H) until all the nuclei are tuned to their resonance frequencies and the reflected power is minimized in all channels.

(3) Locking and shimming magnetic field. Lock the field to a proper state and optimize it to the topmost level. It is advisable to perform gradient shimming to optimize high-order shims efficiently whenever available.

(4) Acquiring spectra. Load the target experimental setup. After calibration of pulse durations and power levels, set acquisition parameters and start experiment.

(5) Acquiring ligand titration spectra. Dissolved ligand in the same NMR buffer as used for RNA sample at relatively high concentration. Add the ligand into RNA sample carefully and slightly centrifuge after mixing them together. For example, one can use a TDZ5-WS/TDZ5WS centrifuge machine (CENCE, China) with No. 2 rotor matched with

NMR tube adaptor and spin at 1000 rpm for 1 min. One can also remove the sample from the NMR tube to add the ligand and reload the sample after mixing.

(6) The molecular ratio values between RNA and ligand vary depend on binding affinity. One or more spectra will be recorded at each ratio.

5. Structure analysis

The secondary structure of most reported riboswitches were predicted based on comparative sequence analysis by Dr. Breaker's lab (Breaker, 2011) and then validated by the experiments such as in-line probing, Selective 2′-Hydroxyl Acylation analyzed by Primer Extension (SHAPE) (Wilkinson, Merino, & Weeks, 2006) or other methods (Weeks, 2015). However, the secondary structure was limited to provide the structural information including multiple base-pairing interactions, the non-canonical base pairing long-range interaction, the metal involved interaction and so on (Zheng et al., 2017). Comparing to the secondary structure, the tertiary structure could provide detailed structural information for riboswitches. In tertiary structure, not only the canonical base pairs can be formed, but also the non-canonical base pairs such as G:U wobble base pair, G:A base pair and many others (Leontis, 2002). Variety of long-range interaction in RNA tertiary structure involves base pairing interaction, and can also be mediated by stacking interaction and by metal ions (Ren, Rajashankar, & Patel, 2012). These long-range interactions usually function in stabilizing the overall tertiary structure or forming binding pocket for proteins and some small molecule ligands. The metal ions, especially divalent cations such as magnesium ion, are essential for RNA to fold into the correct tertiary structure and sometimes even affect their specific activities. As we know, some riboswitches would not bind to their corresponding ligand without magnesium (Ren et al., 2012) and the small self-cleavage ribozymes also need magnesium (Ren et al., 2014) to start self-cleaving reaction. So in analysis of the RNA structure, we should pay more attention to these features to avoid missing critical structural information.

6. Dynamics

6.1 Chemical shift assignment

In general, H1 or H3 chemical shift of imino (HN) groups which exist only in base-paired guanine and uridine is around 9–15 ppm, while H2, H6 and H8 of bases around 6.5–8.5 ppm and H5 of bases around 5–6 ppm (Furtig et al., 2003). Except for H1′ which is around 5–6 ppm, the other ribose protons are around 3–5 ppm, which is a region severely overlapped in 1D ^{1}H spectrum. In NMR study of RNAs, the imino proton resonances are very useful, because they are well resolved, less crowded and very sensitive to structure rearrangement and conformational change (Bothe et al., 2011).

^{15}N-edited NOESY is a commonly used three-dimensional experiment for isotope-labeled RNAs, which records cross peaks with space distance within approximately 4–6 Å depending on mixing time (Furtig et al., 2004). Through a sequential walk, the A-form helix region can be assigned as the distance of imino protons from a base pair or adjacent base pairs are around 4 Å apart. With imino proton (H1) within guanine and imino proton (H3) within uridine both involved, G:U wobble base pairs are easy to be identified with intense cross peaks present from each other with the chemical shift around 10–12 ppm for guanine and 11–12.5 ppm for uridine. The imino protons of Watson–Crick G:C base pairs or A:U base pairs present around 12–14 ppm or 13–15 ppm, respectively, while imino protons of the non-canonical G:A base pairs appear more downfield (Furtig et al., 2003; Ren et al., 2015; Xue et al., 2015).

Based on abundant experimental resonance assignment deposited in biological magnetic resonance bank (BMRB) database, multiple methods for RNA chemical shift prediction with the solved crystal structures or NMR structures can be much helpful for resonance assignment (Frank, Bae, & Stelzer, 2013).

6.2 NMR titration

^{1}H, ^{15}N imino HSQC spectrum serves as a well-resolved “fingerprint” of an RNA for each peak within the spectrum can be traced to one corresponding imino group (HN) (Bothe et al., 2011). The chemical shift in either proton or nitrogen dimension is sensitive to the chemical environment surrounding the imino group. Thus, analysis of a series of ^{1}H, ^{15}N imino HSQC spectra recorded under different condition normally presents chemical shift perturbation and/or changes in peak intensity (Lu et al., 2011).

The chemical shift variation takes place under two conditions (Dagenais et al., 2017). One is the nucleotides involved in the binding interface toward additives. The other is the nucleotides involved in conformation rearrangement. Analysis of the chemical shift changes and mapping of these shifts on the riboswitch structures provide structural information of the binding interface and the induced conformational changes in the riboswitch. The additives for RNA NMR titration vary due to diverse purpose (Lu et al., 2011). As mentioned before, magnesium ion in most cases is important for tertiary structure folding. Titrated with magnesium ion is a powerful method to characterize this dynamic. And of course, to detect ligand binding dynamic, the corresponding ligand is a good choice for titration (Ren et al., 2015).

6.3 RDC data analysis

In RDC data analysis, the ^{1}H, ^{15}N imino HSQC spectra in solution and in alignment medium are recorded using IPAP approach (Kern et al., 2008). The value of J coupling together with dipolar coupling (J + D) is acquired with RNA sample in alignment medium, while in solution without alignment media only J coupling (J) is recorded. The observed dipolar coupling (D_{obs}) is obtained by subtracting J from J + D (Kern et al., 2008; Ottiger, Delaglio, & Bax, 1998).

In this way, D_{obs} of riboswitches in free state or ligand-bound state or other states can be obtained. Comparison of these different D_{obs} datasets, usually presented as a correlation plot, is often discrepant, revealing orientation motion during ligand binding. With the solved crystal structure available, D_{obs} can be fitted to the free state structure and/or ligand-bound state structure *via* programs such as RAMAH (Hansen & Al-Hashimi, 2006) or PALES (Zweckstetter & Bax, 2000). The agreement between experimental RDCs and calculated RDCs based on a crystal structure indicates the consistence of riboswitches structure in solution and the solved structure in crystalline environment (Ren et al., 2015).

7. Summary

In this chapter, we described two principal techniques that are used for studying riboswitch recognition mechanism. Start with a brief introduction, we show the possible challenges and solutions during riboswitch molecules crystallization and NMR spectroscopy. Following a detailed protocol, we additionally discuss the possible recognition mechanism of riboswitches and the data process and analysis details for riboswitches dynamics. The experience summarized here is useful not only for research of riboswitches recognition mechanism but also for structural and dynamic study of other small RNA molecules.

Acknowledgments

We thank Dr. Yi Xue (School of Life Sciences, Tsinghua University), Dr. Pengzhi Wu (ETH Zurich) and Dr. Dinshaw J. Patel (Memorial Sloan-Kattering Cancer Center) for their valuable comments and suggestions in the manuscript revision. The research was funded with grants from the Natural Science Foundation of China (91640104, 31670826 and 31870810), the outstanding youth fund of Zhejiang Province (LR19C050003), the Fundamental Research Funds for the Central Universities (2017QN81010), the new faculty start-up funds from Zhejiang University.

References

Afonine, P. V., Grosse-Kunstleve, R. W., Echols, N., Headd, J. J., Moriarty, N. W., Mustyakimov, M., et al. (2012). Towards automated crystallographic structure refinement with phenix.refine. *Acta Crystallographica. Section D, Biological Crystallography*, *68*, 352–367. Pt. 4.

Batey, R. T. (2011). Recognition of S-adenosylmethionine by riboswitches. *Wiley Interdisciplinary Reviews. RNA*, *2*(2), 299–311.

Bothe, J. R., Nikolova, E. N., Eichhorn, C. D., Chugh, J., Hansen, A. L., & Al-Hashimi, H. M. (2011). Characterizing RNA dynamics at atomic resolution using solution-state NMR spectroscopy. *Nature Methods*, *8*(11), 919–931.

Breaker, R. R. (2011). Prospects for riboswitch discovery and analysis. *Molecular Cell*, *43*(6), 867–879.

Cheong, H. K., Hwang, E., Lee, C., Choi, B. S., & Cheong, C. (2004). Rapid preparation of RNA samples for NMR spectroscopy and X-ray crystallography. *Nucleic Acids Research*, *32*(10), e84.

Dagenais, P., Girard, N., Bonneau, E., & Legault, P. (2017). Insights into RNA structure and dynamics from recent NMR and X-ray studies of the Neurospora Varkud satellite ribozyme. *Wiley Interdisciplinary Reviews. RNA*, *8*(5), 1–18.

de Alba, E., & Tjandra, N. (2004). Residual dipolar couplings in protein structure determination. *Methods in Molecular Biology*, *278*, 89–106.

Duss, O., Maris, C., von Schroetter, C., & Allain, F. H. (2010). A fast, efficient and sequence-independent method for flexible multiple segmental isotope labeling of RNA using ribozyme and RNase H cleavage. *Nucleic Acids Research*, *38*(20), e188.

Eddy, S. R. (2001). Non-coding RNA genes and the modern RNA world. *Nature Reviews. Genetics*, *2*(12), 919–929.

Flores, J. K., Walshe, J. L., & Ataide, S. F. (2014). RNA and RNA–protein complex crystallography and its challenges. *Australian Journal of Chemistry*, *67*(12), 1741.

Frank, A. T., Bae, S. H., & Stelzer, A. C. (2013). Prediction of RNA 1H and 13C chemical shifts: A structure based approach. *The Journal of Physical Chemistry. B*, *117*(43), 13497–13506.

Furtig, B., Richter, C., Bermel, W., & Schwalbe, H. (2004). New NMR experiments for RNA nucleobase resonance assignment and chemical shift analysis of an RNA UUCG tetraloop. *Journal of Biomolecular NMR*, *28*(1), 69–79.

Furtig, B., Richter, C., Wohnert, J., & Schwalbe, H. (2003). NMR spectroscopy of RNA. *Chembiochem*, *4*(10), 936–962.

Getz, M., Sun, X., Casiano-Negroni, A., Zhang, Q., & Al-Hashimi, H. M. (2007). NMR studies of RNA dynamics and structural plasticity using NMR residual dipolar couplings. *Biopolymers*, *86*(5–6), 384–402.

Gilbert, S. D., Montange, R. K., Stoddard, C. D., & Batey, R. T. (2006). Structural studies of the purine and SAM binding riboswitches. *Cold Spring Harbor Symposia on Quantitative Biology*, *71*, 259–268.

Golden, B. L., & Kundrot, C. E. (2003). RNA crystallization. *Journal of Structural Biology, 142*(1), 98–107.
Greenlee, E. B., Stav, S., Atilho, R. M., Brewer, K. I., Harris, K. A., Malkowski, S. N., et al. (2018). Challenges of ligand identification for the second wave of orphan riboswitch candidates. *RNA Biology, 15*(3), 377–390.
Hansen, A. L., & Al-Hashimi, H. M. (2006). Insight into the CSA tensors of nucleobase carbons in RNA polynucleotides from solution measurements of residual CSA: Towards new long-range orientational constraints. *Journal of Magnetic Resonance, 179*(2), 299–307.
Huang, L., & Lilley, D. M. J. (2018). Structure and ligand binding of the SAM-V riboswitch. *Nucleic Acids Research, 46*(13), 6869–6879.
Ke, A., & Doudna, J. A. (2004). Crystallization of RNA and RNA-protein complexes. *Methods, 34*(3), 408–414.
Kern, T., Schanda, P., & Brutscher, B. (2008). Sensitivity-enhanced IPAP-SOFAST-HMQC for fast-pulsing 2D NMR with reduced radiofrequency load. *Journal of Magnetic Resonance, 190*(2), 333–338.
Leontis, N. B. (2002). The non-Watson-crick base pairs and their associated isostericity matrices. *Nucleic Acids Research, 30*(16), 3497–3531.
Liberman, J. A., Suddala, K. C., Aytenfisu, A., Chan, D., Belashov, I. A., Salim, M., et al. (2015). Structural analysis of a class III preQ1 riboswitch reveals an aptamer distant from a ribosome-binding site regulated by fast dynamics. *Proceedings of the National Academy of Sciences of the United States of America, 112*(27), E3485–E3494.
Liu, Y., Holmstrom, E., Zhang, J., Yu, P., Wang, J., Dyba, M. A., et al. (2015). Synthesis and applications of RNAs with position-selective labelling and mosaic composition. *Nature, 522*(7556), 368–372.
Lu, K., Heng, X., Garyu, L., Monti, S., Garcia, E. L., Kharytonchyk, S., et al. (2011). NMR detection of structures in the HIV-1 5'-leader RNA that regulate genome packaging. *Science, 334*(6053), 242–245.
Lu, C., Smith, A. M., Fuchs, R. T., Ding, F., Rajashankar, K., Henkin, T. M., et al. (2008). Crystal structures of the SAM-III/S(MK) riboswitch reveal the SAM-dependent translation inhibition mechanism. *Nature Structural & Molecular Biology, 15*(10), 1076–1083.
Murshudov, G. N., Skubak, P., Lebedev, A. A., Pannu, N. S., Steiner, R. A., Nicholls, R. A., et al. (2011). REFMAC5 for the refinement of macromolecular crystal structures. *Acta Crystallographica. Section D, Biological Crystallography, 67*(Pt. 4), 355–367.
Nudler, E., & Mironov, A. S. (2004). The riboswitch control of bacterial metabolism. *Trends in Biochemical Sciences, 29*(1), 11–17.
Ottiger, M., Delaglio, F., & Bax, A. (1998). Measurement of J and dipolar couplings from simplified two-dimensional NMR spectra. *Journal of Magnetic Resonance, 131*(2), 373–378.
Pape, T., & Schneider, T. R. (2004). HKL2MAP: A graphical user interface for macromolecular phasing with SHELXprograms. *Journal of Applied Crystallography, 37*(5), 843–844.
Pikovskaya, O., Serganov, A. A., Polonskaia, A., Serganov, A., & Patel, D. J. (2009). Preparation and crystallization of riboswitch-ligand complexes. *Methods in Molecular Biology, 540*, 115–128.
Ren, A., Kosutic, M., Rajashankar, K. R., Frener, M., Santner, T., Westhof, E., et al. (2014). In-line alignment and Mg(2)(+) coordination at the cleavage site of the env22 twister ribozyme. *Nature Communications, 5*, 5534.
Ren, A., Rajashankar, K. R., & Patel, D. J. (2012). Fluoride ion encapsulation by Mg^{2+} ions and phosphates in a fluoride riboswitch. *Nature, 486*(7401), 85–89.
Ren, A., Xue, Y., Peselis, A., Serganov, A., Al-Hashimi, H. M., & Patel, D. J. (2015). Structural and dynamic basis for low-affinity, high-selectivity binding of L-glutamine by the glutamine riboswitch. *Cell Reports, 13*(9), 1800–1813.

Serganov, A., & Nudler, E. (2013). A decade of riboswitches. *Cell*, *152*(1–2), 17–24.
Serganov, A., & Patel, D. J. (2012). Metabolite recognition principles and molecular mechanisms underlying riboswitch function. *Annual Review of Biophysics*, *41*, 343–370. Vol. 41. D.C. Rees.
Weeks, K. M. (2015). Review toward all RNA structures, concisely. *Biopolymers*, *103*(8), 438–448.
Wilkinson, K. A., Merino, E. J., & Weeks, K. M. (2006). Selective 2'-hydroxyl acylation analyzed by primer extension (SHAPE): Quantitative RNA structure analysis at single nucleotide resolution. *Nature Protocols*, *1*(3), 1610–1616.
Wunderlich, C. H., Juen, M. A., LeBlanc, R. M., Longhini, A. P., Dayie, T. K., & Kreutz, C. (2015). Stable isotope-labeled RNA phosphoramidites to facilitate dynamics by NMR. *Methods in Enzymology*, *565*, 461–494.
Xue, Y., Kellogg, D., Kimsey, I. J., Sathyamoorthy, B., Stein, Z. W., McBrairty, M., et al. (2015). Characterizing RNA excited states using NMR relaxation dispersion. *Methods in Enzymology*, *558*, 39–73.
Zheng, L., Mairhofer, E., Teplova, M., Zhang, Y., Ma, J., Patel, D. J., et al. (2017). Structure-based insights into self-cleavage by a four-way junctional twister-sister ribozyme. *Nature Communications*, *8*(1), 1180.
Zweckstetter, M., & Bax, A. (2000). Prediction of sterically induced alignment in a dilute liquid crystalline phase: Aid to protein structure determination by NMR. *Journal of the American Chemical Society*, *122*(15), 3791–3792.

CHAPTER TWELVE

Sensitive detection of structural features and rearrangements in long, structured RNA molecules

Rebecca L. Adams[a], Nicholas C. Huston[a], Rafael C.A. Tavares[a,b], Anna M. Pyle[a,b,c,*]

[a]Department of Molecular, Cellular and Developmental Biology, Yale University, New Haven, CT, United States
[b]Department of Chemistry, Yale University, New Haven, CT, United States
[c]Howard Hughes Medical Institute, Chevy Chase, MD, United States
*Corresponding author: e-mail address: anna.pyle@yale.edu

Contents

Abstract

Technical innovations in structural probing have drastically advanced the field of RNA structure analysis. These advances have led to parallel approaches developed in separate labs for analyzing RNA structure and dynamics. With the wealth of methodologies available, it can be difficult to determine which is best suited for a given application. Here, using a long, highly structured viral RNA as an example (the positive strand genome of Hepatitis C Virus), we present a semi-comprehensive analysis and describe

Methods in Enzymology, Volume 623
ISSN 0076-6879
https://doi.org/10.1016/bs.mie.2019.04.002

the major approaches for analyzing the architecture of RNA that is modified with structure-sensitive probes. Additionally, we present an updated method for generating *in vitro* transcribed and folded RNA that maintains native secondary structures in long RNA molecules. We anticipate that the methods described here will streamline the use of current approaches and help investigators who are unfamiliar with structure probing, obviating the need for time-consuming and expensive optimization.

1. Introduction

The ability of RNA to form specific structures is essential for its broad functionality (Dethoff, Chugh, Mustoe, & Al-Hashimi, 2012). Although this has been appreciated for short RNAs such as riboswitches and thermometers, long RNA molecules also contain functional structures, many of which occur at long-range. For example, interactions between the 5′ and 3′ ends of flaviviral genomes, which are >10 kb long, are known to regulate replication of the viral genome (Villordo & Gamarnik, 2009). Until recently, direct detection of conformational changes has not been technically feasible. Innovations in sequencing have provided the ability to assess RNA structure transcriptome-wide and at single-nucleotide resolution, but these developments have been less applicable for structural analysis of low-abundance, long RNAs in cells, and for establishing RNA structural features that are suitable for modeling the architecture of specific RNA structures. We anticipate that dynamic RNA structures play a key role in mediating transitions between phases of the viral life cycle, and the increased incidence of RNA-based viral outbreaks make exploration of viral genome structure essential for understanding viral biology. Additionally, the mechanistic function of long non-coding RNAs in development and disease has become an area of intense interest. It is clear that these RNAs are rich in functional structures, and structural analysis will be paramount for understanding their function.

In this study, we have analyzed the positive sense genome of Hepatitis C Virus (HCV, strain Jc1 Pietschmann et al., 2006), a virus with a long, unsegmented RNA genome (9.6 kb) as a model for calculating specific structures and how they change in response to conditions. Because HCV contains a combination of regions with known structure and less-well characterized regions (Adams, Pirakitikulr, & Pyle, 2017), it provides an ideal proof-of-principle context in which to assess the strengths and weaknesses of available chemical probing methods, library preparation and analyses. Additionally, HCV contains a well-characterized interaction site with the microRNA miR-122 for replication *in vivo* (Luna et al., 2015; Mortimer & Doudna, 2013). We therefore used this interaction as a test case to assess whether current structure probing approaches are sensitive enough to identify structural changes in the HCV genome.

2. RNA structure probes

Historically, RNA structures have been predicted using sequence alignments from diverse homologous sources, using structure-compatible co-variant nucleotides as support for functional base-paring interactions. For this approach, accurate structure prediction requires a "sweet spot" of sequence divergence: there must be enough similarity for accurate primary sequence alignment, but enough diversity so that co-variant base pairs can be identified. For viruses, stringent evolutionary constraints can severely restrict sequence diversity, and the recent emergence of a particular virus can result in sparse sequence information. As a result, viral sequence alignments which may be highly similar are unlikely to aid in structure prediction. At the opposite end of the spectrum, long non-coding RNAs have low primary sequence conservation, making sequence alignment challenging and prone to mistakes. In these cases, *de novo* structure prediction is ideally supplemented by empirical data on RNA secondary structure, obtained using RNA structure-sensitive probes.

The combination of classic and cutting-edge RNA structure probes permits analysis of diverse RNA structural states. The probes covalently modify individual nucleotides in a structure-sensitive manner, resulting in non-natural adducts that perturb reverse-transcriptase enzymes. Historically, DMS has been used to report on base-pairing *in vitro* and *in vivo*, as it modifies the Watson-Crick face of unpaired A and C nucleotides (Wells, Hughes, Igel, & Ares, 2000). To complement this nucleotide selectivity, EDC has been developed to modify unpaired G and U nucleotides *in vitro* and *in vivo* (Mitchell et al., 2019; Wang, Sexton, Culligan, & Simon, 2019). Reporting on nucleotide flexibility as a proxy for strandedness, SHAPE (selective 2′-hydroxyl acylation analyzed by primer extension) modifies the 2′-OH groups of conformationally flexible nucleotides with low apparent nucleotide preference (McGinnis, Dunkle, Cate, & Weeks, 2012; Wilkinson et al., 2009). SHAPE reagents include 1M7 (Mortimer & Weeks, 2007), NMIA (Wilkinson, Merino, & Weeks, 2006), and BzCN (Mortimer & Weeks, 2008) in addition to the probes specifically designed for *in vivo* probing, NAI and FAI (Spitale et al., 2013), and their azide derivatives (NAI-N_3 and FAI-N_3, Chan, Beasley, Zhen, & Spitale, 2018; Flynn et al., 2016). Finally, NAz was recently developed to report on RNA solvent accessibility *in vitro* and *in vivo*, and has the added benefit of specific activation upon exposure to UV_{310}, permitting sensitive control of probing time (Feng et al., 2018). As of the writing of this manuscript, DMS, BzCN, and NAI are currently commercially available (Sigma-Aldrich D186309; Sigma-Aldrich 115959; Millipore Sigma 03-310). Other probes, such as psoralen and its derivative AMT, have been adopted for direct detection of nucleotide base-pairing, and cleavage-based methods, including enzymes, terbium, and hydroxy-radical probing, report on complementary aspects of RNA folding, but these are outside the scope of this discussion.

2.1 HiTS analysis of RNA modification: An overview

With the advent of high-throughput sequencing technologies (HiTS), several labs have developed methods for the read-out of RNA modification. These approaches mostly rely on Illumina-based short-read sequencing of libraries generated from reverse transcription (RT) products and subsequent computational analysis with tailor-made software. The conversion of RNA to DNA and subsequent amplification permits analysis of specific and low-abundance RNAs from *in vivo* or *in vitro* probed sources and parallel analysis of long RNAs in a high-throughput manner. Despite the richness of available approaches, it is difficult to determine which is best suited for each researcher's needs and which is most sensitive to detecting structural dynamics. Because HiTS library generation and sequencing is inherently time-consuming and expensive, each researcher will want to ensure she is getting the best quality data for her efforts.

There are two major ways to read out RNA modification with reverse transcription (RT). Classically, modified nucleotides induce premature termination of cDNA by halting reverse transcription at the nucleotide 3′ to the modified nucleotide. RT stops have historically been detected by electrophoretic separation of short cDNA products (50–200 nucleotides) using labeled primers (Probe-Stop), and the advent of HiTS permits complete analysis of long and low-abundant RNAs in one experiment. Because the position of any given RT stop is unknown, Probe-Stop-HiTS requires ligation of a sequence adaptor during library generation. The most commonly used enzyme for this purpose is CircLigase, an enzyme adapted from a thermostable bacteriophage RNA ligase (Blondal et al., 2005) and which is commercially available (Lucigen, CL9021K). Despite the utility of CircLigase, it is known to have bias in sequence preference for ligated nucleotides (Kwok, Ding, Sherlock, Assmann, & Bevilacqua, 2013). Because the terminal nucleotide of cDNA serves as the acceptor during adaptor ligation and is the informative nucleotide in RT stop-based methods, this bias confounds determination of which nucleotides are most frequently modified. The Bevilacqua lab has developed an approach using T4 DNA ligase to overcome this sequence bias using a non-standard sequencing adaptor (Kwok et al., 2013). Adaptor ligation methods do not discriminate between extended cDNA and unextended primers, so primer must be removed prior to adaptor ligation to eliminate uninformative primer-only products that are preferentially amplified during library preparation and sequencing. The RT-stop method has been validated for all structure probes discussed above, but it suffers from inefficient enzymatic steps, bias during library preparation, and necessary removal of unwanted side-products.

To overcome these experimental difficulties, the Weeks lab developed the mutational profiling (MaP) approach for detecting modified nucleotides (Siegfried, Busan, Rice, Nelson, & Weeks, 2014). In MaP library generation, non-standard RT conditions (Mn^{2+} instead of Mg^{2+} as the divalent cation) are used to induce the MMLV-based SuperScript II reverse transcriptase to incorporate a mutation, rather than stop, at modified nucleotides. This generates cDNA products that can be amplified with specific primers, or used with commercially available second-strand synthesis kits to generate double-stranded DNA for subsequent steps in library generation. Currently, MaP has been used with 1M7, NMIA, BzCN, 1M6, DMS, and NAz probed RNA (Siegfried et al., 2014; Zinshteyn et al., 2019; Zubradt et al., 2017). Note, however, that MaP requires relatively high read depth for accurate quantification of modification rates, and there is a small bias in the types of nucleotides that are substituted during read-through of modified nucleotides.

2.2 Software to calculate nucleotide reactivity and predict RNA structure from HiTS data

With the adoption of HiTS to measure nucleotide reactivities, custom-made software has been created for calculating RT stops or mutation rates and reporting a normalized reactivity. With several labs adopting different approaches in parallel, there is not a standard suite of tools that is used universally in the field. One benefit of RNA probing is to aid in more accurate prediction of RNA structure. Therefore, software designed to output reactivities with normalized values that can be input directly to structure prediction software provides the highest utility and are the only tools detailed here. StructureFold2 has been developed to calculate reactivities from RT stops (Tack, Tang, Ritchey, Assmann, & Bevilacqua, 2018). Similarly, ShapeMapper2 contains the full suite of tools necessary to calculate reactivities from mutation rates (Busan & Weeks, 2018).

Probe reactivities can be used to constrain structure prediction in two broad ways (Lorenz, Hofacker, & Stadler, 2016). First, because high DMS and SHAPE reactivity indicate that a position is unlikely to be base-paired, nucleotides with reactivities above a certain threshold can be forced to be single-stranded during subsequent structure prediction steps. This use of reactivities as hard constraints will not be discussed further here. Alternatively, reactivity values can be used as soft constraints, biasing an ensemble of predicted structures such that structures that correlate with experimental data are scored highest. In the thermodynamic-based structure prediction of RNAStructure, which calculates a minimum free energy (MFE) structure using nearest-neighbor rules for base-pairing, reactivities are converted to pseudo-free energy constraints that bias thermodynamic-based structure prediction (Eq. 1) (Deigan, Li, Mathews, & Weeks, 2009).

$$\Delta G(i) = \boldsymbol{m} \ln[\mathit{reactivity}(i) + 1] + \boldsymbol{b} \tag{1}$$

Using this equation, the slope $\boldsymbol{m}$ and intercept $\boldsymbol{b}$ terms can be adjusted to change how reactivity values penalize base-paring of highly reactive nucleotides and reward base-pairing of nucleotides with low reactivity. This allows flexibility in how reactivities are used during MFE calculations, tuning the bias for prediction of structures that are consistent with experimental observations. RNAStructure (https://rna.urmc.rochester.edu/RNAstructure.html) has the added benefit of being available for local use by command line or GUI for users with limited text-interface experience (Reuter & Mathews, 2010). Additionally, the ViennaRNA package (https://www.tbi.univie.ac.at/RNA/) permits the use of soft constraints with a variety of algorithms (Lorenz, Luntzer, Hofacker, Stadler, & Wolfinger, 2016). Finally, TurboFold II conducts an iterative process that incorporates evolutionary and experimental data to predict structure using a maximum expected accuracy calculation (Tan, Sharma, & Mathews, 2017). To visualize how predicted structures match reactivity calculations, we typically use Varna (Darty, Denise, & Ponty, 2009), Assemble2 (http://www.bioinformatics.org/assemble/index.html), and the Draw tools from RNAStructure because they are user-friendly and facile.

2.3 Preparing samples of folded RNA *in vitro*

While structure probing of RNA *in vivo* can reveal key architectural elements and show how they respond in a biological setting, the presence of bound proteins and other factors often result in sparse data that is unsuitable for generating a *de novo* structural model that can serve as a starting point. To obtain a working architectural model, it is often useful to map the secondary structure of *in vitro* transcribed RNA that has been natively folded (Somarowthu et al., 2015), or to map RNA that has been isolated and purified gently from cells (*ex vivo* RNA) (Smola et al., 2016). Analysis of transcribed and folded RNA provides a simplified system (i.e., absence of protein interactions, nucleotide modification, quaternary RNA interactions) to obtain high resolution information on RNA secondary structure, to understand the motifs that are possible within a transcript, and to test the differential effect of binding partners on RNA folding. However, long RNA transcripts can aggregate and mis-fold, so care must to be taken to produce biologically relevant structures. Because local elements of RNA structure usually fold co-transcriptionally *in vivo* and *in vitro*, our lab has developed approaches to purify RNA such that co-transcriptional folding patterns are preserved, maintaining potentially functional structural elements. For the generation of long RNAs, we recommend the following semi-native purification protocol, which allows analysis of RNA folding with standard lab biochemical techniques. After semi-native purification, *in vitro* transcripts can be studied in much the same way as isolated, *ex vivo* samples of RNA.

2.4 PROTOCOL: Semi-native purification and modification of *in vitro* transcribed long RNA

We have adapted a native purification procedure that was developed for studying long noncoding RNAs (Chillón et al., 2015), and we included an updated analysis of RNA compaction and RNA structure probing. Before getting started, we recommend treating all surfaces (benchtop, pipettes, equipment) with RNAseZap and wearing gloves through every step of the procedure to limit RNase contamination. In an abundance of caution, we also use different pipets and benchtops for plasmid prep procedures, which often include RNase steps, and RNA work. Finally, all buffers are made in autoclaved glassware or RNase-free tubes.

Equipment

- FPLC system with UV_{260} detection
- Sephacryl S1000 SF, 24 mL column (self-packed or pre-packed)
- Benchtop centrifuge
- Nanodrop or alternate spectrophotometer to measure nucleic acid concentration
- 100 μL and 1 mL Hamilton syringes
- Heat block set to 37 and 70 °C
- Electrophoresis Apparatus for nucleic acid sequencing
- 500 mL 0.2 μm Filtration Unit (Fisher, 0974025A)
- 100 kDa NMWL/MWCO 0.5 mL Amicon Ultra filtration unit (MilliporeSigma, UFC510096)
- MicroSpin G-25 Columns (Sigma-Aldrich, GE27-5325-01)
- Vacuum Gel Dryer
- Phosphor Screens and Imager
- Agarose Electrophoresis System
- UV Transilluminator
- Thermocycler

Chemicals

- K-HEPES 1 M solution, pH 7.5
- Potassium Chloride, 1 M solution
- DTT, 1 M solution, stored in small aliquots at −20 °C
- EDTA, 500 mM solution, pH 8.0
- Tris-HCl, 1 M solution, pH 7.5
- Tris-HCl, 1 M solution, pH 8.0
- Calcium Chloride, 1 M solution
- Magnesium Chloride, 1 M solution
- MOPS, 1 M solution, pH 6.5
- Urea (Fisher, 15548274)
- Tris Base
- Boric Acid
- Lithium Chloride
- 5 M Betaine solution (Sigma, B0300)
- Sodium Acetate, 3 M solution, pH 5.2
- Spermidine (Sigma-Aldrich, 85558)
- Sodium Chloride, 5 M solution
- Triton X-100 solution (Sigma, G2025)
- 200 proof Ethanol, stored at −4 °C
- Sucrose

- Ammonium Persulfate
- *N,N,N′,N′*-Tetramethylethylenediamine (TEMED)
- Deionized Formamide
- Formaldehyde
- Xylene Cyanol
- Bromophenol Blue
- Sodium Hydroxide, 2 M
- Hydrochloric Acid, 1 M

Reagents

- Appropriate Restriction Enzyme to linearize plasmid for run-off transcription
- SUPERase In RNase Inhibitor (LifeTech, AM2694)
- Thermostable Inorganic Pyrophosphatase (NEB, M0296S)
- T7 RNA Polymerase (P266L mutation to reduce abortive transcripts, Tang, 2014)
- Nuclease-free Water (ThermoFisher, 10977015)
- ATP, CTP, GTP, and UTP, 100 mM, pH 7.0–7.5
- RQ1 RNase-Free DNase (Promega, M6101)
- Proteinase K Powder (ThermoFisher, 17916)
- 40% 29:1 Acrylamide (National Diagnostics, EC-852)
- SuperScript III Reverse Transcriptase Kit (ThermoFisher, 18080093 or 18080044)
- T4-PNK (NEB, M0201S)
- γ-^{32}P ATP (PerkinElmer, NEG035C001MC)
- dNTP Mix, 10 mM each (ThermoFisher, R0192)
- Dideoxynucleoside Triphosphate Set, 10 mM each (Sigma-Aldrich, 03732738001)
- Thermo Sequenase Cycle Sequencing Kit (ThermoFisher, 785001KT)
- RNaseZap (Life Technologies, AM9780)
- RNA structure probe: DMS (Sigma, D186309), NAI (MilliporeSigma, 03-310), or other desired probe, and appropriate solvent
- ssRNA Ladder (NEB, N0362S)
- SYBR Gold Nucleic Acid Stain (ThermoFisher, S11494)
- Lonza SeaKem LE Agarose (Fisher, BMA500002)

2.4.1 Construct

We generate long RNAs from a linearized plasmid containing a T7 RNA polymerase promoter sequence (Gößringer et al., 2014). For run-off transcription, we follow the desired RNA sequence immediately with a restriction site, requiring only a simple restriction digest and clean-up to generate the template. Note that some viral RNA sequences are toxic to bacteria and must be maintained on low-copy plasmids.

Protocol

A. Prepare Buffers and Solutions

Filter all buffers through a 0.2-μm filter to remove contaminating RNases.

10× Transcription Buffer

Component	Final concentration	Stock concentration	Volume added
Tris-HCl pH 8.0	400 mM	1 M	400 μL
NaCl	100 mM	5 M	20 μL
$MgCl_2$	120 mM	1 M	120 μL
Spermidine	20 mM	2 M	10 μL
Triton X-100	0.1%	100%	1 μL
Deionized H_2O			to 1 mL

Store at −20 °C.

Filtration Buffer

Component	Final concentration	Stock	Volume added
K-HEPES pH 7.5	50 mM	1 M	25 mL
KCl	150 mM	1 M	75 mL
EDTA pH 8.0	100 μM	500 mM	100 μL
Deionized H_2O			to 500 mL

Store at room-temperature, degas before each use.

Proteinase K Storage Buffer

Component	Final concentration	Stock	Volume added
Tris-HCl pH 7.5	10 mM	1 M	100 μL
$CaCl_2$	1 mM	1 M	10 μL
Glycerol	40%	100%	4 mL
Deionized H_2O			to 10 mL

Store at room-temperature.

LiCl solution

Component	Final concentration	Stock	Added
LiCl	7.5 M		15.89 g
EDTA, pH 8.0	50 mM	500 mM	5 mL
Deionized H_2O			to 50 mL

Store at room-temperature.

ME Buffer

Component	Final concentration (mM)	Stock	Added
MOPS, pH 6.5	8	1 M	400 µL
EDTA, pH 8.0	0.1	500 mM	10 µL
Deionized H_2O			to 50 mL

Store at room-temperature.

2× Urea loading dye

Component	Final concentration	Stock	Added
Urea			13.6 g
Tris-HCl pH 7.5	80 mM	1 M	800 µL
EDTA, pH 8.0	1.6 mM	500 mM	32 µL
Sucrose	16%	50%	3.2 mL
Xylene Cyanol/Bromophenol Blue	0.15%	2%	800 µL

Heat to 65 °C to dissolve all components and store at −20 °C in 1 mL aliquots.

10× TBE

Component	Final concentration (mM)	Added
Tris Base	890	108 g
Boric Acid	890	55 g
EDTA	2	7.4 g
Deionized H_2O		to 1 L

pH to 8.3 ± 0.3, store at room-temperature.

10% Polyacrylamide Denaturing Gel

Component	Final concentration	Stock	Added
Urea			84 g
Acrylamide	10%	40%	50 mL
TBE	1 ×	10 ×	40 mL
Deionized H_2O			to 200 mL

Weigh out urea, add TBE, and H_2O to approximately 120 mL. Heat at 65 °C to dissolve urea. Add acrylamide. Filter and store at 4 °C.

10× MOPS Running Buffer

Component	Final concentration (mM)	Stock	Added (mL)
MOPS, pH 6.5	200	1 M	100
NaOAc, pH 5.3	20	3 M	3.3
EDTA, pH 8.0	10	500 mM	10
Deionized H_2O			to 500

Dilute MOPS with H_2O to ~400 mL, adjust pH to 7.0 and add remaining components and H_2O to 500 mL.

10× Gel Loading Buffer

Component	Final concentration	Stock	Added
Glycerol	50%	100%	500 μL
Xylene Cyanol/Bromophenol Blue	0.15%	2%	75 μL
EDTA, pH 8.0	10 mM	500 mM	20 μL
Deionized H_2O			to 1 mL

B. Linearize Plasmid Template

A large amount of template is required to produce enough long RNA for downstream steps. We aim for >50 μg purified final template.

1. Incubate >50 μg plasmid with 100 U of appropriate restriction enzyme in 1 × final appropriate buffer for 250 μL final volume.
2. Incubate reaction overnight at 37 °C in heat block.

3. Run 5 μL plasmid on an agarose gel to ensure complete digestion.
4. Heat-inactivate at 70 °C for 20 min.
5. Add 25 μL 3 M NaOAc pH 5.2, 1 mL 100% EtOH, mix, and precipitate at −80 °C for 30 min.
6. Centrifuge for 30 min at full speed at 4 °C with bench-top centrifuge, pipette out all visible liquid while avoiding the pellet.
7. Wash with freshly prepared 70% EtOH, being careful to not dislodge pellet during washes.
8. Centrifuge for 10 min at full speed at 4 °C with bench-top centrifuge, pipette out all visible liquid while avoiding the pellet and air-dry for ~5 min.
9. Resuspend in ~50 μL nuclease-free H_2O and measure concentration with Nanodrop.

C. *In Vitro* Transcription

We express and purify T7 RNA polymerase in-house for a final concentration of ~2 mg/mL. We have observed that T7 RNA polymerase can be a source of RNase contamination, so a stringent purification protocol that eliminates nucleases is important.

The volumes listed here have been optimized to produce RNA for subsequent FPLC purification steps and result in enough RNA for multiple folding reactions. You may scale down if desired, but this should provide a good starting point.

We typically add NTPs in ratios that correspond to the sequence transcribed, and these volumes should be adjusted according to each target RNA.

With these conditions, the DNA template is the limiting reagent and may be titrated to increase the product yield.

1. Prepare *in vitro* transcription reaction

In vitro transcription

Component	Final concentration	Stock concentration	Volume added
Transcription Buffer	1×	10×	200 μL
ATP	2 mM	100 mM	40 μL
UTP	2.2 mM	100 mM	44 μL
CTP	2.8 mM	100 mM	56 μL
GTP	3 mM	100 mM	60 μL

DTT	10 mM	1 M	20 μL
T7 RNA Polymerase (P266L)	2.5 μg/mL	2 mg/mL	25 μL
Pyrophosphatase	25 U/mL	2000 U/mL	25 μL
SUPERase In	0.25 U/μL	20 U/μL	25 μL
DNA Template			50 μg
Nuclease-free H_2O			up to 2000 μL

2. Incubate at 37 °C for 2–4 h.
   We do not recommend incubating longer than 4 h.
   During transcription, we recommend pre-equilibrating the FPLC column in step D.1.
3. Add DNase mixture to *in vitro* transcription reaction degrade template, pipetting up and down rather than vortexing to mix.

DNase reaction

Component	Final concentration	Stock concentration	Volume added (μL)
10× DNase Buffer	1×	10×	400
DNase	25 mU/μL	1 U/μL	100
Nuclease-free H_2O			1500

4. Incubate at 37 °C for 30 min.
5. During DNase incubation, make >40 μL 30 mg/μL Proteinase K in Proteinase K Storage Buffer.
6. Add 40 μL Proteinase K to reaction to degrade all proteins in the *in vitro* transcription reaction.
7. Incubate at 37 °C for 30 min.
8. Add 200 μL 500 mM EDTA to chelate divalent cations.
   We have observed that very long (> 2 kb) viral genomic RNA aggregates in the presence of magnesium when concentrated in the subsequent step. Therefore, we have included a chelation step to limit aggregation. Note that this chelation step is expected to maintain stable secondary structures formed during transcription, but disrupts magnesium-dependent tertiary interactions. Therefore, we refer to this protocol as semi-native purification.

The transcription reaction contained 12 mM $MgCl_2$, and the DNase reaction contained 10 mM $MgSO_4$, so the resulting EDTA concentration of 25 mM is in excess.

D. Purify RNA *via* Size-Exclusion Chromatography

The following steps purify RNA from free nucleotides, aggregated RNA, protein and DNA fragments, and provide a buffer exchange for subsequent RNA folding steps. Size-exclusion chromatography (SEC) also permits assessment of overall conformational homogeneity with simultaneous isolation of a relatively homogenous population of RNA conformations (Fig. 1A). Note that decreased temperatures cause RNA mis-folding and aggregation, so all steps should be carried out at room-temperature.

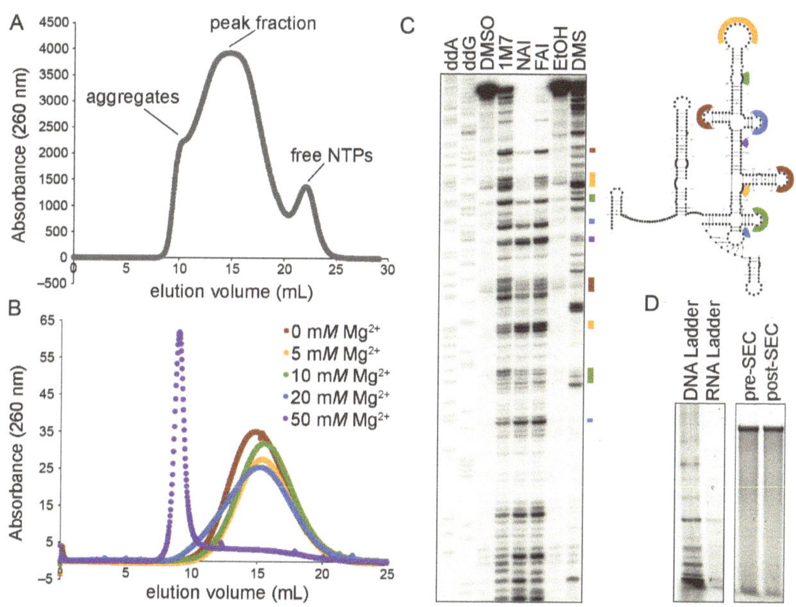

Fig. 1 Semi-Native purification and analysis of the HCV genome. (A) Chromatogram generated during SEC purification of HCV genome with S1000 column. The peak fraction is collected for downstream folding steps. (B) Chromatogram comparing HCV compaction after folding with increasing Mg^{2+} concentrations. Note that 5 mM and 10 mM Mg^{2+} samples are right-shifted relative to 0 mM Mg^{2+}, indicating magnesium-dependent compaction; the 20 mM Mg^{2+} sample is left-shifted indicating partial multimerization/aggregation; and the 50 mM Mg^{2+} sample is almost fully aggregated. (C) ^{32}P-labeled primer extension and PAGE analysis of RNA modification. Note that probe-treated RNA results in an increase of truncated cDNA. Colored regions with high reactivity indicated on the HCV IRES structure on the right. (D) Agarose gel electrophoresis of SEC-purified HCV genome. Denatured RNA is intact and the expected length.

We recommend packing SEC columns at room-temperature, following manufacturer suggestions (http://ge-catalog.campaignhosting.se/Documents/Tricorn/TricornInstructionsAC.pdf). Prior to column packing, we recommend rinsing every part with RNAseZap, and rinsing the packed column with 5 column volumes of filtered Milli-Q water. We store the column in filtered Milli-Q water at 4 °C between uses, and allow equilibration to room-temperature for >30 min prior to use. We use an AKTA pure 25 M FPLC system set to a flow rate of 0.4 mL/min.

1. Prepare FPLC system with 1 mL injection loop. Pre-equilibrate the SEC column with 1.2 column volumes of Filtration Buffer, rinsing the 1 mL Hamilton syringe and all system tubing.
2. Equilibrate 4 Amicon Ultra filters with 500 μL Filtration Buffer. Centrifuge for 5 min at 4000 rcf, and discard flow-through.
3. Add 500 μL RNA sample to each Amicon column and centrifuge to half-volume (250 μL).

   We recommend centrifuging for 1 min and checking the remaining volume until half-volume is reached. It is important not to over-concentrate in order to prevent aggregation.
4. Repeat step 3 until all RNA has been concentrated with ~250 μL remaining in each column.
5. Combine all RNA samples into one Eppendorf tube, and centrifuge on bench-top centrifuge at full speed for 30 s to pellet aggregated RNA and thereby remove it. Move supernatant to new tube.

   It is important to remove large RNA aggregates at this step so that the SEC column does not get clogged.
6. Inject RNA sample and begin the FPLC run, collecting 0.5 mL fractions.
7. Measure RNA concentration of the peaks with the highest UV_{260} absorbance, and continue with the peak fraction.

   At this point, we routinely obtain fraction concentrations of 200–300 ng/μL.

   The three fractions with highest concentration can be combined if more sample is desired, but we do not recommend exceeding three sequential fractions.
8. Remove 5 μL sample and store at −20 °C for up to several days for subsequent analysis by agarose gel electrophoresis.

E. Fold and modify RNA

1. Add $MgCl_2$ and Filtration Buffer to reach desired final concentration of Mg^{2+} and RNA for tertiary folding.

We routinely fold long viral RNA at a concentration of 100–250 ng/μL. This provides enough material for subsequent steps while limiting aggregation. We do not recommend exceeding this concentration due to the likelihood of aggregation.

Note that filtration buffer contains physiological levels of potassium ions, and magnesium can then be titrated in to induce compaction *via* tertiary interactions. Optimal Mg^{2+} concentration must be empirically determined for each RNA. We aim for maximal compaction with no aggregation (Fig. 1B). To determine the optimal Mg^{2+} concentration, we recommend following the folding steps with a range of $MgCl_2$ concentrations (including 0) and assessing compaction/aggregation by SEC with small (100 μL) folding reactions as a first-pass experiment. Note that RNA should be stable in folding buffer at room-temperature for ~12 h at this point, but $MgCl_2$ should be added just prior to folding so that all samples are equilibrated for the same length of time. Finally, once optimal Mg^{2+} concentration is determined, RNA should be folded at the same concentration in all subsequent experiments.

Note that we typically mix Filtration Buffer and $MgCl_2$ prior to adding RNA sample so that the $MgCl_2$ can be quickly mixed after addition of RNA.

RNA folding reaction

Component	Final concentration	Stock concentration	Volume added
RNA (in Filtration Buffer)	100–250 ng/μL		Variable
Filtration Buffer			Variable
$MgCl_2$	Variable	1 M	Variable
Final Volume			500 μL

2. Incubate at 37 °C for 30 min to allow RNA to reach its equilibrated structure, splitting the sample into desired number of tubes for subsequent labeling reactions.
3. Add probe and appropriate solvent, and incubate at 37 °C for 10 min.

   Note that each probe has distinct rates of modification or hydrolysis. For experimental simplicity, we incubate for 10 min with all probes, but it is reasonable to vary the incubation time in order to optimize the experimental outcome. Use an equivalent volume of the appropriate solvent as a negative control for each probing reagent (i.e., DMSO for NAI, FAI, and 1M7, EtOH for DMS, DMSO + UV_{310} for NAz). We additionally recommend that each probe be titrated to ensure that RNA is properly modified. For long RNAs, we recommend averaging "single-hit" kinetics over a range of 200 nucleotides, as can be assessed by primer extension using a reverse transcription primer located 200 nucleotides from the 5′ terminus (step F). In this case, use a concentration/incubation time that results in ~70–80% full-length product for treated sample relative to untreated. We recommend the following concentration ranges as a starting point: 1M7, 2.5–10 mM; NAI, 50–200 mM; FAI, 50–300 mM; DMS, 0.05–0.4%; NAz, 200 mM and exposure to UV_{310} for 30 s to 2 min.

For fast and even distribution of probe in solution with RNA, we usually add the probe to the top of our opened Eppendorf tube, close the tube carefully, and invert the tube several times followed by pulse centrifugation.

Before proceeding with library preparation, we highly recommend performing a primer extension experiment to ensure that RNA is modified (step F). We have observed considerable lot-to-lot variability of modification with probes from commercial sources.

4. Add a half volume of LiCl solution to precipitate probed RNA, mix, and incubate at −80 °C for 30 min to overnight.

5. Centrifuge the precipitating RNA at full speed (≥13,000 rcf) at 4 °C for 30 min, discarding the supernatant.

Note that the pelleted RNA may be difficult to see, so be sure to pipette from the side of the tube away from where the RNA pelleted.

6. Rinse RNA with ice-cold 80% EtOH, centrifuge at full speed (≥13,000 rcf) at 4 °C for 10 min, carefully pipette and discard all visible liquid, and air-dry for ~5 min.

7. Resuspend in ME buffer to a final concentration of 1 μg/μL.

At this point, modified RNA can be stored at −20 °C for weeks or at −80 °C for months.

F. Primer extension and PAGE to assess RNA modification

Follow this protocol to perform primer extension with ^{32}P end-labeled primer followed by electrophoresis to assess RNA modification (Fig. 1C). Under the standard RT conditions described here, the RT enzyme will stop at modified nucleotides, allowing visualization of truncated cDNA products by radiography.

Take appropriate precautions when working with radioactivity, including wearing and using appropriate personal protective equipment, and proper disposal of exposed waste.

We have found success in using OligoWalk (Mathews, 2010) to design RT primers as recommended in (Fang, Moss, Rutenberg-Schoenberg, & Simon, 2015).

1. Prepare 5′ end labeling reaction for reverse transcription (RT) primer, and incubate at 70 °C for 10 min to denature primer. Cool to room-temperature.

5′ End labeling reaction

Component	Volume added (μL)
RT primer (at 10 μM)	2
10 × PNK Buffer	2
Nuclease-free H_2O	14.5

2. Add 0.5 μL γ-^{32}P ATP, 1 μL T4 PNK, mix, and incubate at 37 °C for 1 h.
3. Purify with G-25 column according to manufacturer protocol to remove free γ-^{32}P ATP.
4. Add 2.5 μL NaOAc, vortex to mix, 100 μL EtOH, mix, and precipitate at −80 °C for 30 min.
5. Centrifuge for 30 min at full speed with bench-top centrifuge, pipette all visible liquid while avoiding the pellet, air-dry ~5 min, and resuspend in 20 μL nuclease-free H_2O.
6. Prepare primer annealing mixture, and incubate at 70 °C for 10 min to melt RNA structures. Cool to room-temperature.

Primer annealing reaction

Component	Volume added (μL)
Labeled RT primer	1
Modified RNA (1 μg/μL)	1
5 M Betaine	4
SUPERase In	0.25
Nuclease-free H_2O	7.5

7. Prepare first strand RT reaction and add to cooled primer annealing mixture.

First-strand reaction

Component	Volume added (μL)
5 × First-Strand Buffer	4
10 mM dNTP mix	1
100 mM DTT	1
SuperScript III	0.25

8. Incubate reaction at 50 °C for 10 min, 55 °C for 50 min, and 70 °C for 15 min.
9. Add 2 μL 2 M NaOH to degrade RNA, neutralize with 2 μL 1 M HCl, add 2 μL NaOAc, mix, add 100 μL EtOH and precipitate at −80 °C for 30 min.
10. Centrifuge for 30 min at full speed 4 °C with bench-top centrifuge, pipette out all visible liquid while avoiding the pellet.
11. Wash with freshly prepared 70% EtOH, being careful to not dislodge pellet during washes.
12. Centrifuge for 10 min at full speed at 4 °C with bench-top centrifuge, pipette out all visible liquid while avoiding the pellet and air-dry for ~5 min.
13. Resuspend in 10 μL 20 mM K-HEPES pH 7.5, mix well to ensure pellet is fully dissolved, add 10 μL 2 × Urea loading dye and mix.

    cDNA can be stored for days to weeks at −20 °C, taking into account the 14-day half-life of ^{32}P.
14. Prepare sequencing ladder reaction.

    We typically generate two sequencing ladders, using di-deoxynucleotides complementary to the most abundant nucleotides in the region of interest. Thermo sequenase is appropriate for ladder generation as it uses dideoxynucleotides as efficiently as deoxynucleotides, thus requiring lower ddNTP concentration; however, because it is a DNA-dependent DNA polymerase, the relative amounts of cDNA from the reverse transcription and DNA ladder loaded on the sequencing gel might require optimization. We typically generate ddNTP:dNTP mixtures at 1:100 M ratios (for example, 1 μL 1 mM ddNTP of interest + 10 μL 10 mM dNTP + 989 μL H_2O).

Sequenase ladder reaction

Component	Volume added (μL)
Plasmid with target sequence (100 ng/μL)	0.8
Sequenase reaction buffer	0.92
Labeled primer	1
ddNTP mix (1:100)	8
Sequenase	0.92
Nuclease-free H_2O	2.76

15. Cycle in a thermocycler with the following steps:

Sequenase cycling conditions

Number of cycles	Temperature (°C)	Time
1	95	1 min
40	95	20 s
	55	20 s
	72	1 min
1	72	5 min
	4	Hold

16. Add 16.4 μL Stop Mix, mix.

Sequenase stop mix

Component	Volume added (μL)
3 M NaOAc pH 5.2	50
500 mM EDTA pH 8.0	50
Nuclease-free H_2O	400

17. Add 100 μL EtOH, and precipitate at −80 °C for 30 min.

18. Centrifuge for 30 min at full speed with bench-top centrifuge, pipette out all visible liquid while avoiding the pellet, and air-dry for ~5 min.

19. Resuspend in 10 μL 20 mM K-HEPES pH 7.5, mix well to ensure pellet is fully dissolved, add 10 μL 2× Urea loading dye and mix.

20. Cast 0.4 mm thick × 40 cm tall polyacrylamide gel with one siliconized glass plate and let it polymerize for 1.5–2 h.

Polyacrylamide gel mix

Component	Volume added
10% Polyacrylamide Denaturing Gel Mix	100 mL
10% APS	400 μL
TEMED	40 μL

21. Pre-run gel for ~30 min to equilibrate temperature and current.
22. Denature DNA samples at 70 °C for 10 min and cool to room-temperature.
23. Load samples, run gel at 25–30 watts until bromophenol blue dye is at the bottom (~2–3 h with our apparatus).
24. Dry gel, expose to phosphor screen overnight and image using appropriate detection system.

G. Assess RNA quality by Agarose Gel Electrophoresis

The majority of RNA produced with the above protocol is full-length; a minimal proportion of our sample is degraded or truncated RNA. However, for RNA probing experiments, it is essential to ensure that the final probed RNA is full-length and folding as such. Therefore, we carefully assess RNA integrity by agarose gel electrophoresis of denatured samples (Fig. 1D), as adapted from Sambrook, Fritsch, and Maniatis (1989).

1. Rinse gel casting tray and comb with RNAse Zap.
2. Make a 1% agarose gel by melting 1.5 g agarose in 150 mL 1 × MOPS Running Buffer and cast the gel.
3. Prepare samples in Formaldehyde Gel Loading Buffer by mixing the following components. Incubate at 70 °C for 15 min and cool to room-temperature.

Formaldehyde Gel Loading Buffer

Component	Volume added (μL)
RNA sample/Ladder	5
10 × MOPS Running Buffer	2
Formaldehyde	4
Formamide	10
10 × Gel Loading Buffer	2

4. Load the samples and run gel at 5 V/cm until bromophenol blue has run at least 8 cm (≥3 h with our apparatus).
5. Stain by shaking in SYBR-Gold diluted 1:10,000 in H_2O covered in foil for 30 min and image with UV Transilluminator.

2.5 Probe-Stop-HiTS to calculate nucleotide reactivity

Because RT stop-based methods are the classic way to analyze RNA modification, we first used this approach with HiTS to calculate reactivity (here generically referred to as Probe-Stop-HiTS) in our lab. We have followed several methods for stop-based library preparation (Fang et al., 2015; Flynn et al., 2016; Watters & Lucks, 2016). Specifically, RT primers were designed as in using Illumina sequence followed by a random dodecamer (5′ CAGACGTGTGCTCTTCCGATCTNNNNNNNNNNNN 3′) or sequence complementary to HCV (targeted sequence; 5′ CAGACGTGTGCTCTTCCGATCT— 3′, with — indicating complementary sequence). For targeted primers, RT reactions were carried out in three reactions with primers set >600 nt apart. Adaptor ligation was performed by CircLigase II with 20% PEG 8000 with a PAGE-purified oligonucleotide with random hexamer sequence (5′/5Phos/NNNNNNAGATCGGAAGAGCGTCGTGTAG/3Bio/3′) to remove PCR duplicates computationally. Excess primers, adaptors, and adaptor-ligated primers were removed by silica membrane-based columns, SPRI magnetic beads, or PAGE purification, and libraries amplified with NEB-Next Multiplex Oligos using NEB Q5 polymerase.

2.6 Primer bias for Probe-Stop-HiTS RT

In the first step of library preparation, the researcher has a choice of using primers complementary to the RNA of interest or random primers to non-specifically prime cDNA synthesis. For *in vivo* probing of an RNA of interest, targeted primers are required to ensure that the resulting libraries cover the specific transcript (Fang et al., 2015). In some cases, when probing a short RNA generated by *in vitro* transcription, the RNA can be ligated to a 3′ adaptor, which can be primed from to generate data across the entire sequence (Watters & Lucks, 2016). In any case, Illumina sequence is added to the 3′ end of the RT primer for subsequent library preparation steps.

Using targeted primers to generate Probe-Stop-HiTS libraries from *in vitro* transcribed HCV genome, we have observed a striking peak-and-valley pattern of reactivity that corresponds to primer binding sites

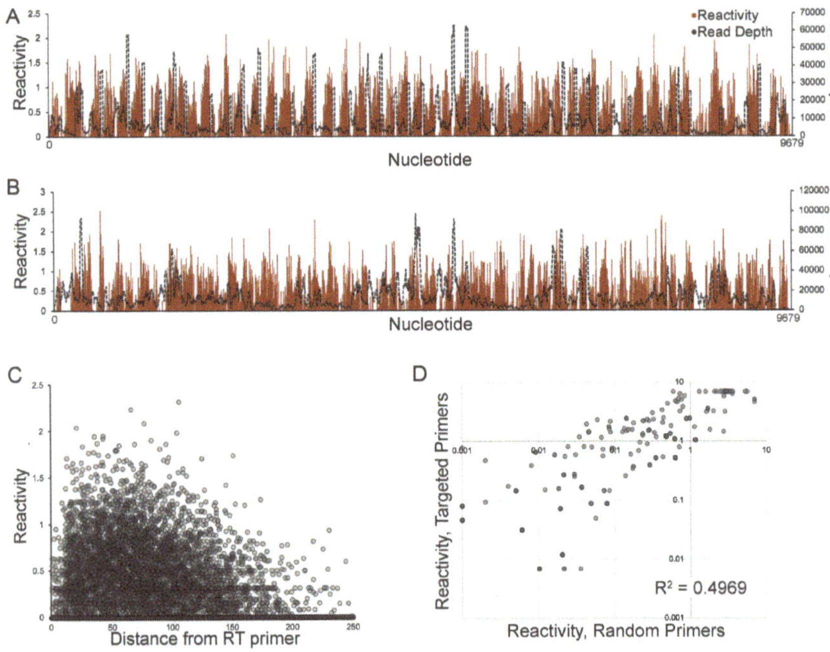

Fig. 2 Primer-dependent bias in Probe-Stop-HiTS reactivity. (A) Reactivity and read depth across the HCV genome using targeted primers for RT. Reactivity calculated from NAI-probed HCV using StructureFold2, with default normalization and natural log transformation. Read depth calculated with SamTools mpileup. (B) Reactivity and read depth across the HCV genome using random primers for RT. Reactivity and read depth calculated as in (A). (C) Reactivities calculated in (A) plotted relative to distance from RT primer. (D) Correlation of reactivities for the HCV IRES from libraries generated with targeted or random primers. Reactivities were generated as in (A) without normalization and natural log transformation.

(observed as regions with relatively high read depth) (Fig. 2A). Furthermore, targeted primers have different priming efficiency, leading to regions with very low read coverage. For probing an *in vitro* generated RNA, an alternative method to generate cDNA is to use random primers during reverse transcription. Using random dodecamer containing Illumina sequence, we observe a decreased apparent correlation between read depth and reactivity (Fig. 2B). We observed an increase in reactivities calculated between 50 and 100 nt 5′ of the primer binding site (Fig. 2C), as has been previously observed (Fang et al., 2015), suggesting bias in reactivity calculations that is especially apparent using targeted primers. Comparison of reactivities obtained from targeted versus random priming resulted in low correlation, highlighting the bias in how reactivities are calculated in relation to the primer binding site (Fig. 2D).

3. Software for Probe-Stop-HiTS computational analysis

StructureFold2 (https://github.com/StructureFold2), developed in a collaboration of the Bevilacqua and Assmann labs, calculates reactivity based on RT stop counts and therefore only requires single-end sequencing with relatively short reads (only ≤35 nt read-length for accurate read mapping). Furthermore, StructureFold2 is available for use on Galaxy (https://usegalaxy.org/ NGS: RNA Structure section), making it user-friendly for researchers who are not comfortable with a command-line interface. We observed that StructureFold2 was robust in reactivity calculation relative to read-depth, with high correlation of reactivity to ~25 reads/nucleotide (Fig. 3A, 250,000 reads), but observed reproducibility was low between biological replicates (Fig. 3B). Reactivities calculated with StructureFold2 correlate with the HCV IRES structure (Fig. 3C), and contributed to accurately predicting the IRES structure and pseudoknot (Fig. 3D), with increased accuracy using non-default slope and intercept values for the pseudo free-energy calculation. Finally, using miR-122 interaction with HCV as a positive control to detect structural changes of the HCV genome, the noise of stop-based library generation made it difficult to distinguish significant differences in reactivity (Fig. 3E).

3.1 Probe-MaP-HiTS

A major technical innovation in RNA structure probing came from recognizing that RT enzymes can incorporate mutations rather than halt synthesis when encountering a nucleotide with an adduct (Siegfried et al., 2014). Specifically, using manganese in the RT buffer rather than magnesium induces read-through of probed nucleotides, and the resulting mutation rates measured at each nucleotide (mutational profiling, MaP) by HiTS recapitulate known RNA structures. This means that every nucleotide of a sequencing read provides information about reactivity, not just the terminal nucleotide of a read that corresponds to the 3′ end of the cDNA.

Here again, the researcher has the choice to use targeted (amplicon workflow) or random (randomer workflow) primers during reverse-transcription, depending on the experimental needs (Smola, Rice, Busan, Siegfried, & Weeks, 2015). When using targeted primers, an amplicon covering the region of interest can be PCR amplified, either incorporating Illumina sequence in this step or in a subsequent tagmentation step. The length of cDNA produced is limited by the activity of the RT enzyme under MaP conditions (6 mM $MnCl_2$), which are sub-optimal for SSII, the

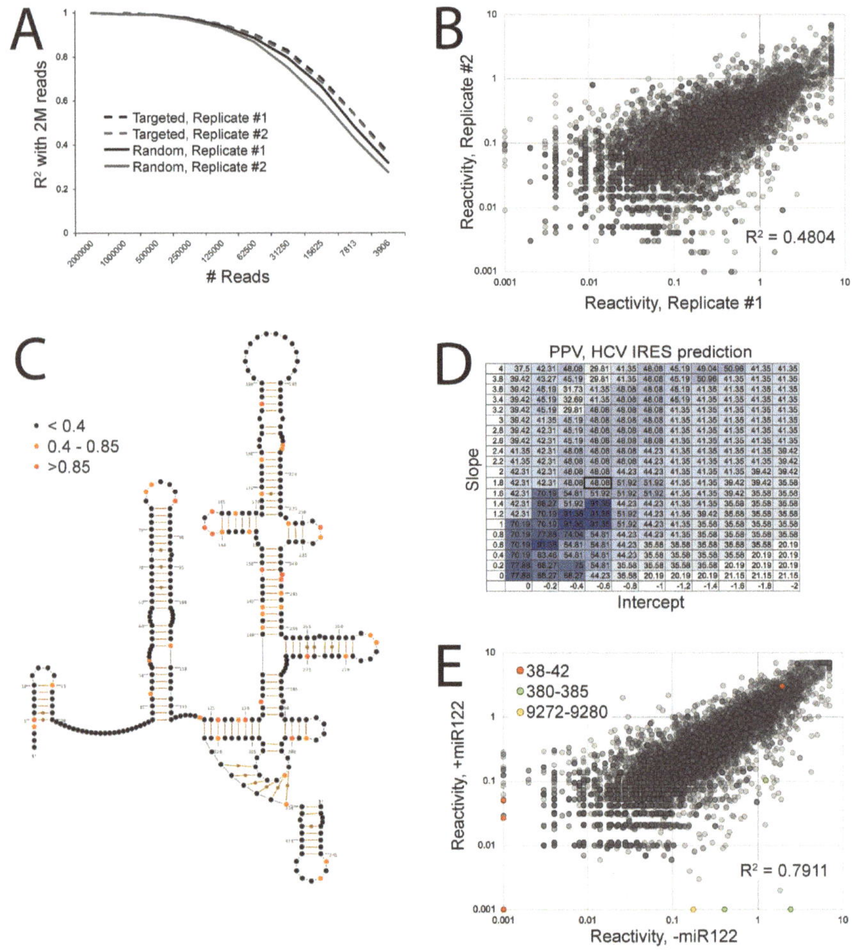

Fig. 3 Utility of StructureFold2-calculated reactivity from Probe-Stop-HiTS. (A) Robustness of StructureFold2 reactivities to down-sampling. FASTQ sequences were iteratively down-sampled, and reactivities were calculated with StrutureFold2 without normalization and natural log transformation and compared to reactivities generated with 2M reads. (B) Comparison of biological replicates of Probe-Stop-HiTS. Independently generated samples were reverse transcribed with random primers, and reactivity was calculated with StructureFold2 without normalization and natural log transformation. (C) Reactivity mapped to the HCV IRES structure. Reactivities generated from randomly-primed samples from NAI-probed HCV, calculated with StructureFold2 with normalization and natural log transformation. Drawn with Assemble2. (D) Positive predictive value of the HCV IRES. Reactivities generated with StructureFold2 (normalized and log-transformed) from NAI-probed HCV were used to constrain pseudoknot prediction from nucleotides 43–353 using ShapeKnots from the RNAStructure package over a range of slope and intercept values. PPV was calculated using Scorer from the RNAStructure package, compared to the known HCV IRES structure. Boxed square indicates default slope and intercept values. (E) Comparison of miR-122-dependent reactivities. HCV RNA was folded in the absence or presence of 4mM miR-122 (Eurofins). Resulting reactivities were computed with StructureFold2 without normalization and natural log transformation. For colored values, zero was plotted as 0.001 to enable visualization.

standard RT enzyme used for these experiments. Therefore, to probe a long RNA, several RT primers tiled across the RNA of interest must be used and several short amplicons generated from the resulting cDNA. Alternatively, for *in vitro* probing of a long RNA, random primers can be used to generate cDNA, with double-stranded DNA generated by second-strand synthesis primed from RNA-DNA hybrids. In our hands, the resulting reactivities map well to the HCV IRES (Fig. 4A and B), regardless of workflow. Furthermore, the reactivities for each workflow correlate (Fig. 4C), suggesting that the calculated mutation rate is not affected by the primer binding site.

The Weeks lab has developed user-friendly software (ShapeMapper2) for calculating mutation rates and resulting normalized reactivities from MaP libraries (Busan & Weeks, 2018). Using Probe-MaP-HiTS, with reactivities calculated using ShapeMapper2, we have observed a very high correlation of reactivity calculated between biological replicates (Fig. 4D). Additionally, calculated reactivities were robust to read-depth, with high correlation of reactivity to 200 reads/nucleotide (Fig. 4E). Finally, Probe-MaP-HiTS data using NAI aided in constraining HCV IRES prediction to high accuracy over a range of slope and intercept values (Fig. 4F).

3.2 PROTOCOL: Library preparation for Probe-MaP-HiTS

This protocol was adapted from the Weeks lab SHAPE-MaP protocol (Smola, Rice, et al., 2015). However, many RNA biochemists have not performed HiTS experiments; therefore, we have included additional information about pooling and sequencing libraries to make it easier to adopt the approach. Additionally, we describe the use of smaller volumes during the final library fragmentation and amplification steps (step C), as we find that carrying out these steps in smaller volumes generates high-quality libraries at reduced cost.

Equipment

- Thermocycler
- Monarch PCR and DNA Cleanup Kit (NEB, T1030S)
- Qubit Fluorometer
- Agilent Bioanalyzer 2100 or TapeStation
- NextSeq 500/550 System or alternate Illumina sequencer
- PCR strips and caps (Dot scientific, 401-8PCR and 400-8PCR)

Chemicals

- Tris-HCl, 1 M solution, pH 7.5
- Potassium Chloride, 1 M solution

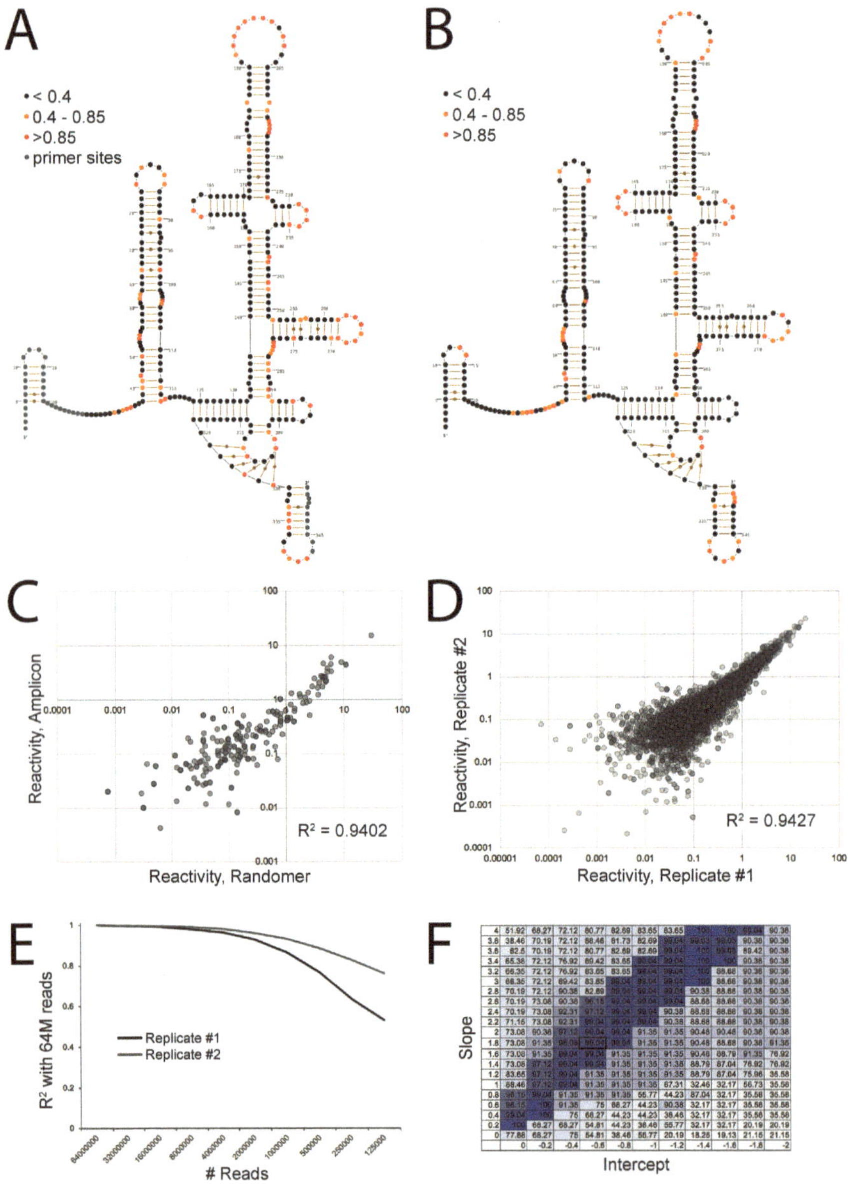

Fig. 4 Utility of ShapeMapper2-calculated reactivity from Probe-MaP-HiTS. (A) Reactivity from targeted primer mapped to the HCV IRES structure. Reactivities generated from targeted primer from NAI-probed HCV, calculated with ShapeMapper with default settings. Drawn with Assemble2. (B) Reactivity from random primers mapped to the HCV IRES structure. Reactivities generated from random primers, calculated as in (A). (C) Correlation of reactivities for the HCV IRES from libraries generated with targeted or random primers. Reactivities were generated as in (A). (D) Comparison of biological

- DTT, 1 M solution
- Manganese (II) chloride solution, 1 M (Sigma-Aldrich, M1787-10X1ML)

Reagents

- dNTP Mix, 10 mM each (ThermoFisher, R0192)
- SuperScript II (ThermoFisher, 18064014)
- Random Nonamer (NEB, S1254S) (randomer workflow)
- NEBNext Second Strand Synthesis Kit (NEB, E6111S) (randomer workflow)
- Q5 High-Fidelity Polymerase (or other high-fidelity DNA Polymerase) (NEB, M0491S) (amplicon workflow)
- Nextera XT DNA Library Preparation Kit (Illumina, FC-131-1024)
- Qubit dsDNA High Sensitivity Assay Kit (ThermoFisher, Q32851)
- Nextera XT Index Kit (Illumina, FC-131-1001)
- Agilent High Sensitivity DNA Kit (Agilent, 5067-4626)
- NextSeq 500/550 High Output v2 Kit (Illumina, FC-404-2005)
- PhiX Control v3 (Illumina, FC-110-3001)

Protocol

A. Order oligonucleotides for reverse transcription.

Illumina adaptors are added during the Tagmentation step, so they are not required during reverse transcription in the MaP protocol.

1. RT primers:

For *in vitro* transcribed long RNA, the randomer workflow is appropriate for generating cDNA, and we have used random nonamers as recommended. However, if analyzing a specific region of interest or analyzing RNA generated from cellular RNA, we recommend the amplicon workflow. In this case, we have successfully used OligoWalk from the RNAStructure package (Mathews, 2010), to design targeted RT primers.

---

replicates of Probe-MaP-HiTS. Independently generated samples were reverse transcribed with random primers, and reactivity was calculated with ShapeMaper2. (E) Robustness of ShapeMapper2 reactivities to down-sampling. FASTQ sequences were iteratively down-sampled, and reactivities were calculated with ShapeMapper2 and compared to reactivities generated with 64 M reads. (F).

Positive predictive value of the HCV IRES. Reactivities generated with ShapeMapper2 from NAI-probed HCV were used to constrain pseudoknot prediction from nucleotides 43–353 using ShapeKnots from the RNAStructure package over a range of slope and intercept values. PPV was calculated using Scorer from the RNAStructure package, compared to the known HCV IRES structure. Boxed square indicates default slope and intercept values.

2. PCR primers:

If following the amplicon workflow, we recommend careful design of PCR primers. Specifically, we have had success in amplifying low abundance RNA from *in vivo* probing by designing primers with a Tm of ~65 °C, according to the NEB calculator for Q5 (https://tmcalculator.neb.com/#!/main). Additionally, due to the low processivity of SuperScript II under MaP RT conditions, amplicons should not be larger than ~700 nt. Finally, we design reverse PCR primers to anneal at least a few nucleotides 5′ to the RT primer binding site for added specificity. Note that PCR primers are not required for the randomer workflow.

B. Reverse Transcription and Second Strand Synthesis

1. Prepare fresh 2.5 × MaP Buffer.

2.5× MaP Buffer

Component	Final concentration (mM)	Stock	Amount
Tris-HCl, pH 8.0	125	1 M	125 μL
KCl	187.5	1 M	187.5 μL
DTT	25	1 M	25 μL
dNTPs	1.25	10 mM	125 μL
$MnCl_2$	15	1 M	15 μL
H_2O			up to 1 mL

It is essential to make this buffer fresh. Upon addition of $MnCl_2$, SSII buffer may momentarily turn light brown.

2. Set up primer annealing mixture in PCR tube by aliquoting 1 μL RNA (1 μg), 8 μL nuclease-free H_2O, and 1 μL random nonamer at 200 ng/μL (for randomer workflow) or targeted RT primer at 1 μM (for amplicon workflow). Incubate at 65 °C for 5 min, and cool to room-temperature

   1 μg *in vitro* transcribed RNA is more than sufficient to generate enough cDNA for downstream steps.

3. Set up first strand RT reaction by adding 8 μL 2.5 × MaP Buffer, pipetting to mix, followed by addition of 2 μL SuperScript II. Incubate at 42 °C 3 h.

4. For randomer workflow, dilute cDNA samples from previous step by adding 48 μL nuclease-free H_2O, followed by addition of 8 μL

$10\times$ NEBNext Second Strand Synthesis Buffer and 4 μL NEBNext Second Strand Synthesis Enzyme Mix. Incubate at 16 °C for 2.5 h, and hold at 4 °C.

It is essential to proceed directly to this step without degrading DNA-RNA hybrids, as intact RNA is required to prime second strand synthesis.

If using the amplicon workflow, continue from step B.3 with the following steps:

a. Add 1 μL 500 mM EDTA, 2 μL 3 M NaOAc, mix, add 100 μL EtOH, and precipitate at −80 °C for 30 min.
b. Centrifuge for 30 min at full speed 4 °C with bench-top centrifuge, pipette out all visible liquid while avoiding the pellet.
c. Wash with freshly prepared 70% EtOH, being careful to not dislodge pellet during washes.
d. Centrifuge for 10 min at full speed at 4 °C with bench-top centrifuge, pipette out all visible liquid while avoiding the pellet, air-dry for ~5 min, and resuspend in 10 μL nuclease-free H_2O
e. Set up PCR using Q5 Polymerase.

Note: any high-fidelity polymerase can be used at this step.

Q5 amplicon PCR

Component	Volume added (μL)
cDNA	1
$5\times$ Q5 Buffer	10
10 mM dNTP mix	1
25 μM Forward Primer	1
25 μM Reverse Primer	1
Q5 Polymerase	0.5
Nuclease-free H_2O	35.5

Cycle in a thermocycler with the following steps:

We have observed that <20 cycles total are required to generate enough amplicon product from *in vitro* transcribed RNA for downstream steps.

Q5 touch-down PCR cycling

Number of cycles	Temperature	Time
1	98 °C	30 s
10	98 °C	10 s
	70 °C, reducing by 1 °C each cycle	30 s
	72 °C	30 s
Variable	98 °C	10 s
	60 °C	30 s
	72 °C	30 s
1	72 °C	5 min
	4 °C	Hold

5. Purify double-stranded cDNA with Monarch PCR and DNA Cleanup Kit, following the manufacturer's recommendations for double-stranded DNA ≤ 2 kb. Elute in 10 μL nuclease-free H_2O.

C. Fragmentation and Library Amplification

Successful fragmentation of dsDNA relies on sensitive quantification of dsDNA concentrations, which can be accomplished using the Qubit dsDNA High Sensitivity Assay Kit to specifically quantify dsDNA concentrations. Note that measuring UV_{260} absorbance using a Nanodrop or spectrophotometer is not sensitive enough and does not distinguish ssDNA from dsDNA.

1. Quantify dsDNA concentration of each sample using the Qubit dsDNA High Sensitivity Assay Kit. Dilute to 0.2 ng/μL with nuclease-free H_2O.
2. Prepare fragmentation reaction with 1 μL dsDNA (0.2 ng/μL), 2 μL Tagment DNA Buffer, and 1 μL Amplicon Tagment Mix from Nextera XT DNA Library Preparation Kit. Incubate in thermocycler at 55 °C for 5 min, then cool to 10 °C. Once cool, immediately add 1 μL NT buffer.

We have found that using 1/5th the reaction volume suggested by the manufacturer's protocol results in high-quality libraries while allowing researchers to defray high reagent costs by increasing the number of libraries prepared with a single kit (i.e., 120 libraries instead of 24 from a > $800 kit). Note that using this smaller volume requires tight capping of PCR tubes to reduce sample evaporation in this and PCR steps, using the recommended strips and caps.

3. Prepare PCR reactions using Nextera XT DNA Library Preparation Kit and Nextera XT Index Kit.

   Take careful notes of each corresponding sample and index number.

Nextera PCR mixture

Component	Volume added (μL)
Tagmented DNA	5
PCR Master Mix	3
Index 1	1
Index 2	1

4. Cycle in a thermocycler with the following steps:

Nextera PCR cycling

Number of cycles	Temperature (°C)	Time
1	72	3 min
1	95	30 s
12	95	10 s
	55	30 s
	72	30 s
1	72	5 min
	4	Hold

5. Purify final libraries with column or SPRI-based purification kit, eluting in 10–12 μL H_2O.

   We use the NEB Monarch PCR and DNA Cleanup Kit for final library purification because of the small volumes used for elution. Use manufacturers recommendations for dsDNA $\leq$2 kb.

D. Library Quantification and Dilution

Although most researchers will submit their final libraries to a core for quality control (QC) and sequencing, we have found that performing our own QC analysis has improved our troubleshooting and resulted in better sequencing results. Library QC requires careful quantification of concentration and size distribution so that molar concentrations can be calculated for pooling.

1. Quantify the concentration of each sample using the Qubit dsDNA High Sensitivity Assay Kit as in step C.1.
2. Calculate library size distribution with BioAnalyzer or TapeStation. From the "Regions" section, determine average library size from a range of 100–1000 bp.
3. Calculate the molar concentration of each library using the following equation (Eq. 2), as recommended by Illumina (https://support.illumina.com/bulletins/2016/11/converting-ngl-to-nm-when-calculating-dsdna-library-concentration-.html).

$$\text{Molarity}\,(\text{n}M) = 10^6 \times \left(\frac{\text{concentration}\left(\frac{\text{ng}}{\mu L}\right)}{660\frac{\text{g}}{\text{mol}} \times \text{average size (bp)}} \right) \tag{2}$$

4. Combine, denature, and dilute as described in https://support.illumina.com/content/dam/illumina-support/documents/documentation/system_documentation/nextseq/nextseq-denature-dilute-libraries-guide-15048776-09.pdf using the Nextera XT DNA Library Preparation Kit. We have found that for each sample, 5% of a lane provides more than enough read-depth to calculate HCV reactivity (9.6 kb) using the NextSeq 500/550 High Output v2 kit (75 cycles). Each researcher will need to determine how many samples to combine depending on the sequencing length and output level of her specific kit and sequencer. We also recommend including a 1% spike-in of PhiX library for calculating read quality.
5. Set up sequencing run parameters on BaseSpace (https://basespace.illumina.com) using Manual Prep with Nextera XT indices.
6. Log in and load onto sequencer as described in each kit's manual.

3.3 Monitoring differential effects on structure: miR-122 and HCV RNA as a test case

Because of the high reproducibility of Probe-MaP-HiTS data, comparing reactivities across the HCV genome folded with or without miR-122 resulted in high correlation between samples, with relatively few nucleotides falling outside of the line of correlation, including the known miR-122 binding site (Fig. 5A). In addition to nucleotide reactivities, ShapeMapper2 outputs statistical measurements of calculated reactivities based on read depth at the specified nucleotide. This allows calculation of statistically significant differences between samples, and the Weeks lab developed DeltaShape for automated analysis (Smola, Calabrese, & Weeks, 2015). Using this tool, we identified several regions with apparent changes in probe reactivity (Fig. 5B). We confirmed changes in select regions by PAGE analysis following RT with ^{32}P-labeled primers (Fig. 5C). Upon examination of the sequence of these regions, we observed that they all displayed complementarity to miR-122 (Fig. 5D). Therefore, the structural rearrangements observed from this analysis likely represent direct interaction between HCV and miR-122.

4. Discussion

This protocol has been written with a focus on analyzing structural changes of RNA under different conditions. However, it should be noted that RNA exists in a dynamic equilibrium of conformational states. Thus, despite using SEC to isolate the most abundant conformation in a population, there is likely not a single homogenous RNA structure probed in the experiments described here, and the most abundant stable structure(s) likely contributes the majority of the probe reactivity signal. Some studies have attempted to decipher relative contributions of each structure among a conformationally heterogenous population from Probe-HiTS data (Li & Aviran, 2018; Spasic, Assmann, Bevilacqua, & Mathews, 2018). Additionally, Probe-MaP-HiTS data can be used to detect co-occurrence of mutations within a single sequencing read, enabling identification of multiple conformational states that co-exist within a single sample of RNA. RING-MaP, developed in the Weeks lab, detects correlated mutations, permitting prediction of distinct conformational states as well as long-range interactions from long-read sequencing (Homan et al., 2014).

In our hands at this point, the noise in calculated reactivities has made sensitive detection of structural changes difficult. We hypothesize that a large part of this experimental variability results from inefficiency of adaptor ligation by CircLigase. We anticipate that this approach (and many HiTS experiments that require ssDNA-ssDNA ligation) will be drastically improved with better molecular tools. Additionally, averaging several technical replicates using an appropriate statistical method might improve signal:noise

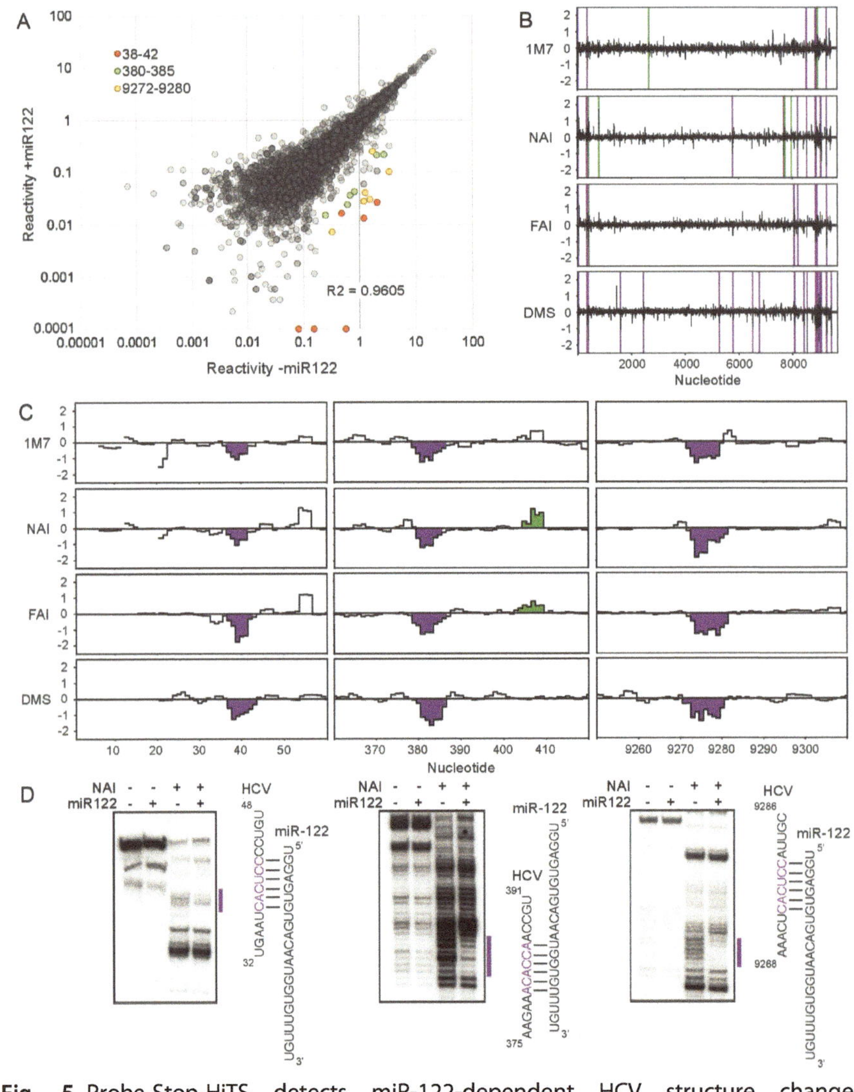

Fig. 5 Probe-Stop-HiTS detects miR-122-dependent HCV structure changes. (A) Comparison of miR-122-dependent reactivities. HCV RNA was folded in the absence or presence of 4 mM miR-122 (Eurofins). Resulting reactivities were computed with ShapeMapper2. For colored values, negative numbers were plotted as 0.0001 to enable visualization. (B) DeltaShape comparison of reactivities with and without miR-122. Results from ShapeMapper2 with indicated probes were input to DeltaShape with the standard deviation threshold set to two and FindSite threshold set to 2.5. (C) DeltaShape detailed results for indicated regions. Note that the very 5′ end had low read depth, resulting in discontinuous comparison. (D) Primer extension of indicated regions. Primer extension with ^{32}P-labeled primers was performed on untreated and NAI-treated RNA for the indicated regions. Complementarity with miR-122 depicted to the right.

when comparing conditions (Vaziri, Koehl, & Aviran, 2018). Although we have also observed bias in reactivity calculations of Probe-Stop-HiTS based on the location of the primer binding site, it is possible that calculating nucleotide reactivity using a more statistical approach, as is done with Spats (Mortimer, Trapnell, Aviran, Pachter, & Lucks, 2012) will decrease this bias. Additionally, including stop and mutational information from Probe-Stop-HiTS sequences may be informative for calculating reactivity and predicting structure, since RT enzymes also incorporate mutations at modified nucleotides under standard conditions (Sexton, Wang, Rutenberg-Schoenberg, & Simon, 2017). Finally, researchers have developed probes that can be conjugated to biotin, enabling purification of modified RNA (Flynn et al., 2016) and cDNA fully extended to the probed nucleotide (Poulsen, Kielpinski, Salama, Krogh, & Vinther, 2015). These probes may increase Probe-Stop-HiTS sensitivity by reducing background premature truncation signal. Therefore, although, Probe-MaP-HiTS is currently a more sensitive method for comparing long RNA structure in different conditions, reactivities calculated from Probe-Stop-HiTS do improve structure prediction (Fig. 3D) and stop-based methods have the potential for technical improvements.

When implementing Probe-HiTS experiments we have found it useful to first analyze an RNA with well-characterized structural elements (example: the HCV IRES). As a test case for evaluating structural changes that can be induced by a change in conditions, we added miR-122, as this miRNA is known to base-pair with the positive strand of the HCV genome (Jopling, Yi, Lancaster, Lemon, & Sarnow, 2005). Under these conditions, we observed that miR-122 interacts with the HCV genome at three sites. One of these is at the canonical binding site (site 2, located at nucleotides 34–42) and two other sites, which are located at positions 380–385 and 9272–9280. Importantly for potential function, the region 9272–9280 binds to the miR-122 seed sequence. However, miR-122 has only been observed to functionally bind the extreme 5′-end (nucleotides 1–42) of genomic HCV in cells (Bernier & Sagan, 2019; Luna et al., 2015). Therefore, probing a transcript in the presence of a microRNA has the potential to uncover all potential miRNA target binding sites, whether functional or not, and subsequent genetic experiments can assess functionality. Ultimately, we anticipate that Probe-MaP-HiTS will be helpful in distinguishing interactions and structural rearrangements *in vivo*.

Acknowledgments

We thank Fei Liu, Shivali Patel, and the Pyle lab for helpful discussions and troubleshooting help. This work was supported by NIH RO1 AI089826 (to A.M.P.) and NIH F32 AI126660 (to R.L.A.). A.M.P. is an investigator of the Howard Hughes Medical Institute.

References

Adams, R. L., Pirakitikulr, N., & Pyle, A. M. (2017). Functional RNA structures throughout the hepatitis C virus genome. *Current Opinion in Virology*, *24*, 79–86. https://doi.org/10.1016/j.coviro.2017.04.007.

Bernier, A., & Sagan, S. M. (2019). Beyond sites 1 and 2, miR-122 target sites in the HCV genome have negligible contributions to HCV RNA accumulation in cell culture. *The Journal of General Virology*. https://doi.org/10.1099/jgv.0.001217.

Blondal, T., Thorisdottir, A., Unnsteinsdottir, U., Hjorleifsdottir, S., Aevarsson, A., Ernstsson, S., et al. (2005). Isolation and characterization of a thermostable RNA ligase 1 from a Thermus scotoductus bacteriophage TS2126 with good single-stranded DNA ligation properties. *Nucleic Acids Research*, *33*(1), 135–142. https://doi.org/10.1093/nar/gki149.

Busan, S., & Weeks, K. M. (2018). Accurate detection of chemical modifications in RNA by mutational profiling (MaP) with ShapeMapper 2. *RNA (New York, N.Y.)*, *24*(2), 143–148. https://doi.org/10.1261/rna.061945.117.

Chan, D., Beasley, S., Zhen, Y., & Spitale, R. C. (2018). Facile synthesis and evaluation of a dual-functioning furoyl probe for in-cell SHAPE. *Bioorganic & Medicinal Chemistry Letters*, *28*(4), 601–605. https://doi.org/10.1016/j.bmcl.2018.01.042.

Chillón, I., Marcia, M., Legiewicz, M., Liu, F., Somarowthu, S., & Pyle, A. M. (2015). Native purification and analysis of long RNAs. *Methods in Enzymology*, *558*, 3–37. https://doi.org/10.1016/bs.mie.2015.01.008.

Darty, K., Denise, A., & Ponty, Y. (2009). VARNA: Interactive drawing and editing of the RNA secondary structure. *Bioinformatics (Oxford, England)*, *25*(15), 1974–1975. https://doi.org/10.1093/bioinformatics/btp250.

Deigan, K. E., Li, T. W., Mathews, D. H., & Weeks, K. M. (2009). Accurate SHAPE-directed RNA structure determination. *Proceedings of the National Academy of Sciences of the United States of America*, *106*(1), 97–102. https://doi.org/10.1073/pnas.0806929106.

Dethoff, E. A., Chugh, J., Mustoe, A. M., & Al-Hashimi, H. M. (2012). Functional complexity and regulation through RNA dynamics. *Nature*, *482*(7385), 322–330. https://doi.org/10.1038/nature10885.

Fang, R., Moss, W. N., Rutenberg-Schoenberg, M., & Simon, M. D. (2015). Probing Xist RNA structure in cells using targeted structure-Seq. *PLoS Genetics*, *11*(12), e1005668. https://doi.org/10.1371/journal.pgen.1005668.

Feng, C., Chan, D., Joseph, J., Muuronen, M., Coldren, W. H., Dai, N., et al. (2018). Light-activated chemical probing of nucleobase solvent accessibility inside cells. *Nature Chemical Biology*, *14*(3), 276–283. https://doi.org/10.1038/nchembio.2548.

Flynn, R. A., Zhang, Q. C., Spitale, R. C., Lee, B., Mumbach, M. R., & Chang, H. Y. (2016). Transcriptome-wide interrogation of RNA secondary structure in living cells with icSHAPE. *Nature Protocols*, *11*(2), 273–290. https://doi.org/10.1038/nprot.2016.011.

Gößringer, M., Helmecke, D., Köhler, K., Schön, A., Kirsebom, L. A., Bindereif, A., et al. (2014). Enzymatic RNA synthesis using bacteriophage T7 RNA polymerase. In *Handbook of RNA biochemistry* (pp. 1–28). John Wiley & Sons, Ltd. https://doi.org/10.1002/9783527647064.ch1.

Homan, P. J., Favorov, O. V., Lavender, C. A., Kursun, O., Ge, X., Busan, S., et al. (2014). Single-molecule correlated chemical probing of RNA. *Proceedings of the National Academy of Sciences of the United States of America, 111*(38), 13858–13863. https://doi.org/10.1073/pnas.1407306111.

Jopling, C. L., Yi, M., Lancaster, A. M., Lemon, S. M., & Sarnow, P. (2005). Modulation of hepatitis C virus RNA abundance by a liver-specific MicroRNA. *Science (New York, N.Y.), 309*(5740), 1577–1581. https://doi.org/10.1126/science.1113329.

Kwok, C. K., Ding, Y., Sherlock, M. E., Assmann, S. M., & Bevilacqua, P. C. (2013). A hybridization-based approach for quantitative and low-bias single-stranded DNA ligation. *Analytical Biochemistry, 435*(2), 181–186. https://doi.org/10.1016/j.ab.2013.01.008.

Li, H., & Aviran, S. (2018). Statistical modeling of RNA structure profiling experiments enables parsimonious reconstruction of structure landscapes. *Nature Communications, 9*(1), 606. https://doi.org/10.1038/s41467-018-02923-8.

Lorenz, R., Hofacker, I. L., & Stadler, P. F. (2016). RNA folding with hard and soft constraints. *Algorithms for Molecular Biology, 11*, 8. https://doi.org/10.1186/s13015-016-0070-z.

Lorenz, R., Luntzer, D., Hofacker, I. L., Stadler, P. F., & Wolfinger, M. T. (2016). SHAPE directed RNA folding. *Bioinformatics, 32*(1), 145–147. https://doi.org/10.1093/bioinformatics/btv523.

Luna, J. M., Scheel, T. K. H., Danino, T., Shaw, K. S., Mele, A., Fak, J. J., et al. (2015). Hepatitis C virus RNA functionally sequesters miR-122. *Cell, 160*(6), 1099–1110. https://doi.org/10.1016/j.cell.2015.02.025.

Mathews, D. H. (2010). Using OligoWalk to identify efficient siRNA sequences. *Methods in Molecular Biology (Clifton, N.J.), 629*, 109–121. https://doi.org/10.1007/978-1-60761-657-3_8.

McGinnis, J. L., Dunkle, J. A., Cate, J. H. D., & Weeks, K. M. (2012). The mechanisms of RNA SHAPE chemistry. *Journal of the American Chemical Society, 134*(15), 6617–6624. https://doi.org/10.1021/ja2104075.

Mitchell, D., Renda, A. J., Douds, C. A., Babitzke, P., Assmann, S. M., & Bevilacqua, P. C. (2019). In vivo RNA structural probing of uracil and guanine base-pairing by 1-ethyl-3-(3-dimethylaminopropyl)carbodiimide (EDC). *RNA (New York, N.Y.), 25*(1), 147–157. https://doi.org/10.1261/rna.067868.118.

Mortimer, S. A., & Doudna, J. A. (2013). Unconventional miR-122 binding stabilizes the HCV genome by forming a trimolecular RNA structure. *Nucleic Acids Research, 41*(7), 4230–4240. https://doi.org/10.1093/nar/gkt075.

Mortimer, S. A., Trapnell, C., Aviran, S., Pachter, L., & Lucks, J. B. (2012). SHAPE-Seq: High-throughput RNA structure analysis. *Current Protocols in Chemical Biology, 4*(4), 275–297. https://doi.org/10.1002/9780470559277.ch120019.

Mortimer, S. A., & Weeks, K. M. (2007). A fast-acting reagent for accurate analysis of RNA secondary and tertiary structure by SHAPE chemistry. *Journal of the American Chemical Society, 129*(14), 4144–4145. https://doi.org/10.1021/ja0704028.

Mortimer, S. A., & Weeks, K. M. (2008). Time-resolved RNA SHAPE chemistry. *Journal of the American Chemical Society, 130*(48), 16178–16180. https://doi.org/10.1021/ja8061216.

Pietschmann, T., Kaul, A., Koutsoudakis, G., Shavinskaya, A., Kallis, S., Steinmann, E., et al. (2006). Construction and characterization of infectious intragenotypic and intergenotypic hepatitis C virus chimeras. *Proceedings of the National Academy of Sciences of the United States of America, 103*(19), 7408–7413. https://doi.org/10.1073/pnas.0504877103.

Poulsen, L. D., Kielpinski, L. J., Salama, S. R., Krogh, A., & Vinther, J. (2015). SHAPE selection (SHAPES) enrich for RNA structure signal in SHAPE sequencing-based probing data. *RNA, 21*(5), 1042–1052. https://doi.org/10.1261/rna.047068.114.

Reuter, J. S., & Mathews, D. H. (2010). RNAstructure: Software for RNA secondary structure prediction and analysis. *BMC Bioinformatics, 11*(129). https://doi.org/10.1186/1471-2105-11-129.

Sambrook, J., Fritsch, E. F., & Maniatis, T. (1989). *Molecular cloning: A laboratory manual* (2nd ed.). Cold Spring Harbor Laboratory.

Sexton, A. N., Wang, P. Y., Rutenberg-Schoenberg, M., & Simon, M. D. (2017). Interpreting reverse transcriptase termination and mutation events for greater insight into the chemical probing of RNA. *Biochemistry*, *56*(35), 4713–4721. https://doi.org/10.1021/acs.biochem.7b00323.

Siegfried, N. A., Busan, S., Rice, G. M., Nelson, J. A. E., & Weeks, K. M. (2014). RNA motif discovery by SHAPE and mutational profiling (SHAPE-MaP). *Nature Methods*, *11*(9), 959–965. https://doi.org/10.1038/nmeth.3029.

Smola, M. J., Calabrese, J. M., & Weeks, K. M. (2015). Detection of RNA-protein interactions in living cells with SHAPE. *Biochemistry*, *54*(46), 6867–6875. https://doi.org/10.1021/acs.biochem.5b00977.

Smola, M. J., Christy, T. W., Inoue, K., Nicholson, C. O., Friedersdorf, M., Keene, J. D., et al. (2016). SHAPE reveals transcript-wide interactions, complex structural domains, and protein interactions across the Xist lncRNA in living cells. *Proceedings of the National Academy of Sciences of the United States of America*, *113*(37), 10322–10327. https://doi.org/10.1073/pnas.1600008113.

Smola, M. J., Rice, G. M., Busan, S., Siegfried, N. A., & Weeks, K. M. (2015). Selective 2′-hydroxyl acylation analyzed by primer extension and mutational profiling (SHAPE-MaP) for direct, versatile and accurate RNA structure analysis. *Nature Protocols*, *10*(11), 1643–1669. https://doi.org/10.1038/nprot.2015.103.

Somarowthu, S., Legiewicz, M., Chillón, I., Marcia, M., Liu, F., & Pyle, A. M. (2015). HOTAIR forms an intricate and modular secondary structure. *Molecular Cell*, *58*(2), 353–361. https://doi.org/10.1016/j.molcel.2015.03.006.

Spasic, A., Assmann, S. M., Bevilacqua, P. C., & Mathews, D. H. (2018). Modeling RNA secondary structure folding ensembles using SHAPE mapping data. *Nucleic Acids Research*, *46*(1), 314–323. https://doi.org/10.1093/nar/gkx1057.

Spitale, R. C., Crisalli, P., Flynn, R. A., Torre, E. A., Kool, E. T., & Chang, H. Y. (2013). RNA SHAPE analysis in living cells. *Nature Chemical Biology*, *9*(1), 18–20. https://doi.org/10.1038/nchembio.1131.

Tack, D. C., Tang, Y., Ritchey, L. E., Assmann, S. M., & Bevilacqua, P. C. (2018). StructureFold2: Bringing chemical probing data into the computational fold of RNA structural analysis. *Methods (San Diego, Calif.)*, *143*, 12–15. https://doi.org/10.1016/j.ymeth.2018.01.018.

Tan, Z., Sharma, G., & Mathews, D. H. (2017). Modeling RNA secondary structure with sequence comparison and experimental mapping data. *Biophysical Journal*, *113*(2), 330–338. https://doi.org/10.1016/j.bpj.2017.06.039.

Vaziri, S., Koehl, P., & Aviran, S. (2018). Extracting information from RNA SHAPE data: Kalman filtering approach. *PLoS One*, *13*(11), e0207029. https://doi.org/10.1371/journal.pone.0207029.

Villordo, S. M., & Gamarnik, A. V. (2009). Genome cyclization as strategy for flavivirus RNA replication. *Virus Research*, *139*(2), 230–239. https://doi.org/10.1016/j.virusres.2008.07.016.

Wang, P. Y., Sexton, A. N., Culligan, W. J., & Simon, M. D. (2019). Carbodiimide reagents for the chemical probing of RNA structure in cells. *RNA (New York, N.Y.)*, *25*(1), 135–146. https://doi.org/10.1261/rna.067561.118.

Watters, K. E., & Lucks, J. B. (2016). Mapping RNA structure in vitro with SHAPE chemistry and next-generation sequencing (SHAPE-Seq). *Methods in Molecular Biology (Clifton, N.J.)*, *1490*, 135–162. https://doi.org/10.1007/978-1-4939-6433-8_9.

Wells, S. E., Hughes, J. M., Igel, A. H., & Ares, M. (2000). Use of dimethyl sulfate to probe RNA structure in vivo. *Methods in Enzymology*, *318*, 479–493.

Wilkinson, K. A., Merino, E. J., & Weeks, K. M. (2006). Selective 2′-hydroxyl acylation analyzed by primer extension (SHAPE): Quantitative RNA structure analysis at single nucleotide resolution. *Nature Protocols*, *1*(3), 1610–1616. https://doi.org/10.1038/nprot.2006.249.

Wilkinson, K. A., Vasa, S. M., Deigan, K. E., Mortimer, S. A., Giddings, M. C., & Weeks, K. M. (2009). Influence of nucleotide identity on ribose 2′-hydroxyl reactivity in RNA. *RNA (New York, N.Y.)*, *15*(7), 1314–1321. https://doi.org/10.1261/rna.1536209.

Zinshteyn, B., Chan, D., England, W., Feng, C., Green, R., & Spitale, R. C. (2019). Assaying RNA structure with LASER-Seq. *Nucleic Acids Research*, *47*(1), 43–55. https://doi.org/10.1093/nar/gky1172.

Zubradt, M., Gupta, P., Persad, S., Lambowitz, A. M., Weissman, J. S., & Rouskin, S. (2017). DMS-MaPseq for genome-wide or targeted RNA structure probing in vivo. *Nature Methods*, *14*(1), 75–82. https://doi.org/10.1038/nmeth.4057.

CHAPTER THIRTEEN

A fluorescent aminosugar to rapidly screen and study RNA binders

Nihar Ranjan[a], Dev P. Arya[b,*]
[a]Department of Medicinal Chemistry, National Institute of Pharmaceutical Education and Research (NIPER)-Raebareli, New Transit Campus, Lucknow, India
[b]Laboratory of Bioorganic and Medicinal Chemistry, Department of Chemistry, Clemson University, Clemson, SC, United States
*Corresponding author: e-mail address: dparya@clemson.edu

Contents

Abstract

RNA targeted high-throughput assays that allow for rapid detection of high affinity binding ligands are important in RNA recognition studies. A number for fluorescent dyes have been reported that can assist in rapidly identifying nucleic acid (RNA) binding

Methods in Enzymology, Volume 623
ISSN 0076-6879
https://doi.org/10.1016/bs.mie.2019.05.027

elements without the need for immobilization of RNA or the ligand. A number of these dyes are planar aromatic molecules that bind non-specifically to nucleic acids and often distort their parent nucleic acid structures leading to ambiguity in the interpretation of results. In this light, we report here, the use of an aminoglycoside (neomycin) based fluorescent probe (F-Neo) which can reversibly bind to different RNA motifs and help identify ligands with needed affinity and selectivity, without any immobilization of the probe or the target. In this chapter, we provide the details of the assay development, experimental considerations and data analysis to use the probe and identify novel ligands. We then provide a brief introduction to calorimetry (ITC) and circular dichroism (CD) spectroscopy based methods in validating the binding of such identified compounds.

1. Introduction

RNA occupies central space in the so called "central dogma of life" as evidenced by its multi-faceted role in the cell survival. The proponents of the "RNA world" theory suggest that RNA played a pivotal role in the evolution of life on earth (Cech & Golden, 1999). The requirements of RNA both during pre- and posttranslational events highlight its importance in the cellular processing. RNA serves as a validated target of drug design in the bacterial and viral life cycles. In eukaryotic systems ncRNAs, miRNAs, four strand RNA G-quadruplexes (Neidle, 2016; Oyaghire et al., 2016), and DNA:RNA hybrid regions of telomeric ends are some potential RNA structures being investigated as drug intervention strategies.

RNA structures are highly dynamic and diverse leading to a structural landscape that is full of canonical and non-canonical forms of nucleic acid architecture (Dethoff, Petzold, Chugh, Casiano-Negroni, & Al-Hashimi, 2012). For example, structures of tRNA, ribozymes and several pre-miRNAs give a glimpse of hairpin loop, bulge, and mismatch sites commonly found in RNAs of various kinds. It is widely believed that this flexibility of frequently adopting non-canonical forms has empowered RNA with catalytic and processing functions unlike its structural counterpart DNA. This flexibility offers unique challenges and opportunities when targeting RNA with small molecules and developing probes for sensing its functions. This is especially true for small molecules that recognize nucleic acids by intercalation or by making interactions through their major and minor grooves. Part of this difference in RNA versus DNA binding is attributed to changes in their structure (A-form to B-form) which leads to sharp differences in their major and minor grooves (Bhaduri, Ranjan, & Arya, 2018; Willis & Arya, 2014). For instance, while the duplex regions of RNA (which adopt A-form) have narrow major grooves and wide minor grooves, the same in the DNA (which predominantly adopt B-form) have wide major groove and narrow minor groove. As a consequence, ligands that have been selected to bind RNA major grooves do not bind or weakly bind the DNA major groove and ligands that have been selected to bind DNA minor grooves strongly are poor binders of RNA. Therefore, discovery of new RNA binding motifs remains an exciting area of research for the discovery of future therapeutic agents.

In this regard, herein, we present methods related to the screening of ligand libraries (Watkins, Norris, Kumar, & Arya, 2013) of small molecules to identify RNA binding molecules using the bacterial rRNA A-site as a "template RNA" which can be easily adapted to screen molecules for other RNA platforms such as ncRNAs and other types of eukaryotic rRNA A-sites (bacterial, human and fungal rRNA) (Degtyareva et al., 2018). We subsequently provide details of other biophysical methods which are commonly used in characterizing and validating interactions of RNA with small molecules. These non-structural characterizations will be limited to circular dichroism (CD) spectroscopy and microcalorimetry (ITC) based characterization of rRNA A-site and aminoglycoside binding which validate the screening data and provide a number of thermodynamic parameters.

The bacterial rRNA A-site is a proven and well-characterized RNA structure which is home to a number of antibiotics belonging to the aminoglycoside class (Arya, 2007). In this chapter, we provide detailed description of a screening platform which provides rapid access to screen hit molecules and preliminary follow-up biophysical characterization using spectroscopic and calorimetry methods.

2. High-throughput fluorescence screening

Fluorescence based high throughput screening is popular because of several reasons: (a) it is a highly sensitive technique which can sense changes in the chemical environment at very low concentrations levels (b) it can be easily adapted to 96, 384 or 1536 well plate readers to enable rapid screening (c) very low levels of analyte concentration requirements reduce or eliminate many unwanted ligand self-association events (d) assays can be tailor-made to screen both fluorescent and non-fluorescent molecules if proper precautions are taken (e) either the receptor or ligand or both can be chosen to sense the changes in the chemical and physical micro-environment upon binding.

Receptor (RNA) modifications with fluorescent tags, FRET pairs, or immobilization are useful complementary tools for screening ligand libraries but require the modification of each individual receptor to be screened. A reversible assay using a small dye molecule has the advantage of probing numerous nucleic acids at the same time. Fluorescence based screening methods such as Competition Dialysis Assay and Fluorescent Intercalator Displacement (FID) (Boger, Fink, Brunette, Tse, & Hendrick, 2001; Tse & Boger, 2004) have been successfully demonstrated to screen a large number and variety of molecules against a range of nucleic acid targets. The competition dialysis assay (Ren & Chaires, 1999) is an extension of equilibrium dialysis assay which can allow detection of structure selective nucleic acid binders. However, this assay requires the ligand to be fluorescent/UV active which often acts as an impediment in the wide applicability of this method. The FID assay overcomes this limitation and allows screening of both fluorescent and non-fluorescent molecules. FID assay relies on the changes in fluorescence emission signal of the probe upon binding of the ligand under investigation. Probes such as Thiazole Orange (TO) have been used routinely for nucleic acid screening but these intercalating probes bind to numerous sites on a nucleic acid leading to alteration of its solution structure. The use of groove binding probes such as Fluorescein-neomycin (F-Neo) offers a useful alternative as F-Neo does not alter the nucleic acid structure by binding in a much lower stoichiometry than TO or ethidium bromide (Watkins, Gong, Kellish, & Arya, 2017). Small molecule assays and leads can, in principle, be generated for any given nucleic acid target of interest. Moreover, with non-specific nucleic acid intercalator dyes such as ethidium bromide or TO, very little information about the nature (groove binding or intercalation) of molecules that displace them can be obtained. This happens partly because the FID assay requires the fluorescent probe to be a moderate binder of the nucleic acid under investigation, with multiple binding sites so that even the allosteric changes in the nucleic acid structure can lead to displacement of the fluorescent probe. These properties of the fluorescent probes (low K_a and high stoichiometry) are useful such that any weak or strong binder can displace them, but are also limiting in the information they provide about the site of binding of the small molecule(s) being investigated. While these issues cannot be completely eliminated with any dye probe, a much better correlation is seen in displacement properties of groove binding F-Neo such that the ligands that directly bind to the site of the probe show the highest ability to displace the probe.

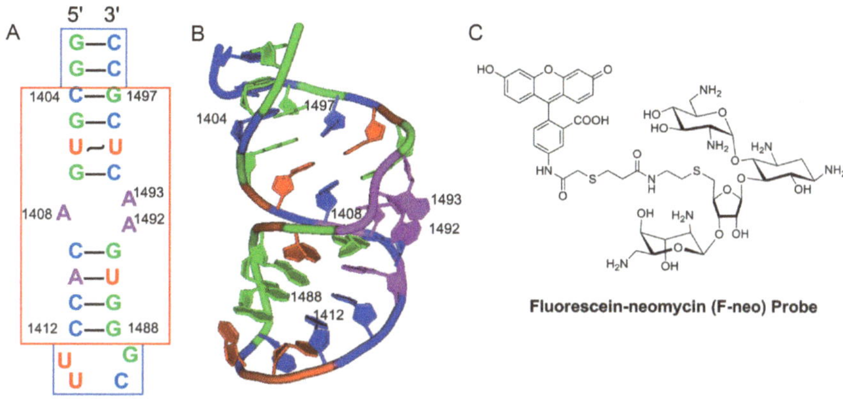

Fig. 1 (A) Base sequence and (B) structure of the ribosomal rRNA A-site (C) chemical structure of F-Neo conjugate. *Figure adapted with permission from Watkins, D., Norris, F. A., Kumar, S., & Arya, D. P. (2013). A fluorescence-based screen for ribosome binding antibiotics.* Analytical Biochemistry, 434*(2), 300–307. Copyright 2013 Elsevier.*

We describe a fluorescence based screening assay to rapidly identify small molecule binders of RNA (Watkins et al., 2013). In contrast to competition dialysis and FID assays, this method utilizes fluorescence quenching of the reporter (a fluorescein-aminoglycoside, F–Ag, hybrid molecule) molecule (Fig. 1) and is very sensitive to changes in the electrostatic potential of the RNA groove. The fluorescent probes used in this assay are typically neomycin-fluorescein or kanamycin fluorescein hybrid molecules which allow identification of both strong and moderate binders of the bacterial rRNA A-site. Outlined below are methods to (a) synthesize the aminoglycoside-fluorescein hybrid reporter probes (b) conduct the F–Ag displacement screen assay (c) adaptation of the assay to other nucleic acid targets, (d) validation of binding and thermodynamic parameters using ITC and (e) assessment of binding and ligand effects on target structure using Circular Dichroism (CD) studies.

3. Fluorescein-neomycin (F-Neo) competition displacement assay

Fluorescein-neomycin (F-Neo) is a hybrid molecule that consists of a fluorescent compound (fluorescein) and an aminosugar (neomycin) linked together by a covalent bond. Fluorescein displays several protonation states of which the monoanion and dianion states that exist near biological pH range are of relevance in this assay. Fluorescein absorption is highly sensitive to changes in the electrostatic potential of the solution, such that the negatively charged polymer like DNA or RNA can shift the equilibrium of the anion to the protonated phenol leading to a quenching of fluorescein fluorescence. The electrostatic sensitivity of fluorescein can be gauged from the fact that in nucleic acids (such as peptide nucleic acids) bereft of phosphate backbones, no quenching is observed (Lavis, Rutkoski, & Raines, 2007; Sjoback, Nygren, & Kubista, 1998). F-Neo has been shown to bind to the bacterial A-site rRNA sequence in manner similar to the binding of neomycin alone. The binding of aminoglycosides to the bacterial rRNA A-site is highly conserved where it binds to the trinucleotide bulge region with high affinity (Fourmy, Recht, & Puglisi, 1998; Lynch & Puglisi, 2001). The binding of F-Neo probe to the bacterial A-site rRNA is also driven by forces similar to non-conjugated neomycin where the fluorescein part of the molecule, because of its spatial positioning, just acts as a signal transmitter to the local changes in the groove micro environment. The fluorescence changes are directly reflective of association or dissociation events and the change in F-Neo fluorescence is proportional to the affinity of the ligand displacing it. Another robust advantage of using the F-Neo probe is that it does not perturb the nucleic acid structure under investigation unlike fluorescent dyes such as thiazole orange, thus one can probe binding to a target RNA in solution very close to its native structure (Watkins et al., 2017).

The binding affinity of F-Neo to the bacterial A-site rRNA is quite similar to neomycin's binding to the same RNA sequence where they both show K_ds in the nanomolar range. Ideally, fluorescent probes are desired to have moderate affinities (typically in the micromolar range or lower) to their cognate nucleic acid sequences so that their dissociation or displacement from the nucleic acid complexes are possible even with low affinity ligands. The range of affinities of F-neo (K_d in nM range) and F-kana (K_d in μM range), allow screening platforms to be tailored for the affinities of interest (Watkins et al., 2013). Outlined below (Fig. 2) is a schematic that represents this assay.

3.1 The assay

The F-Neo probe is fluorescent in water and several other types of buffer systems that have pH close to 7.0. Addition of the bacterial rRNA A-site oligonucleotide sequence causes fluorescence quenching of the fluorescein

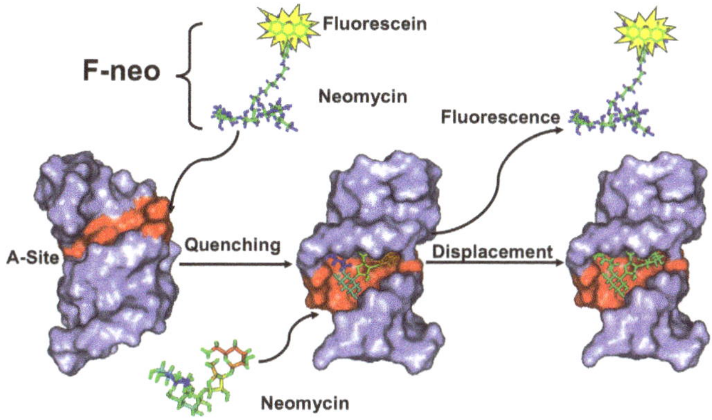

Fig. 2 A schematic showing the F-Neo displacement assay. *Figure adapted with permission from Jiang, L., Watkins, D., Jin, Y., Gong, C., King, A., Washington, A. Z., et al. (2015). Rapid synthesis, RNA binding, and antibacterial screening of a peptidic-aminosugar (PA) library.* ACS Chemical Biology, 10*(5), 1278–1289. Copyright 2015 American Chemical Society.*

moiety. A titration experiment revealed that this quenching is complete when the F-Neo probe and the bacterial rRNA A-site oligonucleotide sequence are at equimolar ratio (Watkins et al., 2013). Further addition of the RNA oligonucleotide to excess amounts (sevenfold higher RNA concentration than the F-Neo probe) caused negligible change in the fluorescence emission signal intensity. Therefore, a 1:1 probe to RNA ratio can be used in the assay. The premise of the assay is that when the F-Neo probe is complexed with the bacterial rRNA A-site sequence at 1:1 ratio, large quenching of the fluorescence emission signal will occur which can be regained by successive displacement of the probe with a competing ligand. Neomycin was chosen as a test ligand from which the probe is derived. As expected, successive addition of neomycin to F-Neo complexed with the rRNA A-site completely displaced the F-Neo probe and brought back the fluorescence emission levels close to the level found in the unbound state (Watkins et al., 2013). The applicability of this assay in distinguishing the strong and weak binders was then checked with a small set of aminoglycoside antibiotics (neomycin, paromomycin, gentamycin, neamine, ribostamycin and streptomycin). The competition displacement assay was done by addition a threefold excess of test ligands (at 0.3 μM) to a 1:1 complex of F-Neo with the rRNA A-site (at 0.1 μM). The displacement assay produced a ranking in the order neomycin > gentamycin > paromoycin > neamine > ribostamycin > streptomycin which was very much in line with the independently generated thermodynamic parameters

(ITC generated K_a values in the range of 4.9×10^9–$1.6 \times 10^6\,M^{-1}$ (Kaul & Pilch, 2002). These results established that both strong and moderate binders can be ranked rapidly and reliably with this assay.

3.2 Adaptability of F-Neo competition assay to 96-well platform

One of the crucial features of a reversible fluorescence based assay is its adaptability to 96- or 384-well plate formats to rapidly screen large libraries of ligands. The initial optimization of this assay was performed on a cuvette based spectrofluorimeter, followed by a well plate format to enable high-throughput screening of ligand libraries. The first step toward this check was reproducing the F-Neo displacement by neomycin in a 96-well plate. The IC_{50} (IC_{50} refers to the concentration of neomycin required to displace half of the bound F-Neo) value of neomycin obtained through the plate reader based displacement was 89 nM, a value which was very close to the same obtained with the spectrofluorimeter (78 nM).

An important statistical parameter in assessing the quality of the assay is the so called Z′-factor which indicates the statistical significance between the positive and negative controls. In this assay, these controls are related to the unbound (free F-Neo) and bound (1:1 F-Neo to RNA oligonucleotide complex) states of the fluorescent probe. The Z′ factor is defined by the following mathematical equation:

$$Z' = 1 - \frac{3(\sigma_p + \sigma_n)}{\mu_p + \mu_n} \qquad (1)$$

where σ_p and σ_n represent the standard deviations in the positive and negative controls whereas μ_p and μ_n represent the mean values of the positive and negative controls. A Z′ value of 1.0 is considered to represent an "ideal" assay whereas any Z′ value between 0.5 and 1.0 is considered as an "excellent" assay.

The Z′ value for F-Neo probe was determined in three separate experiments on 3 different days. In each experiment the Z′ value was ≥ 0.80. Typically, each data in this assay represents an average of three readings from three different wells of the plate.

3.3 Screening of various types of ligands

To establish the applicability of this assay in identifying hits, we have assayed a variety of aminoglycoside derived compounds to search new antibacterial agents. We first used a series of dimeric neomycin (King et al., 2013) and dimeric neomycin-arginine (Jin et al., 2016) ligand libraries whose sample size was below 30 compounds. Subsequently, using solid-phase synthesis approach, a 215 member peptidic-aminosugar (PA) library (Jiang et al., 2015) was screened to identify potent and selective RNA binders. Similarly peptidic–kanamycin conjugates (Kukielski, Maiti, Bhaduri, Story, & Arya, 2018) and several other types of molecules were also screened for their RNA and DNA binding properties (Ghosh et al., 2018; Watkins, Jiang, Nahar, Maiti, & Arya, 2015).

3.4 Applications of F-Neo assay in other nucleic acid systems

F-Neo's utility as a probe is not limited to rRNA sequences. We checked if this assay can be translated to other RNA model systems. The 215 member PA library was screened against variety of miRNA sequences (Watkins et al., 2015). More recently, the F-Neo assay has been used to detect hit molecules for miR-21 binding (Ghosh et al., 2018). Not limiting to RNA, this assay can also be used to identify duplex DNA binding agents (Watkins et al., 2017). These examples highlight that F-Neo can be easily and effectively translated to screen for ligand binding to any nucleic acid of interest.

4. Experimental setup of the F-Neo competition displacement assay

The assay requires one to have access to the following to successfully complete the screening:

(a) F-Neo probe
(b) Appropriate oligonucleotides
(c) Compound library
(d) Data analysis software

F-Neo probe can be synthesized chemically (Scheme 1) (Xi et al., 2011) or it can also be purchased through commercial supplier (https://nubadllc.com/). Similarly, oligonucleotides can be synthesized in the laboratory or purchased from commercial vendors. The compound library of interest is usually self-prepared, however, the available drug libraries from other sources can also be used. Finally, a software capable of performing mathematical operations and plotting (Origin, Kaleidagraph) is required to analyze the data.

4.1 Equipment

1. Analytical balance
2. Glass column (approximately 20mm inner diameter and 600mm length)

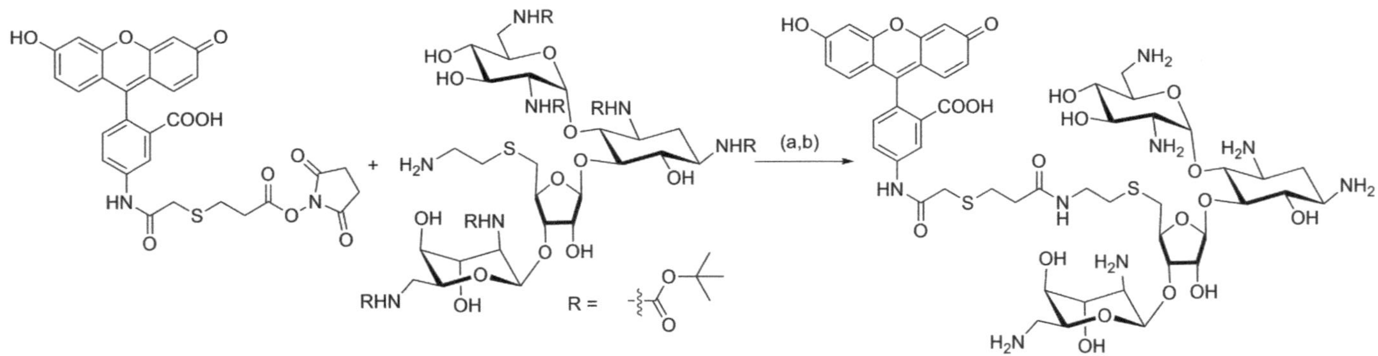

Scheme 1 Reagents and conditions (A) Dimethylformamide (DMF), 4-dimethylaminopyridine (DMAP), room temperature; (B) TFA, dichloromethane (DCM), room temperature. *Scheme adapted with permission from Xi, H., Davis, E., Ranjan, N., Xue, L., Hyde-Volpe, D., & Arya, D. P. (2011). Thermodynamics of nucleic acid "shape readout" by an aminosugar.* Biochemistry, *50(42), 9088–9113. Copyright 2011 American Chemical Society.*

3. Rotary evaporator
4. Vortex
5. Lyophilizer
6. Analytical HPLC
7. UV–Visible spectrophotometer
8. Cuvette based spectrofluorimeter
9. Multi-mode plate reader
10. Circular Dichroism spectropolarimeter
11. Isothermal Calorimetry (ITC) instrument

4.2 Chemicals/consumables

1. Fluorescein Succinimidyl Ester (Thermofisher)
2. Boc protected neomycin amine (Starting materials through MP Biomedicals, Sigma, Acros)
3. DMAP (Acros)
4. Dichloromethane (VWR)
5. Methanol (VWR)
6. Dry dimethyl formamide (Sigma)
7. Trifluoroacetic acid (Sigma)
8. Silica gel (5 μm size, Sorbtech)
9. Pre-coated thin layer chromatography plate (Sorbtech)
10. Flat bottom black 96-well plates (Greener Bioone)
11. Sodium chloride, molecular biology grade (Sigma)
12. Potassium Chloride, molecular biology grade (Sigma)
13. Sodium Cacodylate (Acros)
14. EDTA (Sigma)
15. MOPSO (Sigma)
16. Sea Sand (Acros)

5. Protocol

5.1 Synthesis of F-Neo probe

1.1 In an oven dried 25 mL round bottom flask, add Boc-protected neomycin amine (0.0076 mmol) in dry DMF (1 mL) under Agron atmosphere.

1.2 Add Fluorescein-succinimidyl ester (0.0086 mmol) into the round bottom flask and follow it by adding catalytic amount of DMAP (~0.5 mg).

1.3 Cover the round bottom flask with aluminum foil to protect it from light and stir at room temperature for 12 h.
1.4 Check the completion of reaction using a pre-coated TLC plate which is indicated by disappearance of starting materials and appearance of a new spot with an R_f 0.21 (dichloromethane in methanol 9:1 v/v).
1.5 After the reaction is complete, remove DMF under high vacuum.
1.6 Pack the column with silica (5 μm) in DCM (approximately 10 cm height) and wet load the crude reaction mixture.
1.7 Flush out extra solvent (DCM) until top layer of the packed silica gel is just visible. At this stage, add sea sand on top of silica gel (~3 cm high).
1.8 Elute the crude mixture with a gradient of methanol in dichloromethane (0.5–10% methanol in dichloromethane, 100 mL each) until the desired compound is obtained.
1.9 Pool the fractions containing the desired compound together, and evaporate the volatiles using a rotary evaporator.
1.10 Add DCM (1 mL) to dried Boc-protected F-Neo conjugate followed by addition of TFA (1 mL). **CAUTION!!** TFA is highly corrosive, causes skin burns and problem in inhalation. It must be handled very carefully.
1.11 Put a magnetic stir bar in this mixture and stir at room temperature for 3 h.
1.12 Remove the volatiles under reduced pressure and dissolve the residue in distilled water (0.5 mL).
1.13 Transfer the aqueous F-Neo in a microfuge tube and add ethyl acetate (0.5 mL) to it.
1.14 Vortex it well and remove the ethyl acetate layer. Repeat this step two more times.
1.15 Lyophilize the aqueous layer and store the yellowish-white F-Neo conjugate in an amber vial in freezer.
1.16 Characterize a part of conjugate with NMR, Mass spectrometry, HPLC and UV–Visible spectroscopic methods (Xi et al., 2011).

Tip: The best way to visualize Boc protected F-Neo conjugate is by staining it with ninhydrin. The developed TLC plate is soaked in the ninhydrin stain and the TLC plate is heated gently using a heat gun. The product appears as a brown spot on the TLC plate.

Additionally, a handheld UV-machine can also be used to visualize the product. However, with the handheld UV, the starting material

Boc-protected neomycin cannot be viewed. Therefore, ninhydrin based staining is the best method to isolate the Boc protected neomycin amine and the Boc protected F-Neo since the two spots run very close to each other on silica gel and care must be taken in isolating the two spots.

5.2 F-Neo displacement assay using a model rRNA (96 well format)

2.1 Determine the stock bacterial A-site rRNA (r 5′-GGC GUC ACA CCU UCG GGU GAA GUC GCC-3′) concentration on a UV–Visible spectrophotometer using an extinction coefficient of 2,53,300 $L\,mol^{-1}\,cm^{-1}$.

2.2 Prepare 9.6 mL bacterial A-site rRNA complexed with F-Neo probe (at 0.2 μM concentration each) in buffer 10 mM MOPSO, 0.4 mM EDTA, 50 mM NaCl at pH 7.0.

2.3 Separately, prepare 0.6 μM ligand solutions (300 μL) for each compound in the library in buffer and also 600 μL of 0.1 μM F-Neo solution.

2.4 In wells A1–A3 (Fig. 3), pipette out 200 μL of 0.1 μM F-Neo solution in each well.

2.5 In wells B1–B3 (Fig. 3), pipette out 100 μL of 0.2 μM A-site rRNA-F-Neo complex solution. Following this, add 100 μL buffer solutions (10 mM MOPSO, 0.4 mM EDTA, 50 mM NaCl at pH 7.0) in each of the wells so that the final concentration of the RNA-fluorescent probe complex is 0.1 μM. Mix well.

2.6 In rest of the wells, first add 100 μL of 0.2 μM A-site rRNA-F-Neo complex solution (1:1 ratio) and then add 100 μL of 0.6 μM ligand solution (1:3 ratio) for each compound in triplicates. Mix well. In this way, 30 compounds can be screened in one plate in which the final concentration of the RNA-fluorescent probe complex is 0.1 μM and the final concentration of ligand to be tested is 0.3 μM in each well with 200 μL total volume. One of the tested compounds should be a reference compound (ideally a compound that is capable of close to 100% displacement of the F-Neo probe).

2.7 For additional compounds in the library, a separate plate is prepared making sure that a reference compound is included in each plate. In essence, wells A1–A3, B1–B3, C1–C3 should be kept identical for each plate.

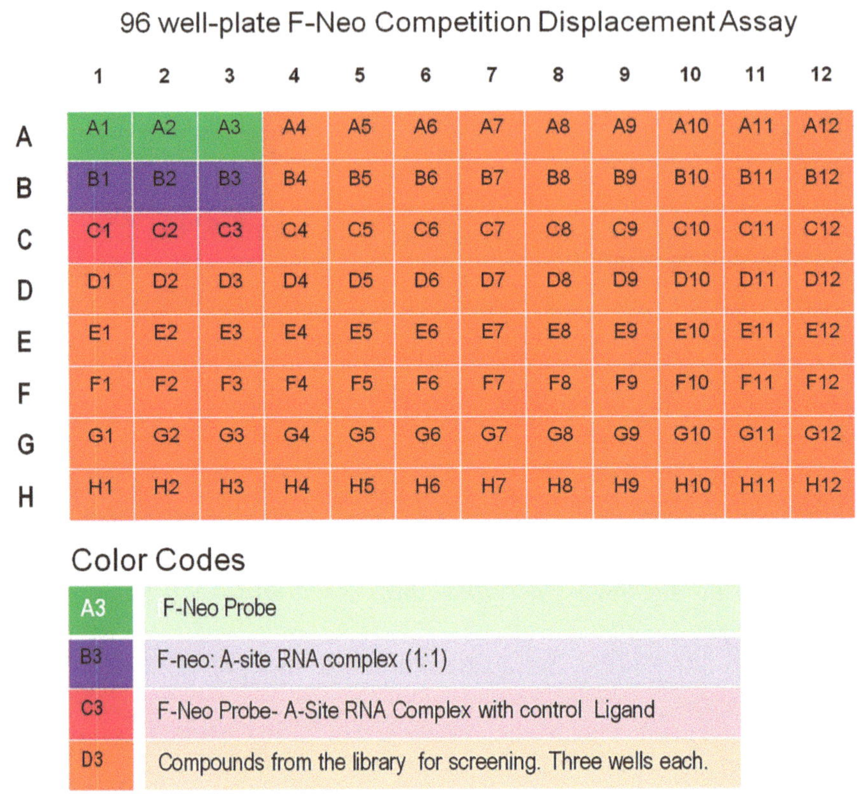

Fig. 3 Typical setup of a 96-well in the F-Neo displacement assay.

2.8 The 96-well plate is then placed in a multi-mode plate reader and fluorescence emission is read under following conditions:
Excitation wavelength: 485 nm
Emission wavelength: 535 nm
Number of reads/well: 100

2.9 The resulting data is then analyzed and plotted using appropriate processing software such as Microsoft Excel, Origin, Kaleidagraph etc.

Tip: Dummy experiments should be done to optimize the "gain" in the plate reader otherwise oversaturation of the fluorescence emission signal may happen.

Tip: A good mixing of the ligand with the A-site RNA complexed with the F-Neo probe is essential to the success of this experiment. To ensure good mixing, each reservoir in the 96 well-plate should be mixed very well with the pipette/multi-pipette initially and then it can additionally be shaken in the multi-mode plate reader instrument.

5.3 Isothermal calorimetry titration (ITC) experiment

3.1 Clean the reference and sample cells with copious amounts of buffer (10 mM sodium cacodylate, 0.5 mM EDTA, 60 mM KCl at pH 7.0).

3.2 In the reference cell, fill up 10 mM sodium cacodylate, 0.5 mM EDTA, 60 mM KCl at pH 7.0 buffer and draw off any excess buffer in the cell.

3.3 In the sample cell, fill up 1.8 mL A-site rRNA solution and draw off any excess solution above the "step" in the cell. Make sure there are no bubbles in either the reference or sample cell.

3.4 Wash the syringe with buffer and fill up the ligand solution in the syringe.

3.5 Insert the syringe needle inside the sample cell and close the sample chamber.

3.6 Start the ITC experiments under following conditions:
Concentration of A-site RNA: 10 μM/strand
Concentration of neomycin: 250 μM
Injection spacing: 240 s
Syringe rotation: 250 rpm
Injection duration: 20 s
Number of injections: 58
Temperature: 20 °C

3.7 Once the experiment is over, perform a control experiment in which neomycin is titrated into a sample cell containing buffer (10 mM sodium cacodylate, 0.5 mM EDTA, 60 mM KCl at pH 7.0) only under identical conditions as used in the previous experiment.

3.8 Integrate the heat burst curves obtained from the experiment outlined in step 3.6 after subtracting the buffer dilution heats obtained in step 3.7.

3.9 The resulting enthalpy versus mole fraction data can be fitted with an appropriate model and the binding parameters can be obtained.

Note: For a successful ITC titration, experiment should be designed in such a way so that the Wiseman value (c) (Turnbull & Daranas, 2003), which is the product of A-site RNA concentration and expected association constant multiplied by the number of ligands at the binding site, is >10 and <1000.

Tip: ITC measurements often require one to conduct experiments at more than one temperature to obtain different thermodynamic parameters. In such cases, experiments can be run under identical conditions by simply changing the temperature to the desired value.

5.4 Circular dichroism spectroscopy based titration

4.1 Turn on the nitrogen gas and the lamp and wait for shutter to open. Once the shutter is open, wait for 15 min before the experiment is started.

4.2 Set the desired parameters using the instrument software.

4.3 Following are the typical parameters of a CD titration:
Scan speed: 500 nm/min
Wavelength range: 200–350 nm
Data acquisition: 1 nm interval
Number of repeats: 3
Temperature: 20 °C

4.4 Once the desired parameters are set, the instrument is ready for the wavelength scan experiment.

4.5 In a quartz cuvette, take 1.8 mL buffer (10 mM sodium cacodylate, 0.5 mM EDTA, 100 mM NaCl at pH 6.8) and perform the wavelength scan experiment.

4.6 Follow this experiment by taking 1.8 mL of bacterial A-Site rRNA (10 μM/strand) solution in buffer and perform wavelength scan experiment.

4.7 To the same solution, add concentrated (millimolar) neomycin solution sequentially so that both pre- (6–8 points) and postsaturation (6–8 points) binding sections are adequately covered. After each addition (typically ~2 μL), the solution in the cuvette is mixed well using a magnetic stirrer and the scan is taken.

4.8 Plot the resulting spectrum after subtracting the buffer using a plotting software.

4.9 Look for the regions with prominent changes and choose the wavelength.

4.10 Plot the graph between CD intensity and molar ratio of ligand to RNA at this ratio.

6. Analysis of results

6.1 Analyzing F-Neo competition displacement assay results

The aim of this assay to rapidly identify compounds that bind to a target of interest with varying affinities and for this purpose, a control ligand is usually employed. The role of control ligand is to serve as a standard compound which displaces most of the bound F-Neo probe from the nucleic acid target. The displacement afforded by the control ligand is taken as 100% displacement and all other ligands screened in the assay are then ranked relative to this control ligand. The percent binding is calculated from the ratio of the change in fluorescence by addition of the test compound (ΔF_{Test}) with the change in fluorescence by addition of the control ligand ($\Delta F_{Control}$), using the following equation:

$$\%\ \text{Displacement} = (\Delta F_{Test}/\Delta F_{Control}) \times 100\% \quad (2)$$

Outlined below are the results of the F-Neo competition displacement screening assay for a 215 member PA library against bacterial rRNA A-site sequence (Jiang et al., 2015). As shown in Fig. 4, out of the 215 compounds screened, 14 compounds showed >100% displacement (as given by the standard used in the assay: neomycin) of the F-Neo probe which identifies them as better binders than the control ligand neomycin. Thirty one compounds showed negative percent binding which indicate that these compounds are incapable of displacing the bound F-Neo from the RNA target. The remaining ligands may be termed as "intermediate" binders. In additional to these interpretations, the binders of needed affinities can be coupled to further properties and analysis such as RB-BIP (King et al., 2013) or principal component analysis (Del Villar-Guerra, Gray, Trent, & Chaires, 2018; Patwardhan, Cai, Newson, & Hargrove, 2018) with other types of biological activity data or other supporting results. As can be seen in Fig. 4, the best binders in this particular rRNA screening were compounds in which neomycin is coupled to a positively charged amino acid such as arginine or lysine together with aromatic amino acids or cysteine.

6.2 Analyzing the results from the ITC titration

The uniqueness of ITC is in its operational simplicity and multitude of information that it can provide in a single experiment, along with the ΔH of an interaction (Buurma & Haq, 2007). Fig. 5 represents ITC titration of neomycin into the bacterial A-site rRNA sequence. The top panel represents the heat burst curves which results from 5 μL injection of neomycin. The interaction of aminoglycosides with the A-site rRNA is an exothermic process (Kaul & Pilch, 2002). The lower panel of the graph in Fig. 5 shows the enthalpy versus molar ratio plot of neomycin interacting with the A-site rRNA. The titration shows that there are two distinct binding events during this interaction: one that saturates at ~1:1 binding ratio while the other at

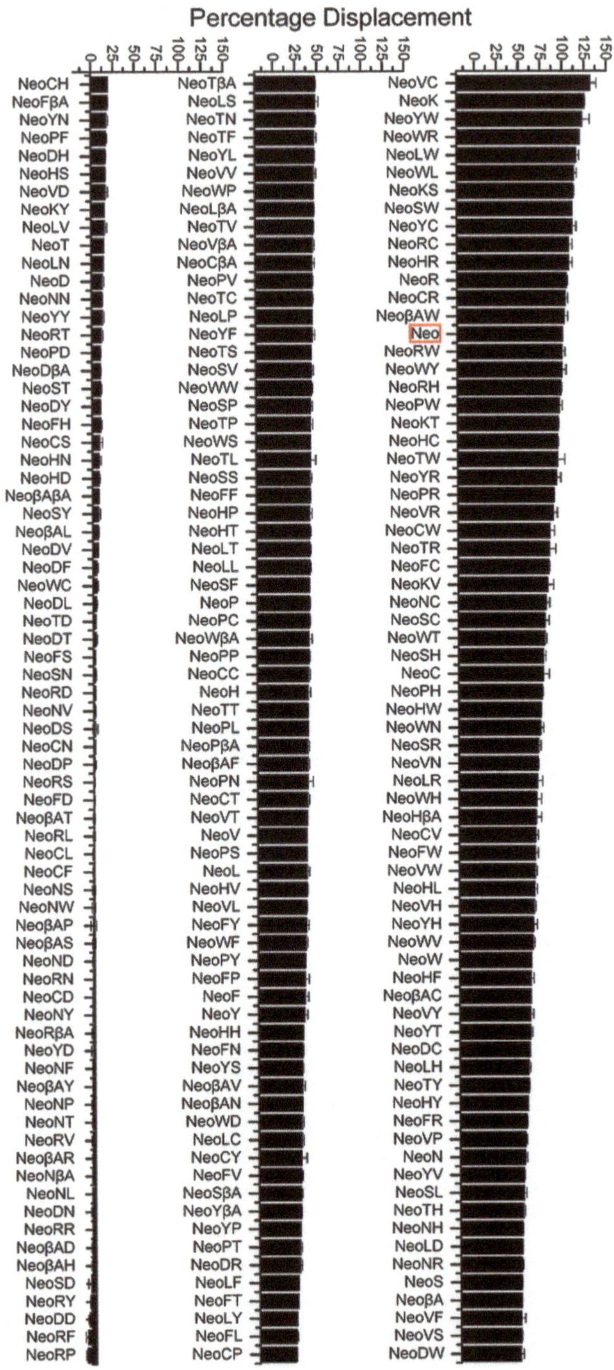

Fig. 4 Results from the 215-member peptidic aminosugar library screen using F-Neo displacement assay. *Figure adapted with permission from Jiang, L., Watkins, D., Jin, Y., Gong, C., King, A., Washington, A. Z. et al. (2015). Rapid synthesis, RNA binding, and antibacterial screening of a peptidic-aminosugar (PA) library.* ACS Chemical Biology, 10(5), 1278–1289. *Copyright 2015 American Chemical Society.*

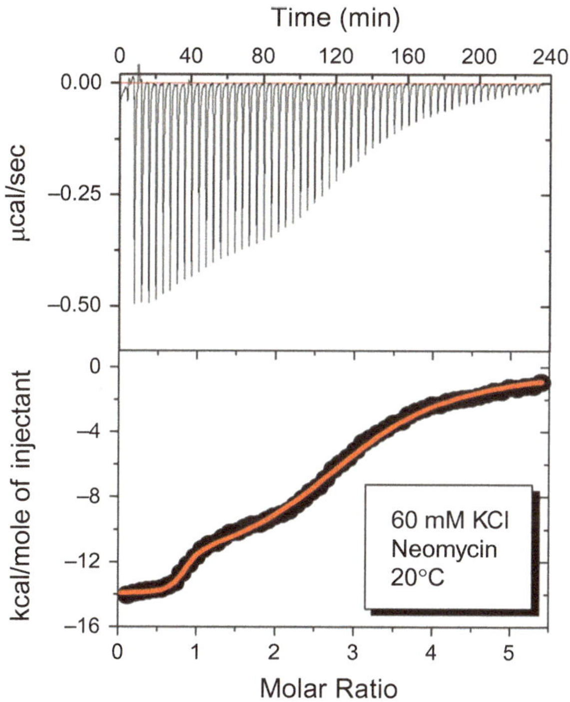

Fig. 5 ITC titration results of rRNA A-site sequence being titrated with neomycin. The experiments were performed at 10 μM/strand RNA concentration in buffer 10 mM sodium cacodylate, 0.5 mM EDTA, 60 mM KCL at PH 7.0. The concentration of neomycin in the syringe was 300 μM and volume of each injection was 5 μL. The experiment was performed at 20 °C.

~2:1 binding ratio. The resulting binding isotherm can be fitted with appropriate mathematical equation representing the equilibrium. In Fig. 5, the binding isotherm was fitted with "two binding site model" given in the instrument software (Orgin 7.0 for Microcal VP-ITC) which assumes that the two binding sites are independent. The fitting results, apart from binding stoichiometry (N), gave the association constants for both binding events ($K_{a1} = 1.21 \times 10^8 M^{-1}$; $K_{a2} = 3.67 \times 10^5 M^{-1}$), the binding enthalpies ($\Delta_{H1} = -13.62$ kcal/mol; $\Delta_{H2} = -12.67$ kcal/mol) and the entropy changes associated with it ($\Delta_{S1} = -9.48$ kcal/mol/deg; $\Delta_{S2} = -17.7$ kcal/mol/deg) which can easily be used to calculate the binding free energy (ΔG) using the relationship $\Delta G = \Delta H - T \cdot \Delta S$. In essence, a single experiment can give all essential thermodynamic parameters and binding events (single or multiple) that may not be visible from a single point fluorescent displacement assay. Thus, ITC serves as a useful validating technique to follow-up in ligand-RNA interactions identified from a high-throughput screen.

6.3 Analysis of circular dichroism experiments

Circular Dichroism spectroscopy allows one to quickly observe the global structural features of a nucleic acid under investigation (Norden, Rodger, & Dafforn, 2010). These features are so distinct and discernible that they serve as diagnostics for different forms (A, B, Z and G-quadruplex) (Ranjan & Arya, 2016) and types (parallel, antiparallel) of nucleic acids. CD titration experiments enables us to observe the ligand binding induced changes in the structure of the nucleic acid under investigation as well as appearance of new induced ligand CD signals which can add more light into the nature of binding process (Garbett, Ragazzon, & Chaires, 2007). The changes in the nucleic acid CD signal at a particular wavelength or induced CD signal with respect to the molar ratio of the ligand to nucleic acid can be used to obtain binding stoichiometry (Ranjan & Arya, 2016). Fig. 6 shows the CD titration results of neomycin being titrated into the bacterial A-site rRNA structure. In the absence of neomycin, the A-site rRNA displays a maximum at 263 nm which is the typical signature of A-form nucleic acids such as RNA. As neomycin was added incrementally to saturation levels, negligible change was observed in CD signal at 263 nm. However, clear changes were observed in the CD intensity at 209 nm which became more negative with the addition of incremental neomycin. These changes show a clear complexation of neomycin to the A-site rRNA structure. The changes observed at 210 nm was further utilized to determine the binding stoichiometry of the interaction by plotting the CD signal intensity changes with the molar ratio of ligand to the RNA. As shown in Fig. 6, the plot shows a clear intersection of lines at 1.0 indicating a 1:1 complex being formed between the RNA and the neomycin and higher concentrations lead to precipitation of the complex so it was not possible to clearly see any higher order complexes (see ITC). Since, neomycin lacks a chromophore in its structure, no induced CD was observed. Induced CD signals are also harbingers of ligand binding modes as groove binding and intercalating molecules can give differences in induced CD signals, although these data should be confirmed by linear dichroism studies (Eriksson & Norden, 2001; Garbett et al., 2007; Ranjan & Arya, 2016). CD is also is very helpful in studying binding induced

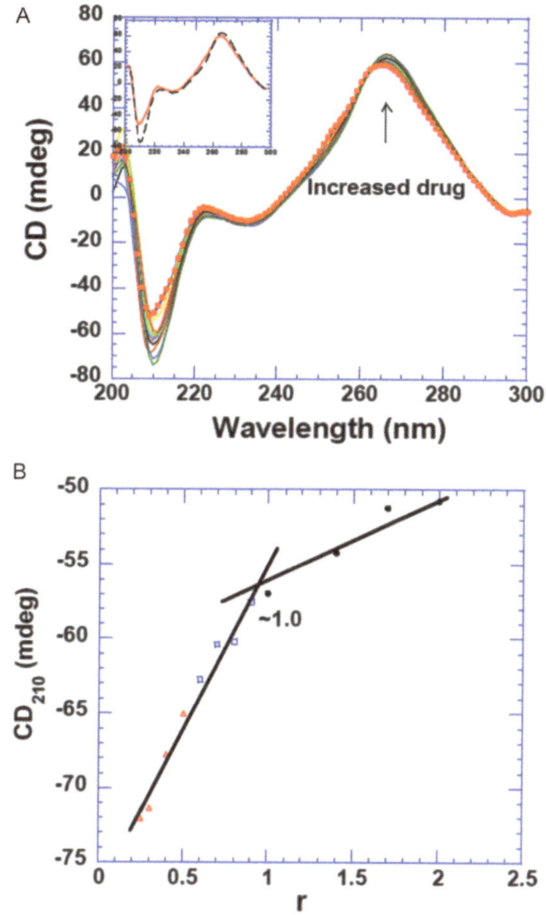

Fig. 6 (A) Circular dichroism studies of rRNA A-site sequence with neomycin. (B) A plot showing the binding stoichiometry of rRNA A-site being titrated with neomycin. *Figure adapted with permission from Xi, H., Davis, E., Ranjan, N., Xue, L., Hyde-Volpe, D., & Arya, D. P. (2011). Thermodynamics of nucleic acid "shape readout" by an aminosugar.* Biochemistry, *50(42), 9088–9113. Copyright 2011 American Chemical Society.*

structural change such as A to B transitions or vice versa. For brevity, we will not discuss the other uses of CD in studying ligand-nucleic acid interactions such as assessing thermal stability effects upon ligand binding and kinetic studies. Nevertheless, experiments discussed here outline the importance of CD in studying ligand-nucleic acid interactions to assess the global effects of ligand binding and to confirm the stoichiometries obtained from fluorescence and ITC titrations.

7. Conclusions

New infections by microbial pathogens and increasing cases of resistance to existing drug regimen have mandated the search of new chemical entities targeting validated or exploratory targets which can be further developed as eventual therapeutic agents. Along these lines, the current efforts to explore RNA targeted therapies have warranted operationally simple, robust and sensitive high-throughput screening techniques to enable rapid discovery of hit molecules. The groove binding nucleic acid (RNA and DNA) probe F-Neo can rapidly assist in the discovery of molecules which bind in the nucleic acid grooves with varying affinities and selectivities. Importantly, the fluorescent probe described here does not significantly perturb the native nucleic acid structure allowing for an assessment of the free target (RNA, DNA) close to its native solution state.

Acknowledgment

This work was supported by National Institutes of Health grants GM097917 to D.P.A.

References

Arya, D. P. (2007). *Aminoglycoside antibiotics: From chemical biology to drug discovery*. John Wiley & Sons.

Bhaduri, S., Ranjan, N., & Arya, D. P. (2018). An overview of recent advances in duplex DNA recognition by small molecules. *Beilstein Journal of Organic Chemistry*, *14*, 1051–1086.

Boger, D. L., Fink, B. E., Brunette, S. R., Tse, W. C., & Hendrick, M. P. (2001). A simple, high-resolution method for establishing DNA binding affinity and sequence selectivity. *Journal of the American Chemical Society*, *123*, 5878–5891.

Buurma, N. J., & Haq, I. (2007). Advances in the analysis of isothermal titration calorimetry data for ligand-DNA interactions. *Methods (Oxford, United Kingdom)*, *42*(2), 162–172.

Cech, T. R., & Golden, B. L. (1999). *The RNA world Cold*. Spring Harbor Laboratory Press.

Degtyareva, N. N., Gong, C., Story, S., Levinson, N. S., Oyelere, A. K., Green, K. D., et al. (2018). Antimicrobial activity, AME resistance, and A-site binding studies of anthraquinone-neomycin conjugates. *ACS Infectious Diseases*, *3*(3), 206–215.

Del Villar-Guerra, R., Gray, R. D., Trent, J. O., & Chaires, J. B. (2018). A rapid fluorescent indicator displacement assay and principal component/cluster data analysis for determination of ligand–nucleic acid structural selectivity. *Nucleic Acids Research*, *46*(7), E41.

Dethoff, E. A., Petzold, K., Chugh, J., Casiano-Negroni, A., & Al-Hashimi, H. (2012). Visualizing transient low-populated structures of RNA. *Nature*, *491*(7426), 724–728.

Eriksson, M., & Norden, B. (2001). Linear and circular dichroism of drug-nucleic acid complexes. *Methods in Enzymology*, *340*, 68–98.

Fourmy, D., Recht, M. I., & Puglisi, J. D. (1998). Binding of neomycin-class aminoglycoside antibiotics to the A-site of 16 S rRNA. *Journal of Molecular Biology*, *277*, 347–362.

Garbett, N. C., Ragazzon, P. A., & Chaires, J. B. (2007). Circular dichroism to determine binding mode and affinity of ligand-DNA interactions. *Nature Protocols*, *2*(12), 3166–3172.

Ghosh, A., Degyatoreva, N., Kukielski, C., Story, S., Bhaduri, S., Maiti, K., et al. (2018). Targeting miRNA by tunable small molecule binders: Peptidic aminosugar mediated interference in miR-21 biogenesis reverts epithelial to mesenchymal transition. *MedChemComm*, *9*(7), 1147–1154.

Jiang, L., Watkins, D., Jin, Y., Gong, C., King, A., Washington, A. Z., et al. (2015). Rapid synthesis, RNA binding, and antibacterial screening of a peptidic-aminosugar (Pa) library. *ACS Chemical Biology*, *10*(5), 1278–1289.

Jin, Y., Watkins, D., Degtyareva, N. N., Green, K. D., Spano, M. N., Garneau-Tsodikova, S., et al. (2016). Arginine-linked neomycin B dimers: Synthesis, rRNA binding, and resistance enzyme activity. *Medchemcomm*, 7(1), 164–169.

Kaul, M., & Pilch, D. S. (2002). Thermodynamics of aminoglycoside—rRNA recognition: The binding of neomycin-class aminoglycosides to the a site of 16s rRNA. *Biochemistry*, *41*(24), 7695–7706.

King, A., Watkins, D., Kumar, S., Ranjan, N., Gong, C., Whitlock, J., et al. (2013). Characterization of ribosomal binding and antibacterial activities using two orthogonal high throughput capable screens. *Antimicrobial Agents and Chemotherapy*, *57*(10), 4717–4726.

Kukielski, C., Maiti, K., Bhaduri, S., Story, S., & Arya, D. P. (2018). Rapid solid-phase syntheses of a peptidic-aminoglycoside library. *Tetrahedron*, *74*(33), 4418–4428.

Lavis, L. D., Rutkoski, T. J., & Raines, R. T. (2007). Tuning the pKa of fluorescein to optimize binding assays. *Analytical Chemistry*, *79*, 6775–6782.

Lynch, S. R., & Puglisi, J. D. (2001). Structural origins of aminoglycoside specificity for prokaryotic ribosomes. *Journal of Molecular Biology*, *306*(5), 1037–1058.

Neidle, S. (2016). Quadruplex nucleic acids as novel therapeutic targets. *Journal of Medicinal Chemistry*, *59*(13), 5987–6011.

Norden, B., Rodger, A., & Dafforn, T. (2010). *Linear dichroism and circular dichroism*. Cambridge, UK: RSC Publishing.

Oyaghire, S. N., Cherubim, C. J., Telmer, C. A., Martinez, J. A., Bruchez, M. P., & Armitage, B. A. (2016). RNA G-quadruplex invasion and translation inhibition by antisense gamma-peptide nucleic acid oligomers. *Biochemistry*, *55*(13), 1977–1988.

Patwardhan, N. N., Cai, Z., Newson, C. N., & Hargrove, A. E. (2018). Fluorescent peptide displacement as a general assay for screening small molecule libraries against RNA. *Organic & Biomolecular Chemistry*, *17*(7), 1778–1786.

Ranjan, N., & Arya, D. P. (2016). Linker dependent intercalation of bisbenzimidazole-aminosugars in an RNA duplex; selectivity in RNA vs. DNA binding. *Bioorganic & Medicinal Chemistry Letters*, *26*(24), 5989–5994.

Ren, J., & Chaires, J. B. (1999). Sequence and structural selectivity of nucleic acid binding ligands. *Biochemistry*, *38*(49), 16067–16075.

Sjoback, R., Nygren, J., & Kubista, M. (1998). Characterization of fluorescein-oligonucleotide conjugates and measurement of local electrostatic potential. *Biopolymers*, *46*, 445–453.

Tse, W. C., & Boger, D. L. (2004). A fluorescent intercalator displacement assay for establishing DNA binding selectivity and affinity. *Accounts of Chemical Research*, *37*, 61–69.

Turnbull, W. B., & Daranas, A. H. (2003). On the value of c: Can low affinity systems be studied by isothermal titration calorimetry? *Journal of the American Chemical Society*, *125*(48), 14859–14866.

Watkins, D., Gong, C., Kellish, P., & Arya, D. P. (2017). Probing A-form DNA: A fluorescent aminosugar probe and dual recognition by anthraquinone-neomycin conjugates. *Bioorganic & Medicinal Chemistry*, *25*(4), 1309–1319.

Watkins, D., Jiang, L., Nahar, S., Maiti, S., & Arya, D. P. (2015). A Ph sensitive high-throughput assay for miRNA binding of a peptide-aminoglycoside (Pa) library. *PLoS One*, *10*(12), E0144251.

Watkins, D., Norris, F. A., Kumar, S., & Arya, D. P. (2013). A fluorescence-based screen for ribosome binding antibiotics. *Analytical Biochemistry*, *434*(2), 300–307.

Willis, B., & Arya, D. P. (2014). Recognition of RNA duplex by a neomycin-hoechst 33258 conjugate. *Bioorganic & Medicinal Chemistry*, *22*(7), 2327–2332.

Xi, H., Davis, E., Ranjan, N., Xue, L., Hyde-Volpe, D., & Arya, D. P. (2011). Thermodynamics of nucleic acid "shape readout" by an aminosugar. *Biochemistry*, *50*(42), 9088–9113.

CHAPTER FOURTEEN

The use of electrospray ionization mass spectrometry to monitor RNA-ligand interactions

Danielle N. Dremann, Christine S. Chow*
Wayne State University, Detroit, MI, United States
*Corresponding author: e-mail address: cchow@wayne.edu

Contents

Abstract

RNAs are drawing increasing attention as potential therapeutic targets. A significant challenge in the RNA drug discovery process is identification of compounds that not only disrupt the natural functions of RNA by binding with high affinity, but also do so selectively. Assessing the binding mode of small molecules with RNA is important for understanding how they select their binding site and impart their mechanism of action. A number of complementary assays are often employed for analysis of the binding mode and to determine selectivity. One important technique that gives information about the binding affinity and stoichiometry is electrospray ionization mass spectrometry (ESI MS). More recent methods have also revealed the usefulness of ESI MS in determining the binding loci of small molecules on RNA.

Methods in Enzymology, Volume 623
ISSN 0076-6879
https://doi.org/10.1016/bs.mie.2019.05.013

1. Introduction

1.1 RNA as a drug target

Small molecules that target RNA have a long history dating back to the 1940s (Thomas & Hergenrother, 2008; Waksman, Frankel, & Graessle, 1949; Waksman & Lechevalier, 1949; Waksman, Lechevalier, & Harris, 1949). For example, about half of the known antibiotics—including macrolides, aminoglycosides, chloramphenicol, tetracyclines, and oxazoladinones—target ribosomal RNA (rRNA) (Steitz, 2005). While many of these compounds are natural products, others have been discovered by screening synthetic libraries (Guan & Disney, 2012). More recently developed RNA-targeting agents include flavin mononucleotide mimics (Ribocil) and pyridazines (Branaplam) (Warner, Hajdin, & Weeks, 2018). These newer efforts also expand the range of RNA targets to include non-coding RNAs (ncRNA), messenger RNA (mRNA), viral RNA, riboswitches, and splicing RNAs, among others (Guan & Disney, 2012). These studies take advantage of the rich diversity of RNA structures and functions, as well as the ability of RNA to undergo conformational changes that are important for its function (Aboul-Ela, 2010; Chow & Bogdan, 1997; Connelly, Moon, & Schneekloth, 2016; Thomas & Hergenrother, 2008; Tor, 2003). For the cases in which rRNA-targeting drugs exist, resistance is an ongoing problem (Overbye & Barrett, 2005; Projan, 2003; Quadri, 2007; Tu, Blaha, Moore, & Steitz, 2005) as well as off-target effects such as mitochondrial rRNA binding that leads to hearing loss (Böttger, Springer, Prammananan, Kidan, & Sander, 2001; Xie, Talaska, & Schacht, 2011). Therefore, the methods used to study these systems need to be able to distinguish small differences in binding modes, such as change in affinity and/or contact sites on the RNA. The resistance or off-target effects may be the result of RNA mutations or modifications, so the methods need to be universal and not depend on nucleotide composition, which is the case for traditional chemical probing methods to map binding of small molecules on large RNAs within biological contexts (Waduge, Sakakibara, & Chow, 2019).

1.2 Aminoglycosides and peptides as examples of rRNA-targeting molecules

Although numerous RNAs have been shown to be potential drug targets (Chow & Bogdan, 1997; Connelly et al., 2016; Guan & Disney, 2012;

Thomas & Hergenrother, 2008; Warner et al., 2018), the focus here will be rRNA motifs. The ribosome is a key antibiotic target because of its major role in protein synthesis (Cate, Yusupov, Yusupova, Earnest, & Noller, 1999; Moore & Steitz, 2005; Steitz, 2005; Tu et al., 2005; Yusupov et al., 2001). Although the bacterial and eukaryotic ribosomes are similar in structure and function, there are key differences between them such as their subunit sizes, or composition of their initiation complexes, elongation factors, and recycling steps, which impact specific steps of protein synthesis and allow for selective targeting (Wilson & Doudna Cate, 2012). Despite the complexity of the ribosome, a number of studies have shown that small model RNAs about 20 to 40 nucleotides in length provide useful information about drug binding modes that correlates well with interactions observed with complete ribosomes (Thomas & Hergenrother, 2008). One example is the decoding region A site located in helix 44 (h44) of the bacterial 16S rRNA in the 30S subunit (Fig. 1A) (Recht, Fourmy, Blanchard, Dahlquist, & Puglisi, 1996). This site, which is responsible for monitoring base pairing between the codon of mRNA and the anticodon of the corresponding aminoacyl-transfer RNA (aa-tRNA), is also a well-known binding site for aminoglycoside antibiotics (Moazed & Noller, 1987; Wang et al., 2012). Aminoglycosides also target helix 69 (H69) of the bacterial 23S rRNA in the 50S subunit (Fig. 1A) and interfere with

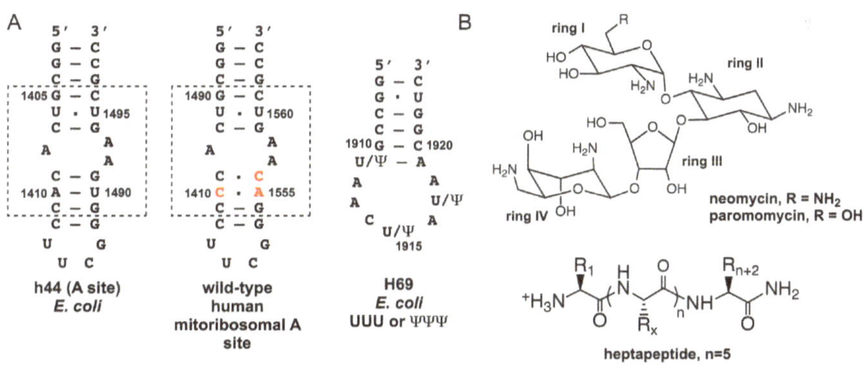

Fig. 1 Target rRNAs and examples of RNA-binding ligands. (A) The secondary structures of h44 (A site) from *E. coli* and wild-type human mitochondrial ribosomes (the boxed area represents the native rRNA sequence, whereas the nucleotides outside the box serve to stabilize the structure), and H69 from *E. coli* (Ψ = pseudouridine; the UUU RNA contains U at three positions and the ΨΨΨ RNA contains Ψ at the corresponding positions) are shown. (B) The chemical structures of aminoglycosides neomycin and paromomycin are shown, along with a generic amidated peptide in which R represents any amino acid side chain (in the case shown, n = 5 to give a heptapeptide).

ribosome recycling (Borovinskaya et al., 2007). One distinguishing feature of H69 is the presence of modified nucleotides, namely pseudouridine (Ψ), at three positions in the loop region (Ofengand & Bakin, 1997). Although aminoglycosides are effective antibiotics with a broad spectrum of activity against gram-positive and gram-negative bacteria, they have several disadvantages such as oto- and nephrotoxicity at high doses, leading to negative impacts on hearing and kidney function (Xie et al., 2011). One of the reasons for this toxicity is close resemblance between the mitochondrial 12S rRNA and bacterial A site (Fig. 1A) (Böttger et al., 2001).

Aminoglycosides belong to a large group of antibiotics containing amino sugars as their main components (Fig. 1B) (Guan & Disney, 2012; Thomas & Hergenrother, 2008; Tor, 2003). They are water soluble, cationic molecules (at physiological pH) with broad antimicrobial activity. A number of aminoglycoside derivatives have been synthesized in order to overcome antibiotic resistance or ototoxicity (Böttger et al., 2001). They also serve as ideal compounds for better understanding RNA-ligand interactions, such as roles of electrostatics and RNA dynamics, as well as for comparative studies and method development. Their high charge makes them particularly well suited for mass spectrometry applications, which will be discussed below. In addition to rRNA, aminoglycosides target many other RNA molecules such as tRNA, viral RNA, and catalytic RNA (Tor, 2003).

Peptide ligands are attractive drug leads because of ease of synthesis and the range of both natural and unnatural amino acids available to diversify their structures (Lau & Dunn, 2018). In addition, they are comparable in size and composition to some natural products that bind to RNA. Peptides can be modified to contain charged residues, which is beneficial for RNA binding as well as analysis by mass spectrometry. Cyclic peptides with high affinity (nM) to viral RNAs have been developed (Davidson et al., 2009). Similarly, amidated linear peptides (Fig. 1B) have been shown to have moderate affinity (μM) to both h44 and H69, which is comparable to the affinity of aminoglycosides but with higher selectivity for the target RNA over other RNAs (Dremann & Chow, 2016; Kaur, Rupasinghe, Klosi, Spaller, & Chow, 2013; Li et al., 2009).

1.3 Use of electrospray ionization mass spectrometry (ESI MS) to monitor RNA-ligand interactions

Experimental methods to study RNA-ligand interactions serve two main purposes. The first is to screen and identify compounds with high affinity and selectivity, and the other is to characterize the binding modes.

There are a number of high- and low-throughput methods that have been used for these purposes (Blakeley et al., 2012; Connelly et al., 2016; Guan & Disney, 2012). Computational studies have guided the search for novel and selective compounds. Other methods that have been used include dynamic combinatorial chemistry, in vitro selection, X-ray crystallography, NMR spectroscopy, fluorescence spectroscopy (including single molecule methods), surface plasmon resonance (SPR), isothermal calorimetry (ITC), UV melting, electrochemistry, electrophoretic mobility shift assays (EMSA), and filter binding assays (Blakeley et al., 2012). Biochemical and functional assays have also been useful to identify compounds that inhibit a range of RNA functions, including splicing (Chow, Mahto, & Lamichhane, 2008; Guan & Disney, 2012; Thomas & Hergenrother, 2008; Tor, 2003).

The primary use of mass spectrometry in RNA binding has been to obtain information about ligand affinity and stoichiometry of the bound ligand. Matrix-assisted laser desorption ionization time-of-flight mass spectrometry (MALDI-TOF MS) has been applied to determine the stoichiometry of oligonucleotide complexes, but the MALDI process is not ideal for generating intact non-covalent complexes in a quantitative manner (Tang, Callahan, Zhou, & Vertes, 1995). Previous studies revealed that ESI MS is a useful technique for characterizing non-covalent complexes between small molecules and RNA because it is a gentle ionization technique (Sannes-Lowery, Griffey, & Hofstadler, 2000). Furthermore, the target RNA and potential ligands do not require any tagging or labeling, except for certain applications in which RNAs of the same mass are being compared (Hofstadler et al., 1999). ESI MS analysis only requires a small amount of material (ng), which is important when employing RNAs that are difficult to obtain such as modified H69. The major advantage of ESI MS is that it reveals the stoichiometry of the interaction, which is important when studying highly charged molecules with a high level of non-specific interactions. ESI MS can also be used to determine binding affinities (Sannes-Lowery et al., 2000) and locations of binding sites (Griffey, Greig, An, Sasmor, & Manalili, 1999; Schneeberger & Breuker, 2017; Vusurovic, Schneeberger, & Breuker, 2017). Mass spectrometry is particularly useful for comparative studies in which relative binding affinities of closely related species are desired. For example, the relative binding affinities of aminoglycoside analogues to a single RNA motif such as h44 or H69 (Li et al., 2009; Sakakibara, Abeysirigunawardena, Duc, Dremann, & Chow, 2012), or binding of a single compound to the unmodified and modified H69 variants (Dremann & Chow, 2016; Kaur et al., 2013), can be determined.

2. Background, sample preparation, and ESI MS methods

2.1 ESI MS

The electrospray ionization process converts molecules (e.g., RNAs and ligands) present in solution into gas-phase ions that can be detected and identified (Cole, 2010; Fenn, Mann, Meng, Wong, & Whitehouse, 1989). Most standard ESI MS instrumentation can handle a wide array of sample types ranging from small organic and inorganic compounds to large proteins and nucleic acids (typically from 100 Da to 80 kDa). The solution will contain electrolyte ions that are converted into charged droplets at the heated electrospray capillary tip. These charged droplets are further consolidated by solvent evaporation, and repeated droplet disintegration forces the charged droplets into gas-phase ions (Silverstein, Webster, & Kiemle, 2005). In order to convert solution-phase ions into gas-phase ions, the solution needs volatile solvents such as methanol, ethanol, isopropanol, or acetonitrile incorporated into the sample preparation. Titrations of the sample material (ligand) to be tested for target RNA interactions can provide important binding information. In situations involving equal ionization amongst all components in the sample, dissociation constants can be determined directly (Bligh, Haley, & Lowe, 2003; Sannes-Lowery et al., 2000). In cases in which ionization of the samples is not equal, relative intensities of all charged species can be used to determine relative dissociation constants, which will be discussed in more detail for peptide-RNA interactions (Section 3.2) (Dremann & Chow, 2016; Li et al., 2009; Nordhoff et al., 1993).

One of the biggest challenges in ESI MS of RNA is efficient ionization of both the free RNA and RNA-ligand complex. Cation adducts with ions such as K^+ and Na^+ are also problematic because they lead to charge neutralization of the RNA, and thus reduce ion signals (Gaston & Limbach, 2014; Wetzel & Limbach, 2016) and sensitivity. Because of the high negative charge of the phosphodiester backbone of RNA and high affinity for cations, focusing on clean sample preparation and optimizing the sample compositions (i.e., buffer components, pH, additives) as well as instrument parameters are important in order to achieve efficient and reproducible ionization.

2.2 Materials and general procedures

The laboratory space used for all RNA preparations must be free of ribonucleases (RNases). Ribonucleases are ubiquitous; therefore, equipment and supplies should be dedicated for RNA work and/or cleaned thoroughly to ensure success of the experiments. Other precautionary measures include wiping down the lab bench with ethanol, dedicating pipets for RNA use only, and using tips and polypropylene tubes that are autoclaved and/or pre-packaged under "RNase-free" conditions. Gloves and a facemask are recommended in order to prevent RNase contamination of the samples.

RNase-free polypropylene sample tubes are used for all RNA and ligand preparations and autoclaved prior to use. Double-deionized water (ddH_2O), pH 7.0, that is RNase free is used for all experiments. All solvents used for reverse-phase high performance liquid chromatography (RP HPLC) are HPLC grade and pre-filtered (0.2 μm nylon ZAPCAP® bottle-top filter, 500 mL capacity, GE Whatman) and degassed prior to use. Degassed solvents are stored in autoclaved, clear glass bottles that are wrapped in foil when working with light-sensitive compounds or reagents.

2.3 RNA preparation

Regardless of which method is used to obtain RNA, purification and desalting are imperative for the success of ESI MS experiments. For example, the presence of non-volatile cations (e.g., Na^+, K^+, Mg^{2+}) can lead to signal suppression and complex spectra that are difficult to analyze. There are several methods that can be used to purify and desalt RNA in preparation for ESI MS analysis that will be described below.

2.3.1 RNA synthesis

RNA can be produced synthetically or by transcription with T7 RNA polymerase (Milligan, Groebe, Witherell, & Uhlenbeck, 1987). RNAs made synthetically with the 2′-*O*-ACE chemistry (Dharmacon) (Hartsel, Kitchen, Scaringe, & Marshall, 2005) are generally less salty than RNAs made with 2′-*O*-silyl phosphoramidites (Ogilvie, Beaucage, Schifman, Theriault, & Sadana, 1978) due to the mild deprotection conditions. For our purposes, the following RNAs were obtained from Dharmacon (Lafayette, CO; 1 μmol scale) and deprotected according to the protocols provided by the company (Hartsel et al., 2005):

h44: 5′-GGCGUCACACCUUCGGGUGAAGUCGCC-3′
H69 UUU: 5′-GGCCGUAACUAUAACGGUC-3′
H69 ΨΨΨ: 5′-GGCCGΨAACΨAΨAACGGUC-3′ (Ψ = pseudouridine)
H69 biotin-ΨΨΨ: 5′-Biotin-GGCCGΨAACΨAΨAACGGUC-3′

2.3.2 RNA purification

Gel purification and HPLC are two methods that can be used to prepare RNA for mass spectrometric analysis. HPLC is performed using an X-Terra MS C18 (2.5 μm, 10 × 50 mm, Waters Corporation, Milford, MA) column. HPLC-grade acetonitrile (ACN) in 25 mM triethylammonium acetate (TEAA) at pH 7.0 is used as the mobile phase with a linear gradient from 6% to 11% ACN over 20 min at a flow rate of 4.0 mL/min. The fractions (typically 250–500 μL volumes) containing the desired RNA are collected and lyophilized to dryness. Minimal amounts of RNase-free ddH_2O (about 10–20 μL) are used to dissolve the RNA sample so that the final concentration is about 300 μM. The fractions are analyzed with MALDI-TOF MS to identify those containing the full-length RNA product.

2.3.3 RNA desalting

There are several methods that can be used for desalting RNA. An ethanol (EtOH) precipitation with ammonium acetate (NH_4OAc) is acceptable in most cases. The purified RNA samples with the correct masses undergo two precipitations. To form the pellet, RNA is placed in a 1.7 mL micro-centrifuge tube. One volume (V) of RNA to 3 V of chilled (–20 °C) EtOH and 0.1 V of 4 M NH_4OAc are mixed. The sample is placed on dry ice for 30–40 min, and then centrifuged for 30 min at 13,000 rpm. The supernatant is carefully removed from the RNA pellet, the pellet is redissolved in ddH_2O (about 20 μL), and the process is repeated. After the supernatant is removed the second time, 500 μL of cold EtOH (–20 °C) is used to wash the pellet, which is then dried in a speed vac and stored at –20 °C until use. Typically MALDI-TOF MS analysis of the sample will reveal if salt (e.g., Na^+, K^+, Mg^{2+}) is present. Desalting can also be accomplished by dialyzing the sample against RNAse-free ddH_2O, typically using membranes such as Spectra/Por7 (1000 MWCO, Spectrum, Fisher Scientific). A 200 μL RNA sample (maximum 200 μg) is dialyzed against 4 L of ddH_2O for 3 days with a complete change of ddH_2O each day. The RNA is lyophilized to dryness, then redissolved in ddH_2O to give a concentration of approximately 300 μM.

2.3.4 RNA concentration determination and renaturation

The dried, purified, RNA pellets are dissolved in a minimal amount of ddH_2O (typically about 25 μL), and 1 μL of RNA is used to determine the concentration by measuring the absorbance (e.g., Beckman Coulter DU800 UV–Vis spectrophotometer; Beckman Coulter, Brea, CA) at 260 nm and using the calculated extinction coefficient (Fasman, 1975).

The RNA is then distributed into 100 μL microcentrifuge tubes (typically 50 μL of 100–300 μM stock solutions) with at least three sets of samples for repetitions of experiments, and lyophilized to dryness. Aliquoting the RNA stock and drying helps avoid potential contamination or degradation through repetitive handling and freeze-thaw cycles. RNA is stored at −20 °C until use in the binding experiments. Upon use, a fixed amount of ddH_2O or buffer (depending on previous concentration determination and aliquot preparation) is added to the RNA. Renaturation of the RNA is done by heating the sample for 5 min at 90 °C followed by slow cooling to room temperature. When working with small amounts of RNA, it is ideal to produce master mixes (RNA, NH_4OAc, and isopropanol in ddH_2O) with a constant concentration of RNA.

2.4 Peptide and small molecule preparation

Manual solid phase peptide synthesis (SPPS) is carried out using fmoc (Anthis & Clore, 2013; Carpino & Han, 1972; Merrifield, 1963) as the temporary protective group, and varying solid supports are utilized to generate small peptides of about 7–12 amino acids as shown in Fig. 1B (Dremann & Chow, 2016; Kaur et al., 2013). Peptides used for ESI MS of RNA are typically amidated at the C-terminus in order to avoid a negatively charged carboxylate that would have unfavorable interactions with the target (using RINK Amide AM 200–400 mesh solid support, Novabiochem, San Diego, CA) (Rink, 1987).

The permanent protective groups and solid support are cleaved from the peptide with 94:2:2:2 trifluoroacetic acid (TFA):triisopropylsilane:anisole: thioanisole for 3–6 h at room temperature (Dremann & Chow, 2016). The workup following chemical cleavage involves washing with cold diethyl ether. The liquid contents are drained into 50 mL centrifuge tubes containing diethyl ether that is pre-chilled at −20 °C, vortexed, and centrifuged at 8000 RPM at 4 °C for 6 min to pellet the peptide. The diethyl ether is drained and replaced with fresh cold diethyl ether. The solution is vortexed and centrifuged, and this washing step is repeated three times. A minimal amount of ddH_2O with 0.1% TFA is added to dissolve the peptide pellet, followed by HPLC purification.

Peptide purification is typically done by HPLC with a Waters 600 controller, 2996 detector, and autosampler, using a Phenomenox® C18 Luna Column (C18, 10.00 × 250 mm, 5 μm, 100 Å) and a flow rate of 5 mL/min. Buffer A contains ddH_2O + 0.1% TFA and buffer B contains

HPLC-grade ACN + 0.08% TFA. All solvents are degassed for a minimum of 1 h. Over the course of 84 min, 3 column volumes (CVs) of 100% buffer A (12 min at 5 mL/min) is followed by a 0–5% buffer B over 3 CVs (12 min 100% → 95% buffer A). This gradient is held for 3 CV (12 min). A 30% ACN gradient is attained over 3 CVs (5–30% buffer B; 95–70% buffer A) and then maintained for 12 min because products with permanent protective groups, particularly those with aromatic functionalities, tend to elute off the column at this percentage. Buffer B is then returned to 0% (100% buffer A) over 3 CVs and maintained for an additional 12 min. Please note that gradients may need to be adjusted. The peptides used in the described experiments lacked aromatic side chains, which led to short elution times from the column at a very low concentration of ACN. Fractions (0.5–1 mL each) containing the desired peptide, as determined by MALDI-TOF MS analysis, are lyophilized to dryness to give a white powder. Peptides that contain aromatic side chains or still contain permanent protection groups with aromaticity tend to yield yellowish compounds. Quantification is done by pre-weighing the tubes (typically 1.7 mL polypropylene microcentrifuge tubes) used to collect the HPLC fractions. Fractions that contain the desired peptide are lyophilized and then weighed again to determine the quantity of peptide available for binding experiments. Based on the molecular weight of the peptide, ddH_2O with 0.1% TFA is added to generate 500 μM stocks to use in ESI MS binding experiments with RNA. In addition, the peptide concentration is verified using a UV–Vis spectrophotometer (e.g., Beckman Coulter DU800 spectrophotometer) at a wavelength of 205 nm (for peptides that contain histidine, but do not contain aromatic side chains) or 254 nm (for compounds that contain aromatic side chains such as tyrosine) (Anthis & Clore, 2013). Samples are read in triplicate and absorbance values are averaged prior to using Beer's law for concentration determination.

Peptide stock solutions (500 μM) in ddH_2O with 0.1% TFA are used for binding experiments. RNA is resuspended in ddH_2O at pH 7.0 to make a 300 μM stock solution and renatured as described in Section 2.3.4. Microcentrifuge tubes (100 μL) are prepared with varying amounts of ddH_2O in each tube to give a final volume of 50 μL. A master mix is produced so that the final concentrations are 2.7 μM RNA, 25% isopropanol, and 150 mM NH_4OAc at pH 7.0. Peptide (from the 500 μM stock) is then added (0–120 μM) and the sample is incubated at room temperature for 15 min. Based on the number of titration points (typically 10–20), the amount of added peptide will vary. The total volume for infusion is 50 μL (Dremann & Chow, 2016).

Neomycin sulfate (Sigma-Aldrich, St. Louis, MO) and paromomycin sulfate (Sigma-Aldrich) stock solutions (100 μM) are prepared in ddH_2O, pH 7.0. RNA is resuspended in 100 mM NH_4OAc, pH 7.0, to create a 100 μM RNA stock and renatured as described in Section 2.3.4. A master mix is also created for this titration set (10–12 titrations) so that the final concentrations are 1 μM RNA, 25% isopropanol, and 150 mM NH_4OAc (pH 7.0). Aminoglycoside antibiotics (neomycin, 0–6 μM; paromomycin, 0–14 μM) are added to the RNA and the 50 μL samples are incubated at room temperature for 15 min (Sakakibara et al., 2012). The RNA and ligand concentrations, as well as the additives (e.g., isopropanol), need to be optimized for each case, since ionization efficiencies vary with instrumentation, ligand composition, and the RNA being tested.

2.5 Instrument preparation

Our binding studies are performed on a Micromass QuattroLC triple quadrupole mass spectrometer with an electrospray/APCI source and Waters Alliance 2695 LC, autosampler and photodiode array UV detector (Manchester, UK), but the method can be adapted for other instruments. The Micromass QuattroLC is useful for analyzing polar and high molecular weight compounds ranging from less than 100 Da to proteins over 80 kDa. The m/z range covers 2–4000 Da, but higher MW species can be measured since they take on multiple charges. The QuattroLC has an accurate mass option, which allows acquisition of mass data at low resolution (0.15 Da, 128 points/Da) but high accuracy (within 5 ppm).

Instrument preparation includes a thorough wash in 50% isopropanol in ddH_2O and cone voltage of 40 V over the course of 3–12 h. This washing step helps eliminate possible contaminants from previous experiments and prepares the instrument for RNA experiments. Care is taken to ensure that any components used in connection with the mass spectrometer are cleaned prior to experiments with RNA. Many compounds, including aminoglycosides and peptides, have a tendency to leave trace amounts in the system long after the infusion has occurred. To maintain accurate titrations, the ESI MS instrumentation undergoes a wash step with 50% isopropanol in ddH_2O between each infusion. This procedure takes place over several minutes and is done for both RNA-aminoglycoside as well as RNA-peptide experiments. The wash is monitored for any residual traces of the previous infusion and the next infusion does not begin until the running spectrum is clean. While this step takes place, the syringe

(Hamilton®, 100 μL) is also washed with 50% isopropanol, followed by ddH_2O, and once more with 50% isopropanol. The solvents are freshly filtered into clear, glass bottles that have been autoclaved. Brown bottles tend to leach salts into the solutions over time, which can complicate the spectra. All samples to be infused are prepared in autoclaved polypropylene tubes or class A clear glass vials (Fisher Scientific, 03-339-25A, ion-free borosilicate).

3. Data collection and analysis

3.1 RNA-aminoglycoside binding experiments

In our studies, aminoglycosides (up to 120 μM) are titrated into renatured synthetic RNAs (e.g., H69 or h44) (1–3 μM is required for the Micromass Quattro, but ideally the RNA concentrations should be below the μM range if the dissociation constant for the ligand complex is also in the μM range) in 25% isopropanol in ddH_2O and NH_4OAc (150 mM; this concentration will likely need to be reduced for nanospray applications) and incubated for 15 min at room temperature prior to infusion. Experiments are carried out in the negative ion mode. The spectrum of the infused sample is obtained for the mass-to-charge ratio (m/z) range of 1300–2400 to reveal complexes of RNA and aminoglycoside, as well as free RNA, in various charges states (in our case, the (3-) and (4-) charge states are observed for H69, and (4-) and (5-) charge states for h44). For the Micromass Quattro, the sample (50 μL) is infused with a Hamilton® 100 μL glass syringe at a rate of 6 μL/min. A cone voltage of 40 for ΨΨΨ and UUU H69 RNAs and 50 for 5′-Biotin-ΨΨΨ H69 and h44 RNAs was maintained. Typically 60–70 spectra are averaged for each measurement. A thorough wash with 50% isopropanol in ddH_2O is done between samples. When traces of the aminoglycoside are no longer detected, the next sample is infused. Several washes of the syringe also take place between infusions. Experiments are performed sequentially in duplicate or triplicate with 10–20 different ligand concentrations over a 10-fold or greater range.

3.2 RNA-peptide binding experiments

Unless otherwise noted, the peptide binding experiments use the same instrumentation and setup used for the aminoglycoside binding experiments. In our studies, peptides (up to 120 μM) are titrated into renatured synthetic H69 or h44 RNAs (2.7 μM) in 25% isopropanol in ddH_2O and NH_4OAc (150 mM) and incubated for 15 min at room temperature prior to infusion.

The spectrum of the infused sample is obtained for the m/z range of 1100–2400 to reveal complexes of RNA and peptide, as well as free RNA, in various charges states (in our case, the (3-), (4-), and (5-) charge states are observed).

To illustrate some other design features of the experiment, the H69 RNA construct has a mass of 6058 Da. The aminoglycosides and peptides have masses ranging from 600 to 1000 Da (ideally less than 15% difference between the mass of free RNA and RNA-ligand complex). Another consideration is whether the free ligand can be observed in the spectra. Under certain conditions, the ligand is observed in the (1-) charge state within the mass range (600–1000 Da) used to observe the free RNA and RNA complexes. The (2-) charge state for the ligand would be in a lower mass range (i.e., 300–500 Da for aminoglycosides or peptides). Initial optimization of the ESI MS experiments may reveal that the free ligand (1-) will severely reduce the intensity of the free RNA and corresponding complexes, making it impossible to do the binding experiment. In those cases, tagging the RNA to increase its mass may resolve the problem, but the tag may then interfere with the binding event. The presence of free ligand in the spectra makes it challenging to observe the binding interactions, particularly at higher concentrations and/or if the ionization efficiency of the ligand is significantly greater than the free RNA or RNA-ligand complex. Therefore, the chosen mass range of the experiment is typically outside the range of the ligand mass. In some cases, as the ligand concentration increases, ligand dimers are observed, which occurs frequently with hydrophobic peptides.

3.3 Representative data and analysis

During the binding experiment, the fraction of RNA bound typically increases as the concentration of ligand increases, which can be observed by ESI MS if the dissociation constant is in the μM range or lower. For a simple binding mode (1:1), the RNA-ligand complex is observed along with free RNA. If non-specific binding or more complex binding modes occur, then 2:1 and higher-order complexes (3:1, 4:1, etc.) will be observed as the concentration of ligand increases. Once saturation is achieved, no more change in intensity of either the free RNA or RNA-ligand complex will be observed. For reliable K_d determination, enough titration points need to be included to observe saturation for the 1:1 complex before non-specific binding occurs. We note, however, that it is often difficult to achieve saturation for ligands with both specific and non-specific binding modes for their target. In addition, the ESI MS spectra often display increased background peaks and RNA salt adducts as the ligand concentration increases. Thus, the stoichiometry of the complex is typically determined at concentrations below 50 μM.

As shown in the ESI MS spectra in Fig. 2, the binding of neomycin or peptide (RQVANHQ-NH_2) to unmodified (UUU) H69 RNA or

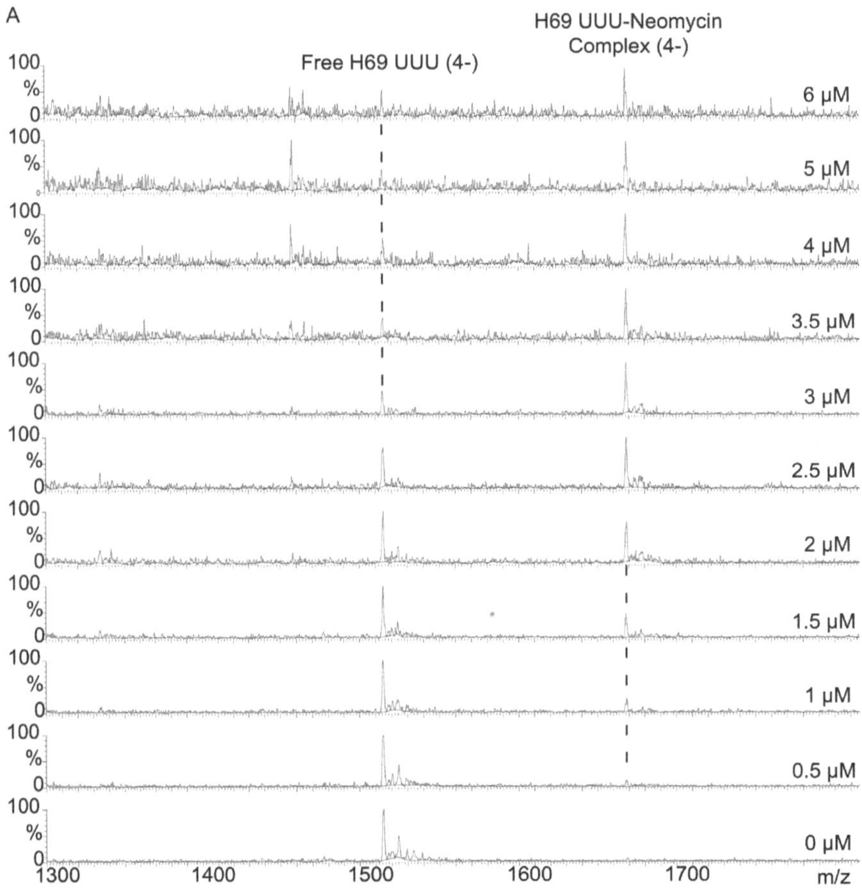

Fig. 2 Representative ESI MS spectra for the binding of ligands to H69 RNAs. (A) Neomycin binding to unmodified (UUU) H69 and (B) peptide (RQVANHQ-NH_2) binding to unmodified (UUU) and Biotin(Bi)-ΨΨΨ H69 RNAs are shown (150 mM NH_4OAc, pH 7.0). The free RNA and ligand-RNA 1:1 complexes in the (4-) and (5-) charge states are shown. Smaller peaks correspond to the Na^+ and K^+ adducts. The free RNA peak intensity decreases as the ligand concentration increases, and conversely the 1:1 complex peak intensity increases with increasing ligand concentration (up to 6 μM for neomycin and 50 μM for peptide).

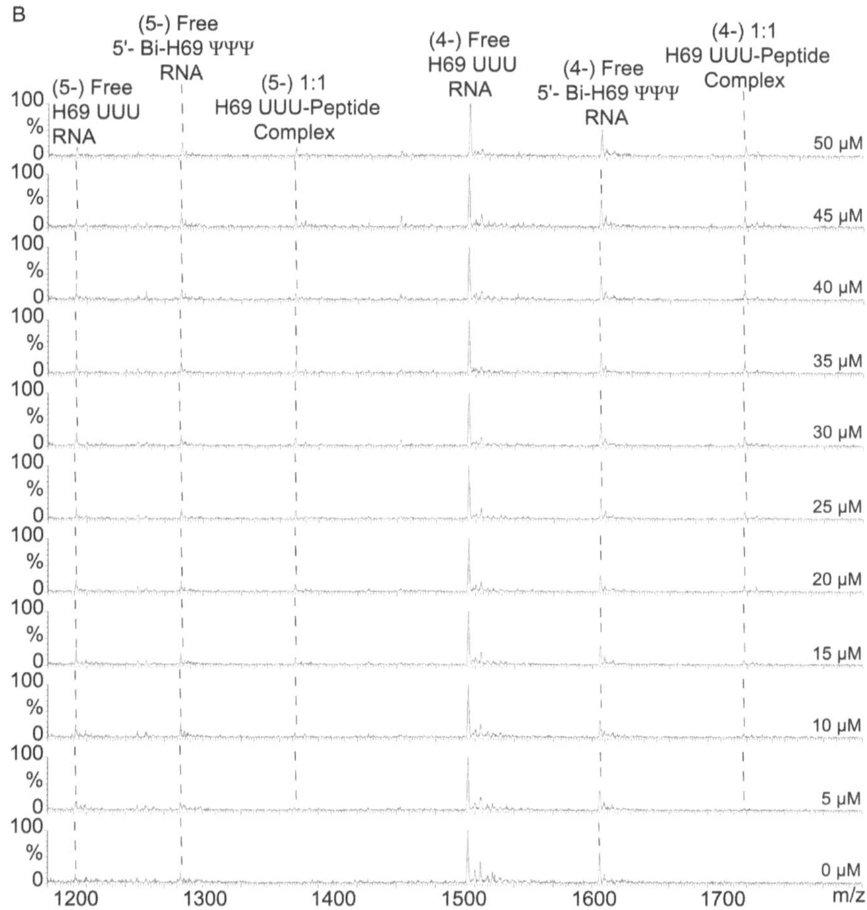

Fig. 2—Cont'd

biotinylated modified (ΨΨΨ) H69 causes a change in peak intensity of the free RNA (4- or 5- charge state). As the ligand concentration increases, the intensity of the peak corresponding to the 1:1 ligand-RNA complex in the (4-) or (5-) charge state also increases. Smaller peaks can be observed, which correspond to Na^+ and K^+ adducts. In the titration shown in Fig. 2A, saturation was achieved above 10 μM of ligand. The 50% fraction bound is reached at 2 μM neomycin, suggesting an apparent dissociation constant, or K_d value, in the low μM range. For both aminoglycoside and peptide complexes with h44 or H69 RNAs, the peak areas are analyzed and used to calculate the relative dissociation constant of the ligand to RNA, which will be discussed further below (Dremann & Chow, 2016). It is important to note that if complex formation in a given charge state is not observed, then information from the same charge state for the free RNA is not included in the analysis.

As seen in Fig. 2, ESI MS experiments of small RNAs with cationic molecules such as aminoglycosides (panel A) show better ionization efficiencies compared to peptides (panel B) (Dremann & Chow, 2016; Sakakibara et al., 2012; Sannes-Lowery et al., 2000). This effect is particularly noticeable with complexes involving hydrophobic peptides, in which the ionization efficiencies are often not equivalent between the RNA and peptide. Despite having lower ionization efficiencies, consistent data are obtained with the peptides. The apparent K_d values derived from ESI MS correlate well with those obtained by solution methods such as ITC, fluorescence spectroscopy, or chemical probing (Dremann & Chow, 2016; Kaur et al., 2013; Li et al., 2009; Sakakibara et al., 2012; Waduge et al., 2019). Consideration must be taken into account when analyzing ESI MS data, however, as the decrease in the free RNA peak area may be quite small as the concentration of ligand is increased; therefore, K_d values obtained from ESI MS are considered to be "relative" and typically only useful for comparisons between related ligands or RNAs.

Peak areas are obtained using Mass Lynx® software V. 4.0 (Waters, Milford, MA). Smoothing takes place for each experiment by selecting the spectrum associated with an RNA-only run (in absence of ligand). The peak associated with the dominant charge state of free RNA (e.g., (4-) or (5-) for H69 RNA) is selected and expanded if its presence is dominant and consistent throughout all ESI MS experimentation with both free RNA and the selected ligand. In most cases more than one charge state is selected and results are averaged. The value associated with the peak width at 50% of the peak height is inserted into the Savitzky-Golay algorithm (Savitzky & Golay, 1964). The result for the algorithm is then applied toward smoothing the rest of the spectra associated with the experiment. The relative fraction bound of RNA (F_r) is determined by taking the peak intensities associated with the dominant charge states of the complex (e.g., (4-) and (5-) for H69 RNA-peptide complexes) and dividing those values by the total intensities associated with the binding complex and free RNA at the same charge states. Peaks correlating with salt adducts associated with the binding complex and free RNA are included in this calculation. For these studies, it is assumed that the peak area is proportional to the concentration of the RNA in solution, and the apparent K_d value for each ligand with RNA is obtained by plotting F_r vs. [ligand, μM] using a non-linear curve fitting with the quadratic equation (Eq. 1) (Bligh et al., 2003; Sannes-Lowery et al., 2000). Plotting of the data can be done with software such as Kaleidagraph (version 4.0). Eq. (1) correlates the data obtained from free RNA (R) relative to the RNA-ligand complex (RL). Ligand (L) is used in the quadratic portion of this equation (Bligh et al., 2003; Sannes-Lowery et al., 2000). If only a relative binding affinity for all of the bound species is desired, then the total intensity of ligand-RNA complexes at all stoichiometries (i.e., 1:1, 1:1 + 2:1, 1:1 + 2:1 + 3:1, etc.) and salt adducts (Na^+, K^+, etc.) are summed, but this approach removes key information about the binding mode of the ligand to the chosen RNA target.

$$\frac{\Sigma RL}{\Sigma R + \Sigma RL} = \frac{([R]_0 + [L]_0 + K_d) - \left(([R]_0 + [L]_0 + K_d)^2 - 4[R]_0[L]_0\right)^{\frac{1}{2}}}{2[R]_0} \quad (1)$$

Results for RNA binding experiments with aminoglycosides (Li et al., 2009; Sakakibara et al., 2012) and peptides (Dremann & Chow, 2016; Kaur et al., 2013; Li et al., 2009) have been described previously, and a representative binding curve for the ESI MS data is shown in Fig. 3. The plot for the data shown in Fig. 2B using Eq. (1), gives an apparent K_d of

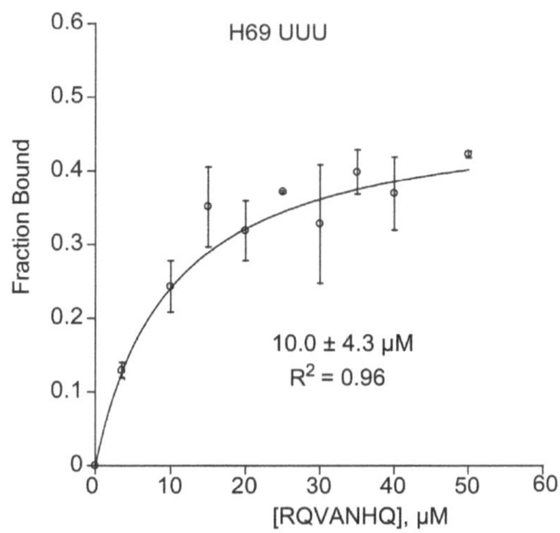

Fig. 3 Binding curve for peptide binding to unmodified (UUU) H69 RNA. Curve fitting of the fraction of UUU H69 bound vs. peptide (RQVANHQ-NH_2) concentration gives an apparent K_d value of 10.0 μM (average of three independent experiments).

10.0 μM. Similarly, the plot for the data shown in Fig. 2A using Eq. (1), gives an apparent K_d of 1.2 μM (Sakakibara et al., 2012). The peak areas for free and complexed RNA were calculated and the fits to the data for the multiple charge states were averaged.

The concentration of NH_4OAc in the ESI MS experiments as well as additives such as methanol and piperidine are important. These reagents will not only impact the binding affinity of the ligand (due to changes in ionic strength), but will also affect ionization efficiencies and charge state distribution (Schneeberger & Breuker, 2017; Vusurovic, et al., 2017). For example, NH_4OAc helps stabilize ion formation; however, the concentration used for each binding experiment may need to be adjusted depending on the type of ligand being tested. Too high of a concentration can lead to greater noise in the spectra. Too low of a concentration may lead to ion instability and hinder the observation of binding interactions. Some of these variables also depend on the instrumentation and settings used to collect the spectra (e.g., cone voltage). For these reasons, we suggest that it is better to analyze the relative changes in F_r rather than using peak areas at single titration points to obtain K_d values.

The advantage of ESI MS binding analyses is that multiple RNAs or ligands can be examined simultaneously. As shown in Fig. 2B, two different RNAs can be tested for peptide binding. In that case, the modified (ΨΨΨ) and unmodified (UUU) H69 have the same mass, therefore one RNA species was tagged with biotin in order to compare relative affinities of the amidated heptamer peptide RQVANHQ-NH_2 (Dremann & Chow, 2016). In a related study, binding of aminoglycosides to 2-aminopurine (2AP)-labeled RNA was compared to the corresponding unlabeled RNA (Sakakibara et al., 2012). The 2AP is a fluorescent adenine analogue that has been used to monitor ligand-induced RNA structural changes. ESI MS studies revealed that changes in 2AP fluorescence do not necessarily correlate with binding affinities, highlighting the need for multiple methods to determine ligand binding modes with structurally dynamic RNAs (Sakakibara et al., 2012).

Finding molecules that bind to RNA with moderate affinity is actually not that difficult, but discovery of molecules with a high selectivity for RNA over DNA, or for certain RNA motifs relative to sequence mutations or modified variants (e.g., bacterial vs. mitochondrial h44 rRNAs shown in Fig. 1), is considerably more challenging. ESI MS allows for multiple RNAs or compounds to be analyzed simultaneously such that specificity can be assessed. In some cases, the ESI MS analysis will reveal more complex

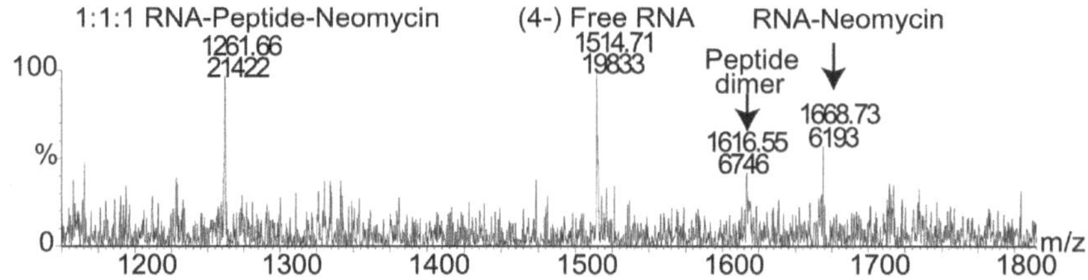

Fig. 4 Dual binding mode of neomycin (10 μM) and NQVANHQ-NH_2 (60 μm) with unmodified (UUU) H69 RNA (2.7 μM). This spectrum reveals a profile in which free H69, peptide dimer, and 1:1 H69-neomycin in the (4-) charge state, and 1:1:1 H69-neomycin-NQVANHQ-NH_2 complex in the (6-) charge state, are observed.

binding modes with dual compound binding, which suggests that conjugates of those two ligands may have desirable biological properties with increased selectivity. The example given in Fig. 4 shows that neomycin, peptide NQVANHQ-NH_2, and unmodified (UUU) H69 RNA have the ability to form a 1:1:1 binding complex.

The ability to monitor non-covalent interactions between RNA and ligands by ESI MS nicely set the stage for more advanced binding site mapping studies. Work from the laboratories of Loo (Yin & Loo, 2010; Yin, Xie, & Loo, 2008) and Fabris (Turner, Hagan, Kohlway, & Fabris, 2006) showed that in top-down ESI MS studies, some ligand interactions are sufficiently strong enough to survive phosphodiester cleavage by collisionally activated dissociation (CAD). This method therefore allows the binding sites of ligands on targets such as RNA to be determined. Schneeberger and Breuker used top-down MS to map wild-type tat peptide on TAR RNA, and their results indicate that information on both complex assembly and stoichiometry can be obtained in a time-resolved manner (Schneeberger & Breuker, 2017). These studies show that ESI MS is able to provide high-resolution information about the interaction sites of a variety of ligands on RNA. In order to improve on current compounds that target RNA, information obtained by using advanced ESI MS approaches can reveal binding affinity, loci of binding, stoichiometry, specificity for the target over closely related variants, and kinetics of binding.

4. Summary

Investigators who seek to determine ligand binding modes with RNA have a variety of methods from which to choose. Each method has unique advantages and disadvantages, so multiple methods will typically be employed in order to get a more complete picture of the binding modality. Some specific challenges include determining the binding stoichiometry, loci, relative affinity, and level of selectivity of a small molecule bound non-covalently to an RNA target. This chapter describes the use of ESI MS to determine binding stoichiometry and affinity with short RNAs, with an emphasis on sample preparation and data analysis. While the ESI MS method is best applied to charged ligands, we also provide an example of peptides with hydrophobic character binding to RNA. Future studies that employ more advanced methods will include top-down MS methods (Schneeberger & Breuker, 2017; Vusurovic, et al., 2017) and MS-MS with larger RNAs (Wetzel & Limbach, 2016). The ESI MS method provides key information that can be used to design new ligands with the potential for greater affinity and selectivity for the target RNA.

Acknowledgment

We are grateful for a Schaap Faculty Scholar Award to support our work.

References

Aboul-ela, F. (2010). Strategies for the design of RNA-binding small molecules. *Future Medicinal Chemistry*, *2*, 93–119.

Anthis, N. J., & Clore, G. M. (2013). Sequence-specific determination of protein and peptide concentrations by absorbance at 205 nm. *Protein Science*, *22*, 851–858.

Blakeley, B. D., DePorter, S. M., Mohan, U., Burai, R., Tolbert, B. S., & McNaughton, B. R. (2012). Methods for identifying and characterizing interactions involving RNA. *Tetrahedron*, *68*, 8837–8855.

Bligh, S. W., Haley, T., & Lowe, P. N. (2003). Measurement of dissociation constants of inhibitors binding to Src SH2 domain protein by non-covalent electrospray ionization mass spectrometry. *Journal of Molecular Recognition*, *16*, 139–148.

Borovinskaya, M. A., Pai, R. D., Zhang, W., Schuwirth, B. S., Holton, J. M., Hirokawa, G., et al. (2007). Structural basis for aminoglycoside inhibition of bacterial ribosome recycling. *Nature Structural & Molecular Biology*, *14*, 727–732.

Böttger, E. C., Springer, B., Prammananan, T., Kidan, Y., & Sander, P. (2001). Structural basis for selectivity and toxicity of ribosomal antibiotics. *EMBO Reports*, *2*, 318–323.

Carpino, L. A., & Han, G. Y. (1972). The 9-fluorenylmethoxycarbonyl amino-protecting group. *Journal of Organic Chemistry*, *37*, 3404–3409.

Cate, J. H., Yusupov, M. M., Yusupova, G. Z., Earnest, T. N., & Noller, H. F. (1999). X-ray crystal structures of 70S ribosome functional complexes. *Science*, *285*, 2095–2104.

Chow, C. S., & Bogdan, F. M. (1997). A structural basis for RNA-ligand interactions. *Chemical Reviews*, *97*, 1489–1514.

Chow, C. S., Mahto, S. K., & Lamichhane, T. N. (2008). Combined approaches to site-specific modification of RNA. *ACS Chemical Biology*, *3*, 30–37.

Cole, R. B. (2010). *Electrospray and MALDI mass spectrometry: Fundamentals, instrumentational practicalities, and biological applications*. Hoboken, NJ: John Wiley & Sons, Inc.

Connelly, C. M., Moon, M. H., & Schneekloth, J. S. (2016). The emerging role of RNA as a therapeutic target for small molecules. *Cell Chemical Biology*, *23*, 1077–1090.

Davidson, A., Leeper, T. C., Athanassiou, Z., Patora-Komisarska, K., Karn, J., Robinson, J. A., et al. (2009). Simultaneous recognition of HIV-1 TAR RNA bulge and loop sequences by cyclic peptide mimics of Tat protein. *Proceedings of the National Academy of Sciences of the United States of America*, *106*, 11931–11936.

Dremann, D. N., & Chow, C. S. (2016). The development of peptide ligands that target helix 69 rRNA of bacterial ribosomes. *Bioorganic and Medicinal Chemistry*, *24*, 4486–4491.

Fasman, G. (1975). *Handbook of biochemistry and molecular biology*. Cleveland, OH: CRC Press.

Fenn, J. B., Mann, M., Meng, C. K., Wong, S. F., & Whitehouse, C. M. (1989). Electrospray ionization for mass spectrometry of large biomolecules. *Science*, *246*, 64–71.

Gaston, K. W., & Limbach, P. A. (2014). The identification and characterization of non-coding and coding RNAs and their modified nucleosides by mass spectrometry. *RNA Biology*, *11*, 1568–1585.

Griffey, R. H., Greig, M. J., An, H. Y., Sasmor, H., & Manalili, S. (1999). Targeted site-specific gas-phase cleavage of oligoribonucleotides. Application in mass spectrometry-based identification of ligand binding sites. *Journal of the American Chemical Society*, *121*, 474–475.

Guan, L. R., & Disney, M. D. (2012). Recent advances in developing small molecules targeting RNA. *ACS Chemical Biology*, *7*, 73–86.

Hartsel, S. A., Kitchen, D. E., Scaringe, S. A., & Marshall, W. S. (2005). RNA oligonucleotide synthesis via 5′-silyl-2′-orthoester chemistry. *Methods in Molecular Biology*, *288*, 33–50.

Hofstadler, S. A., Sannes-Lowery, K. A., Crooke, S. T., Ecker, D. J., Sasmor, H., Manalili, S., et al. (1999). Multiplexed screening of neutral mass-tagged RNA targets against ligand libraries with electrospray ionization FTICR MS: A paradigm for high-throughput affinity screening. *Analytical Chemistry*, *71*, 3436–3440.

Kaur, M., Rupasinghe, C. N., Klosi, E., Spaller, M. R., & Chow, C. S. (2013). Selection of heptapeptides that bind helix 69 of bacterial 23S ribosomal RNA. *Bioorganic and Medicinal Chemistry*, *21*, 1240–1247.

Lau, J. L., & Dunn, M. K. (2018). Therapeutic peptides: Historical perspectives, current development trends, and future directions. *Bioorganic and Medicinal Chemistry*, *26*, 2700–2707.

Li, M., Duc, A. C. E., Klosi, E., Pattabiraman, S., Spaller, M. R., & Chow, C. S. (2009). Selection of peptides that target the aminoacyl-tRNA site of bacterial 16S ribosomal RNA. *Biochemistry*, *48*, 8299–8311.

Merrifield, R. B. (1963). Solid phase peptide synthesis. 1. Synthesis of a tetrapeptide. *Journal of the American Chemical Society*, *85*, 2149–2154.

Milligan, J. F., Groebe, D. R., Witherell, G. W., & Uhlenbeck, O. C. (1987). Oligoribonucleotide synthesis using T7 RNA polymerase and synthetic DNA templates. *Nucleic Acids Research*, *15*, 8783–8798.

Moazed, D., & Noller, H. F. (1987). Interaction of antibiotics with functional sites in 16S ribosomal RNA. *Nature*, *327*, 389–394.

Moore, P. B., & Steitz, T. A. (2005). The ribosome revealed. *Trends in Biochemical Sciences*, *30*, 281–283.

Nordhoff, E., Cramer, R., Karas, M., Hillenkamp, F., Kirpekar, F., Kristiansen, K., et al. (1993). Ion stability of nucleic acids in infrared matrix-assisted laser desorption/ionization mass spectrometry. *Nucleic Acids Research*, *21*, 3347–3357.

Ofengand, J., & Bakin, A. (1997). Mapping to nucleotide resolution of pseudouridine residues in large subunit ribosomal RNAs from representative eukaryotes, prokaryotes, archaebacteria, mitochondria and chloroplasts. *Journal of Molecular Biology*, *266*, 246–268.

Ogilvie, K. K., Beaucage, S. L., Schifman, A. L., Theriault, N. Y., & Sadana, K. L. (1978). The synthesis of oligoribonucleotides. II. The use of silyl protecting groups in nucleoside and nucleotide chemistry VII. *Canadian Journal of Chemistry*, *56*, 2768–2780.

Overbye, K. M., & Barrett, J. F. (2005). Antibiotics: Where did we go wrong? *Drug Discovery Today*, *10*, 45–52.

Projan, S. J. (2003). Why is big Pharma getting out of antibacterial drug discovery? *Current Opinion in Microbiology*, *6*, 427–430.

Quadri, L. E. (2007). Strategic paradigm shifts in the antimicrobial drug discovery process of the 21st century. *Infectious Disorders Drug Targets*, *7*, 230–237.

Recht, M. I., Fourmy, D., Blanchard, S. C., Dahlquist, K. D., & Puglisi, J. D. (1996). RNA sequence determinants for aminoglycoside binding to an A-site rRNA model oligonucleotide. *Journal of Molecular Biology*, *262*, 421–436.

Rink, H. (1987). Solid-phase synthesis of protected peptide-fragments using a trialkoxy-diphenyl-methylester resin. *Tetrahedron Letters*, *28*, 3787–3790.

Sakakibara, Y., Abeysirigunawardena, S. C., Duc, A. C. E., Dremann, D. N., & Chow, C. S. (2012). Ligand- and pH-induced changes of RNA domain helix 69 revealed by 2-aminopurine fluorescence. *Angewandte Chemie International Edition*, *51*, 12095–12098.

Sannes-Lowery, K. A., Griffey, R. H., & Hofstadler, S. A. (2000). Measuring dissociation constants of RNA and aminoglycoside antibiotics by electrospray ionization mass spectrometry. *Analytical Biochemistry*, *280*, 264–271.

Savitzky, A., & Golay, M. J. E. (1964). Smoothing and differentiation of data by simplified least squares procedures. *Analytical Chemistry*, *36*, 1627–1639.

Schneeberger, E. M., & Breuker, K. (2017). Native top-down mass spectrometry of TAR RNA in complexes with wild-type tat peptide for binding site mapping. *Angewandte Chemie International Edition*, *56*, 1254–1258.

Silverstein, R. M., Webster, F. X., & Kiemle, D. J. (2005). *Spectrometric identifcation of organic compounds* (7th ed.). . Hoboken, NJ: John Wiley & Sons.

Steitz, T. A. (2005). On the structural basis of peptide-bond formation and antibiotic resistance from atomic structures of the large ribosomal subunit. *FEBS Letters*, *579*, 955–958.

Tang, X., Callahan, J. H., Zhou, P., & Vertes, A. (1995). Noncovalent protein—Oligonucleotide interactions monitored by matrix-assisted laser desorption/ionization mass spectrometry. *Analytical Chemistry*, *67*, 4542–4548.

Thomas, J. R., & Hergenrother, P. J. (2008). Targeting RNA with small molecules. *Chemical Reviews*, *108*, 1171–1224.

Tor, Y. (2003). Targeting RNA with small molecules. *Chembiochem*, *4*, 998–1007.

Tu, D., Blaha, G., Moore, P. B., & Steitz, T. A. (2005). Structures of MLSBK antibiotics bound to mutated large ribosomal subunits provide a structural explanation for resistance. *Cell*, *121*, 257–270.

Turner, K. B., Hagan, N. A., Kohlway, A. S., & Fabris, D. (2006). Mapping noncovalent ligand binding to stemloop domains of the HIV-1 packaging signal by tandem mass spectrometry. *Journal of the American Society for Mass Spectrometry*, *17*, 1402–1411.

Vusurovic, J., Schneeberger, E. M., & Breuker, K. (2017). Interactions of protonated guanidine and guanidine derivatives with multiply deprotonated RNA probed by electrospray ionization and collisionally activated dissociation. *ChemistryOpen*, *6*, 739–750.

Waduge, P., Sakakibara, Y., & Chow, C. S. (2019). Chemical probing for examining the structure of modified RNAs and ligand binding to RNA. *Methods*, *156*, 110–120.

Waksman, S. A., Frankel, J., & Graessle, O. (1949). The in vivo activity of neomycin. *Journal of Bacteriology*, *58*, 229–237.

Waksman, S. A., & Lechevalier, H. A. (1949). Neomycin, a new antibiotic active against streptomycin-resistant bacteria, including tuberculosis organisms. *Science*, *109*, 305–307.

Waksman, S. A., Lechevalier, H. A., & Harris, D. A. (1949). Neomycin-production and antibiotic properties. *The Journal of Clinical Investigation*, *28*, 934–939.

Wang, L., Pulk, A., Wasserman, M. R., Feldman, M. B., Altman, R. B., Doudna Cate, J. H., et al. (2012). Allosteric control of the ribosome by small-molecule antibiotics. *Nature Structural & Molecular Biology*, *19*, 957–963.

Warner, K. D., Hajdin, C. E., & Weeks, K. M. (2018). Principles for targeting RNA with drug-like small molecules. *Nature Reviews Drug Discovery*, *17*, 547–558.

Wetzel, C., & Limbach, P. A. (2016). Mass spectrometry of modified RNAs: Recent developments. *The Analyst*, *141*, 16–23.

Wilson, D. N., & Doudna Cate, J. H. (2012). The structure and function of the eukaryotic ribosome. *Cold Spring Harbor Perspectives in Biology*, *4*, a011536.

Xie, J., Talaska, A. E., & Schacht, J. (2011). New developments in aminoglycoside therapy and ototoxicity. *Hearing Research*, *281*, 28–37.

Yin, S., & Loo, J. A. (2010). Elucidating the site of protein-ATP binding by top-down mass spectrometry. *Journal of the American Society for Mass Spectrometry*, *21*, 899–907.

Yin, S., Xie, Y., & Loo, J. A. (2008). Mass spectrometry of protein-ligand complexes: Enhanced gas-phase stability of ribonuclease-nucleotide complexes. *Journal of the American Society for Mass Spectrometry*, *19*, 1199–1208.

Yusupov, M. M., Yusupova, G. Z., Baucom, A., Lieberman, K., Earnest, T. N., Cate, J. H., et al. (2001). Crystal structure of the ribosome at 5.5 Å resolution. *Science*, *292*, 883–896.

CHAPTER FIFTEEN

Design of RNA-targeting macrocyclic peptides

Matthew J. Walker, Gabriele Varani*
Department of Chemistry, University of Washington, Seattle, WA, United States
*Corresponding author: e-mail address: varani@chem.washington.edu

Contents

Abstract

RNA structures play a pivotal role in many biological processes and the progression of human disease, making them an attractive target for therapeutic development. Often RNA structures operate through the formation of complexes with RNA-binding proteins, however, much like protein-protein interactions, RNA-protein interactions span large surface areas and often lack traditional druggable properties, making it challenging to target them with small molecules. Peptides provide much greater surface areas and therefore greater potential for forming specific and high affinity interactions with

Methods in Enzymology, Volume 623
ISSN 0076-6879
https://doi.org/10.1016/bs.mie.2019.04.029

RNA. In this chapter, we discuss our approach for engineering peptides that bind to structured RNAs by highlighting methods and design strategies from previous successful projects aimed at inhibiting the HIV Tat-TAR interaction and the biogenesis of oncogenic microRNAs.

1. Introduction

RNA molecules play crucial roles in regulating healthy and diseased cellular processes (e.g., transcription, splicing, mRNA transport, translation, etc) and are therefore tightly regulated transcriptionally and post-transcriptionally, often through their interactions with other RNAs and with RNA binding proteins. RNAs and RNA-protein interactions regulate viral replication or the expression of proto-oncogenes, and are mis-regulated in many infectious and chronic diseases (Cooper, Wan, & Dreyfuss, 2009; Esteller, 2011) making these RNA structures and RNA-protein surfaces an untapped source of potential drug targets (Burnett & Rossi, 2012; Ling, Fabbri, & Calin, 2013). However, RNA-protein interactions are much more challenging to target with small molecules than traditional enzymatic active sites (Warner, Hajdin, & Weeks, 2018). These interactions span large surface areas and often lack structural complexity (Jones, Daley, Luscombe, Berman, & Thornton, 2001; Lunde, Moore, & Varani, 2007). Therefore, it is more challenging, though not impossible (Afshar, Prescott, & Varani, 1999; Bower et al., 2003; Davis et al., 2004; Howe et al., 2015; Murchie et al., 2004; Palacino et al., 2015; Ratni et al., 2016), to discover small molecules that compete with much larger proteins and stabilize the often dynamic single stranded regions of the RNA. Intermediate molecular weight (1.5–2 kDa) peptides can provide much greater surface area and therefore have greater potential to form high affinity and specific complexes (Battiste et al., 1996; Puglisi, Chen, Blanchard, & Frankel, 1995). Thus, our group has used peptides to discover RNA-binding ligands for probing structural and mechanistic aspects of RNA-protein interactions and investigating possible new RNA inhibitors.

Here, we discuss our approach to engineering peptides that bind RNAs by highlighting methods and design strategies (Fig. 1). In Section 2, we describe limitations of targeting RNA with linear peptides (Leulliot & Varani, 2001) and how conformationally constrained peptide mimetics address some of these issues (Robinson, 2008). Section 3 discusses the design process for building mimetics from protein structure and sequence, based on

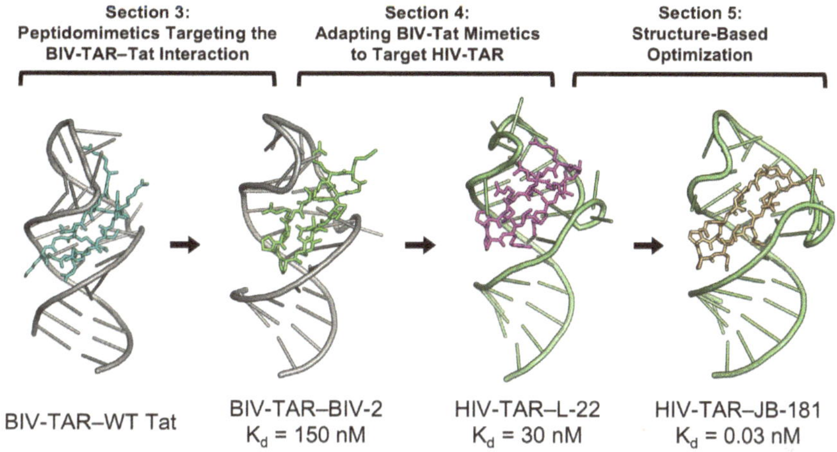

Fig. 1 Schematic depicting the three design stages described in Sections 3–5 that led to the discovery of very potent RNA-targeting macrocyclic peptides. From left to right: Bovine immunodeficiency virus trans-activation response (BIV-TAR) is a structured *cis*-regulatory element in the BIV 5′ UTR mRNA that is recognized by BIV Tat to stimulate viral transcription (PDB ID: 1MNB; Puglisi et al., 1995); Section 3 overviews the process of using peptidomimetics to target the BIV-TAR—Tat interaction by grafting the wildtype Tat sequence onto a β-hairpin scaffold. This strategy led to the discovery of peptide BIV-2 and its structure determination by NMR (PDB ID: 2A9X; Leeper, Athanassiou, Dias, Robinson, & Varani, 2005); Section 4 describes how a large positional scanning peptide library was organized to generate structure activity relationship (SAR), from which a BIV-Tat mimetic was adapted to target human immunodeficiency virus TAR (HIV-TAR) (PDB ID: 2KDQ; Davidson et al., 2009); Section 5 describes the structure-based optimization process and peptide design strategies that incorporated non-canonical amino acid substitution (e.g., backbone extension, side-chain truncation) to achieve low picomolar binding, JB-181 (PDB ID: 6D2U; Shortridge et al., 2018).

our successful targeting of the interaction between viral trans-activator of transcription (Tat) and trans-activating response element (TAR) (Athanassiou et al., 2004; Leeper et al., 2005). Section 4 describes the use of positional scanning libraries to discover a high affinity peptide capable of binding human immunodeficiency virus TAR RNA (Athanassiou et al., 2007; Davidson et al., 2009). Section 5 is dedicated to explaining how structure-based optimization can lead to the discovery of peptides with low pM affinity and exquisite specificity (Davidson, Patora-Komisarska, Robinson, & Varani, 2011; Shortridge et al., 2018). In Section 6, we evaluate how β-turn mimetics can be adapted to target other pharmaceutically relevant RNA stem-loop structures by two cases as examples (Moehle et al., 2007; Shortridge et al., 2017).

2. Targeting RNA with macrocyclic β-hairpin peptide mimetics

2.1 Challenges in targeting RNA with peptides

Engineering peptides to bind RNA requires an understanding of several fundamental challenges inherent to targeting RNA. First, RNA structures are not abundant in the PDB. Crystallization can be difficult and sometimes impossible because of RNA's intrinsic flexibility (Leulliot & Varani, 2001; Williamson, 2000). Our group uses NMR spectroscopy to circumnavigate this issue as it provides characterization of RNA and peptide conformation, even when structures are flexible (Varani, Aboula-ela, & Allain, 1996). Structure determination of peptide-RNA complexes produce atomic resolution accuracy and provides critical insight regarding key peptide-RNA interactions, yet only becomes accessible once a potent enough peptide is identified. Furthermore, overlapped NMR spectra reduce resolution, making unambiguous chemical shift assignment impossible for weakly bound peptides discovered early on in the design process. Second, binding interfaces are sometimes so large that it can be difficult to target them with peptides, particularly for larger multimeric RNA-protein complexes. Third, for any RNA binding ligand, it is difficult to differentiate using only biochemical assays (K_d) how stable the RNA-ligand complex actually is. Even when a molecule is bound, conformational flexibility might remain, or the molecule might bind in multiple locations on the RNA, boosting affinity through avidity and giving the illusion of potency. It is often the case that the resulting affinity results from non-specific interactions from electrostatic interactions between the phosphate backbone or electronegative major groove with basic side chains of peptides, or charged small molecules. Maintaining a single RNA-ligand conformation is also inherently energetically unfavorable because of the entropic penalty associated with rigidification of flexible RNA molecules by induced fit (Leulliot & Varani, 2001; Williamson, 2000). The penalty is magnified when RNA is targeted with linear peptides with poor conformational stability, since short peptide fragments lack the tertiary interactions that buttress protein binding sites in folded proteins. This is the reason we prefer to use highly structured peptide mimics with stable secondary structure in our targeting efforts.

2.2 Macrocyclic β-hairpin scaffolds as an effective strategy for targeting RNA

Structure-based peptidomimetics is a peptide design strategy where inspection of a protein-bound structure provides a rational to design a minimal peptide that mimics the activity of the full protein (Avan, Hall, & Katritzky, 2014; Mason, 2010; Pelay-Gimeno, Glas, Koch, & Grossmann, 2015). The protein structure provides a starting point for the design of structurally rigid peptides that recapitulate intermolecular interactions within a minimal framework and inhibit biological activity (Muppidi et al., 2012; Rosenström et al., 2006; Werder, Hauser, Abele, & Seebach, 1999; Whitby et al., 2011). It is crucial for the mimetic to match both sequence and structure, and this is done most effectively when using secondary structure motifs that stabilize a peptide fold through local intramolecular interactions while allowing incorporation of the desired sequence into small peptide fragments (Fujii et al., 2003; Pelay-Gimeno et al., 2015). The two most common protein secondary structures are β-hairpins and α-helices, whose backbones act as scaffolds upon which energetically favorably sidechains can be pre-presented for binding in position precisely defined by intramolecular hydrogen bonding and packing interactions. β-hairpins are of particular interest for targeting RNA as many RNA-binding proteins (RBP's) exploit β-sheet structures (Cléry, Blatter, & Allain, 2008; Sibanda & Thornton, 1985).

Our group has targeted RNA using macrocyclic β-hairpin peptidomimetic chemistry. Head-to-tail cyclization rigidifies the structure (Pelay-Gimeno et al., 2015), and cyclized peptides are advantageous from a pharmacological standpoint as well because they are exonuclease resistant (Robinson, 2000) and have increased cellular uptake when arginine-rich (Lättig-Tünnemann et al., 2011). Their structure is comprised of two anti-parallel β-strands stabilized by two β-hairpin favoring turn inducers located at the tip of both strands. β-hairpins strongly favor Type I′ ($\varphi_{i+1} = 60$, $\psi_{i+1} = 30$, $\varphi_{i+2} = 90$, $\psi_{i+2} = 0$) and Type II′ ($\varphi_{i+1} = 60$, $\psi_{i+1} = -120$, $\varphi_{i+2} = -80$, $\psi_{i+2} = 0$) turn inducing residues (Madan, Seo, & Lee, 2014). Most of our published work uses 14-mer β-hairpins with a Type II′ D-Pro/L-Pro turn inducer as it readily favors β-hairpin folding (Fig. 2A) (Bean, Kopple, & Peishoff, 1992; Nair, Vijayan, Venkatachalapathi, & Balaram, 1979). The D-Pro/L-Pro turn developed in the Robinson group acts as a template for grafting mimicked sequences onto the β-hairpin scaffold (Robinson, 2008), favoring formation of stable β-hairpins.

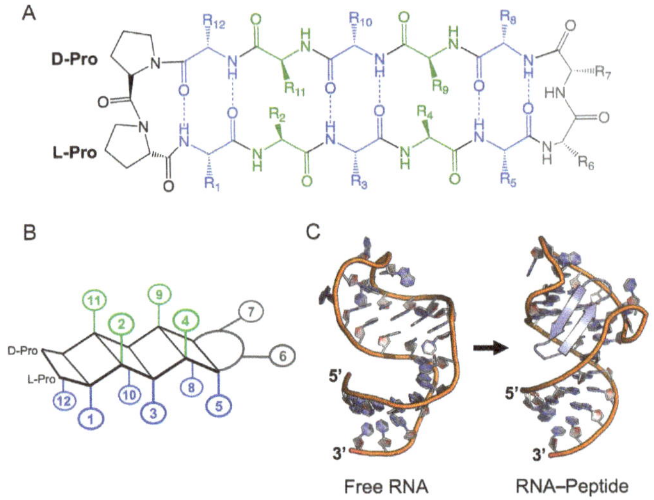

Fig. 2 (A) Generic 2D structure of a 12-mer β-hairpin macrocycle with a D-Pro/L-Pro template. (B) Representation of a 12-mer β-hairpin macrocycle with sidechains grouped by orientation relative to the backbone plane. Positions 2, 4, 9, 11 point "up" (green); while positions 1, 3, 5, 8, 10, 12 point "down" (blue). (C) β-hairpin scaffolds match the shape and size of the RNA major groove and mold the RNA structure. From left to right: (PDB ID: 1ANR; Aboul-ela, Karn, & Varani, 1996), (PDB ID: 6D2U; Shortridge et al., 2018).

The second turn varies in sequence depending on the protein being mimicked with the only requirement that it favors Type I′ or Type II′ (Robinson, 2008).

If properly designed, cyclic β-hairpins will possess a rigid backbone structure, stabilized by multiple NH to CO backbone hydrogen bonds. A rigid β-hairpin provides a reliable scaffold for mimicking a protein and makes amino acid substitution less likely to affect the β-hairpin character. A 14-mer cyclic beta-hairpin favors a 2:2 sidechain orientation (Sibanda, Blundell, & Thornton, 1989) presenting 6 residues in one direction (for the purpose of this chapter, positions: 1, 3, 5, 8, 10, 12) and 4 residues in the opposite direction (positions: 2, 4, 9, 11) (Fig. 2B). Considering this positioning is essential during optimization and highlighted in Section 4. This class of cyclic β-hairpins also has size and shape to match major groove RNA features found at apical loop and/or bulge junctions (Fig. 2C), allowing sidechains to make stabilizing contacts with RNA bases and the phosphate backbone. The combination of major groove fit and optimizable intermolecular interactions gives the β-hairpin a higher probability of inducing RNA conformation changes that favor formation of both canonical and non-canonical base-pairs, even when these are not present in the unbound state of the RNA (Bardaro, Shajani, Patora-Komisarska, Robinson, & Varani, 2009;

Borkar et al., 2016). Formation of these base-pairs is influential in developing high affinity RNA-targeted peptides.

A detailed description of synthetic methods used to prepare these peptides can be found in previous publications (Athanassiou et al., 2004, 2007). Briefly, cyclic β-hairpin peptides can be generated following solid-state peptide synthesis followed by a standard cyclization procedure, using Fmoc chemistry on acid sensitive 2-chlorotrityl resin in order to retain protecting groups during resin cleavage and HoBt/HBTU head-to-tail cyclization. HPLC purification is necessary after cyclization and before side-chain deprotection in order to separate cyclized product from unreacted linear peptide. It should be noted that due to the highly constrained nature of β-hairpins, some sequences have unexpected poor yields, likely due to unfavorable reaction geometries. In addition, the cyclization reaction is sensitive to sequences that produce flexible backbones (e.g., incorporation of glycine, backbone kinks) which may inhibit the cyclization reaction and/or require longer reaction times.

3. Peptidomimetic targeting the BIV-TAR–Tat interaction

Mimicking the TAR–Tat interaction of the Bovine Immunodeficiency Virus (BIV) provides an illustration of how to mimic an RBP by grafting amino acids onto a cyclic β-hairpin peptidomimetic scaffold (Athanassiou et al., 2004; Leeper et al., 2005), and in turn shows the feasibility of using β-hairpin scaffolds to target RNA.

3.1 Design rational

Regulatory RNAs in viral RNA genomes provide attractive targets for discovery of new antiviral drugs because these elements are essential for viral replication, structured, and highly conserved between isolates (Gallego & Varani, 2001; Kirk, Luedtke, & Tor, 2000; Wilson & Li, 2000). A well-studied example is provided by the Tat protein, which plays a critical role lentiviral life cycle and contributes significantly to HIV pathogenesis by increasing transcription of the integrated pro-virus by recognition of TAR RNA, a structured *cis*-regulatory element in viral 5′ UTR mRNA (Gonda, Luther, Fong, & Tobin, 1994; Jones & Peterlin, 1994; Wei, Garber, Fang, Fischer, & Jones, 1998). The TAR-Tat interaction is essential for viral replication and driven by TAR recognition of the highly basic Tat Arginine Rich Motif (ARM), which, in the BIV protein (though not in the human counterpart) forms a β-hairpin that binds to BIV TAR within the major groove at the apical loop-stem junction (Puglisi et al., 1995; Ye, Kumar, & Patel, 1995). Single nucleotide or amino acid substitutions within either the protein or the RNA severely affect transcriptional levels and hence viral replication (Fischer, Huber, Boelens, Mattaj, & Luhrmann, 1995). BIV and HIV-TAR have similar structure and function, but at the time structural information was limited to the BIV TAR–Tat complex, to which we directed our mimicking efforts.

The starting point for this project was the structure of a Tat 14-mer linear peptide (residues 68-RPRGTRGKGRRIRR-81) bound to the BIV TAR stem-loop (residues 4–31) (Fig. 3A) (Puglisi et al., 1995). In the structure,

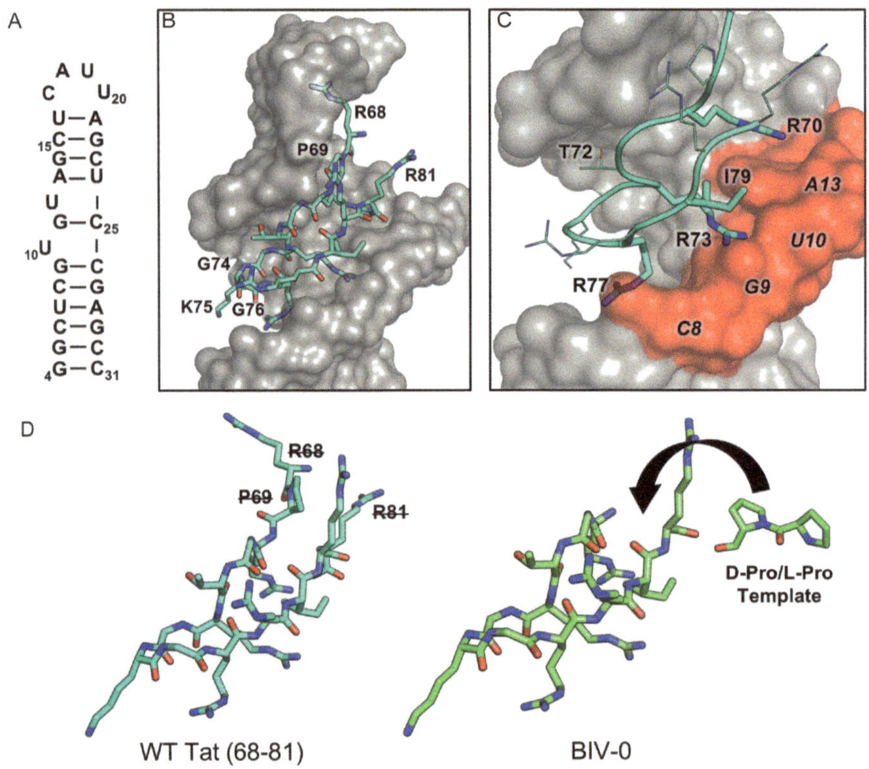

Fig. 3 (A) Secondary structure of BIV-TAR. (B) BIV Tat arginine-rich motif (cyan, 68–81) binds to within the BIV-TAR (gray) major groove at the stem-bulge and stem-apical loop interface (PDB ID: 1MNB; Puglisi et al., 1995). (C) Arg-77, Arg-73, and Arg-70 are critical recognition elements for BIV-TAR binding by sandwiching between nucleotides C8, G9, U10, G11 forming arginine sandwich motifs (ASM). (D) BIV-0 cyclization was designed by eliminating R68, P69, and R81 and grafting a D-Pro/L-Pro template (Athanassiou et al., 2004).

TAR grasps the Tat peptide in a glove-like manner within a widened major groove pocket, sandwiching it between the apical loop and thumb-like bulge (Fig. 3B). The 14-mer peptide bound to the RNA forms a β-hairpin that is kinked due to flexible Gly-71, pushing Thr-72 to bulge upward toward G22 within the BIV-TAR apical loop. The β-hairpin is favored by a Gly-Lys-Gly γ-turn facing downward and away from the apical loop. Significant contact points include: hydrophobic contacts between the methyl group of Thr-72 and the ribose ring of G22; phosphate H-bonding and π-stacking interactions between Arg-70/G14 and Arg-77/C8 and G9; a triple base pair induced by Arg-73 positioned between G9 and U10 and stabilized by hydrophobic packing of Ile-79 neighboring U11 (Fig. 3C). Together, sandwiching of three arginines (Arg-77, Arg-73, Arg-70) between the neighboring bulge nucleotides (C8, G9, U10, G11) are critical elements for BIV-TAR recognition forming arginine sandwich motifs (ASM) (Puglisi et al., 1995).

Since BIV Tat forms a β-hairpin when bound to BIV TAR, design of the first mimetic (BIV-0) was straightforward. Mutagenesis data shows that residues critical for binding (Arg-70, Gly-71, Thr-72, Arg-73, Gly-74, Arg-77, and Ile-79) are all located close to the reverse turn formed by residues Gly-74, Lys-75, and Gly-76 (Chen & Frankel, 1995). Therefore, the peptide was cyclized by replacing Arg-68, Pro-69, and Arg-81 with the D-Pro/L-Pro Type II' turn template (Fig. 3D). However, this mutagenesis and sequence-based approach to graft the wildtype Tat peptide fragment onto a beta-hairpin scaffold was unsuccessful, since BIV-0, containing an almost identical sequence to wild-type Tat showed very poor, probably non-specific binding ($K_d \sim 40\,\mu M$) (Athanassiou et al., 2004).

This result prompted us to re-consider the importance of peptide stability for β-hairpin design. When we investigated by NMR the structure of BIV-0, we observed essentially a random coil conformation (Athanassiou et al., 2004). In retrospect, a regular 2:2 β-hairpin conformation (Sibanda et al., 1989) requires a particular number of residues on each strand for ideal hydrogen bonding (Griffiths-Jones, Maynard, & Searle, 1999). Turn chemistry may have also destabilized the peptide since the rigid turn-inducing template D-Pro/L-Pro near Gly-71 can induce an irregular backbone conformation. We therefore redesigned the peptide mimic by using 12-residue peptides to promote 2:2 β-hairpin formation and mounting the sequence on a D-Pro/L-Pro template (Favre, Moehle, Jiang, Pfeiffer, & Robinson, 1999; Jiang, Moehle, Dhanapal, Obrecht, & Robinson, 2000). Eight peptides were synthesized (BIV-1–BIV-8, Table 1) with residues Thr-4, Arg-5, Gly-6, Lys-7,

Table 1 Sequences of the BIV-Tat ARM (68–81), peptide mimic BIV-0, and β-hairpin mimetics BIV-1–BIV-8. Reported K_ds were measured by electromobility shift assays (EMSA) (nb = no binding) (Athanassiou et al., 2004). Position numbering based on the schematic representation shown below.

BIV Tat R_{68} P R G **T R G K G R R I** R R_{81}

BIV-0 ^{D}P R G **T R G K G R R I** R ^{L}P

	1	2	3	4	5	6	7	8	9	10	11	12	K_d (uM)
BIV-1	I	R	G	T	R	G	K	R	R	I	R	V	30
BIV-2	R	V	R	T	R	G	K	R	R	I	R	V	0.15
BIV-3	I	Y	R	T	R	G	K	R	R	I	R	T	nb
BIV-4	Y	R	G	T	R	G	K	R	R	I	Y	V	>50
BIV-5	R	R	G	T	R	G	K	R	R	I	G	R	1–2
BIV-6	V	R	G	T	R	G	K	R	R	I	K	Y	>50
BIV-7	V	R	R	T	R	G	K	R	R	I	K	Y	nb
BIV-8	K	R	G	T	R	G	K	R	R	I	G	Y	>50

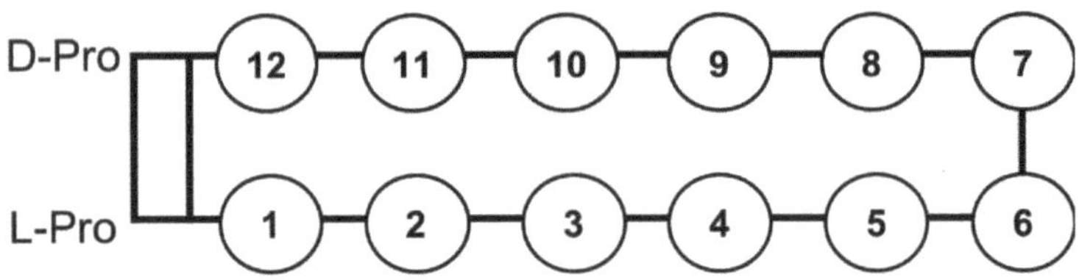

Arg-8, Arg-9, and Ile-10 kept constant, corresponding to residues Thr-72, Arg-73, Gly-74, Lys-75, Arg-77, Arg-78, and Ile-79 in BIV Tat, which make critical interactions with BIV-TAR (Athanassiou et al., 2004). The sequence near the apical loop was trimmed by changing positions 1–3, 11, and 12 (positions 68–70, 80, and 81 in wildtype Tat) to residues with charged and aromatic sidechains. Since β-hairpins prefer type I′ and type II' chemistry, Gly-76 from the γ-turn was also removed to favor a Type II' turn of Gly-74 and Lys-75. Although all peptides are basic and bind with low μM affinity, only of the eight mimetics (BIV-2) bound with nM affinity (K_d = 150 nM) (Athanassiou et al., 2004). BIV-2 is arginine-rich and lacks a mid-strand glycine between turns, further supporting our hypothesis that formation of a stable and regular β-hairpin is essential for peptides to bind RNA.

Next, we investigated which BIV-2 structural features made it bind BIV-TAR strongly by evaluating each peptide structure by NMR. BIV-0, for example, had random coil characteristics, but NMR data for BIV-2 showed it had a single set of peaks (Athanassiou et al., 2004). Structures were determined for BIV-2 and BIV-3, and both adopted β-hairpin structures, whereas the other mimetics adopted multiple conformations, yet BIV-2 had a lower RMSD compared to BIV-3, suggesting BIV-2 was more rigid and that peptide stability contributed to binding (Athanassiou et al., 2004).

3.2 Structure and SAR development

The structure of HIV-TAR bound to BIV-2 was determined using NMR (Leeper et al., 2005). Inspection of the BIV-TAR–BIV-2 complex reveals that BIV-2 is bound in the same major groove pocket as BIV Tat, but in a completely different orientation, being upside down relative to what was expected from the targeted structure with the D-Pro/L-Pro template positioned downward into the major groove and the Gly-6/Lys-7 turn upward into the apical loop (Fig. 4A). Yet, despite the flipped orientation, many sidechain interactions with the RNA are similar to BIV Tat. Arg-1, Arg-3, and Arg-5 are sandwiched between bases forming ASMs mimicking BIV Tat sidechain positioning for Arg-70, Arg-73, and Arg-77, respectively. Intercalation occurs between C8, G9, U10, G11, and G14 through a combination of cation-π interactions and hydrogen bonding with purine N7's and guanosine O2's (Fig. 4B). Arg-8 is also buried relative to the β-hairpin, yet positioned near flexible apical loop nucleotides and only making transient contacts with multiple apical loop nucleotides. Mimicry is nearly exact between the two structures for Ile-10 and Ile-79, both are positioned deep within the base stack. Ile-10 stabilizes the U10-A13-U24 base triple through van der Waal interactions with A13 and U24 and is a major driving force for TAR recognition (Fig. 4C and D). Solvent exposed Arg-9 and Arg-11 make no hydrogen bonds or cation-π interactions, yet contacts with the phosphate backbone of A21 and G22 favor the apical loop to fold over the peptide (Fig. 4E). Val-2 and Thr-4 point away from the RNA and into the solvent, making little contact with the RNA.

In summary, the structure revealed that, once a stable β-hairpin conformation is achieved, residue changes that increase or decrease affinity with the RNA are likely to arise from changes in side-chain to RNA interactions. We identified key stabilizing interactions, which contributed to future efforts to characterize the different binding potentials for each position (see Section 4).

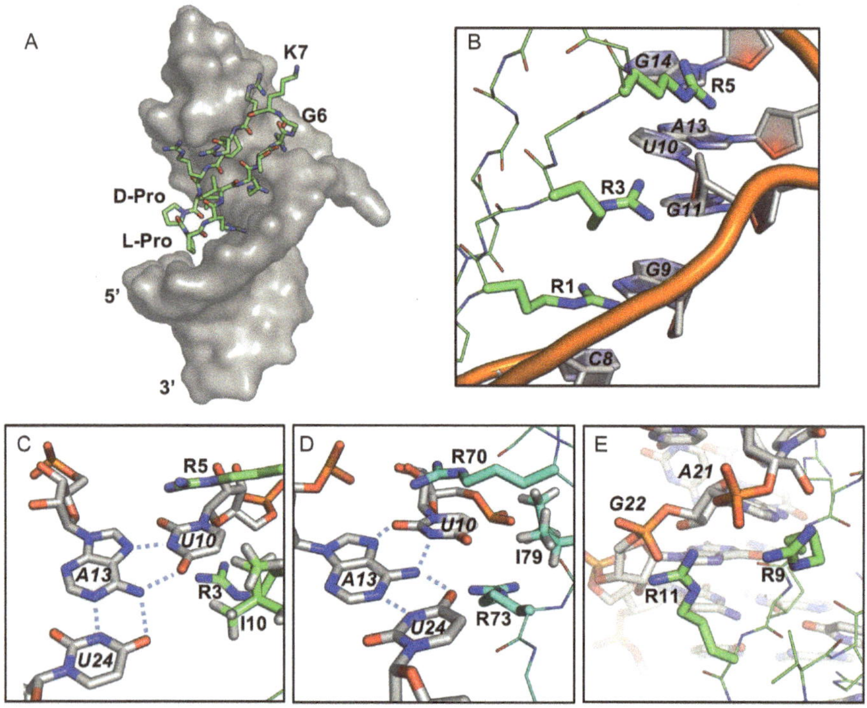

Fig. 4 (A) BIV-2 (green) bound to BIV-TAR (gray) upside down relative to wildtype BIV-Tat. (PDB ID: 2A9X; Leeper et al., 2005). (B) ASM formed by Arg-1, Arg-3 and Arg-5 intercalating between C8, G9, U10, and G11. (C and D) The base triple induced by BIV-2 mimics that induced by BIV Tat (cyan) (PDB ID: 1MNB; Puglisi et al., 1995); Arg-5, Arg-3, Ile-10 mimic Arg-70, Arg-73, Ile-79, respectively. (E) Solvent exposed Arg-9 and Arg-11 make electrostatic interactions with backbone phosphates at A21 and G22.

4. Discovery of peptide mimetics targeting HIV-TAR

The long-term focus of targeting BIV-TAR project discussed in Section 3 was to discover peptide inhibitors targeting Human Immunodeficiency Virus (HIV) TAR, located at positions 17–45 within the 5′-UTR of the HIV genome (Fig. 5A). Its trinucleotide bulge (U23, C24, U25) nucleotides are functionally analogous to BIV TAR U10, G11, and U12 in that they are necessary for Tat recognition and have been shown to reduce replication rates in mutational studies (Dingwall et al., 1990; Keen, Gait, & Karn, 1996; Roy, Delling, & Rosen, 1990). We used the SAR developed for the BIV-TAR–BIV-2 complex to generate a large positional scanning peptide library, synthesized both to optimize binding affinity, but also to

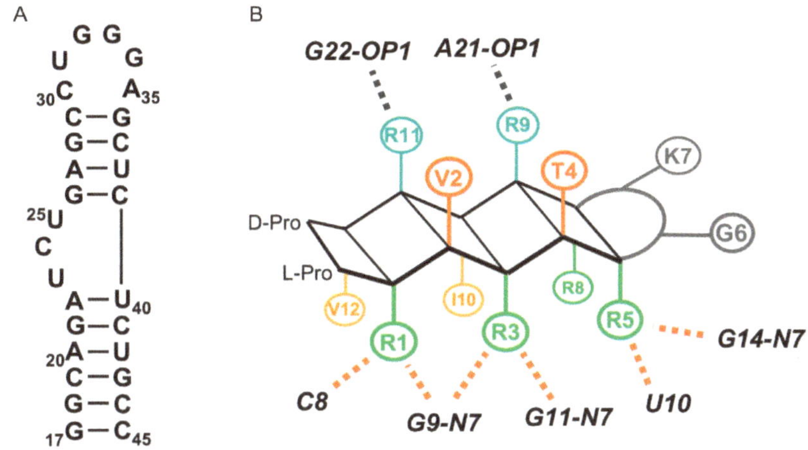

Fig. 5 (A) Secondary structure of HIV-TAR. (B) Cartoon depicting SAR for BIV-2 binding to BIV-TAR. Orange dotted lines indicate hydrogen bonding and/or π-stacking interactions, gray dotted indicate electrostatic interactions (Athanassiou et al., 2007).

discover new HIV-TAR specific peptides (Athanassiou et al., 2007). Our aim of discovering a peptide which bound to HIV-TAR strongly enough for structure determination was successful (Davidson et al., 2009), which was used for developing an HIV-TAR specific SAR utilized in future rounds of optimization.

4.1 Discovery of low nM peptide ligands for HIV-TAR using a large positional scanning peptide library

The structure and initial SAR for the BIV-2 peptide led to the following summary of its requirement for high affinity binding:

- Positions 1, 3, 5, 8 are buried against the RNA and exhibit ASM sandwiching with C8, G9, G11, U10, and G14. Arginines are likely favored at these positions because of their multiple hydrogen bond acceptors and charged protonation state. Hydrogen bonding was observed between purine N7's and guanosine O2's, along with cation-π and π-π interactions with various bases. Arg-5 contributes significantly to triple basepair formation based on its location in the binding pocket.
- Positions 6 and 7 are non-template turn inducing residues required to be Type I′ and Type II' turn inducers, since these turns favor β-hairpin formation.
- Positions 2 and 4 are solvent exposed and make negligible interactions with the RNA and minor contacts with neighboring side-chains.

- Positions 9 and 11 are solvent exposed but make interactions with A21 and G22 that support the RNA conformation through induced fit.
- Positions 10 and 12 are deeply buried into the RNA interface and form hydrophobic interactions. Ile-10 contributes significantly to triple base-pair formation by packing against A13 and U24.

These interactions are represented on a schematic of a β-hairpin (Fig. 5B) and provide a design rational for the positional scanning peptide library (L-1–L-86) that led to the discovery of HIV-TAR binding peptides L-22, L-50 and L-51 (Athanassiou et al., 2007). Since optimization was carried out for both BIV and HIV-TAR, patterns discussed are explained as observations for both RNAs unless otherwise specified.

4.1.1 Ala scanning

In a pre-panel screen, each position except the D-Pro/L-Pro template was changed to alanine to identify which residues contributed to binding. Remarkably, every alanine substitution reduced binding significantly, with the exception of solvent exposed Val-2 and Thr-4 where K_ds only increased twofold relative to BIV-2 (Athanassiou et al., 2007). Changing the buried Ile-10 and Val-12 to alanine also led to low μM binding, likely because of a reduced number of hydrophobic interactions near the base triple. Decreased binding affinity for each residue was also proposed to be correlated to the loss of cross-strand packing with arginine methylene chains. This observation invited us to consider peptide intramolecular stability through side-chain to side-chain interactions when making design decisions.

4.1.2 ASM-forming residues: Arg-1, Arg-3, Arg-5, Arg-8

Arg-1, Arg-3, Arg-5 essentially mimic BIV Tat Arg-77, Arg-73 and Arg-70, while Arg-8 interacts with both A21 associated base and phosphate in the apical loop. Arginines at positions 1,3,5, and 8 were substituted with both charged (Lys, L-ornithine Orn); polar (Asn, Gln, L-citrulline Cit); and aromatic (Tyr, Trp) residues. Arginines are less likely to make non-specific phosphate salt bridges than lysine and its analogue ornithine (Rohs et al., 2010). Therefore, when affinity is retained by mutating an arginine to lysine, the interaction is potentially non-specific. Positions Arg-1, Arg-3, Arg-5 all show no binding when replaced with Lys, but substitution of position 8 to lysine and ornithine only increased its K_d to 2.5 μM suggesting some degree of non-specific electrostatic backbone interaction (Athanassiou et al., 2007).

4.1.3 Non-template beta-hairpin turn inducers: Gly-6, Lys-7

The D-Pro/L-Pro template and Gly-6, Lys-7 form Type II' turns. Their identity is important (Chen, Lin, Lee, Jan, & Chan, 2001); BIV-0 bound poorly likely due to conformational instability from dis-favorable turn chemistry (Athanassiou et al., 2004). This hypothesis was tested with L-22 and L-51 by switching Gly-6 and Lys-7 to residues favoring a Type I′ turn. This resulted in a significant increase in affinity and selectivity for HIV-TAR over BIV-TAR, and equally important, produced cleaner NMR data for structure determination even when compared to peptides with higher binding affinities (Athanassiou et al., 2007). Lys-7 was also changed to polar, aromatic, and charged residues to improve binding, yet lysine appeared to be the preferred amino acid here. How turn sequences can be used to modulate specificity to other RNAs is an ongoing subject of investigation.

4.1.4 Solvent exposed Val-2, Thr-4

Solvent exposed residues were expected to make negligible contributions to binding. Changing Val-2 and Thr-4 to hydrophobic (Thr, Leu); polar (Asn, Gln); and aromatic (Tyr, Trp) produced mixes results. Val-2 to Thr and Thr-4 to Gln, Val, Tyr increased binding by two- to threefold (as ordered by largest to smallest increase in affinity) (Athanassiou et al., 2007). Increase in affinity was speculated to be due to changes in peptide conformational stability or lower solvation free energy, but careful analysis of these unbound peptides would be needed to confirm this conjecture.

4.1.5 Apical loop binding residues Arg-9, Arg-11

Residues that might make contacts with the apical loop because of their location on the peptide structure are difficult to optimize. Structures tend to converge poorly in these regions due to the inherent apical loop flexibly and non-specific interactions with the phosphate backbone might drive any energetic contribution. Arg-9 and Arg-11 were substituted, similarly to Arg-1, Arg-3, Arg-5, and Arg-8, by changing arginine to lysine. For BIV-TAR, lysine and ornithine substitutions retained binding ($K_d = 150\,nM$) (Athanassiou et al., 2007). When the same peptides were assayed against HIV-TAR, changes at position 9 retained binding, yet changes at position 11 led to two- to threefold loss in binding affinity suggesting a greater and unexpected dependence on arginine at position 11.

4.1.6 Major groove buried hydrophobic residues: Ile-10 and Val-12

Inspection of the BIV-2 bound structure shows both of these hydrophobic residues are buried deep within the major groove and contribute a significant faction of the buried interfacial surface. There residues were altered to optimize hydrophobic packing, and for Ile-10, increased stabilization for the triple base pair. Both were substituted to hydrophobic (Leu, cyclohexylalanine); aromatic (Phe, Tyr); and polar (Asn, Gln). Cyclohexylalanine was used as a control due to its hydrophobicity, yet sterically unfavorable size resulting in low μM affinity at each position. All hydrophobic changes retained or increased K_d two- to threefold, likely due to their increased size leading to a looser fit of the peptide in the major groove (Athanassiou et al., 2007). Val-12 to Ile-12 increased binding affinity ($K_d = 100\,nM$), however, and was incorporated in conjunction with changes at other positions (e.g., in L-22). Surprisingly, Ile-10 could be changed to different hydrophobic residues for BIV-TAR without significant changes in binding, yet for HIV-TAR substitutions to other residue led to loss of binding, revealing that isoleucine is optimal at this position for HIV-TAR recognition.

4.1.7 Combining changes at multiple positions

Residue changes that resulted in increased binding affinity were introduced in combination aiming for additive effects. Table 2 provides examples of peptides with several single residue changes (L-42, L-46, L-51, L-59) with improved binding affinity against BIV-TAR. L-72 was designed to incorporate both modifications at position 11 (Arg-to-Orn, $K_d = 100\,nM$) and

Table 2 Combination of sequences and K_ds of peptides from the L-1–L-86 peptide where single residue changes improved binding affinity relative to BIV-2 (L-42, L-46, L-51, L-59). Single residue changes were incorporated together as peptides with multiple position changes (L-72, L-76, L-78) (Athanassiou et al., 2007).

	1	2	3	4	5	6	7	8	9	10	11	12	K_d BIV-TAR (nM)
BIV-2	R	V	R	T	R	G	K	R	R	I	R	V	150
L-42	R	V	R	T	R	G	K	R	R	I	R	I	100
L-46	R	V	R	T	R	G	K	R	R	I	O	V	100
L-51	R	I	R	T	R	G	K	R	K	I	R	V	50
L-59	R	V	R	Q	R	G	K	R	R	I	R	V	50
L-72	R	V	R	T	R	G	K	R	R	I	O	I	50
L-76	R	I	R	T	R	G	K	R	R	I	O	I	20
L-78	R	I	R	Q	R	G	K	R	R	I	O	I	50

12 (Val-to-Ile, $K_d = 100\,nM$), which together increased binding twofold ($K_d = 50\,nM$). L-76 incorporated Thr at position 2 for L-72, decreasing the K_d further ($K_d = 20\,nM$) (Athanassiou et al., 2007). Not all changes were additive, however, since L-78 incorporated a glutamate into position 4 and reduced binding affinity 2.5-fold ($K_d = 50\,nM$). L-76 and L-78 together demonstrates incorporation of multiple positional changes for additivities effects requires careful step-wise design as well, since the effects are not simply additive.

4.2 HIV-TAR–L-22 structure and SAR

Peptides L-22, L-50 and L-51 were discovered as the highest-affinity peptides targeting HIV-TAR (Davidson et al., 2009) (Table 3). L-51 also demonstrated 10-fold specificity, with binding affinities of 5 nM for HIV-TAR and 50 nM for BIV-TAR, but competition assays showed that L-22 inhibited a 37-mer Tat-derived linear peptide better, since its IC_{50} was only two- to threefold above its K_d (Davidson et al., 2009). This promising inhibition data, combined with cleaner NMR data compared to L-50 and L-51, prompted us to determine the structure of HIV-TAR bound to L-22, even if L-50 binds most strongly and L-51 is specific. It is commonly the case that the most potent or highest affinity ligand is chosen as the optimal lead, but we strongly believe that the optimal starting structure is that which provides for structure determination, because of more facile optimization. Not only because of the obvious advantage of the availability of a structure, but also because the quality of the NMR data indicates that the peptide binds in a unique rigid orientation, which is not the case for L-50 and L-51.

The structure of the HIV-TAR–L-22 complex was determined using a very high-quality NMR dataset (Davidson et al., 2009). Positioning of L-22 is similar to BIV-2 on BIV TAR, with the D-Pro/L-Pro template facing

Table 3 Sequences and K_ds for peptides with highest-affinity targeting HIV-TAR from the L-1–L-86 library (Davidson et al., 2009).

	1	2	3	4	5	6	7	8	9	10	11	12	K_d (nM), HIV	K_d (nM), BIV
L-22	R	V	R	T	R	K	G	R	R	I	R	I	30	5
L-50	R	V	R	T	R	G	K	R	R	I	R	R	1	1
L-51	R	T	R	T	R	K	G	R	R	I	R	V	5	50

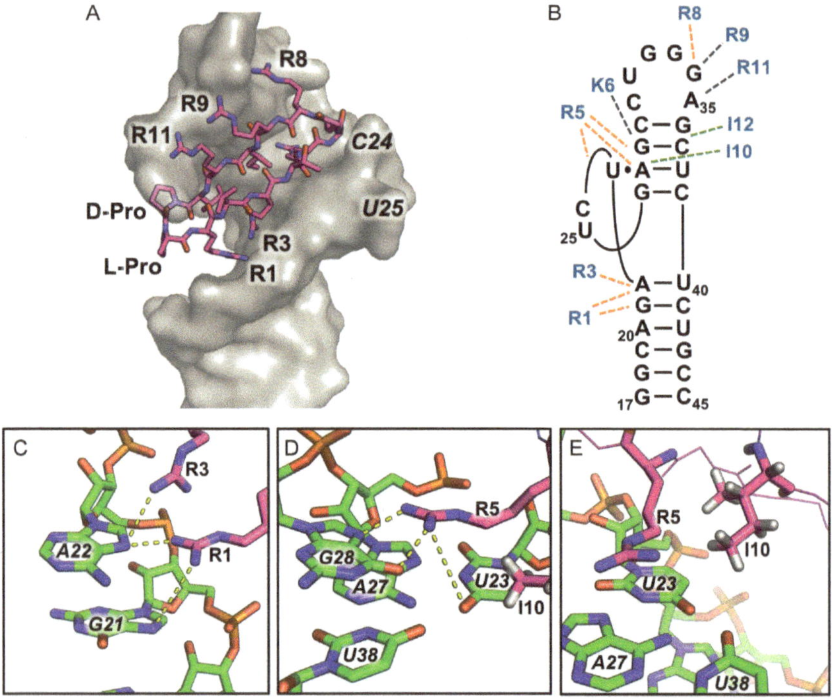

Fig. 6 (A) L-22 (magenta) bound HIV-TAR (gray). (PDB ID: 2KDQ; Davidson et al., 2009). (B) SAR for L-22 binding to HIV-TAR. Orange dotted lines indicate hydrogen bonding and/or π-stacking interactions, green dotted lines indicate hydrophobic interactions, gray dotted indicate electrostatic interactions. (C) Hydrogen bonds between Arg-1 with G21 N7 and Arg-3 with A22 N7 stabilize the stem-bulge interface. (D) Hydrogen bonds between R5 and U23 and G28 induce the U23-A27-U38 base triple. (E) Ile-10 contributes to base triple stabilization by hydrophobic interactions with the H5 and H6 side of U23.

down into the major groove and a base triple forming between U23, A27, and U38 (Fig. 6A). However, L-22 bound is buried even deeper into the major groove than BIV-2 into BIV TAR, sequestering ~50% of the peptide surface from solvent, including most of its hydrophobic side chains. L-22 SAR with HIV-TAR is as follows: residues 1, 3, 5, 8 make hydrogen bond and/or stacking interactions with the RNA; 10 and 12 make hydrophobic interactions; positions 9 and 11 bind through transient electrostatic interactions near the backbone of G34 and A35 (Fig. 6B). Arg-1 and Arg-3 stabilized the stem-bulge interface, with Arg-1 contacting the RNA backbone and making specific hydrogen bonds with G21 N7 and A22 N7 and Arg-3 with A22 (Fig. 6C). Both side chains are proximal to the phosphate backbone making electrostatic interactions as well. As for the BIV Tat and BIV-2 structures, Arg-5 and Ile-10 stabilize the HIV-TAR base triple, with Arg-5 stacking over the three nucleotides. Very high-quality NMR data provided reliable identification of hydrogen bonding between the Arg-5 guanidium group and G28 N7 and O2 (Fig. 6D). Ile-10 is packed deep into the groove and makes hydrophobic interactions with the H5 and H6 side of U23 (Fig. 6E). A hypothesis for why the KG turn was preferred over the GK turn in L-50 was a newly formed electrostatic interaction between Lys-6 and the phosphate backbone of G28. This long-distance interaction with the RNA was considered in future design and optimized in the later generation peptide JB-181 (Section 5.2).

5. Structure-based optimization of L-22 binding affinity and specificity

Further improvements in affinity and specificity were achieved by generating reliable SAR and the structure of the complex as a basis for design hypothesis. In this section, we will dissect these strategies by discussing both a blind alley (Davidson et al., 2011) and success (Shortridge et al., 2018).

5.1 KP-Z-41: Extending cyclic β-hairpin peptides to 15, 16, and 18-mers

Synthesis of the KP-Z peptide series sought to improve specificity and affinity by extending the peptide to fully engage the apical loop. Inspection of the 14-mer L-22 bound to HIV-TAR shows that several apical loop nucleotides make no interactions with it. Peptide libraries based on the L-22 sequence were designed to take advantage of these available interaction sites by extending the peptides to 15, 16 and 18 amino acids (Davidson et al., 2011). Essential structural requirements for specific recognition of HIV-TAR were observed based on the structure of KP-Z-41 bound to TAR, which revealed an unexpected position of the peptide distinct from our original design (Table 4).

We synthesized a medium sized cyclic β-hairpin peptide library (~25–50) of 15, 16, and 18 residues in length (Davidson et al., 2011). The L-22 sequence was conserved in these peptides, yet new amino acids were mainly based on our previously established understanding of buried and solvent exposed residues and model-based predictions of new possible contacts in the apical loop by extension of the peptide. Cross-strand disulfide linkages were also introduced to test whether increased rigidity from

Table 4 Sequences and K_ds for peptides with highest-affinity to HIV-TAR from the KP-Z library (Davidson et al., 2011).

	1	2	3	4	5	6	7	8	9	10	11	12	13	14	15	16	K_d (nM)
L-22	R	V	R	T	R	K	G	R	R	I	R	I					30
KP-Z-40	R	V	R	C	R	Q	R	K	G	R	R	A	C	I	R	I	25
KP-Z-41	R	V	R	C	R	Q	R	K	G	R	R	I	C	I	R	I	75

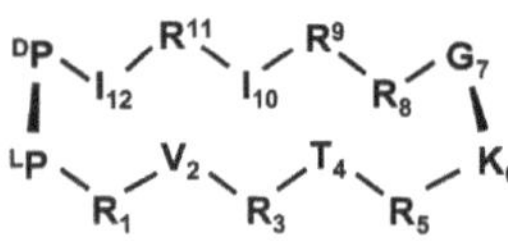
L-22

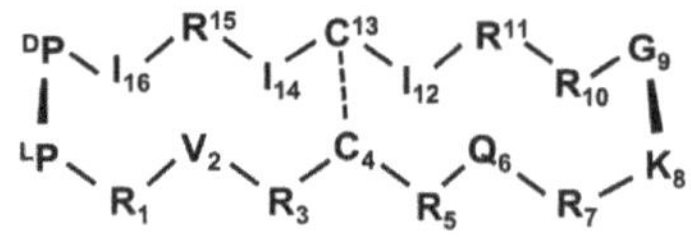
KP-Z-41

additional covalent linkage would enhance binding. 15 and 16 amino acid β-hairpins showed poor binding in all cases, perhaps not unexpectedly since these peptides should be unstably folded, whereas 18 amino acid β-hairpins fold similarly to the 14-mers, following the $n+4$ periodicity for 2:2 beta-hairpins (Griffiths-Jones et al., 1999). Binding affinity was enhanced for some 18-mer β-hairpins, improving binding relative to L-22, likely due to an additional arginine positioned on the major groove side of the peptide (positions 1, 3, 5, 7). KP-Z-40 displayed the highest binding affinity for HIV-TAR with a K_d in the pM range without excess tRNA present, and a 25 nM K_d even in the presence of 10,000-fold excess of tRNA (Table 4) (Davidson et al., 2011). However, quality of NMR data for bound KP-Z-40 was poor. We hypothesized that Ala-12 may be occupying the triple base pair pocket normally occupied by Ile-10. KP-Z-41 was therefore designed to have this position changed to an isoleucine and we observed an unexpected threefold drop in affinity, yet spectral quality improved allowing for structure determination to be executed (Table 4) (Davidson et al., 2011).

The structure for KP-Z-41 bound to HIV-TAR shows that the peptide is positioned completely opposite to our expectations and farther down into the major groove with the D-Pro/L-Pro template sticking out of the complex (Fig. 7). This sliding might be due to the requirement to form arginine-mediated interaction at or near the UGU bulge. Combined with the upside down BIV-2 mimicry of BIV Tat, this result exemplifies how deviations from a scaffold that might seem conservative, can induce drastic changes in binding position and orientation.

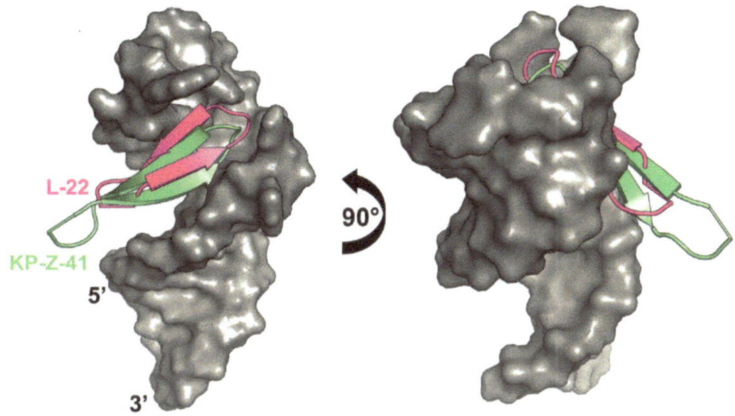

Fig. 7 Overlaid HIV-TAR-bound structures (gray) for L-22 (magenta) and KP-Z-41 (pale green) (PDB ID: 2KX5, Davidson et al., 2011). HIV-TAR backbone phosphates were used as anchor points for the superposition.

5.2 JB-181: Non-canonical amino acids

Peptides L-22, L-50 and L-51 bound to RNA potently, with low nM K_ds, but discriminated effectively only against structurally unrelated RNAs, such as tRNA present in all binding assays to reduce non-specific binding (Davidson et al., 2009). When comparing closely-related stem-loop structures such as HIV-TAR mutants or BIV-TAR, binding affinity only decreased by a few to at most 10-fold (Davidson et al., 2009). A new generation of peptides was rationally designed by inspection of the structure of L-22. Since the KP-Z series indicated that extension of the peptide to make additional specific contacts with the apical loop would not be straightforward or possible, new peptides were designed with the goal of allowing deeper penetration into the major groove by incorporating truncated, non-canonical side chains. Using this approach, JB-181 was discovered, a cyclic β-hairpin mimetic of Tat with low picomolar affinity for HIV-TAR and remarkable specificity.

A new library of ~200 peptides was synthesized based on the L-22 sequence, with single residue changes comprising natural L-amino acids, mixing chirality and non-canonical side chains (Table 5, top panel) (Shortridge et al., 2018). The inclusion of L-2,4-diaminobutryric (Dab) at position 1 (JB-59), and L-2-amino-4-guanidinobutyric acid (noR) (JB-82) at position 11 led to the greatest improvement in binding. When both changes were combined into a single peptide, JB-181, it produced pM binding ($K_d = 28 \pm 4$ pM, as measured by a fluorescence assay) and NMR data of unprecedented quality (Table 5)

Table 5 Top: Most effective non-canonical amino acids used in the JB peptide series. Bottom: Sequences and K_ds for peptides with the highest HIV-TAR affinities from the JB library (Shortridge et al., 2018).

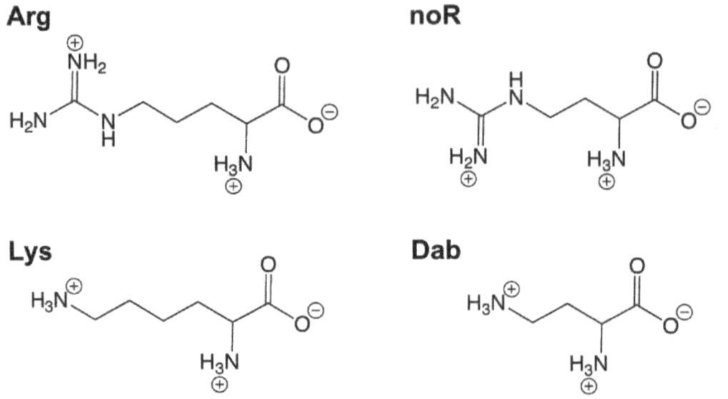

	1	2	3	4	5	6	7	8	9	10	11	12	K_d (nM)
L-22	R	V	R	T	R	K	G	R	R	I	R	I	30
JB-59	**Dab**	V	R	T	R	K	G	R	R	I	R	I	1
JB-82	R	V	R	T	R	K	G	R	R	I	**noR**	I	75
JB-181	**Dab**	V	R	T	R	K	G	R	R	I	**noR**	I	0.03
JB-190	Dab	V	R	T	R	**G**	**K**	R	R	I	noR	I	1.5

(Shortridge et al., 2018). Although the original β-hairpin mimetics were designed using the BIV structure as a model (Athanassiou et al., 2004; Puglisi et al., 1995), binding of JB-181 to BIV-TAR occurs at low nM affinity (Kd = 4 nM) and low μM binding affinity is observed toward the non-coding RNA 7SK-SL4 (Shortridge et al., 2018) that also binds to HIV-Tat, demonstrating that JB-181 has 10^2–10^6-fold discrimination against closely related structures.

The quality of the NMR data revealed new interactions responsible for the large increase in binding affinity. JB-181 binds in the same major groove pocket and orientation as the previous Tat-mimetics (Fig. 8A) (Shortridge et al., 2018). The backbone ψ and ϕ angles for JB-181 and L-22 were nearly identical with the exception of Lys-6, yet their positioning in the TAR binding pocket are different (Fig. 8B). The introduction of truncated side chains (Dab-1, noR-11) essentially allows the D-Pro/L-Pro template and neighboring residues to fit deeper into the major groove pocket. Inspection of the structure shows that substitution to an amine at positioning one maintains a salt bridge with the RNA backbone (2.4 Å from O1P of G21 and

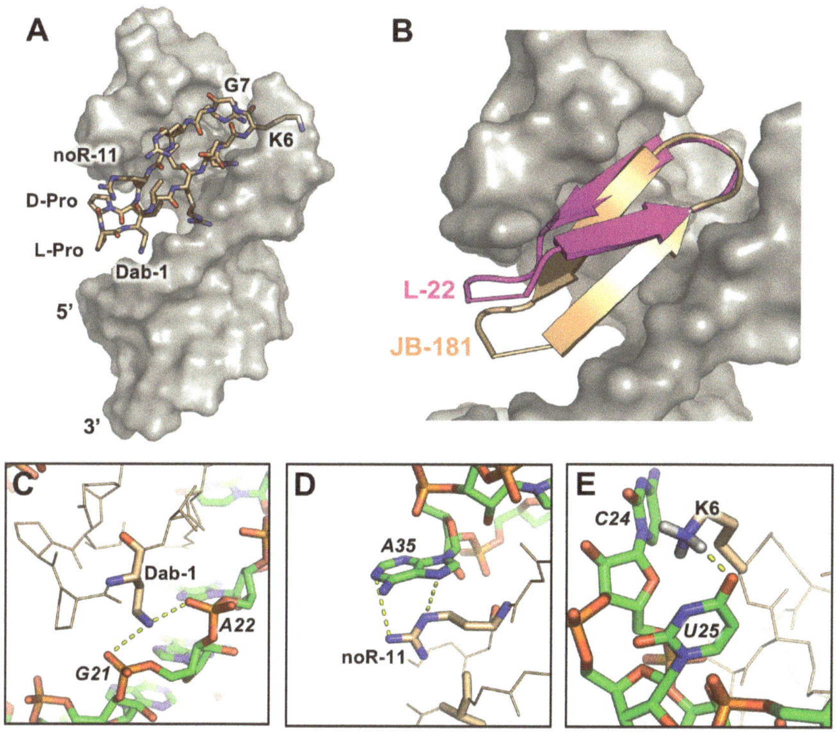

Fig. 8 (A) JB-181 (wheat) bound to HIV-TAR (gray) (PDB ID: 6D2U; Shortridge et al., 2018). (B) Overlaid of HIV-TAR-bound structures (gray) for L-22 (magenta) (PDB ID: 2KDQ; Davidson et al., 2009) and JB-181 (wheat). HIV-TAR backbone phosphates were used as anchor points for the superposition. (C) Dab-1 forms salt bridges with the G21 and A22 phosphates. (D) noR-11 makes a cation-π interaction with A35 favoring a *syn* conformation for the A35 base. (E) Change in peptide position favors formation of a novel hydrogen bond between Lys-6 and U25 O4.

3.6 Å from O1P of A22) (Fig. 8C), similar to Arg-1, but truncation of the side chain pulls the peptide closer to the major groove by about 2.4 Å and allows the KG turn to pitch upward with respect to the double helical axis. NoR-11 reduces affinity more than twofold relative to L-22 (JB-81, $K_d = 75$ pM), yet we detected more intermolecular NOEs from position noR-11 to A35, which now occupies the *syn* conformation allowing formation of a more favorable cation-π interaction with the new sidechain (Fig. 8D). Changes at positions 1 and 11 affected other parts of the structure as well. Deeper burial of the peptide favors improved Arg-3 and Arg-5 sandwiching between bases, increasing polar contacts and closer packing of Ile-10 against the base triple. Ile-10 is stabilized by many new NMR observables never observed by us in >20 years of investigation of multiple complexes of TAR such as a 2′-OH in the base triple protected from exchange with solvent. One of the most notable changes is a new hydrogen bond formed between the amine of Lys-6 and the O4 of U25 (Fig. 8E). This novel interaction is supported by mutational analysis such as swapping of the turn from KG to GK in JB-190 (Table 5), or mutating U25 to C25, which both decrease binding from low pM to low nM (Shortridge et al., 2018). Altogether, the remarkable structural and binding properties of JB-181 demonstrate that high affinity and specific RNA recognition of relatively simple RNA secondary structures, hairpin loops, can be achieved using a structure guided approach and macrocyclic β-hairpin scaffolds.

6. Targeting other RNAs using macrocyclic β-hairpin peptides

In this section, we show how the same class of peptides can be used to target RNAs other than TAR, namely, the Rev-RRE (Rev Response Element) interaction in HIV (Moehle et al., 2007) and pre-microRNA-21 (pre-miR-21) the precursor to the potent oncomir miRNA-21 (miR-21) (Shortridge et al., 2017). These additional studies demonstrate the versatility of cyclic β-hairpin scaffolds for targeting RNA.

6.1 Targeting the Rev-RRE interaction

The virally encoded Rev protein plays a critical role in viral replication by regulating the transport of unspliced and partially spliced viral RNA from the nucleus to the cytoplasm of infected cells (Emerman & Peden, 1989; Malim, Hauber, Le, Maizel, & Cullen, 1989). Rev functions by binding through an α-helical ARM to stem loop IIb (Fig. 9A) of the RRE a structured ~300 mRNA segment located within the Env coding region of HIV-1 (Mann et al., 1994). Following this high affinity binding event, approximately 10 additional Rev molecules oligomerize through protein–protein and protein–RNA interactions and coat the entire RRE (Mann et al., 1994). Because the Rev-RRE interaction is analogous to the TAR-Tat interaction in that it involves a high-affinity RNA binding protein that binds by a major groove recognizing ARM. We elected to target it with our BIV-TAR mimetics even if this interaction is distinct from the TAR-Tat interactions in that Rev binds the RRE major groove in an α-helical conformation (Fig. 9B) (Battiste et al., 1996). Similar to previous studies where β-hairpin scaffolds were used to mimic an α-helices and block HDM2 binding to

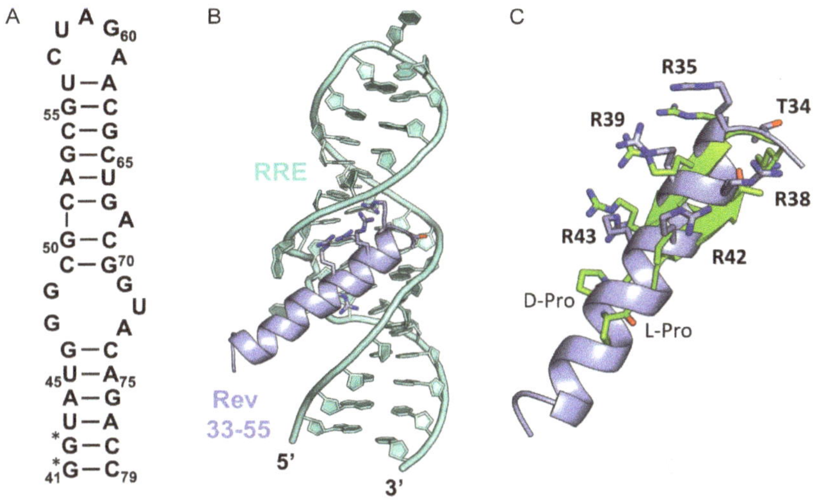

Fig. 9 (A) Secondary structure of stem loop IIb of the Rev Response Element (RRE). (*) indicated next to residues changed for increased T7 transcription efficiency. (B) α-helical Rev ARM (33–55, light blue) bound to the RRE major groove (light cyan) (PDB ID: 1ETF; Battiste et al., 1996). (C) A standard 2:2 β-hairpin mimetic (residues 1–12 with a D-Pro/L-Pro template) (green) superimposed onto the Rev residues critical for RRE binding demonstrating how β-hairpin scaffolds can be adapted to mimic similar α-helix sidechain positioning.

p53 (Fasan et al., 2004), we reasoned our mimetics could be adapted to position sidechains in similar orientations as in the Rev α-helix. Indeed, the β-hairpins can be modeled to show that arginines can be positioned on the peptide to moderately mimic the HIV Rev α-helical ARM (Fig. 9C).

We screened the small peptide library develop for BIV TAR (Table 1) for binding to the RRE. BIV-5, which had only a low μM affinity for HIV-TAR ($K_d = 1–2\,\mu M$) bound to the RRE much more strongly ($K_d = 0.3\,\mu M$) (Moehle et al., 2007). Interestingly, BIV-7 which showed no measurable affinity for TAR, bound with ~1 μM affinity as well, further suggesting that specificity can be found in arginine rich peptides if the basic side chains are presented on a rigid scaffold like a β-hairpin. BIV-5 showed effective inhibition potential in a competition-based assay as well, by displacing the Rev peptide with IC_{50} similar to its K_d (Moehle et al., 2007). Similar optimization as for BIV Tat-TAR was carried out using a medium-size library, which led to the discovery of R-27, a low nM RRE binding peptide ($K_d = 2\,nM$) that was 50-fold more selective for RRE compared to HIV-TAR ($K_d = 100\,nM$) (Moehle et al., 2007). This project demonstrated in a proof of principle manner that molecular interactions observed on the Rev α-helix could be successfully transplanted onto a β-hairpin scaffold by rational design, leading to rapid discovery of cyclic peptides that are more potent than the Rev ARM in binding to the RRE high affinity site.

6.2 Pre-microRNA 21

Lastly, we used this chemistry to develop peptides targeting the potent oncogene miR-21. MicroRNAs (miRNA) are short non-coding RNAs that silence gene translation through sequence specific binding in most cases at complementary 3′ UTR loci inducing mRNA degradation or translational suppression (Carthew & Sontheimer, 2009). Their dysregulation is also often observed in human disease (Esteller, 2011; Ling et al., 2013). The biogenesis of miRNAs begins with expression of a transcript sometimes thousands of nucleotides long called the pri-microRNA (pri-miRNA). The primary transcript undergoes two exonucleolytic cleavage reactions, first in the nucleus by Drosha-DGCR8 to generate the pre-miRNA of 55–75 nucleotides, and second in the cytoplasm by Dicer-TRBP to form the functional mature miRNA of 21–24 nucleotides (Carthew & Sontheimer, 2009). We focused on miR-21 which is upregulated in many human cancers (Asangani et al., 2008; Kjaer-Frifeldt et al., 2012; Meng et al., 2007; Ribas & Lupold, 2010; Yan et al., 2008) and sought to discover a peptide inhibitor of miR-21 maturation by screening our cyclic peptide libraries to find an initial scaffold to subsequently construct pre-miR-21 specific cyclic β-hairpins.

Targeting pre-miR-21 was unlike previous projects because of the lack of structural information and because pre-miR-21, our target, has a large apical loop, that NMR data indicate is largely unstructured and flexible (Fig. 10A). The Dicer cleavage sites are located above the A29 bulge and at the $i + 1$ cross-strand phosphate backbone, between G45 and C46. Therefore, we focused our screen on this apical section of the structure to avoid discovering peptides that bind to other sections of this RNA. Thus, we truncated the full length pre-miR-21 sequence by removing the bottom half of the helix and used this RNA in both NMR experiments and EMSAs when screening our peptide library (Shortridge et al., 2017). Although EMSAs are cumbersome, they allow direct visualization of complexes and distinguish multimeric and non-specific complexes. We prioritized monomeric hits from screening rather than just affinity because of much easier follow-up to optimize affinity and specificity.

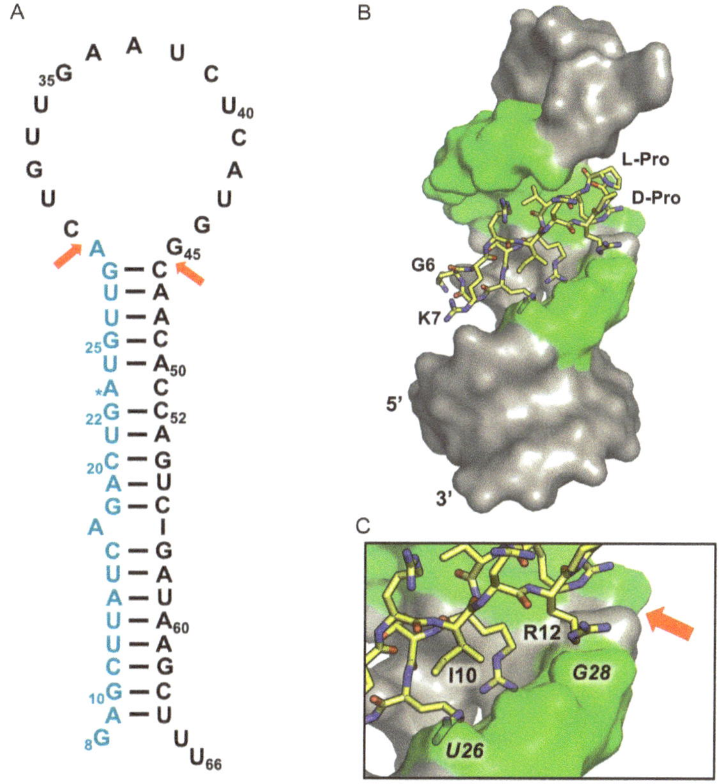

Fig. 10 (A) Secondary structure of pre-miRNA-21 (pre-miR-21; numbers based on miRBASE v21 (Kozomara & Griffiths-Jones, 2014), the Dicer cleavage sites are indicated by red arrows to generate the ~22 nucleotide mature miR-21 (blue). (B) Structure of pre-miR-21-L50 complex with nucleotides that make intermolecular NOEs highlighted in green (PDB ID: 5UZZ; Shortridge et al., 2017). (C) Both Ile-10 and Arg-12 contact the RNA near one of the Dicer cleavage sites.

From the initial screening, L-50 was identified to bind to pre-miR-21 with nanomolar binding affinity even in the presence excess tRNA ($K_d = 200\,nM$) (Shortridge et al., 2017), the only peptide to bind with sub-μM affinity. Follow-up studies revealed that L-50 does not bind the A29 deletion mutant, suggesting L-50 bound near the Dicer cleavage site and it out-competed streptomycin ($K_d = 100\,nM$) (Bose et al., 2012). These results collectively provided enough evidence to warrant structure determination.

Despite poor definition at this early stage of discovery, the structure revealed critical sites of contacts and contributed an initial SAR. Inspection of the lowest energy structure reveals that L-50 binds in the opposite orientation compared to TAR binding predecessors, in that the D-Pro/L-Pro template is facing upward into the apical loop (Fig. 10B), and charged residues appear to make non-specific contacts with the phosphate backbone (Shortridge et al., 2017). Interestingly, L-50 is the only peptide in our screening library with an arginine at position 12, located near the phosphate backbone of G28, suggesting Arg-12 is necessary for pre-miR-21 binding; both Ile-10 and Arg-12 contact the RNA near the Dicer cleavage site (Fig. 10C). When we tested for activity in an in vitro dicer assay using in-house purified Dicer-TRBP, we observed anomalous Dicer products; cellular assays showed modest but specific inhibition of miR-21 production in HEK293 cell (Shortridge et al., 2017). These preliminary results motivated efforts currently ongoing to optimize L-50 for better pre-miR selectivity, and to incorporate new turn chemistry that deviates from the D-Pro-L-Pro template. Current unpublished results have identified peptides with new turn chemistry and sequences that bind much more strongly to pre-miR-21 compared to TAR, inhibit Dicer processing in vitro, and reduce miR-21 levels in cells with greatly improved potency compared to the original hit.

7. Summary

Peptides provide an appealing solution to inhibiting large intermolecular interfaces, such as protein-protein or protein-RNA interactions, that are less amenable to small molecule inhibitors than traditional enzyme targets. While less pharmaceutically attractive than small molecules, they represent a growing area of pharmacology (Fosgerau & Hoffmann, 2015; Pelay-Gimeno et al., 2015; Teixido & Giralt, 2017) that can sometimes lead to bioactive molecules and even, in favorable cases, oral bioavailability (Renukuntla, Dutt, Patel, Boddu, & Mitra, 2013). Past efforts to use peptides to target RNA focused on peptoids (Hamy et al., 1997) or other backbone modification (Niu, Jones, Wu, Varani, & Cai, 2011) invariably led to highly flexible linear structure from which specific binding activity was very difficult to generate, or very short fragments whose binding activity was too weak to observe any biochemical or cellular effects. We show here that cyclic β-hairpin mimics can yield very potent (pM K_d) binding with strong specificity, and that these can be rapidly identified from a combination of conventional library-generated SAR and structure-based design. Success can be found either by mimicking the known interfaces of RNA-binding protein, or by an agnostic process of library screening to identify imperfect hits that can then be rapidly optimized, due to the relative ease of peptide chemistry compared to small molecules. Conformationally restrained peptides are particularly important for RNA, which often have flexible structure. For this reason, peptide folding and binding must be established. Critical information on the characteristics of the intermolecular interface that transcend and extend binding constants can be obtained by NMR. Namely, whether the complex is formed at a unique rigid binding site on the RNA, as opposed to sliding on the structure as observed, for example, with aminoglycosides like streptomycin (Bose et al., 2012; Maiti, Nauwelaerts, & Herdewijn, 2012; Shortridge & Varani, 2015). Inspection of the structure of the complex, which can be generated by NMR even for initial relatively low affinity hits if the peptide is rigid, reveals how the peptide is positioned and identifies the location of key sidechains identified through library-based SAR.

While peptides do not generally match small molecules with regards to in vivo application, they are nevertheless more conductive to optimization of pharmacological properties than oligonucleotide analogues (Juliano, 2016; Stein & Goel, 2011; Wang, Lon, Lee, Burckart, & Pisetsky, 2015). Combined with in vitro evolution, our methods rapidly generate leads that can be optimized to generate peptides with improved pharmacological potential through the introduction of side chain and backbone modifications, leading even, in favorable cases, to oral bio-availability (Adessi & Soto, 2002; Boehm et al., 2017; Chatterjee, Gilon, Hoffman, & Kessler, 2008; Witt, Gillespie, Huber, Egleton, & Davis, 2001). We consider this family of peptides to be very useful alternatives to small molecules in discovering RNA targeting bioactive structures and provide herein protocols for their discovery, characterization and optimization of their binding activity to RNA.

Acknowledgments

Work in the Varani group was supported by NIH grants 1 RO1 GM-110569, 1 RO1 GM-103934 and R35 GM-126942 to G.V.

References

Aboul-ela, F., Karn, J., & Varani, G. (1996). Structure of HIV-1 TAR RNA in the absence of ligands reveals a novel conformation of the trinucleotide bulge. *Nucleic Acids Research*, *24*(20), 3974–3981.

Adessi, C., & Soto, C. (2002). Converting a peptide into a drug: Strategies to improve stability and bioavailability. *Current Medicinal Chemistry*, *9*(9), 963–978.

Afshar, M., Prescott, C. D., & Varani, G. (1999). Structure-based and combinatorial search for new RNA-binding drugs. *Current Opinion in Biotechnology*, *10*(1), 59–63.

Asangani, I. A., Rasheed, S. A., Nikolova, D. A., Leupold, J. H., Colburn, N. H., Post, S., et al. (2008). MicroRNA-21 (miR-21) post-transcriptionally downregulates tumor suppressor Pdcd4 and stimulates invasion, intravasation and metastasis in colorectal cancer. *Oncogene*, *27*, 2128–2136.

Athanassiou, Z., Dias, R., Moehle, K., Dobson, N., Varani, G., & Robinson, J. (2004). Structural mimicry of retroviral tat proteins by constrained beta-hairpin peptidomimetics: Ligands with high affinity and selectivity for viral TAR RNA regulatory elements. *Journal of the American Chemical Society*, *126*, 6906–6913.

Athanassiou, Z., Patora, K., Dias, R. L., Moehle, K., Robinson, J. A., & Varani, G. (2007). Structure-guided peptidomimetic design leads to nanomolar beta-hairpin inhibitors of the Tat-TAR interaction of bovine immunodeficiency virus. *Biochemistry*, *46*, 741–751.

Avan, I., Hall, C. D., & Katritzky, A. R. (2014). Peptidomimetics via modifications of amino acids and peptide bonds. *Chemical Society Reviews*, *43*, 3575–3594.

Bardaro, M. F., Jr., Shajani, Z., Patora-Komisarska, K., Robinson, J. A., & Varani, G. (2009). How binding of small molecule and peptide ligands to HIV-1 TAR alters the RNA motional landscape. *Nucleic Acids Research*, *37*(5), 1529–1540.

Battiste, J. L., Mao, H., Rao, N. S., Tan, R., Muhandiram, D. R., Kay, L. E., et al. (1996). Alpha helix-RNA major groove recognition in an HIV-1 rev peptide-RRE RNA complex. *Science*, *273*, 1547–1551.

Bean, J. W., Kopple, K. D., & Peishoff, C. E. (1992). Conformational analysis of cyclic hexapeptides containing the D-Pro-L-Pro sequence to fix β-turn positions. *Journal of the American Chemical Society*, *114*, 5328–5334.

Boehm, M., Beaumont, K., Jones, R., Kalgutkar, A. S., Zhang, L., Atkinson, K., et al. (2017). Discovery of potent and orally bioavailable macrocyclic peptide–peptoid hybrid CXCR7 modulators. *Journal of Medicinal Chemistry*, *60*(23), 9653–9663.

Borkar, A. N., Bardaro, M. F., Camilloni, C., Aprile, F. A., Varani, G., & Vendruscolo, M. (2016). Structure of a low-population binding intermediate in protein-RNA recognition. *Proceedings of the National Academy of Sciences of the United States of America*, *113*(26), 7171–7176.

Bose, D., Jayaraj, G., Suryawanshi, H., Agarwala, P., Pore, S. K., Banerjee, R., et al. (2012). The tuberculosis drug streptomycin as a potential cancer therapeutic: Inhibition of miR-21 function by directly targeting its precursor. *Angewandte Chemie, International Edition*, *51*, 1019–1023.

Bower, J., Drysdale, M., Hebdon, R., Jordan, A., Lentzen, G., Matassova, N., et al. (2003). Structure-based design of agents targeting the bacterial ribosome. *Bioorganic & Medicinal Chemistry Letters*, *13*(15), 2455–2458.

Burnett, J. C., & Rossi, J. J. (2012). RNA-based therapeutics: Current progress and future prospects. *Chemistry & Biology*, *19*(1), 60–71.

Carthew, R. W., & Sontheimer, E. J. (2009). Origins and mechanisms of miRNAs and siRNAs. *Cell*, *136*, 642–655.

Chatterjee, J., Gilon, C., Hoffman, A., & Kessler, H. (2008). N-methylation of peptides: A new perspective in medicinal chemistry. *Accounts of Chemical Research*, *41*(10), 1331–1342.

Chen, L., & Frankel, A. D. (1995). A peptide interaction in the major groove of RNA resembles protein interactions in the minor groove of DNA. *Proceedings of the National Academy of Sciences of the United States of America*, *92*, 5077–5081.

Chen, P., Lin, C., Lee, C., Jan, H., & Chan, S. I. (2001). Effects of turn residues in directing the formation of the beta-sheet and in the stability of the beta-sheet. *Protein Science*, *10*, 1794–1800.

Cléry, A., Blatter, M., & Allain, F. H. (2008). RNA recognition motifs: Boring? Not quite. *Current Opinion in Structural Biology*, *18*(3), 290–298.

Cooper, T. A., Wan, L., & Dreyfuss, G. (2009). RNA and disease. *Cell*, *136*, 777–793.

Davidson, A., Leeper, T. C., Athanassiou, Z., Patora-Komisarska, K., Karn, J., Robinson, J. A., et al. (2009). Simultaneous recognition of HIV-1 TAR RNA bulge and loop sequences by cyclic peptide mimics of Tat protein. *Proceedings of the National Academy of Sciences of the United States of America, 106*(29), 11931–11936.
Davidson, A., Patora-Komisarska, K., Robinson, J. A., & Varani, G. (2011). Essential structural requirements for specific recognition of HIV TAR RNA by peptide mimetics of Tat protein. *Nucleic Acids Research, 39*(1), 248–256.
Davis, B., Afshar, M., Varani, G., Murchie, A. I., Karn, J., Lentzen, G., et al. (2004). Rational design of inhibitors of HIV-1 TAR RNA through the stabilisation of electrostatic "hot spots" *Journal of Molecular Biology, 336*(2), 343–356.
Dingwall, C., Ernberg, I., Gait, M. J., Green, S. M., Heaphy, S., Karn, J., et al. (1990). HIV-1 tat protein stimulates transcription by binding to a U-rich bulge in the stem of the TAR RNA structure. *The EMBO Journal, 9*(12), 4145–4153.
Emerman, M., & Peden, K. (1989). The rev gene product of the human immunodeficiency virus affects envelope-specific RNA localization. *Cell, 57*, 1155–1165.
Esteller, M. (2011). Non-coding RNAs in human disease. *Nature Reviews Genetics, 12*, 861–874.
Fasan, R., Dias, R. L. A., Moehle, K., Zerbe, O., Vrijbloed, J. W., Obrecht, D., et al. (2004). Using a beta-hairpin to mimic an alpha-helix: Cyclic peptidomimetic inhibitors of the p53–HDM2 protein–protein interaction. *Angewandte Chemie, International Edition, 43*, 2109–2112.
Favre, M., Moehle, K., Jiang, L., Pfeiffer, B., & Robinson, J. A. (1999). Structural mimicry of canonical conformations in antibody hypervariable loops using cyclic peptides containing a heterochiral diproline template. *Journal of the American Chemical Society, 121*, 2679–2685.
Fischer, U., Huber, J., Boelens, W. C., Mattaj, I. W., & Luhrmann, R. (1995). The HIV-1 Rev activation domain is a nuclear export signal that accesses an export pathway used by specific cellular RNAs. *Cell, 82*, 475–483.
Fosgerau, K., & Hoffmann, T. (2015). Peptide therapeutics: Current status and future directions. *Drug Discovery Today, 20*(1), 122–128.
Fujii, N., Oishi, S., Hiramatsu, K., Araki, T., Ueda, S., Tamamura, H., et al. (2003). Molecular-size reduction of a potent CXCR4-chemokine antagonist using orthogonal combination of conformation-and sequence-based libraries. *Angewandte Chemie International Edition, 42*(28), 3251–3253.
Gallego, J., & Varani, G. (2001). Targeting RNA with small-molecule drugs: Therapeutic promise and chemical challenges. *Accounts of Chemical Research, 34*, 836–843.
Gonda, M. A., Luther, D. G., Fong, S. E., & Tobin, G. J. (1994). Bovine immunodeficiency virus: Molecular biology and virus-host interactions. *Virus Research, 32*, 155–181.
Griffiths-Jones, S. R., Maynard, A. J., & Searle, M. S. (1999). Dissecting the stability of a beta-hairpin peptide that folds in water: NMR and molecular dynamics analysis of the beta-turn and beta-strand contributions to folding. *Journal of Molecular Biology, 292*(5), 1051–1069.
Hamy, F., Felder, E. R., Heizmann, G., Lazdins, J., Aboul-ela, F., Varani, G., et al. (1997). An inhibitor of the Tat/TAR RNA interaction that effectively suppresses HIV-1 replication. *Proceedings of the National Academy of Sciences of the United States of America, 94*, 3548–3553.
Howe, J. A., Wang, H., Fischmann, T. O., Balibar, C. J., Xiao, L., Galgoci, A. M., et al. (2015). Selective small-molecule inhibition of an RNA structural element. *Nature, 526*(7575), 672–677.
Jiang, L., Moehle, K., Dhanapal, B., Obrecht, D., & Robinson, J. A. (2000). Combinatorial biomimetic chemistry: Parallel synthesis of a small library of β-hairpin mimetics based on loop III from human platelet-derived growth factor B. *Helvetica Chimica Acta, 83*, 3097–3112.

Jones, S., Daley, D. T. A., Luscombe, N. M., Berman, H. M., & Thornton, J. M. (2001). Protein-RNA interactions: A structural analysis. *Nucleic Acids Research, 29*(4), 943–954.

Jones, K. A., & Peterlin, B. M. (1994). Bovine immunodeficiency virus: Molecular biology and virus-host interactions. *Annual Review of Biochemistry, 63*, 717–743.

Juliano, R. L. (2016). The delivery of therapeutic oligonucleotides. *Nucleic Acids Research, 44*(14), 6518–6548.

Keen, N. J., Gait, M. J., & Karn, J. (1996). Human immunodeficiency virus type-1 Tat is an integral component of the activated transcription-elongation complex. *Proceedings of the National Academy of Sciences of the United States of America, 93*, 2505–2510.

Kirk, S. R., Luedtke, N. W., & Tor, Y. (2000). Neomycin–acridine conjugate: A potent inhibitor of Rev-RRE binding. *Journal of the American Chemical Society, 122*, 980–981.

Kjaer-Frifeldt, S., Hansen, T. F., Nielsen, B. S., Joergensen, S., Lindebjerg, J., Soerensen, F. B., et al. (2012). The prognostic importance of miR-21 in stage II colon cancer: A population-based study. *British Journal of Cancer, 107*, 1169–1174.

Kozomara, A., & Griffiths-Jones, S. (2014). miRBase: Annotating high confidence microRNAs using deep sequencing data. *Nucleic Acids Research, 42*, D68–D73.

Lättig-Tünnemann, G., Prinz, M., Hoffmann, D., Behlke, J., Palm-Apergi, C., Morano, I., et al. (2011). Backbone rigidity and static presentation of guanidinium groups increases cellular uptake of arginine-rich cell-penetrating peptides. *Nature Communications, 2*, 453.

Leeper, T., Athanassiou, Z., Dias, R., Robinson, J. A., & Varani, G. (2005). TAR RNA recognition by a cyclic peptidomimetic of Tat protein. *Biochemistry, 44*, 12362–12372.

Leulliot, N., & Varani, G. (2001). Current topics in RNA-protein recognition: Control of specificity and biological function through induced fit and conformational capture. *Biochemistry, 40*(27), 7947–7956.

Ling, H., Fabbri, M., & Calin, G. A. (2013). MicroRNAs and other non-coding RNAs as targets for anticancer drug development. *Nature Reviews Drug Discovery, 12*, 847–865.

Lunde, B. M., Moore, C., & Varani, G. (2007). RNA-binding proteins: Modular design for efficient function. *Nature Reviews. Molecular Cell Biology, 8*(6), 479–490.

Madan, B., Seo, S. Y., & Lee, S. G. (2014). Structural and sequence features of two residue turns in beta-hairpins. *Proteins, 82*(9), 1721–1733.

Maiti, M., Nauwelaerts, K., & Herdewijn, P. (2012). Pre-microRNA binding aminoglycosides and antitumor drugs as inhibitors of Dicer catalyzed microRNA processing. *Bioorganic & Medicinal Chemistry Letters, 22*(4), 1709–1711.

Malim, M. H., Hauber, J., Le, S. Y., Maizel, J. V., & Cullen, B. R. (1989). The HIV-1 rev trans-activator acts through a structured target sequence to activate nuclear export of unspliced viral mRNA. *Nature, 338*, 254–257.

Mann, D. A., Mikaelian, I., Zemmel, R. W., Green, S. M., Lowe, A. D., Kimura, T., et al. (1994). Co-operative rev binding to stem I of the rev-response element modulates human immunodeficiency virus type-1 late gene expression. *Journal of Molecular Biology, 241*, 193–207.

Mason, J. M. (2010). Design and development of peptides and peptide mimetics as antagonists for therapeutic intervention. *Future Medicinal Chemistry, 2*(12), 1813–1822.

Meng, F., Henson, R., Wehbe-Janek, H., Ghoshal, K., Jacob, S. T., & Patel, T. (2007). MicroRNA-21 regulates expression of the PTEN tumor suppressor gene in human hepatocellular cancer. *Gastroenterology, 133*, 647–658.

Moehle, K., Athanassiou, Z., Patora, K., Davidson, A., Varani, G., & Robinson, J. A. (2007). Design of beta-hairpin peptidomimetics that inhibit binding of alpha-helical HIV-1 rev protein to the rev response element RNA. *Angewandte Chemie, International Edition, 46*, 9101–9104.

Muppidi, A., Doi, K., Edwardraja, S., Drake, E. J., Gulick, A. M., Wang, H. G., et al. (2012). Rational design of proteolytically stable, cell-permeable peptide-based selective Mcl-1 inhibitors. *Journal of the American Chemical Society, 134*(36), 14734–14737.

Murchie, A. I., Davis, B., Isel, C., Afshar, M., Drysdale, M. J., Bower, J., et al. (2004). Structure-based drug design targeting an inactive RNA conformation: Exploiting the flexibility of HIV-1 TAR RNA. *Journal of Molecular Biology*, *336*(3), 625–638.

Nair, C. M., Vijayan, M., Venkatachalapathi, Y. V., & Balaram, P. (1979). X-ray crystal structure of pivaloyl-D-Pro-L-Pro-L-Ala-N-methylamide; observation of a consecutive β-turn conformation. *Journal of the Chemical Society, Chemical Communications*, 1183–1184.

Niu, Y., Jones, A., Wu, H., Varani, G., & Cai, J. (2011). γ-AApeptides bind to RNA by mimicking RNA-binding proteins. *Organic & Biomolecular Chemistry*, *9*, 6604–6609.

Palacino, J., Swalley, S. E., Song, C., Cheung, A. K., Shu, L., Zhang, X., et al. (2015). SMN2 splice modulators enhance U1–pre-mRNA association and rescue SMA mice. *Nature Chemical Biology*, *11*(7), 511.

Pelay-Gimeno, M., Glas, A., Koch, O., & Grossmann, T. N. (2015). Structure-based design of inhibitors of protein-protein interactions: Mimicking peptide binding epitopes. *Angewandte Chemie (International Ed. in English)*, *54*(31), 8896–8927.

Puglisi, J. D., Chen, L., Blanchard, S., & Frankel, A. D. (1995). Solution structure of a bovine immunodeficiency virus Tat-TAR peptide-RNA complex. *Science*, *270*, 1200–1203.

Ratni, H., Karp, G. M., Weetall, M., Naryshkin, N. A., Paushkin, S. V., Chen, K. S., et al. (2016). Specific correction of alternative survival motor neuron 2 splicing by small molecules: Discovery of a potential novel medicine to treat spinal muscular atrophy. *Journal of Medicinal Chemistry*, *59*(13), 6086–6100.

Renukuntla, J., Dutt, A., Patel, A., Boddu, S. H. S., & Mitra, A. K. (2013). Approaches for enhancing oral bioavailability of peptides and proteins. *International Journal of Pharmaceutics*, *447*(1–2), 75–93.

Ribas, J., & Lupold, S. E. (2010). The transcriptional regulation of miR-21, its multiple transcripts, and their implication in prostate cancer. *Cell Cycle*, *9*, 923–929.

Robinson, J. A. (2000). The design, synthesis and conformation of some new β-hairpin mimetics: Novel reagents for drug and vaccine discovery. *Synlett*, *2000*, 429–441.

Robinson, J. A. (2008). Beta-hairpin peptidomimetics: Design, structures and biological activities. *Accounts of Chemical Research*, *41*(10), 1278–1288.

Rohs, R., Jin, X., West, S. M., Joshi, R., Honig, B., & Mann, R. S. (2010). Origins of specificity in protein-DNA recognition. *Annual Review of Biochemistry*, *79*, 233–269.

Rosenström, U., Sköld, C., Lindeberg, G., Botros, M., Nyberg, F., Karlén, A., et al. (2006). Design, synthesis, and incorporation of a β-turn mimetic in angiotensin II forming novel pseudopeptides with affinity for AT1 and AT2 receptors. *Journal of Medicinal Chemistry*, *49*(20), 6133–6137.

Roy, S., Delling, U., & Rosen, C. A. (1990). A bulge structure in HIV-I TAR RNA is required for Tat binding and Tat-mediated trans-activation. *Genes & Development*, *4*, 1365–1373.

Shortridge, M. D., & Varani, G. (2015). Structure based approaches for targeting non-coding RNAs with small molecules. *Current Opinion in Structural Biology*, *30*, 79–88.

Shortridge, M. D., Walker, M. J., Pavelitz, T., Chen, Y., Yang, W., & Varani, G. (2017). A macrocyclic peptide ligand binds the oncogenic microRNA-21 precursor and suppresses Dicer processing. *ACS Chemical Biology*, *12*, 1611–1620.

Shortridge, M. D., Wille, P. T., Jones, A. N., Davidson, A., Bogdanovic, J., Arts, E., et al. (2018). An ultra-high affinity ligand of HIV-1 TAR reveals the RNA structure recognized by P-TEFb. *Nucleic Acids Research*. https://doi.org/10.1093/nar/gky1197. [Epub ahead of print].

Sibanda, B. L., Blundell, T. L., & Thornton, J. M. J. (1989). Conformation of β-hairpins in protein structures: A systematic classification with applications to modelling by homology, electron density fitting and protein engineering. *Journal of Molecular Biology*, *206*, 759–777.

Sibanda, B. L., & Thornton, J. M. (1985). β-Hairpin families in globular proteins. *Nature*, *316*, 170–174.
Stein, C. A., & Goel, S. (2011). Therapeutic oligonucleotides: The road not taken. *Clinical Cancer Research*, *17*(20), 6369–6372.
Teixido, M., & Giralt, E. (2017). Jumping hurdles: Peptides able to overcome biological barriers. *Accounts of Chemical Research*, *50*(8), 1847–1854.
Varani, G., Aboula-ela, F., & Allain, F. H. (1996). NMR investigation of RNA structure. *Progress in Nuclear Magnetic Resonance Spectroscopy*, *29*, 51–127.
Wang, J., Lon, H. K., Lee, S. L., Burckart, G. J., & Pisetsky, D. S. (2015). Oligonucleotide-based drug development: Considerations for clinical pharmacology and immunogenicity. *Therapeutic Innovation & Regulatory Science*, *49*(6), 861–868.
Warner, K. D., Hajdin, C. E., & Weeks, K. M. (2018). Principles for targeting RNA with drug-like small molecules. *Nature Reviews Drug Discovery*, (8), 547–558.
Wei, P., Garber, M. E., Fang, S.-M., Fischer, W. H., & Jones, K. A. (1998). A novel CDK9-associated C-type cyclin interacts directly with HIV-1 Tat and mediates its high-affinity, loop-specific binding to TAR RNA. *Cell*, *92*, 451–462.
Werder, M., Hauser, H., Abele, S., & Seebach, D. (1999). β-Peptides as inhibitors of small-intestinal cholesterol and fat absorption. *Helvetica Chimica Acta*, *82*(10), 1774–1783.
Whitby, L. R., Ando, Y., Setola, V., Vogt, P. K., Roth, B. L., & Boger, D. L. (2011). Design, synthesis, and validation of a β-turn mimetic library targeting protein–protein and peptide–receptor interactions. *Journal of the American Chemical Society*, *133*(26), 10184–10194.
Williamson, J. R. (2000). Induced fit in RNA–protein recognition. *Nature Structural and Molecular Biology*, 7(10), 834.
Wilson, W. D., & Li, K. (2000). Targeting RNA with small molecules. *Current Medicinal Chemistry*, 7, 73–98.
Witt, K. A., Gillespie, T. J., Huber, J. D., Egleton, R. D., & Davis, T. P. (2001). Peptide drug modifications to enhance bioavailability and blood-brain barrier permeability. *Peptides*, *22*(12), 2329–2343.
Yan, L. X., Huang, X. F., Shao, Q., Huang, M. Y., Deng, L., Wu, Q. L., et al. (2008). MicroRNA miR-21 overexpression in human breast cancer is associated with advanced clinical stage, lymph node metastasis and patient poor prognosis. *RNA*, *14*, 2348–2360.
Ye, X., Kumar, R. A., & Patel, D. J. (1995). Molecular recognition in the bovine immunodeficiency virus Tat peptide-TAR RNA complex. *Chemistry & Biology*, *2*, 827–840.

Further reading

Mei, H.-Y., Cui, M., Heldsinger, A., Lemrow, S. M., Loo, J. A., Sannes-Lowery, K. A., et al. (1998). Inhibitors of protein-RNA complexation that target the RNA: Specific recognition of human immunodeficiency virus type 1 TAR RNA by small organic molecules. *Biochemistry*, *37*, 14204–14212.

CHAPTER SIXTEEN

Molecular recognition of HIV-1 RNAs with branched peptides

Ashley N. Peralta[a], Yumin Dai[a], Chringma Sherpa[b], Stuart F.J. Le Grice[b], Webster L. Santos[a,*]

[a]Department of Chemistry and Center for Drug Discovery, Virginia Tech, Blacksburg, VA, United States
[b]Basic Research Laboratory, National Cancer Institute, Frederick, MD, United States
*Corresponding author: e-mail address: santosw@vt.edu

Contents

Abstract

Targeting RNA offers the potential in many diseases of a therapeutic treatment. Due to its large surface area and ability to adopt different conformations, targeting RNA has proven challenging. Medium-sized branched peptides are of the size to competitively bind RNA while remaining cell permeable, stable *in vivo*, and non-toxic. Additionally, the ease in generating a large library followed by high-throughput screening provides a way to suggest a scaffold with high diversity that is capable of targeting the structure and sequence of RNA. The ability to select various types of amino acid modifications in the branched peptide allows for variable structures and interactions of the branched peptide but can result in too large a task if not approached properly. In this chapter, we discuss a strategy to selectively recognize RNAs of interest through high throughput screening of branched peptides, validation of hits and biophysical characterization, leading by example with our experience in targeting HIV-1 RNAs with branched peptides.

Methods in Enzymology, Volume 623
ISSN 0076-6879
https://doi.org/10.1016/bs.mie.2019.04.021

Abbreviations

2-AP	2-amino purine
ANP	3-amino-3-(2-nitrophenyl)propionic acid
ARM	arginine rich motif
BP	branched peptide
BSA	bovine serum albumin
EMSA	electrophoretic mobility shift assay
ITC	isothermal titration calorimetry
MTT	3-(4,5-dimethylthiazol-2-yl)-2,5-diphenyltetrazolium bromide
nt	nucleotide
NF H_2O	nuclease free water
PBS	phosphate buffer solution
PNA	peptide nucleic acid
RT	reverse transcriptase
RRE	Rev response element
SHAPE-MaP	selective 2′-hydroxyl acylation analyzed by primer extension and mutational profiling
SPPS	solid phase peptide synthesis
TAR	transactivation response element

1. Introduction

RNA is pertinent for many biological functions, from genetic transfer to cellular upkeep; this includes transcription, translation, splicing, and catalysis (Draper, 1995). Its critical functions make RNA an interesting drug target, while its unique secondary and tertiary structure offer opportunity for selective targeting (Zaman, Michiels, & van Boeckel, 2003). This is accomplished through designing constructs that achieve molecular recognition of RNA by exploiting the unique potential binding sites afforded by its secondary and tertiary structure. HIV-1 has many potential RNA targets, of which the transactivation response element (TAR) and Rev response element (RRE) RNAs are among the best characterized. Several strategies have been developed and unique agents have been designed to target these two important RNAs, including small molecules (Thomas & Hergenrother, 2008), aminoglycosides (Mei et al., 1997), bi-functional conjugates (Luedtke, Liu, & Tor, 2003), RRE decoys (Symensma, Baskerville, Yan, & Ellington, 1999), TAR decoys (Michienzi, Li, Zaia, & Rossi, 2002), antisense oligonucleotides (Prater, Saleh, Wear, & Miller, 2007), zinc fingers (Friesen & Darby, 2001), metallopeptide complexes (Joyner & Cowan, 2011), peptide nucleic acids (PNAs) (Kumagai, Takahashi, Hamasaki, Ueno, & Mihara, 2001), cyclic peptides (Chaloin et al., 2007; Lalonde et al., 2011), β-hairpin peptidomimetics (Davidson et al., 2009), and α-helical peptidomimetics (Hyun et al., 2007; Lee, Hyun, Kieft, & Yu, 2009; Lee, Hyun, Kim, & Yu, 2008; Mills, Daugherty, Frankel, & Guy, 2006). Unfortunately, there has been limited success, with the exception of antibiotics, in targeting RNA.

Obstacles in RNA drug development are due to many reasons. The large surface area, conformational dynamics, and high binding affinity of the endogenous substrates all contribute to the difficulty in selectively binding RNA (Davidson & McCray, 2011; Lu et al., 2011). These difficulties lead to considerations in designing a scaffold for a novel RNA therapeutic. As with all cellular therapeutics, the designed scaffold is required to exhibit cell permeability, no toxicity, and good *in vivo* stability. To address the difficulty in targeting RNA, the therapeutic also needs to demonstrate good binding affinity and selectivity. Additionally, a well-characterized RNA target will help immensely in the design process of the therapeutic.

As a point of focus on how to overcome the difficulties in targeting RNA, RRE RNA, a highly conserved HIV-1 RNA will be used as an example in this chapter. RRE binds the Rev protein and together they function to transfer unspliced and singly-spliced transcripts from the nucleus to the cytoplasm (Kjems, Brown, Chang, & Sharp, 1991). This process provides the genetic transcript and necessary proteins for a new virion to form. Due to structural rearrangements, 4- and 5-stem models have been proposed for RRE RNA, and common structural elements between the models have been identified (Rausch & Le Grice, 2015). The high-affinity binding site of RRE is the IIB stem loop (RRE IIB), which contains an internal loop and single nucleotide (nt) bulge, offering a unique architecture (Fig. 1) (Rausch & Le Grice, 2015). Rev binds RRE IIB with an arginine rich motif (ARM), an α-helical motif consisting of four base (3 Arg residues and 1 Asn residue) and six sugar-phosphate backbone (1 Thr and 6 Arg) contacts with RRE IIB (Battiste et al., 1996). Sequence, structure, and points of contact within the RNA should be considered in the design for a highly selective inhibitor.

Among agents developed to target the RRE, peptides have been a large area of focus. The use of α-helical peptidomimetics was previously investigated in order to mimic the α-helical character of the Rev ARM (Hyun et al., 2007; Lee et al., 2008; Lee et al., 2009; Mills et al., 2006). Although peptides with good binding affinity were produced, many did not proceed to the next stage of targeting due to a lack of selectivity. Problems with selectivity likely arose due to the commonality of RNA:α-helical interaction. With this, a search focused on differentially structured peptides and proved

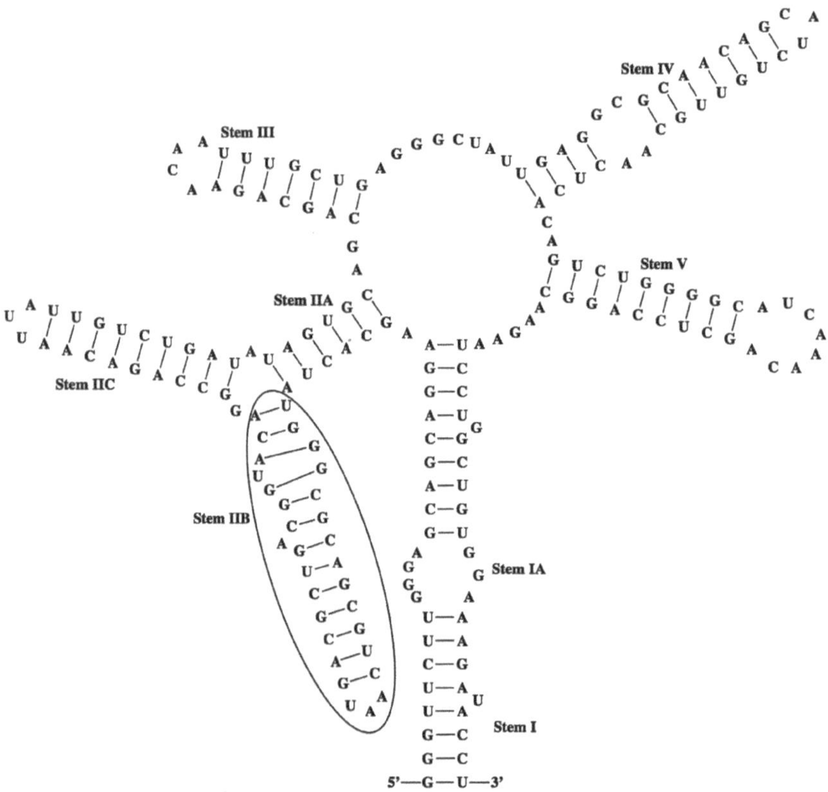

Fig. 1 The 5-stem secondary structure of RRE RNA with the high-affinity binding site, Stem IIB, highlighted with an oval. Stem I is truncated here to show the most pertinent parts of the structure. Stem IA is the secondary Rev binding site. The remainder of Stem I comprises seven additional stem loops and a nucleotide bulge.

that α-helicity was not necessary for targeting RRE (Harada, Martin, Tan, & Frankel, 1997; Zhang, Harada, Cho, Frankel, & Wemmer, 2001). Ultimately, selectivity, as well as *in vivo* stability of the peptides, remained an unsolved topic.

Building off this work, we proposed medium-sized branched peptides (BPs) for targeting RNA. Medium-sized BPs lend many desirable characteristics: their branch promotes *in vivo* stability, their size allows for cell permeability, and their natural constituents results in little to no toxicity (Brown, Madauss, Lian, Beck, Tolbert, & Rodgers, 2001; Bryson, Zhang, McLendon, Reineke, & Santos, 2012). Generating a BP library is, in itself, facile, with the main priority belonging to amino acid selection. In knowing which interactions the RNA favors, the appropriate amino acids can be selected to compete with the cognate protein. Additionally, an array of BPs with different amino acid sequences that give rise to unique structures and non-canonical binding is preferred. This introduces a potential novel set of interaction and conformation within each BP. By screening the BPs, the appropriate intermolecular interactions, as well as tertiary structure, can be investigated to determine what is needed for RNA recognition. The overall effect is creating a construct for RNA with multiple points of interaction and a bias toward the intended tertiary structure.

Herein, we describe the design in targeting HIV-1 RNA through molecular recognition. Our approach involves. (1) Peptide design: selecting amino acids with favorable properties for the RNA of interest; (2) Library generation for screening: selecting potential peptides with ideal properties and structures; (3) Validation: binding affinity and selectivity determination; and (4) Investigation of BP: RNA binding site. The steps outlined in this chapter are the rational design for generating and screening a BP library, validating the hits, and further assessing selected BPs. Thus, RNA recognition is achieved through selectivity filters, binding site identification, and identifying RNA local structural changes to accommodate interactions with branched peptides.

2. Design, synthesis, and high-throughput screening of a branched peptide library

In our discussion on the approach, the selection of amino acids and design of the BP library will be the focus. Emphasis will be placed on lessons learned in each generational library. However, solid phase peptide synthesis (SPPS), synthesis of the library, and synthesis of the individual peptides will not be discussed (Zhang, Bryson, Crumpton, Wynn, & Santos, 2013a). In addition, the design and protocol of the screening conditions for the library will be discussed.

2.1 Considerations in designing a branched peptide library

While using BPs to target RNA, our group has conducted a series of four libraries in the past 10 years. The first library targeted HIV-1 TAR RNA, a highly conserved sequence containing a hexanucleotide loop and a 3-nt bulge, which is located on the 5′ end of all nascent mRNA and is critical for viral transcription (Fig. 2) (Keen, Gait, & Karn, 1996; Rana & Jeang, 1999).

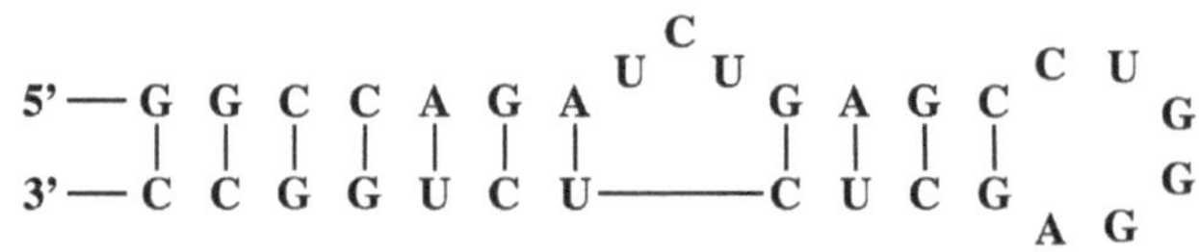

Fig. 2 Secondary structure of TAR RNA.

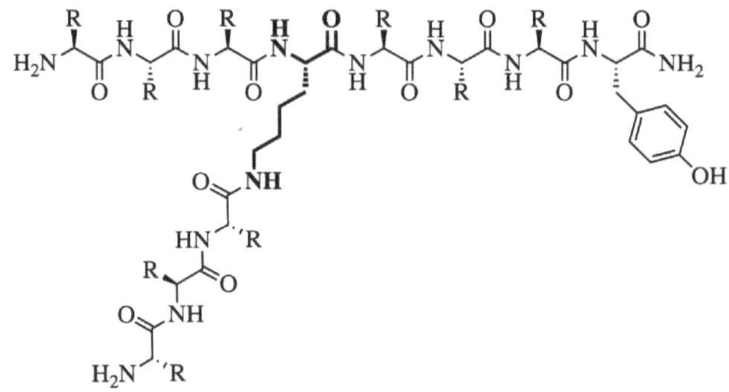

Fig. 3 Structure of a branched peptide. The lysine branching position is bolded. R groups are the variable amino acid positions dependent on what functional groups are chosen for the specific library. The first generation library did not include tyrosine as a spectroscopic handle at the C-terminus, but this was included in all following libraries.

The initial focus was on searching for favorable amino acid interactions for binding RNA. Natural amino acids capable of electrostatic interactions (Arg, His) with the negatively charged backbone, π-π stacking (Phe, Trp, Tyr) with the nucleobases, and hydrogen bonding (Ser, Asn, Gln) with the backbone or nucleobases were largely used (Bryson et al., 2012; Bryson, Zhang, Ray, & Santos, 2009). The most pivotal part of the BP was the branching unit itself, in which the following amino acids were coupled to the α- and ε-amino groups of lysine (Fig. 3). A concern with RNA is that a false positive can be generated by creating a peptide with too many postive charges, with the consequence that the peptide soley interacts with the negatively charged backbone. While the positive charge was found to be favored, it was not the sole determining factor of binding due to peptides with additional charged amino acids having worse binding affinities. The structure was also investigated by scrambling the sequence and systematically removing the C-terminus, α-amino lysine branch, and ε-amino lysine branch, all of which showed a decrease in binding affinity. This showcased that sequence and structure play an important role in binding RNA. In this library low micromolar binding to

TAR RNA was achieved, however, it became clear that specifity in targeting RNA would be the next challenge to overcome.

The next library selected RRE RNA, which includes more complicated secondary structures (stem loops, stem bulge, etc.), with emphasis on improving specificity (Zhang et al., 2013a; Zhang, Bryson, Crumpton, Wynn, & Santos, 2013b). The goal was to increase selectivity of the BPs for RNA over DNA and protein. Encouragingly, enhanced selectivity was acheived by incorporating a boronic acid unit, which can form a reversible covalent bond with the hydroxy group from the sugar backbone. Therapeutics with a boronic acid targeting an active site with a hydroxy group were already approved for use (Velcade®) or in development (now Kerydin®) (Kane, Bross, Farrell, & Pazdur, 2003; Rock et al., 2007). It was hypothesized that a boronic acid would favor binding to the RNA 2′-OH group, making the BP less likely to bind to DNA, which lacks a 2′-OH, and to proteins, which have considerably fewer –OH groups. Two different unnatural amino acids were originally developed, namely phenylalanine (F_{BPA}) and lysine (K_{BBA}) with a boronic acid modification, to interrogate a shorter or longer alkyl linker (Fig. 4). Of these, the longer boronic acid, K_{BBA}, was favored. Not only did the boronic acid play an important role in selectivity, but the BPs also showed high binding affinity to RRE RNA in the low micromolar range. In addition, tyrosine was added at the C-terminus as a quantitative handle for the BPs. Overall, it was reconfirmed that electrostatic interactions were still important, but not the determining factor, and that hydrophobic residues were not beneficial for binding with RNA. More importantly, the second library demonstrated a new mode of binding to increase selectivity toward RNA.

Besides selectivity, binding affinity was another key consideration in order to competitively bind the RNA target. In the third library, with 4096 members, an amino acid containing an acridinyl moiety was selected due to its ability to intercalate into RNA (Fig. 5) (Wynn et al., 2016).

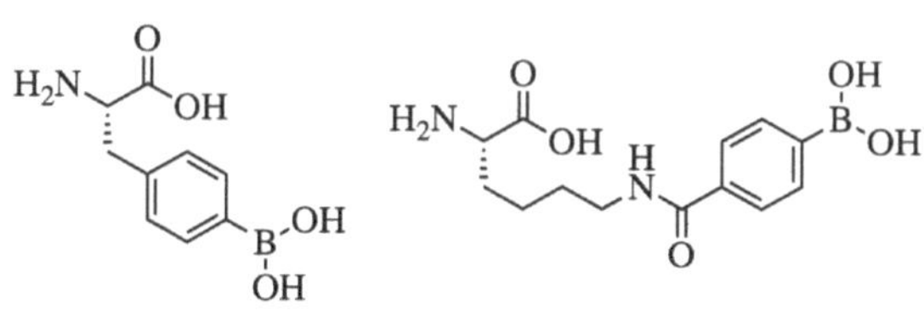

Fig. 4 Structures of boronic acid derivatives tested in the second generation branched peptide library. The longer length K_{BBA} was ultimately used in the remaining libraries.

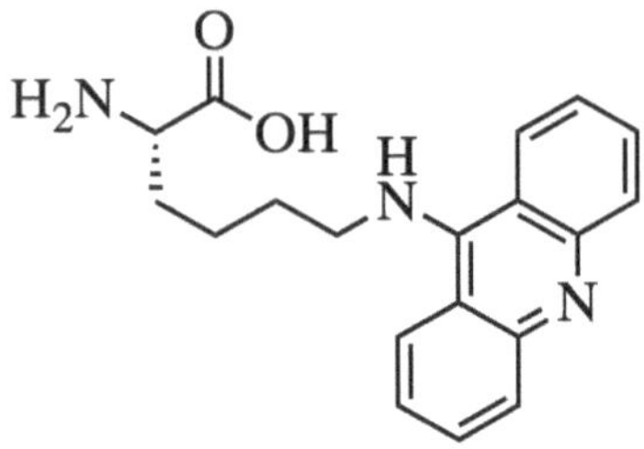

$\mathbf{K_{ACR}}$

Fig. 5 Structure of acridine derivative used in the third generation branched peptide library.

However, improved binding affinity was accompanied by loss of selectivity. It is possible that either acridine had trouble discriminating between RNA and DNA, or the lack of amino acid diversity hindered the peptide interactions and structure, as only one amino acid for each interaction (electrostatic, π-π, hydrogen bonding, hydrophobic) was selected, in addition to tyrosine for BP quantification. Acridine has been a common small molecule in targeting RRE RNA (Thomas & Hergenrother, 2008) and as such may again be worth attempting with a peptide, but within a library of greater diversity. Nevertheless, as acridine is a known intercalator, it will have difficultly differntiating RNA *versus* DNA, thus a peptide that can recognize structure and sequence of the RNA is highly needed.

The previous libraries have demonstrated success in achieving selectivity or affinity. Selectivity was increased with the use of a boronic acid in the second library, whereas the first and third libraries demonstrated the importance of amino acid diversity in a competitive binder. Combining all efforts, a fourth library was designed and constructed. This included six different unnatural amino acid possibilities in each position, resulting in a library with 46,656 BPs (Dai et al., 2018). The boronic acid moiety was still included for selectivity, and five other amino acids with each capable of multiple types of interactions (π-π stacking, electrostatic interactions, hydrogen bonding, or structural rigidity) were investigated, in addition to tyrosine for quantification (Fig. 6). Multiple BPs with sub-micromolar binding affinities were observed, while one of these selectively bound RRE IIB with a promising *in vitro* outcome.

In summary, the lesson to be learned from our many libraries is the importance in (1) designing selectivity toward RNA; and (2) including diverse amino acids to increase interactions with RNA, as well as structural diversity.

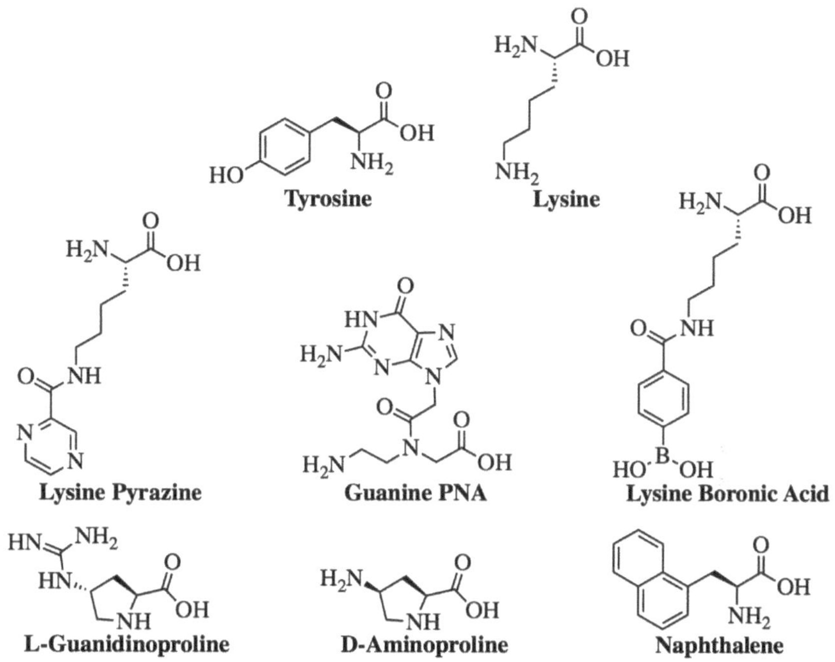

Fig. 6 The amino acid derivatives chosen for the fourth generation branched peptide library.

2.2 Screening and sequencing of a branched peptide library

A high-throughput assay can be employed to screen the BP library. The screening process was developed and modifed based on previous work with small, linear peptides (Hwang et al., 1999). The combinatorial libraries were constructed on the Tentagel resin linked to a photocleavable linker, 3-amino-3-(2-nitrophenyl)propionic acid (ANP), and each sequence was replicated three times. It was important to produce the library in triplicate as a small percentage of resins are prone to loss during the Kaiser test in SPPS. Additionally, during the screening process, replicated hits may be obtained (acheiving the same sequence more than once) which indicates a promising result.

In order to avoid false postives, the screening underwent two incubations: Block and Bind. The first removes promiscous binders, while the second determines any BPs that bind the target RNA. Since the BPs were biased with positive charges, bovine serum albumin (BSA), a negatively charged protein, was selected to remove peptides that would interact solely

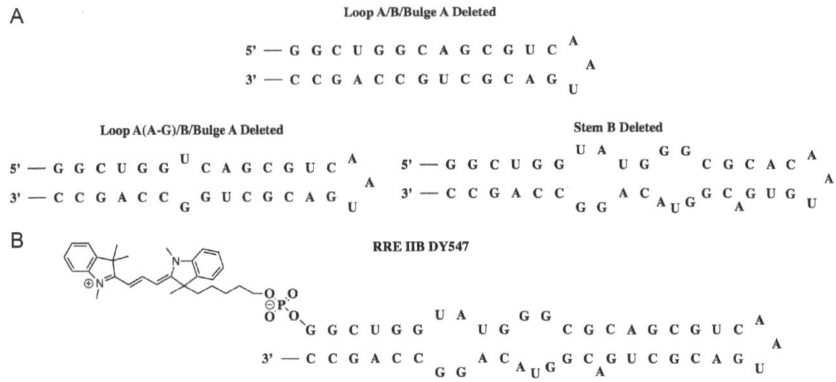

Fig. 7 (A) RRE IIB variants used in the incubation process. (B) RRE IIB with DY547 dye label.

through electrostatics. To remove any BPs that interacted non specifically with RNA, α-synuclein mRNA, a large RNA (340-nt) capable of adopting many conformers, was originally used. Later libraries opted for tRNA, a shorter RNA, but present in high concentrations within the cell. Additionally, to ensure BPs were targeting structure and not sequence, variants of the RNA target (Fig. 7A) were included in the incubation process (i.e., removal of stems, loops, bulges).

The second incubation was with the RNA of interest. The RNA was tagged with the DY547 fluorophore on the 5′-end (Fig. 7B). It is recommended to test a small batch to optimize incubation time for a given library. Rather than 24 h, as suggested from the literature (Hwang et al., 1999), incubation time was set to 3 h, in order to minimize the number of hits, but keep the most potent. Even at 3 h, 6 hits out of approximately 2000 beads (of a total of 12,288) were found. Therefore, the second incubation time was reduced to 1 h. In subsequent libraries, the second incubation time was 3 h for libraries 2 and 3, and 5 h for library 4.

The procedure for the most recent library screening is as follows.

2.2.1 Equipment

- Barnstead/Thermolyne Labquake rotisserie shaker

2.2.2 Buffers and reagents

See Table 1

See Table 2

Table 1 Non-specific binding incubation.

Component	Concentration (μM)	Volume (μL)	Ratio to DY547 RRE IIB
BSA	300	30	150:1
tRNA	400	150	1000:1
Loop A/B/bulge A deleted	35.4	1.69	1:1
Loop A(A-G)/B/bulge A deleted	25.8	2.33	1:1
Stem B deleted	7.77	7.72	1:1
1 × PBS (pH 7.4)		408.3	
Total volume		600	

Table 2 Dye incubation.

Component	Concentration (μM)	Volume (μL)	Final concentration (nM)
DY547 RRE IIB RNA	19	3.16	100
1 × PBS (pH 7.4)		597	
Total volume		600	

2.2.3 Procedure

1. Weigh out 100 mg of resin into a 1.5 mL sterile eppendorf tube.
2. Add components listed in Table 1.
3. With use of a rotisserie shaker, rotate eppendorf tube at room temperature for 3 h.
4. Wash beads five times with 1 × PBS
5. Add componets listed in Table 2.
6. Rotate eppendorf tube at 4 °C for 5 h
7. Wash beads five times with 1 × PBS

2.2.4 Notes

1. Concentrations and ratios of the non-specific binding molecules can vary for individual needs.
2. When washing the resins with 1 × PBS, invert the tube to mix. Allow resins to settle before removing 1 × PBS to avoid losing resins.
3. RNA variants should include removal of important structural features to ensure targeting of the structure and not sequence.

Once incubation was complete, screening was initiated with the use of fluorescence microscopy. Beads were distributed into a 96-well plate, together with a negative and positive control subjected to just the second incubation. Resin alone was used as a negative control and a resin with a BP of all lysine residues was used as a positive control. With lysine being a positively charged residue, a peptide consisting of lysine should heavily bind non-specifically to the negatively charged RNA backbone. Generally, the library is screened under the appropriate excitation wavelength of the dye and any beads that have fluorescence comparable to the positive control are selected (Fig. 8). Due to auto-fluorescence observed from the BPs with solely unnatural amino acids, confocal microscopy was selected for the fourth library in order to distinguish the fluorescence from the dye and auto-fluorescence of the library beads. To optimize screening conditions, the RNA incubated beads were compared to the resin alone and the detector sensitivity was adjusted in order to remove the autofluorescence.

Resins that bound the fluorescent dye were selected from the well plate and subjected to UV irradiation (365 nm, 1 h). This cleaved the peptides from the resin through the use of the photo-cleavable linker, ANP, releasing them into a 1:1 MeOH:H_2O solution. The resulting solution was then directly used for MALDI-TOF analysis where tandem mass spectrometry

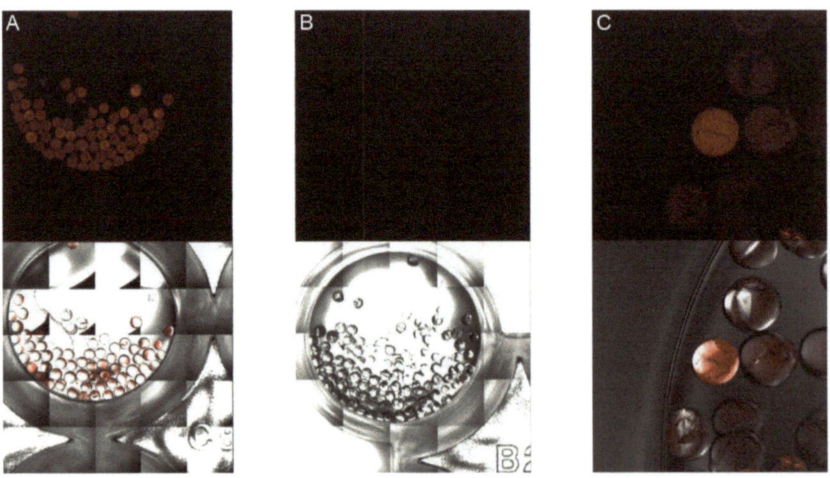

Fig. 8 Comparison of beads under confocal microscopy. Top: View of beads under UV excitation. Bottom: Overlay of beads under UV excitation and no excitation. (A) Positive control consisting of a polylysine BP on resin. (B) Negative control consisting of only the resin. (C) A hit bead among beads that are non-hits. The hit shows fluorescence upon UV excitation whereas the beads that are not hits remain non-fluorescent.

was utilized in order to sequence the BPs based on their consistent cleavage patterns (Crumpton, Zhang, & Santos, 2011). BPs whose sequences were found more than once, or contained specific repeated sequence motifs, were selected as leads. These lead BPs were then synthesized through SPPS and purified through HPLC before continuing to biophysical characterizations.

3. Determination of binding affinity and selectivity

Upon synthesis of the BPs, their bindng affinities to the RNA were validated. In this section, a flourescent titration assay was utilized. Compounds with sub-micromolar binding affinities were then subjected to *in vitro* studies to evaluate their toxicity and inhibition effect. In brief, a standard MTT assay and p24 ELISA assay were performed to evaluate the toxicity and *in vitro* inhibition, respectively. The p24 capsid protein is an important component of HIV-1 and is transcribed by one of the mRNAs that the Rev/RRE complex transports to the nucleus. When a BP successfully binds to RRE the transcript for p24 cannot be transported to the nucleus and a decrease in p24 concentration is seen. Branched peptides that showed no toxicity and inhibition of p24 formation were further tested in a selectivity assay before evaluation of RNA binding.

3.1 Binding affinity assay

Methods to determine binding affinity of the BP to RNA included a dot blot assay, electrophoretic mobility shift assay (EMSA), and isothermal calorimetry (ITC). Initial libraries relied on a dot blot assay for binding affinity. However, by the third library, select BPs would stick to the nitrocellulose which resulted in inconsistent results. For EMSA, there were issues with many of the peptide: RNA complexes entering the gel, rendering the assay unreliable. ITC, while a useful technique, requires a large amount of RNA sample. Consequently, a validated fluorescence assay was employed, offering a quick, cost efficient, and realiable procedure. The fluorescent nucleobase, 2-amino purine (2-AP) (Millar, 1996), was incorporated into the target RNA (Fig. 9) at a

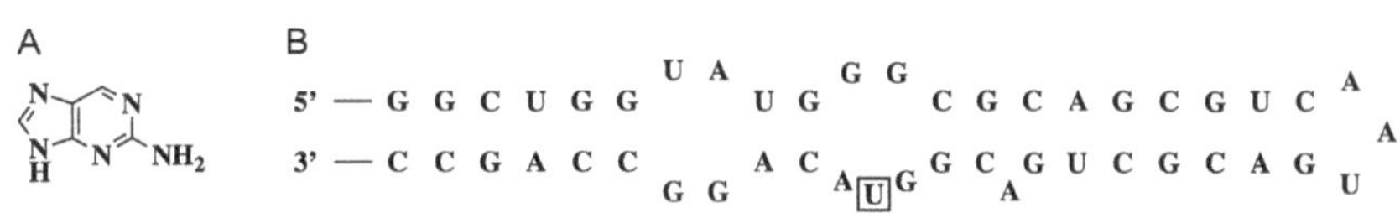

Fig. 9 (A) Structure of 2-AP. (B) RRE IIB secondary structure, the nucleotide replaced by 2-AP is highlighted with a box.

non-base-paired position of the sequence, so that fluorescence was not quenched. This fluorophore was previously incorporated into RRE and shown to be quenched upon binding of Rev (Lacourciere, Stivers, & Marino, 2000). For simiplicity of transcribing the RNA, the high-affinity binding site, RRE IIB, was used in place of the full length RRE RNA. In solution, 2-AP emits a fluorescent signal which is quenched upon peptide binding. By normalizing the fluorescence and plotting the values against the concentration of the peptide, the binding affinity can be determined.

3.1.1 Equipment

- Block heater
- Varian Cary Eclipse fluorescence spectrophotometer using a xenon flash lamp with a thermoelectrically controlled cell holder
- Quartz cell

3.1.2 Buffers and reagents

- 0.2 μm sterile-filtered 1 × PBS—10 mM potassium phosphate, 100 mM KCl, 1 mM $MgCl_2$, 20 mM NaCl, pH 7.0
- 2-AP RNA (IDT)—freeze dry 3 μL (10 μM stock) in a sterile eppendorf tube. Store in −20 °C freezer until ready for use.
- BP—freeze dry 20 μL (200 μM stock). Store in −20 °C freezer until ready for use.

3.1.3 Procedure

1. Add 300 μL of 1 × PBS to the freeze dried 2-AP substituted RNA (0.1 μM working solution).
2. Heat at 95 °C for 3 min (to denature).
3. Cool for 10 min on ice (to refold the RNA).
4. Add 200 μL of the 0.1 μM working solution to the BP (20 μM) and mix by pipette.
5. Include a blank of 80 μL of 1 × PBS.
6. Add 80 μL of the 2-AP RNA to the cleaned quartz cuvette. Obtain fluorescence spectra twice at 10 min, to ensure a stable signal. This is the 100% value.
7. Titrate 2-AP/BP solution in at concentrations from 0 to 20 μM. Obtain the fluorescence spectra twice at 10 min of each titration, until the fluorescence signal is quenched.
8. Analyze binding data using the Hill equation (Eq. 1) with Kaleidagraph (Synergy Software). In this equation, b and y are the fluorescence

emission intensities of the RNA in the absence and presence of peptide, respectively; m is the fluorescence emission intensity of the RNA in the presence of an infinite drug concentration; R and x are the total concentrations of the RNA and peptide, respectively; K is the K_d value of the peptide binding to the RNA; n is the apparent cooperativity.

$$y = b + (m - b) * \left(\frac{1}{1 + \left(\frac{K}{x} \right)^n} \right) \quad (1)$$

3.1.4 Notes

1. Allow the fluorometer to warm up before use.
2. Adding 2-AP RNA to the BP ensures that the concentration of the 2-AP RNA remains consistent throughout the titration.
3. Titrations were carried out in a 50 μL quartz fluorometer cuvette (1 cm path length) with a minimum working volume of 50 μL.
4. 2-AP RNA is fluorescent at 372 nm, the samples were excited at 310 nm and spectra were collected from 340 to 450 nm. The excitation slit width and emission slit width are 10 nm. The fluorescence intensity at 372 nm was used for binding data.
5. Each experiment was performed in at least duplicate with error bars representing the standard deviation calculated over the replicates.
6. The absorption and emission spectra of peptides should be taken to ensure that fluorescence quenching or other fluorescence interference do occur.

3.2 Selectivity assay

The BPs that had good binding affinity, no toxicity, and showed promising *in vitro* activity were selected for more stringent tests. A number of assays could be used to determine the selectivity of the BP. Ultimately, an assay similar to the binding affinity assay was used, with a modification made to the incubation process to screen for selectivity. Selecting potential non-selective targets for the peptides is crucial during the experimental design; tRNA and variants of the RNA, as well as the DNA transcript of the RNA, were used. The DNA was particularly important for accessing whether the peptides targeted structure over sequence.

3.2.1 Equipment

- Equipment listed in Section 3.1.1

3.2.2 Buffers and reagents

- 1 × PBS and BP—Prepared as described in Section 3.1.2
- 2-AP RNA (IDT)—Freeze dry 3 μL (10 μM stock) with the appropriate volume of the off target molecule (tRNA, DNA, or variant RNA, all stock concentrations vary) in a sterile eppendorf tube. Store in −20 °C freezer until ready for use.

3.2.3 Procedure

1. Add 300 μL of 1 × PBS to the freeze dried 2-AP RNA containing the off target molecule (0.1 μM working solution with 1 μM, 10 × excess, of off target molecule).
2. Carry out steps 2–8 as described in Section 3.1.3

3.2.4 Notes

1. Notes from Section 3.1.4 apply.

The same concept is applied with the selectivity assays, however, a non-fluorescently labeled variant RNA (or DNA) is added. As such, if the peptide is not selective for the fluorescent probe it can bind to the off target molecule instead. This results in a much higher concentration of peptide needed to bind the fluorescent probe and cause the signal to decrease, meaning the binding affinity will increase. For a molecule to be considered selective the K_d should increase no more than twofold, or no more than fivefold to be partially selective.

4. Determination of RNA binding sites

Further analysis is needed to ensure that hit BPs target structured RNA. Two complimentary approaches were used for this purpose. A common approach for assessing RNA binding is an RNase protection assay. The RNA binding site can be envisioned by incubation with the peptide and subjecting it to various RNases. The limitation to this assay was that only a fragment of the RNA, the high affinity-binding site RRE IIB, was used. To further validate binding of the RNA, selective 2′-hydroxyl acylation analyzed by primer extension and mutational profiling (SHAPE-MaP) analysis was invoked which uses the full-length wild type RNA to assess binding of the peptide.

4.1 RNase protection assay

Below is described the protocol for the RNase protection assay. RNA must first be dephosphorylated then phosphorylated with ^{32}P-ATP, resulting in a 5′-^{32}P labeled RNA. This is done through standard procedures, as indicated by the provider of the enzymes for these experiments. The RNase is performed as described below.

4.1.1 Equipment

- Vortex
- Centrifuge
- Block heater
- Sequencing gel electrophoresis apparatus
- Gel dryer
- Phosphor screen
- Phosphor imaging system

4.1.2 Buffers and reagents

- BP—Aliquot 10 μL (200 μM stock) into sterile eppendorf tube. Prepare 20 μM solution at 10 μL (from 200 μM stock). Prepare 2 μM solution at 10 μL (from 20 μM solution).
- Alkaline hydrolysis ladder—Add 1.66 μL of 5′-^{32}P-labeled RRE IIB (50 μM stock) to a sterile eppendorf tube followed by 8.33 μL of 1 × alkaline hydrolysis buffer. Heat the sample at 95 °C for 3 min then cool on ice for 10 min.
- RNase A (1 μg/μL stock)—Dilute 50 × for a 20 ng/μL solution
- RNase V1 (0.1 U/μL stock)—Dilute 50 × for a 0.002 U/μL solution
- RNase T1 (1 U/μL stock)
- Inactivation/precipitation buffer
- Nuclease free water (NF H_2O)
- Glycogen
- Tracking dye
- 12% PAGE gel with 7.5 M urea

4.1.3 Procedure

1. Allow peptide, unlabeled RNA, labeled RNA, and structure buffer to thaw. Once solutions are thawed, mix by vortex then centrifuge.
2. To a screw top eppendorf tube, add first three components listed in Table 3 in the following order: NF H_2O, unlabeled RNA, labeled RNA. Mix by pipette.

Table 3 RNA solution.

Component	Concentration (nM)	Volume (μL)	Final concetration (nM)
5′-^{32}P-labeled RRE IIB	50	28	12.5
Dephosphorylated RRE IIB	Varying	X	250
NF H_2O		70–X	
Structure buffer		14	
Total volume		112	

Table 4 Prepared tubes for RNase assay.

Enzyme	Component	Volume (μM)	Final concentration (μM) [BP]
None	H_2O	2	0
T1/A/V1	H_2O	1	0
T1/A/V1	2 μM BP	1	0.2
T1/A/V1	20 μM BP	1	2
T1/A/V1	200 μM BP	1	20

3. Heat at 95 °C for 3 min.
4. Cool on ice for 10 min.
5. While RNA is cooling, prepare 13 screw top eppendorf tubes. One of the tubes will be a control, four will be RNase T1, four will be RNase A, and four will be RNase V1. Add the appropriate component and volume as listed in Table 4:
6. Place prepared tubes on ice.
7. Once RNA is cooled, add 14 μL of structure buffer to the eppendorf tube (from Step 4) and mix by pipette.
8. Add 8 μL of the resulting RNA solution to each of the prepared tubes. Place the tubes on ice for 3.5–4 h.
9. Add 1 μL of the appropriate enzyme to each tube, being careful to note the time for each tube. (Each enzyme has a specified incubation time: RNases T1 and A for 1 h, RNase V1 for 10 min). The control remains on ice during this time.
10. Quench the RNase with 20 μL of precipitation buffer at the end of the specified incubation time and mix by pipette. Also add precipitatation buffer to the control.

11. Add 1 μL of glycogen to each sample.
12. Place the tubes in the −20 °C freezer for 3 h for pellet formation.
13. Centrifuge all tubes at full speed (13,200 rpm) for 15 min.
14. Carefully remove the supernatant from each tube.
15. Add 10 μL of loading dye to the pellet and mix thouroughly. The samples can either be used for the sequencing gel or stored in the −20 °C until required.
16. Load samples onto the sequencing gel (Lanes: alkaline hydrolysis ladder, control, each RNase sample, additional alkaline hydrolysis ladder). Following electrophoresis and once the gel is dry, use a phosphor screen to image the gel.

4.1.4 Notes

1. When dealing with radioisotopes the labeled component is always added last, with the exception of enzymes.
2. The 3.5–4 h period allows peptide binding to the RNA.
3. Once the enzyme is added, for a total of 10 μL in each tube, 5′-^{32}P-labeled RRE IIB has a final concentration of 10 nM. The dephosporylated RRE IIB has a final concentration of 200 nM.
4. Glycogen helps to visualize the RNA pellet once it precipitates.
5. The pellet still may be hard to visualize after being centrifuged, a sharpie can be used to mark the outside of the tube before centrifuging to mark where the pellet will be formed.
6. When removing tubes from centrifuge be careful to do so gently, as knocking the tube can dislodge the pellet, making it harder to remove the supernatant.
7. Try to remove as much of the supernatant as possible without removing any of the pellet.

4.2 SHAPE-MaP based footprinting

Peptide binding to the RNA can be more accurately determined *via* SHAPE. SHAPE measures the reactivity of the 2′-hydroxyl group of the ribose sugar in the RNA backbone to electrophiles. Ribonucleotides that are more accessible to the electrophile, such as single-stranded regions, exhibit higher SHAPE reactivity, whereas ribonucleotides that are rendered inaccessible, such as double-stranded regions, exhibit lower SHAPE reactivity. Thus, when reactive, flexible single-stranded regions exhibit significantly lower reactivity in the presence of the peptide it likely indicates binding sites of the peptide. Such regions are determined by subtracting ribonucleotide reactivities of the RNA obtained in the absence of the peptide from those obtained in its presence.

Several variations of SHAPE exist, depending on the method for identifying the modified/reactive ribonucleotides. One of the most robust is Selective 2′-hydroxyl acylation analyzed by primer extension and mutational profiling (SHAPE-MaP) (Dai et al., 2018; Sherpa, Rausch, & Le Grice, 2018; Smola, Rice, Busan, Siegfried, & Weeks, 2015). In SHAPE-MaP, RNA is probed with an electrophile that selectively acylates accessible ribonucleotides followed by reverse transcription under conditions that promote mutagenesis in cDNA opposite to sites of modification. By high-throughput sequencing, the sites and frequencies of these mutations are determined and ribonucleotides protected by base-pairing or ligands are identified. Herein, we describe the SHAPE-MaP based footprinting method for the determination of peptide binding sites on a 234-nt long HIV-1 RRE RNA.

4.2.1 Equipment

- Thermal cycler
- Heat block
- RNeasy mini kit RNA (Qiagen)
- G50 spin columns
- Electrophoresis unit
- UV transilluminator
- Electroelution system
- Real-time PCR machine
- High-throughput sequencer
- KAPA Universal Library Quantification Kit
- MiniSeq Mid Output Kit

4.2.2 Buffers and reagents

- 234-nt RRE RNA (5 μM)
- 10 × Renaturation buffer—100 mM Tris HCl, 1000 mM KCl, 1 mM EDTA, pH 8
- NF H_2O
- 5 × RNA folding buffer—350 mM Tris, 900 mM KCl, 1.5 mM EDTA, 40 mM $MgCl_2$, pH 8
- 1 × PBS—10 mM potassium phosphate, 100 mM KCl, 1 mM $MgCl_2$, 20 mM NaCl, pH 7.0
- Anhydrous DMSO

- BP—5 μM (for 1:1 RNA:BP) and 50 μM (for 1:10 RNA:BP), solutions prepared in 1 × PBS
- 1-Methyl-7-nitroisatoic anhydride (1M7, 25 mM)
- Formamide
- 10 × Denaturation buffer—500 mM HEPES, 40 mM EDTA, pH 8
- 5 × MaP prebuffer—250 mM Tris, 375 mM KCl, 50 mM DTT, 2.5 mM of each dNTP, pH 8. Store in small aliquots at −20 °C as it should not be subjected to multiple freeze-thaw cycles.
- 2.5 × MaP buffer—125 mM Tris, 187.5 mM KCl, 15 mM $MnCl_2$, 25 mM DTT, 1.25 mM of each dNTPs, pH 8. Combine equal volumes of 5 × Map Prebuffer and 30 mM $MnCl_2$ immediately prior to use to prevent oxidation of manganese, which makes this solution ineffective.
- Superscript II Reverse Transcriptase (200 U/μL)
- 2 N NaOH
- 2 N HCl
- dNTPs (10 mM) (Table 5)
- Q5 Hot Start High-Fidelity DNA polymerase
- 5 × Q5 reaction buffer
- Agarose
- Ethanol
- PhiX control V3 (Illumina)
- 1 × TE buffer—10 mM Tris, 0.1 mM EDTA, pH 7.5.

Table 5 Oligos used in SHAPE-MaP analysis.

Oligo name	Sequence (5′ to 3′)	Purpose
RT oligo	AGGAGCTGTTGATCCTTTAG	Oligo used to reverse transcribe RRE RNA
PCR1_F	GACTGGAGTTCAGACGTGTGCTCTTCC GATCTNNNNNAGGAGCTTTGTTCCTTG	Forward SHAPE-MaP PCR1 oligo for NL4-3234 nt RRE.
PCR1_R	CCCTACACGACGCTCTTCCGATCTNN NNNAGGAGCTGTTGGTCC	Reverse SHAPE-MaP PCR1 oligo for NL4-3234 nt RRE.

Continued

Table 5 Oligos used in SHAPE-MaP analysis.—cont'd

Oligo name	Sequence (5′ to 3′)	Purpose
P7,I-1	CAAGCAGAAGACGGCATACGAGATC GTGATGTGACTGGAGTTCAGAC	Forward SHAPE-MaP PCR2 oligos containing different indices for barcoding
P7,I-2	CAAGCAGAAGACGGCATACGAGATA CATCGGTGACTGGAGTTCAGAC	
P7,I-3	CAAGCAGAAGACGGCATACGAGATG CCTAAGTGACTGGAGTTCAGAC	
P7,I-4	CAAGCAGAAGACGGCATACGAGATT GGTCAGTGACTGGAGTTCAGAC	
P7,I-5	CAAGCAGAAGACGGCATACGAGATC ACTGTGTGACTGGAGTTCAGAC	
P7,I-6	CAAGCAGAAGACGGCATACGAGATA TTGGCGTGACTGGAGTTCAGAC	
P7,I-7	CAAGCAGAAGACGGCATACGAGATG ATCTGGTGACTGGAGTTCAGAC	
P7,I-12	CAAGCAGAAGACGGCATACGAGATT ACAAGGTGACTGGAGTTCAGAC	
P5	AATGATACGGCGACCACCGAGATCTAC ACTCTTTCCCTACACGACGCTCTTCCG	Reverse SHAPE-MaP PCR2 oligo

4.2.3 Procedure

Each SHAPE experiment requires three types of reactions that are performed in parallel: (1) 1M7 +, where the RNA is probed with an electrophile (here 1M7) both in the absence of peptide (negative control for peptide binding) and in the presence of varying peptide concentrations; (2) 1M7 −, where the RNA is treated in the absence of 1M7 and peptide, to account for the intrinsic background, Reverse transcriptase (RT) stop sites, RT mutation sites, and naturally occurring RNA modification events; (3) Denaturation control, which measures sequence and structure specific biases in mutation detection.

1M7 +:

1. To 3 eppendorf tubes, suspend 1 μL of 5 μM 234-nt RRE RNA in 1 × renaturation buffer for a total volume of 5 μL by mixing 0.5 μL of 10 × renaturation buffer, 3.5 μL NF H_2O, and 1 μL of 5 μM of RNA.
2. Using a thermocycler, heat each tube to 85 °C (to denature) and slowly cool (0.1 °C/s) to 25 °C (to renature).
3. Add 2.2 μL NF H_2O and 1.8 μL of 5 × RNA folding buffer to each tube.

4. Incubate at 37 °C for 15 min.
5. To tube one add 1 μL 1 × PBS (negative control for peptide binding), to tube two add 1 μL of 5 μM BP (1:1 RNA:BP), and to tube three add 1 μL of 50 μM BP (1:10 RNA:BP).
6. Incubate at 37 °C for 15 min.
7. Add 1 μL of 25 mM 1M7.

1M7 −:

1. To an eppendorf tube, suspend 1 μL of 5 μM 234-nt RRE RNA in 1 × renaturation buffer for a total volume of 5 μL by mixing 0.5 μL of 10 × renaturation buffer, 3.5 μL NF H_2O, and 1 μL of 5 μM of RNA.
2. Using a thermocycler, heat to 85 °C and slowly cool (0.1 °C/s) to 25 °C.
3. Add 2.2 μL NF H_2O and 1.8 μL of 5 × RNA folding buffer.
4. Incubate at 37 °C for 15 min.
5. Add 1 μL 1 × PBS.
6. Incubate at 37 °C for 15 min.
7. Add 1 μL of DMSO.

Denaturation Control:

1. To an eppendorf tube, mix 1 μL of 5 μM 234-nt RRE RNA, 2 μL of NF H_2O, 5 μL of 100% formamide, and 1 μL of 10 × denaturation buffer.
2. Heat at 95 °C for 1 min.
3. Add 1 μL of 25 mM of 1M7.
4. Heat at 95 °C for 1 min.
5. Place the tube on ice.

1M7 −, 1M7 +, and Denaturation Controls:

1. Recover RNA from all samples by RNeasy mini kit RNA according packaging protocol. Elute in 30 μL of NF H_2O and store samples at −20 °C until further use.
2. Perform reverse transcription of samples to generate cDNA library:
   a. Add 1 μL of the **RT Oligo** (2 μM) to 10 μL of RNA.
   b. Incubate at 65 °C for 5 min.
   c. Hold on ice.
   d. Add 8 μL of 2.5 × MaP buffer.
   e. Add 1 μL of Superscript II RT.
   f. Incubate at 42 °C for 3 h.
   g. Add 1 μL of 2 N NaOH to hydrolyze RNA template.
   h. Incubate at 95 °C for 5 min.
   i. Place on ice, spin briefly, and neutralize with 1 μL of 2 N HCl.
   j. Add 28 μL of NF H_2O to bring the volume to 50 μL.
   k. Purify samples using G50 spin columns following manufacturer's protocol.

3. Convert cDNA to dsDNA containing Illumina adapters:
   a. To each PCR tube add the following: 50 μL of cDNA (from previous step), 1.1 μL of 50 μM **PCR1_F**, 1.1 μL of 50 μM **PCR1_R**, 24.8 μL of NF H_2O, 2 μL of 10 mM dNTPs, 20 μL of 5 × Q5 reaction buffer, and 1 μL of Hot Start High-Fidelity DNA polymerase.
   b. Amplify the product (**PCR1**) using a thermocycler programmed as follows:

Step	Temperature	Time
Initial denaturation	98 °C	30 s
15 cycles	98 °C	10 s
	50 °C	30 s
	72 °C	30 s
Final extension	72 °C	3 min
Hold	4 °C	

   c. Purify **PCR1** products by extracting from a 2% agarose gel, followed by electroelution at room temperature for 2 h, then ethanol precipitation.
   d. Resuspend the **PCR1** amplicons in 20 μL of NF H_2O.
   e. Add 2 μL of each purified **PCR1** product to a PCR tube.
   f. To each tube add the following: 1.1 μL of appropriate 50 μM barcoding index forward oligo (**P7,I1-7,12**), 1.1 μL of 50 μM reverse oligo **P5**, 72.8 μL of NF H_2O, 2 μL of 10 mM dNTPs, 20 μL of 5 × Q5 reaction buffer, and 1 μL of Hot Start High-Fidelity DNA polymerase.
   g. Amplify the product (**PCR2)** using a thermocycler programmed as follows:

Step	Temperature	Time
Initial denaturation	98 °C	30 s
15 cycles	98 °C	10 s
	50 °C	30 s
	72 °C	30 s
Final extension	72 °C	3 min
Hold	4 °C	

   h. Purify **PCR2** products by extracting from 2% agarose gel, followed by electroelution at room temperature for 2 h, then ethanol precipitation.
4. Quantify each **PCR2** product with a real-time PCR according to the protocol for the KAPA Universal Library Quantification Kit.
5. Prepare the DNA library and PhiX control V3 for sequencing on a MiniSeq Mid Output Kit (300-cycles) according to manufacturer's guidelines.
6. Generate 2 × 150 paired-end reads by combining the DNA library and PhiX control V3 followed by loading 500 μL of the mixture (1.8 pM DNA library and 20% Phix control) to the kit cartridge and sequence.
7. Input the FASTA sequence files for each DNA library into the ShapeMapper software (https://weeks.chem.unc.edu/software.html) using the default settings to generate SHAPE reactivity profiles.
8. Subtract the reactivity measurements of the RNA with no peptide from the RNA with peptide. A negative value indicates protection by the peptide.

4.2.4 Notes

1. Ratio of RNA:peptide can vary, ratios of 1:1 and 1:10 were chosen to provide example.
2. The **PCR1** appends partial Illumina adapters to the ends of the RRE DNA. The **PCR2** completes the Illumina adapters while adding appropriate barcoding indices.
3. A decrease in the reactivity values of the RNA incubated with peptide relative to free RNA reactivity measurements can result not only from peptide binding but also from conformational changes induced by peptide binding.
4. Since base-paired ribonucleotides have low reactivity in free RNA, SHAPE cannot detect peptide binding in base-paired regions of the RNA.

5. Conclusions

Overall, there are a number of steps involved in engineering peptides that are capable of selective RNA targeting while also being able to compete with the cognate protein. The first step is to select an RNA target that has been well characterized. By knowing which amino acids of the protein are reported to bind the target, this may facilitate the selecting of favorable amino acids that increase chances of binding the RNA. The same type of amino acid interactions of the peptide as the cognate protein is not enough to ensure selective targeting. A diverse library should be synthesized so that peptides with different structures are generated, so that they may be biased toward binding the RNAs tertiary structure. Once leads are selected, it is necessary to validate them; this involves determining binding affinity, selectivity, and binding location, as described in Sections 3 and 4. Additionally, while not described here, it is necessary to evaluate the leads *in vitro* to ensure that there is no toxicity. In summary, our work has collectively provided important lessons in evaluating RNA recognition with the use of BPs. The techniques presented herein winnow peptides that recognize RNA through selectivity filters while the biophysical techniques identify the specific peptide-RNA interactions. Our studies demonstrate that good binding affinity and/or selectivity, thereby RNA molecular recognition, can be improved with a diverse library ideally with unique features predisposed to RNA interaction.

Acknowledgments

We thank K. Ray and R. Helm of the Virginia Tech Mass Spectrometry Incubator for their assistance with MALDI-TOF analysis, D. Tebit and D. Rekosh of University of Virginia for the initial *in vitro* screening of the BPs, B. Jayaraman and A. Frankel of the University of California San Francisco for conducting the p24 inhibition assay, and K. DeCourcy of the Fralin Life Science Institute for her assistance with confocal microscopy. This work was supported by National Institutes of Health grant RO1 GM093834 (W.L.S.) and R25GM072767 (A.N.P.).

References

Battiste, J. L., Mao, H., Rao, N. S., Tan, R., Muhandiram, D. R., Kay, L. E., et al. (1996). Alpha helix-RNA major groove recognition in an HIV-1 Rev peptide-RRE RNA complex. *Science*, *273*(5281), 1547–1551.

Brown, C. K., Madauss, K., Lian, W., Beck, M. R., Tolbert, W. D., & Rodgers, D. W. (2001). Structure of neurolysin reveals a deep channel that limits substrate access. *Proceedings of the National Academy of Sciences of the United States of America*, *98*(6), 3127–3132.

Bryson, D. I., Zhang, W., McLendon, P. M., Reineke, T. M., & Santos, W. L. (2012). Toward targeting RNA structure: Branched peptides as cell-permeable ligands to TAR RNA. *ACS Chemical Biology*, 7(1), 210–217.

Bryson, D. I., Zhang, W., Ray, W. K., & Santos, W. L. (2009). Screening of a branched peptide library with HIV-1 TAR RNA. *Molecular BioSystems*, *5*(9), 1070–1073.

Chaloin, L., Smagulova, F., Hariton-Gazal, E., Briant, L., Loyter, A., & Devaux, C. (2007). Potent inhibition of HIV-1 replication by backbone cyclic peptides bearing the Rev arginine rich motif. *Journal of Biomedical Science*, *14*(5), 565–584.

Crumpton, J., Zhang, W., & Santos, W. L. (2011). Facile analysis and sequencing of linear and branched peptide boronic acids by MALDI mass spectrometry. *Analytical Chemistry*, *83*(9), 3548–3554.

Dai, Y., Wynn, J. E., Peralta, A. N., Sherpa, C., Jayaraman, B., Li, H., et al. (2018). Discovery of a branched peptide that recognizes the rev response element (RRE) RNA and blocks HIV-1 replication. *Journal of Medicinal Chemistry*, *61*(21), 9611–9620.

Davidson, B. L., & McCray, P. B., Jr. (2011). Current prospects for RNA interference-based therapies. *Nature Reviews Genetics*, *12*(5), 329–340.
Davidson, A., Leeper, T. C., Athanassiou, Z., Patora-Komisarska, K., Karn, J., Robinson, J. A., et al. (2009). Simultaneous recognition of HIV-1 TAR RNA bulge and loop sequences by cyclic peptide mimics of Tat protein. *Proceedings of the National Academy of Sciences of the United States of America*, *106*(29), 11931–11936.
Draper, D. E. (1995). Protein-RNA recognition. *Annual Review of Biochemistry*, *64*, 593–620.
Friesen, W. J., & Darby, M. K. (2001). Specific RNA binding by a single C2H2 zinc finger. *The Journal of Biological Chemistry*, *276*(3), 1968–1973.
Harada, K., Martin, S. S., Tan, R., & Frankel, A. D. (1997). Molding a peptide into an RNA site by *in vivo* peptide evolution. *Proceedings of the National Academy of Sciences of the United States of America*, *94*(22), 11887–11892.
Hwang, S., Tamilarasu, N., Ryan, K., Huq, I., Richter, S., Still, W. C., et al. (1999). Inhibition of gene expression in human cells through small molecule-RNA interactions. *Proceedings of the National Academy of Sciences of the United States of America*, *96*(23), 12997–13002.
Hyun, S., Kim, H. J., Lee, N. J., Lee, K. H., Lee, Y., Ahn, D. R., et al. (2007). Alpha-helical peptide containing N,N-dimethyl lysine residues displays low-nanomolar and highly specific binding to RRE RNA. *Journal of the American Chemical Society*, *129*(15), 4514–4515.
Joyner, J. C., & Cowan, J. A. (2011). Targeted cleavage of HIV RRE RNA by Rev-coupled transition metal chelates. *Journal of the American Chemical Society*, *133*(25), 9912–9922.
Kane, R. C., Bross, P. F., Farrell, A. T., & Pazdur, R. (2003). Velcade: U.S. FDA approval for the treatment of multiple myeloma progressing on prior therapy. *The Oncologist*, *8*(6), 508–513.
Keen, N. J., Gait, M. J., & Karn, J. (1996). Human immunodeficiency virus type-1 Tat is an integral component of the activated transcription-elongation complex. *Proceedings of the National Academy of Sciences of the United States of America*, *93*(6), 2505–2510.
Kjems, J., Brown, M., Chang, D. D., & Sharp, P. A. (1991). Structural analysis of the interaction between the human immunodeficiency virus Rev protein and the Rev response element. *Proceedings of the National Academy of Sciences of the United States of America*, *88*(3), 683–687.
Kumagai, I., Takahashi, T., Hamasaki, K., Ueno, A., & Mihara, H. (2001). HIV Rev peptides conjugated with peptide nucleic acids and their efficient binding to RRE RNA. *Bioorganic & Medicinal Chemistry Letters*, *11*(9), 1169–1172.
Lacourciere, K. A., Stivers, J. T., & Marino, J. P. (2000). Mechanism of neomycin and Rev peptide binding to the Rev responsive element of HIV-1 as determined by fluorescence and NMR spectroscopy. *Biochemistry*, *39*(19), 5630–5641.
Lalonde, M. S., Lobritz, M. A., Ratcliff, A., Chamanian, M., Athanassiou, Z., Tyagi, M., et al. (2011). Inhibition of both HIV-1 reverse transcription and gene expression by a cyclic peptide that binds the Tat-transactivating response element (TAR) RNA. *PLoS Pathogens*, 7(5), e1002038.
Lee, S. J., Hyun, S., Kieft, J. S., & Yu, J. (2009). An approach to the construction of tailor-made amphiphilic peptides that strongly and selectively bind to hairpin RNA targets. *Journal of the American Chemical Society*, *131*(6), 2224–2230.
Lee, Y., Hyun, S., Kim, H. J., & Yu, J. (2008). Amphiphilic helical peptides containing two acridine moieties displace picomolar affinity toward HIV-1 RRE and TAR. *Angewandte Chemie International Edition*, *47*(1), 134–137.
Lu, J., Kadakkuzha, B. M., Zhao, L., Fan, M., Qi, X., & Xia, T. (2011). Dynamic ensemble view of the conformational landscape of HIV-1 TAR RNA and allosteric recognition. *Biochemistry*, *50*(22), 5042–5057.

Luedtke, N. W., Liu, Q., & Tor, Y. (2003). RNA-ligand interactions: Affinity and specificity of aminoglycoside dimers and acridine conjugates to the HIV-1 Rev response element. *Biochemistry*, *42*(39), 11391–11403.

Mei, H. Y., Mack, D. P., Galan, A. A., Halim, N. S., Heldsinger, A., Loo, J. A., et al. (1997). Discovery of selective, small-molecule inhibitors of RNA complexes-1. The Tat protein/TAR RNA complexes required for HIV-1 transcription. *Bioorganic & Medicinal Chemistry*, *5*(6), 1173–1184.

Michienzi, A., Li, S., Zaia, J. A., & Rossi, J. J. (2002). A nucleolar TAR decoy inhibitor of HIV-1 replication. *Proceedings of the National Academy of Sciences of the United States of America*, *99*(22), 14047–14052.

Millar, D. P. (1996). Fluorescence studies of DNA and RNA structure and dynamics. *Current Opinion in Structural Biology*, *6*(3), 322–326.

Mills, N. L., Daugherty, M. D., Frankel, A. D., & Guy, R. K. (2006). An alpha-helical peptidomimetic inhibitor of the HIV-1 Rev-RRE interaction. *Journal of the American Chemical Society*, *128*(11), 3496–3497.

Prater, C. E., Saleh, A. D., Wear, M. P., & Miller, P. S. (2007). Allosteric inhibition of the HIV-1 Rev/RRE interaction by a 3′-methylphosphonate modified antisense oligo-2′-O-methylribonucleotide. *Oligonucleotides*, *17*(3), 275–290.

Rana, T. M., & Jeang, K.-T. (1999). Biochemical and functional interactions between HIV-1 Tat protein and TAR RNA. *Archives of Biochemistry and Biophysics*, *365*(2), 175–185.

Rausch, J. W., & Le Grice, S. F. J. (2015). HIV Rev assembly on the Rev response element (RRE): A structural perspective. *Viruses*, 7(6), 3053–3075.

Rock, F. L., Mao, W., Yaremchuk, A., Tukalo, M., Crepin, T., Zhou, H., et al. (2007). An antifungal agent inhibits an aminoacyl-tRNA synthetase by trapping tRNA in the editing site. *Science*, *316*(5832), 1759–1761.

Sherpa, C., Rausch, J. W., & Le Grice, S. F. J. (2018). Structural characterization of maternally expressed gene 3 RNA reveals conserved motifs and potential sites of interaction with polycomb repressive complex 2. *Nucleic Acids Research*, *46*(19), 10432–10447.

Smola, M. J., Rice, G. M., Busan, S., Siegfried, N. A., & Weeks, K. M. (2015). Selective 2′-hydroxyl acylation analyzed by primer extension and mutational profiling (SHAPE-MaP) for direct, versatile and accurate RNA structure analysis. *Nature Protocols*, *10*(11), 1643–1669.

Symensma, T. L., Baskerville, S., Yan, A., & Ellington, A. D. (1999). Polyvalent Rev decoys act as artificial Rev-responsive elements. *Journal of Virology*, *73*(5), 4341–4349.

Thomas, J. R., & Hergenrother, P. J. (2008). Targeting RNA with small molecules. *Chemical Reviews*, *108*(4), 1171–1224.

Wynn, J. E., Zhang, W., Tebit, D. M., Gray, L. R., Hammarskjold, M.-L., Rekosh, D., et al. (2016). Effect of intercalator and Lewis acid-base branched peptide complex formation: Boosting affinity towards HIV-1 RRE RNA. *Medicinal Chemistry Communications*, 7(7), 1436–1440.

Zaman, G. J. R., Michiels, P. J. A., & van Boeckel, C. A. A. (2003). Targeting RNA: New opportunities to address drugless targets. *Drug Discovery Today*, *8*(7), 297–306.

Zhang, W., Bryson, D. I., Crumpton, J. B., Wynn, J., & Santos, W. L. (2013a). Branched peptide boronic acids (BPBAs): A novel mode of binding towards RNA. *Chemical Communications*, *49*(24), 2436–2438.

Zhang, W., Bryson, D. I., Crumpton, J. B., Wynn, J., & Santos, W. L. (2013b). Targeting folded RNA: A branched peptide boronic acid that binds to a large surface area of HIV-1 RRE RNA. *Organic & Biomolecular Chemistry*, *11*(37), 6263–6271.

Zhang, Q., Harada, K., Cho, H. S., Frankel, A. D., & Wemmer, D. E. (2001). Structural characterization of the complex of the Rev response element RNA with a selected peptide. *Chemistry & Biology*, *8*(5), 511–520.

CHAPTER SEVENTEEN

Sequence-specific recognition of structured RNA by triplex-forming peptide nucleic acids

Dziyana Hnedzko, Eriks Rozners*
Department of Chemistry, Binghamton University, Binghamton, NY, United States
*Corresponding author: e-mail address: erozners@binghamton.edu

Contents

Abstract

While <2% of DNA encodes for functional proteins, >70% is transcribed into RNA. Although the function of most RNA transcripts is unknown, such non-coding RNAs are attractive targets for molecular recognition because of the potentially important roles they play in regulation of gene expression and development of disease. In this chapter, we describe peptide nucleic acids (PNAs) that form sequence-specific triple helices with double-stranded RNA (dsRNA). We provide protocols for sequence design and biophysical characterization of PNAs and discuss first examples where such PNAs have been used for functional modulation of dsRNA. The triplex-forming PNAs represent a new approach for RNA recognition that may find future applications in fundamental science, biotechnology and medicine.

Methods in Enzymology, Volume 623
ISSN 0076-6879
https://doi.org/10.1016/bs.mie.2019.04.003

1. Introduction

Compared to DNA, molecular recognition of double-stranded RNA (dsRNA) has received relatively little attention. However, most recently several large pharmaceutical companies and smaller biotechs have launched new programs on small-molecule drugs that target regulatory RNA motifs in mRNAs and non-coding RNAs (Mullard, 2017). The key challenge for therapeutic targeting of regulatory RNAs is sequence specificity—it is easy to find strong RNA binders, but it is very difficult to find sequence-specific binders (Childs-Disney & Disney, 2016; Morgan, Forte, Hargrove, & Hargrove, 2018; Shortridge & Varani, 2015; Warner, Hajdin, & Weeks, 2018). The regulatory RNA motifs in mRNAs and non-coding RNAs form complex double helical structures. Because of the uniform and polar surface and conformational flexibility, molecular recognition of such structures is a formidable problem. Despite these challenges, several groups have made impressive progress in developing small-molecules that target dsRNAs (Childs-Disney & Disney, 2016; Disney & Angelbello, 2016; Shortridge & Varani, 2015; Velagapudi, Vummidi, & Disney, 2015). However, these studies also revealed that small-molecules will not provide a general solution for targeting every RNA, which leaves a substantial gap in technology and room for development of novel approaches to molecular recognition of dsRNA.

The present chapter describes a relatively new approach to molecular recognition of dsRNA using triplex-forming peptide nucleic acids (PNAs) that achieve sequence specificity using Hoogsteen hydrogen bonding in the major groove of the RNA target (Fig. 1). Following our initial discovery that short PNAs form highly stable and sequence-specific triple helices with dsRNA (Li, Zengeya, & Rozners, 2010), our group (Hnedzko, McGee, Karamitas, & Rozners, 2017; Muse et al., 2013; Zengeya, Gupta, & Rozners, 2012) and others (Devi, Yuan, Lu, Zhao, & Chen, 2014; Sato, Sakamoto, & Nishizawa, 2018; Tähtinen, Granqvist, Murtola, Strömberg, & Virta, 2017; Zhou et al., 2013) have used physical and organic chemistry to study and optimize binding affinity and expand the sequence scope that can be targeted using the triplex-forming PNAs. The most significant development from our group was the replacement of cytosine (pK_a ~4.5) with the more basic (pK_a ~6.7) 2-aminopyridine (M, Fig. 1) nucleobase (Zengeya et al., 2012), which enabled strong binding of PNAs as short as hexamers to dsRNA at physiological salt and pH

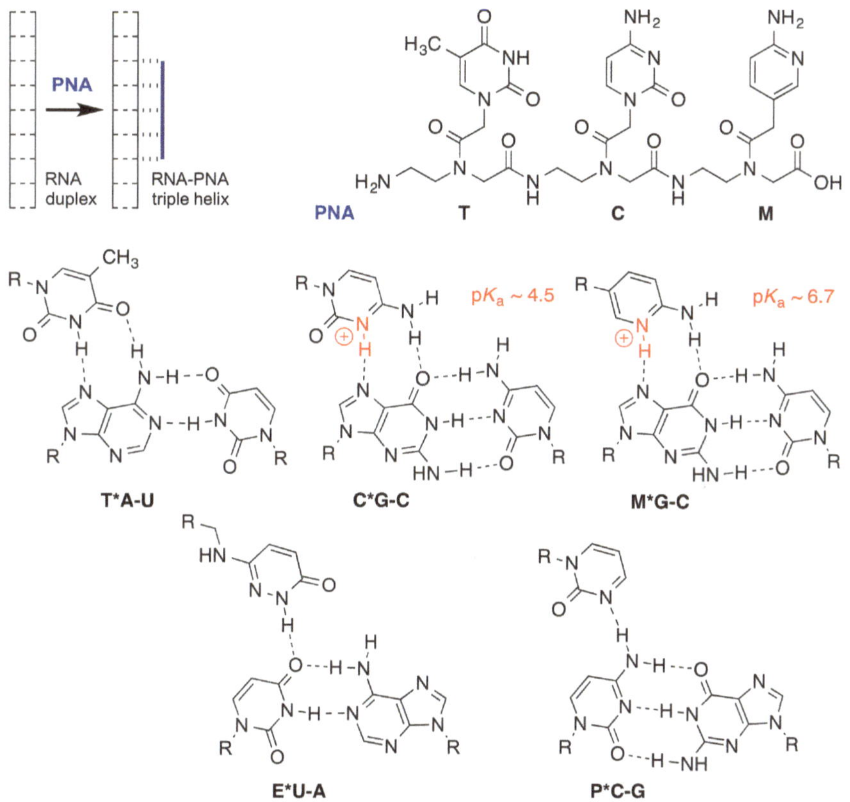

Fig. 1 Structure of PNA and Hoogsteen base triplets used in molecular recognition of dsRNA.

conditions (Muse et al., 2013). Moreover, M-modified PNAs had unique selectivity for binding RNA, as the PNA-dsRNA triplex was significantly more stable than the PNA-dsDNA triplex of the same sequence (Hnedzko et al., 2017; Muse et al., 2013; Zengeya et al., 2012). These properties make M-modified PNAs uniquely suited for sequence-specific molecular recognition of biologically relevant complex RNA species. Herein, we describe the recent progress and remaining challenges in using nucleobase-modified triplex-forming PNAs for this purpose.

2. Sequence-specific formation of PNA-dsRNA triplex

Conceptually our approach rests on the notion that H-bonding to nucleobases is inherently the most efficient means of selective recognition of helical nucleic acids. This concept has been elegantly applied by Dervan and co-workers (Dervan & Edelson, 2003) in targeting DNA by minor groove-binding polyamides. In contrast to DNA, the shallow minor groove of RNA is not a good fit for polyamides (Chenoweth, Meier, & Dervan, 2013). Before our studies, the deep and narrow major groove of RNA was also considered a poor receptor for ligand binding (Thomas & Hergenrother, 2008). Although the sequence-specific recognition of *single stranded* RNA by Watson-Crick base pairing has been well-established (the "antisense" approach), recognition of *double-stranded* RNA had few relevant precedents (Escude et al., 1993; Han & Dervan, 1993; Roberts & Crothers, 1992). Our group was the first to report that PNAs formed triple helix in the major groove of dsRNA with unusually high affinity and an excellent sequence selectivity (Li et al., 2010).

The sequence specificity in traditional RNA triple helices derives from uridine recognition of A-U base pairs (U*A-U triplet) and protonated cytidine recognition of G-C base pairs (C^{+}*G-C triplet) via the Hoogsteen hydrogen-bonding (Fig. 1; Devi, Zhou, Zhong, Toh, & Chen, 2015). However, because cytosine (pK_a ~4.5) protonation is required for formation of the C^{+}*G-C triplet (Fig. 1), unmodified PNAs have low affinity for double-stranded nucleic acids at physiological pH. We discovered that the more basic (pK_a ~6.7) 2-aminopyridine nucleobase M enabled formation of strong and selective PNA-dsRNA triplexes at physiological salt and pH (Zengeya et al., 2012). To evaluate the binding affinity and sequence specificity of modified PNAs, we routinely use a model system of four hairpins (**HRP1–HRP4**, Fig. 2) that have a single variable base pair (color coded in Fig. 2) in the middle of the stem. Paired with four PNAs that bear native or modified nucleobases at the variable position, the 16 PNA-dsRNA

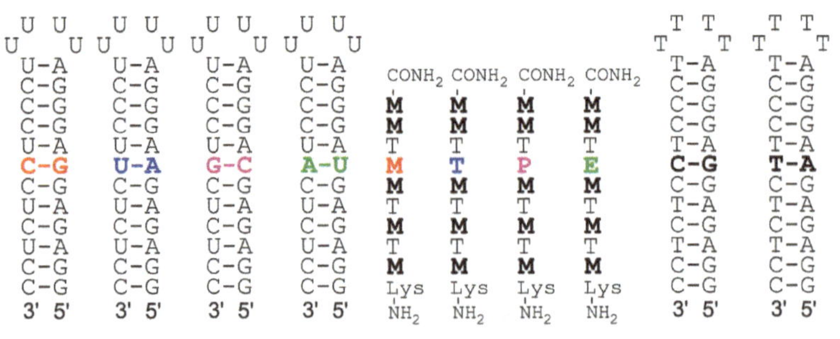

Fig. 2 Sequences of RNA hairpins used to study the binding affinities and sequence specificity of M-modified PNAs.

Table 1 Binding affinities and sequence specificities of M-modified PNAs for dsRNA hairpins.[a]

PNA	HRP1	HRP2	HRP3	HRP4	dsDNA1	dsDNA2
PNA1[b]	**110**	9	6	3	3	ND
PNA2[b]	11	**94**	3	3	ND	0.5
PNA3	0.3	0.9	**3**	0.8	ND	ND
PNA4	2	9	13	**33**	ND	ND

[a]Association constant $K_a \times 10^6\ M^{-1}$, measured by ITC using a Malvern MicroCal iTC200 instrument in 2 mM $MgCl_2$, 90 mM KCl, 10 mM NaCl, 50 mM potassium phosphate buffer pH 7.4 at 25 °C. The bold entries represent the affinities of the matched triple helices, color coded as in Fig. 2.
[b]Data from Hnedzko et al. (2017).
ND—not determined.

combinations represent a rigorous test for binding affinity and sequence specificity of PNAs. As illustrated in Table 1, M-modified 9-mers **PNA1** and **PNA2** (Fig. 2) showed excellent affinity and selectivity for their matched **HRP1** and **HRP2**, respectively, as measured by isothermal titration calorimetry (ITC) (Hnedzko et al., 2017). The superior binding was due to the cationic M nucleobase as our previous study (Zengeya et al., 2012) showed that a PNA with the neutral pseudoisocytosine (J) (Egholm et al., 1995; Hansen, Bentin, & Nielsen, 2009) modification had ~10-fold weaker binding than its M-modified counterpart. The low nanomolar binding affinity and high RNA selectivity at physiologically relevant pH and salt concentration opened the doors for potential applications of M-modified triplex-forming PNAs in biological systems.

Surprisingly, PNA-dsRNA triplexes were significantly more stable than the corresponding PNA-dsDNA triplexes. The stability of triplexes between **PNA1** and **PNA2** and their matched **dsDNA1** and **dsDNA2** (having the same sequences as **HRP1** and **HRP2**) was as low, if not lower, as the stability of mismatched PNA-dsRNA triplexes. Comparing our results to literature (Hansen et al., 2009) we noticed that strong PNA-dsDNA triple helices (low nanomolar affinity) were formed by PNA 15mers, while an 11mer had only low micromolar affinity. Thus, the low affinity of our PNA 9mer for dsDNA was consistent with literature (Hansen et al., 2009) and the surprising result was the high affinity of PNA for dsRNA. It is conceivable that the deep and narrow major groove of RNA presents a better steric and electrostatic fit for the PNA ligands than the wider major groove of DNA. Supporting this notion, our recent NMR study (Kotikam, Kennedy, MacKay, & Rozners, 2019) showed that PNA backbone amides formed hydrogen-bonds to phosphates of dsRNA, which was consistent with the hydrogen bonding scheme observed in a crystal structure (Betts, Josey, Veal, & Jordan, 1995) of A-form PNA-DNA-PNA triplex and may explain the preference of PNA to bind A-form dsRNA over B-form dsDNA where such hydrogen bonding is not feasible due to a larger phosphate-phosphate distance.

Traditional DNA and RNA triple helices are restricted to sequences where one strand of the double helix consists only of purine nucleosides (Fox, Brown, & Rusling, 2018). This is because the purine nucleosides are easily recognized by the common Hoogsteen triplets (Fig. 1), while recognition of pyrimidines is less efficient because they present fewer hydrogen-bonding sites in the major groove. As recently reviewed (Fox et al., 2018), substantial effort has been devoted to development of modified nucleobases that bind to the so-called pyrimidine inversions in DNA triple helices. In earlier studies (Gupta, Zengeya, & Rozners, 2011), we found that PNA modified with 2-pyrimidinone P and 3-oxo-2,3-dihydropyridazine E (Fig. 1) recognized the C-G and U-A inversions in dsRNA, respectively, with mid nanomolar affinity and good sequence selectivity in acetate buffer at pH 6.25. The use of P base in PNA had no precedent before our study, although derivatives of pyrimidinone had been used as oligonucleotide modifications for recognition of C-G inversions in dsDNA (Buchini & Leumann, 2004). The E base was originally designed by Nielsen and co-workers (Eldrup, Dahl, & Nielsen, 1997) as a PNA modification to recognize T-A inversions in dsDNA. Expanding our studies to physiological conditions, we found that the recognition was less efficient than that of uninterrupted purine tracts as **PNA3** (Fig. 2 and Table 1) had relatively weak binding affinity and the selectivity of **PNA4** was moderate. Nevertheless, as illustrated by Table 1 and an example discussed below, the P- and E-modified PNAs provided working solutions for recognition of dsRNA having eight consecutive purines interrupted by an isolated pyrimidine, motifs that are commonly found in complex non-coding RNAs, such as ribosomal RNAs (http://www.rna.ccbb.utexas.edu/) and miRNAs (Griffiths-Jones, Saini, van Dongen, & Enright, 2008).

A recent study (Endoh, Hnedzko, Rozners, & Sugimoto, 2016) demonstrated biological activity of triplex-forming PNAs in cells using reporter mRNAs encoding Renilla luciferase (RLuc) and having either the wild-type (wtHRP) or mutant (mtHRP) hairpin sequences in 5′ UTR (Fig. 3). For these studies, we developed a method for attaching fluorescent dyes (HiLyte Fluor 488, Anaspec) to PNA during solid-phase synthesis and

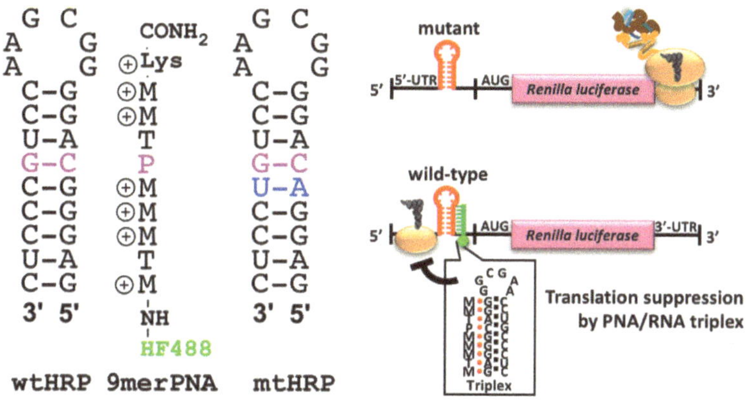

Fig. 3 Sequences of wild-type (**wrHRP**) and mutant-type (**mtHRP**) hairpins that were used in 5′ UTR of ~1200 nt long mRNA to study the biological activity of triplex-forming PNA (**9merPNA**) in cells. HF488 denotes HiLyte Fluor 488 fluorescent dye (Anaspec).

showed that HiLyte Fluor 488-labeled PNAs bind complimentary dsRNA with affinities similar to those of unlabeled PNA (Hnedzko, McGee, & Rozners, 2016). PAGE experiments confirmed the formation of a sequence-specific triplex between **9merPNA** and the ~1200 nt long wild-type mRNA, but not the mutant mRNA (Endoh et al., 2016). Consistent with triplex formation, **9merPNA** suppressed translation of RLuc from wild-type (but not mutant) mRNA in vitro. Most likely, the PNA-dsRNA triplex acted as a roadblock for the small ribosomal subunit. Interestingly, when the hairpin was inserted after AUG codon, the triplex could not block translation by the fully assembled ribosome. When the reporter mRNA was incorporated in a plasmid and electroporated in MCF7 cells together with **9merPNA**, the result was ~60% reduction of translation of the wild-type mRNA (but not mutant) compared to control experiments without PNA. This study (Endoh et al., 2016) was the first demonstration of a biological effect of PNA-dsRNA triplex formation in live cells.

The biological activity of a PNA 9mer containing the P nucleobase that gives lower binding affinity (Table 1) than other PNAs, is important because it demonstrates a strong potential of triplex-forming PNAs to modulate RNA's function. Winssinger and co-workers (Kim, Chang, & Winssinger, 2018) used PNAs modified with M, P, and E nucleobases in a novel fluorescent RNA detection method that targeted purine tracts having up to three isolated pyrimidine inversions. Most remarkably, 2 PNA 9mers, one having two P nucleobases, and the second having 2 Ps and an E nucleobase, efficiently and selectively directed a dsRNA templated reaction that unmasked a profluorophore and produced signal amplification specific to dsRNA (Kim et al., 2018). Taken together these studies (Endoh et al., 2016; Kim et al., 2018) suggest that M-modified triplex-forming PNAs are uniquely suited for sequence-specific recognition of dsRNA and may find various future applications in fundamental studies and biotechnology of functional RNAs.

3. Protocols for synthesis, purification and testing of M-modified PNA

Previous studies by us and others reviewed above suggest that PNA 9mers achieve the required binding affinity and sequence selectivity for modulating structure and function of dsRNA. To improve PNA solubility, it is common to add one lysine residue to the carboxy terminus. If necessary, short cationic peptides can be added to improve the cellular uptake. We recommend adding three lysines to the carboxy terminus (Hnedzko et al., 2017). It is also recommended that the affinity of PNA ligand for the intended RNA receptor is checked using ITC and a model hairpin similar to those in Fig. 2. Below, we provide protocols for synthesis, purification and binding affinity studies of triplex-forming PNA.

3.1 Equipment

1. DNA or peptide synthesizer capable of PNA synthesis. The protocols for PNA synthesis described below are for the Expedite 8909 DNA synthesizer using the Fmoc/Boc protected monomers (Fig. 4). The specific steps may be different if other instruments or PNA synthesis chemistry is used.
2. UV spectrophotometer.
3. HPLC instrument including gradient-capable pumps, column oven, and UV detector, equipped with C-18 reversed phase column (4.6 × 150 mm or larger diameter).
4. Freeze-dry system with SpeedVac capability.
5. Rotary evaporator.
6. Isothermal titration microcalorimeter suitable for small biological samples (cell volume 1 mL or smaller and minimum detectable heat 0.1 μJ or smaller), such as a Malvern MicroCal iTC200 instrument.
7. pH-meter.
8. LC/MS instrument (or MALDI TOF MS) for analysis of PNA.

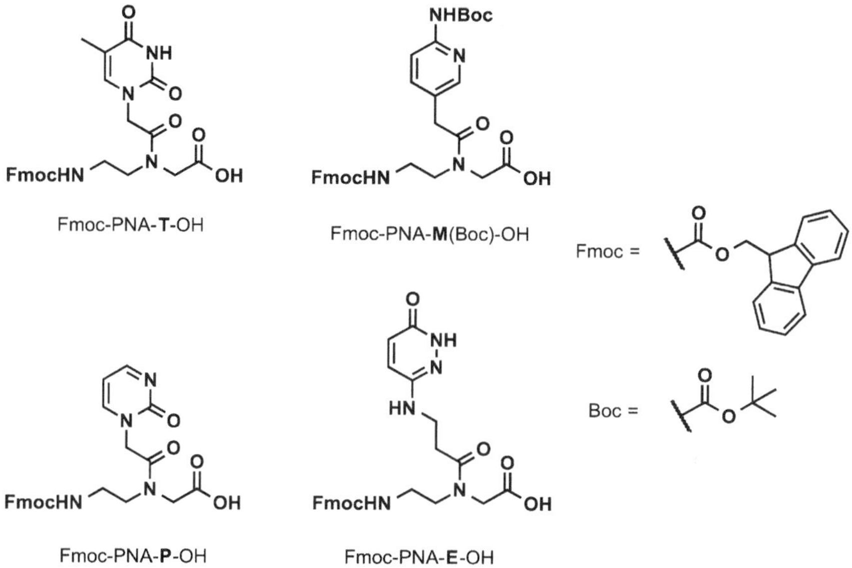

Fig. 4 Structures of the standard T (Link Technologies) and nucleobase modified (Hnedzko, Cheruiyot, & Rozners, 2014) PNA monomers used in the solid-phase synthesis of triplex-forming PNA.

3.2 Chemicals, materials and supplies

1. Fmoc-PNA-**T**-OH PNA monomer commercially available from Link Technologies.
2. PNA monomers carrying the modified nucleobases (Fig. 4): 2-aminopyridine (**M**), 2-pyrimidinone (**P**), 3-oxo-2,3-dihydropyridazine (**E**) synthesized following our reported procedures (Hnedzko et al., 2014).
3. Fmoc-Lys(Boc)-OH (natural L series) commercially available from NovaBiochem.
4. Empty synthesis columns for the Expedite 8909 commercially available from Glen Research.
5. NovaSyn TG Sieber solid support for PNA synthesis (functionalized at 0.2 mmol/g).
6. *N,N*-dimethylformamide (DMF), *N*-methyl-2-pyrrolidone (NMP), and diethyl ether, anhydrous.
7. Piperidine, biotech grade, commercially available from Aldrich.
8. *O*-(7-azabenzotriazol-1-yl)-1,1,3,3-tetramethyluronium hexafluorophosphate (HATU).

9. Diisopropylethylamine (DIPEA) and 2,6-lutidine, reagent grade, distilled over CaH_2.
10. Acetic anhydride and m-cresol, reagent grade.
11. Trifluoroacetic acid (TFA), formic acid, water, and acetonitrile, HPLC grade.
12. Syringe filter, 13 mm with a 0.2 μm nylon membrane.
13. $MgCl_2 \cdot 6H_2O$ and KCl, ACS grade.
14. NaCl, K_2HPO_4, and KH_2PO_4, molecular biology grade.

3.3 PNA synthesis protocol for expedite 8909 instrument

1. Place 10 mg of NovaSyn TG Sieber resin into an empty synthesis column to set up a standard 2 μmol scale PNA synthesis on an Expedite 8909 synthesizer.
2. Prepare 0.2 M solution of PNA monomers (Fig. 4) in anhydrous NMP. Weigh the following amounts of PNA monomers: Fmoc-PNA-**T**-OH (203 mg), Fmoc-PNA-**M**(Boc)-OH (230 mg), Fmoc-PNA-**P**-OH (196 mg), Fmoc-PNA-***E***-OH (202 mg), and Fmoc-Lys (Boc)-OH (178 mg) in oven dried Expedite 8909 monomer vials and add 1.9 mL of anhydrous NMP to each vial. The resulting 2 mL of monomer solution is sufficient for ~14 couplings including ~0.3 mL that will be consumed to prime the Expedite reagent lines.
3. Prepare the deblocking solution (20% (v/v) piperidine in DMF) for PNA synthesis by mixing 20 mL of piperidine and 80 mL of anhydrous DMF.
4. Prepare the activator solution (0.19 M HATU) by dissolving 1.089 g of HATU in 15 mL of anhydrous DMF.
5. Prepare the base solution by mixing 3.3 mL of DIPEA (0.2 M), and 3.5 mL of 2,6-lutidine (0.3 M) in 93.2 mL of anhydrous DMF.
6. Prepare the capping solution (5% (v/v) acetic anhydride and 6% (v/v) 2,6-lutidine in DMF) by mixing (in the following order) 89 mL of anhydrous DMF, 6 mL of 2,6-lutidine, and 5 mL of acetic anhydride.
7. Prime the Expedite 8909 synthesizer with the monomers and reagents (prepared in steps 2–6 in this section), install the synthesis column (prepared in step 1 in this section), and program the sequences to be synthesized.
8. Run the standard PNA synthesis protocol on Expedite 8909 synthesizer.

9. It is recommended to monitor the synthesis efficiency by collecting the deblocking solution in each Fmoc deprotection step, diluting it to 10 mL with the deblocking solution, and measuring the absorbance of the resulting mixture at 300 nm using a UV spectrophotometer. The difference in absorbances will provide a rough estimate of the coupling yield.
10. After completion of PNA synthesis, remove the terminal Fmoc group as described in Expedite 8909 PNA synthesis manual.

3.4 PNA cleavage from solid support and deprotection

1. Prepare the cleavage cocktail (20% (v/v) m-cresol in TFA) by mixing 0.8 mL of TFA and 0.2 mL of m-cresol.
2. As described in Expedite 8909 PNA synthesis manual, remove the synthesis column from synthesizer and attach two 1 mL syringes to each end. Take 0.6 mL of the cleavage cocktail in one of the syringes and move it forth and back through the column several times. Leave the synthesis column with the cleavage cocktail solution for 2 h occasionally moving the solution forth and back through the column.
3. Collect the cleavage cocktail solution in a plastic 2 mL centrifuge vial. Use another 0.3 mL of the cleavage cocktail to treat the synthesis column for another 30 min. Collect the second cleavage cocktail solution in the same vial.
4. Distribute the combined cleavage cocktail solutions evenly into five plastic 2 mL centrifuge vials. Precipitate the crude deprotected PNA by adding 1.8 mL of anhydrous diethyl ether to each of the vials and mixing vigorously.
5. Centrifuge the vials at 14,500 rpm for 15 min. Carefully remove the diethyl ether from the PNA precipitate with a pipette. Add another 1 mL of diethyl ether, mix vigorously, centrifuge and remove the solvent with a pipette.
6. Dissolve the PNA pellets by adding 0.3 mL of HPLC water in each vial. Combine all solutions in one vial and evaporate using a SpeedVac.

3.5 HPLC purification and quantification of PNA

1. Dissolve the PNA sample in 1 mL of HPLC water. Filter the sample through a 13 mm syringe filter.

2. Purify the PNA by reverse phase HPLC using a C-18 column at 55 °C, eluting with a linear gradient of acetonitirile (5–20%, v/v) in water containing 0.1% formic acid over 40 min, flow rate 1 mL/min for a 4.6 mm diameter column. Collect the fractions corresponding to the major peak as detected by UV absorbance monitored at 254 nm.
3. Using rotary evaporator, reduce the volume of the collected fractions to ~50% of the original volume (to remove formic acid) and freeze dry the residual sample.
4. Confirm the purity and identity of PNA by LC/MS or HPLC and MALDI TOF mass spectrometry.
5. Dissolve the dry PNA sample in 1 mL of HPLC water. Take 50 μL of the sample and dilute to 1 mL (dilution factor of 20).
6. Measure the UV absorbance (A) of PNA at 260 nm. If $A > 1$, dilute the sample further or repeat previous step 5 in this section and increase the dilution factor.
7. Calculate the molar extinction coefficient (ε_{260} M^{-1} cm^{-1}) of PNA by summing up the individual extinction coefficients of monomers according to the PNA sequence: 8560 for T, 2212 for **M**, 2150 for **P**, and 5984 for **E**.
8. Determine the quantity of PNA (n) in nmols using the following formula:

$$n = A \times 10^6 \times 20/\varepsilon$$

where A is the UV absorbance of PNA at 260 nm, 20 is the dilution factor, and ε is the molar extinction coefficient.

9. Freeze dry the sample and store as solid at −20 °C. The triplex forming PNA is ready for research applications.

3.6 Isothermal titration calorimetry protocol for Malvern MicroCal iTC200 instrument

1. Prepare a 0.24 mM stock solution of PNA (from Section 3.5 step 9) in HPLC water by adding 1 mL of water for each 240 nmols. Store the stock solution frozen at −20 °C.
2. Prepare a 0.24 mM stock solution of RNA (commercially available from Dharmacon) in HPLC water by adding 1 mL of water for each 240 nmols. Store the stock solution frozen at −20 °C.
3. Prepare ITC buffer (2 mM $MgCl_2$, 90 mM KCl, 10 mM NaCl and 50 mM potassium phosphate, pH 7.4). Prepare buffer A by dissolving 203 mg of $MgCl_2 \cdot 6H_2O$, 3.355 g of KCl, 292 mg of NaCl, and 4.355 g

of K_2HPO_4 in a total volume of 500 mL of HPLC water. Prepare buffer B by dissolving 203 mg of $MgCl_2 \cdot 6H_2O$, 3.355 g of KCl, 292 mg of NaCl and 3.402 g of KH_2PO_4 in a total volume of 500 mL of HPLC water. Slowly add buffer B in buffer A until pH 7.4 is reached.

4. Evaporate 12.5 μL of RNA stock solution (0.24 mM) in a plastic 2 mL centrifuge vial using a SpeedVac. Dissolve the RNA sample in 300 μL of degassed ITC buffer.
5. Evaporate 22.5 μL of PNA stock solution (0.24 mM) in a plastic 2 mL centrifuge vial using a SpeedVac. Dissolve the PNA sample in 60 μL of degassed ITC buffer.
6. Load the RNA sample (250 μL) into the sample cell of MicroCal iTC200 instrument. Load HPLC-grade water into the reference cell of the ITC instrument.
7. Using the automated syringe loading protocol, load the PNA sample (40 μL) in titration syringe of MicroCal iTC200 instrument.
8. Program the experimental parameters in the MicroCal iTC200 software: temperature (25 °C), reference power (5 μcal/s), initial delay (60 s), number of injections (16), injection (1.0 μL of PNA for the first injection, 2.45 μL of PNA for others), interval between injections (500 s), stirring speed (750 rpm), filter period (5 s), and feedback mode (high). The intervals between injections can be either increased or decreased depending on the rate of the PNA binding.
9. Start the ITC experiment and collect the titration data (Fig. 5).
10. Analyze the data using Malvern software MicroCal PEAQ-ITC Analysis to obtain experimental association constant (K_a), binding enthalpy (ΔH), binding entropy (ΔS), free energy (ΔG), and binding stoichiometry (N) (Fig. 5).

4. Remaining challenges and outlook

Specific recognition of any sequence of complex biologically relevant RNA is a highly desirable, yet still an elusive goal. In this chapter, we describe a relatively new approach that uses triple helical binding of cationic PNAs in the major groove of dsRNA. Advances in chemical modification of PNA has allowed us to solve some of the problems of triple helical recognition, such as, the low stability of triplexes at physiological pH, but other challenges remain. Most importantly, the triple helical recognition of dsRNA is limited to continuous tracts of eight or more purines interrupted by one (or possibly two) pyrimidines. While this is still a significant sequence

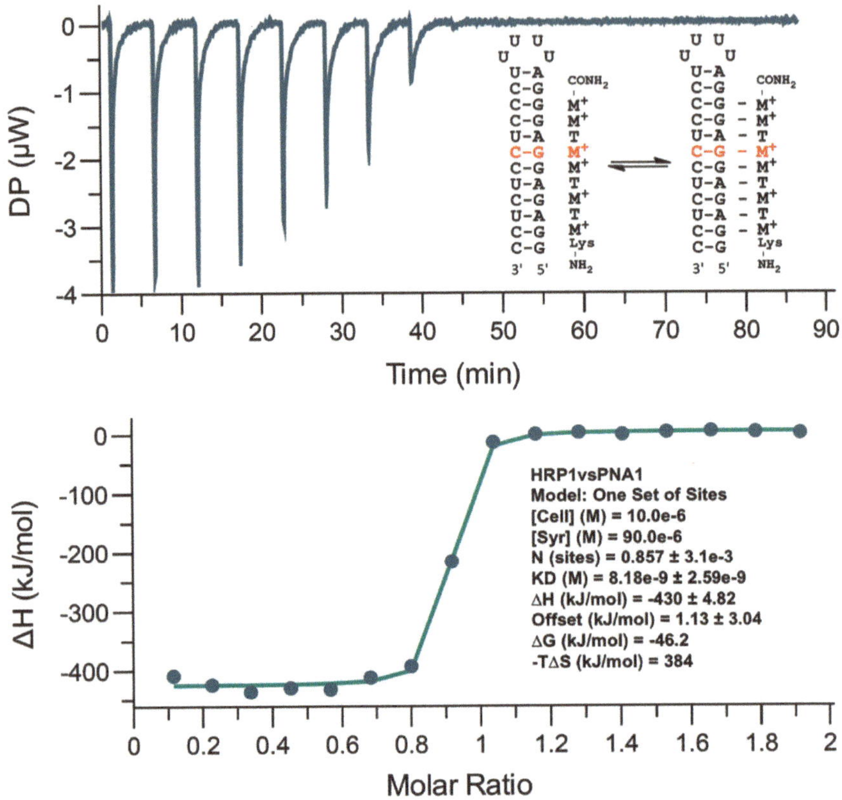

Fig. 5 ITC titration curve of **PNA1** binding to the matched **HRP1** ran on a Malvern MicroCal iTC200 instrument in 2 mM $MgCl_2$, 90 mM KCl, 10 mM NaCl, 50 mM potassium phosphate buffer pH 7.4 at 25 °C. Top panel: raw data plot with baseline subtracted. Bottom panel: integrated heat plot with summary of experimental results.

limitation, such motifs do occur in biologically relevant RNAs. Another possible challenge is the poor cellular uptake of PNA. We have shown that M modification alone or in combination with three lysine residues at carboxy terminus greatly improves PNA's ability to cross cellular membranes, but more work is needed to insure effective delivery of PNAs to intracellular targets. The triple helical recognition of dsRNA is an approach still under active development.

Acknowledgment

This work was supported by NSF (CHE-1406433 and CHE-1708761) and NIH (R35 GM130207).

References

Betts, L., Josey, J. A., Veal, J. M., & Jordan, S. R. (1995). A nucleic acid triple helix formed by a peptide nucleic acid-DNA complex. *Science*, *270*(5243), 1838–1841.

Buchini, S., & Leumann, C. J. (2004). Stable and selective recognition of three base pairs in the parallel triple-helical DNA binding motif. *Angew. Chem., Int. Ed. Engl.*, *43*(30), 3925–3928.

Chenoweth, D. M., Meier, J. L., & Dervan, P. B. (2013). Pyrrole-imidazole polyamides distinguish between double-helical DNA and RNA. *Angewandte Chemie, International Edition*, *52*(52), 415–418.

Childs-Disney, J. L., & Disney, M. D. (2016). Approaches to validate and manipulate RNA targets with small molecules in cells. *Annu. Rev. Pharmcol. Toxicol.*, *56*, 123–140.

Dervan, P. B., & Edelson, B. S. (2003). Recognition of the DNA minor groove by pyrrole-imidazole polyamides. *Curr. Opin. Struct. Biol.*, *13*(3), 284–299.

Devi, G., Yuan, Z., Lu, Y., Zhao, Y., & Chen, G. (2014). Incorporation of thio-pseudoisocytosine into triplex-forming peptide nucleic acids for enhanced recognition of RNA duplexes. *Nucleic Acids Research*, *42*(6), 4008–4018.

Devi, G., Zhou, Y., Zhong, Z., Toh, D.-F. K., & Chen, G. (2015). RNA triplexes: From structural principles to biological and biotech applications. *Wiley. Interdiscip. Rev. RNA*, *6*(1), 111–128.

Disney, M. D., & Angelbello, A. J. (2016). Rational Design of Small Molecules Targeting Oncogenic Noncoding RNAs from sequence. *Accounts of Chemical Research*, *49*(12), 2698–2704.

Egholm, M., Christensen, L., Dueholm, K. L., Buchardt, O., Coull, J., & Nielsen, P. E. (1995). Efficient pH-independent sequence-specific DNA binding by pseudoisocytosine-containing bis-PNA. *Nucleic Acids Research*, *23*(2), 217–222.

Eldrup, A. B., Dahl, O., & Nielsen, P. E. (1997). A novel peptide nucleic acid monomer for recognition of thymine in triple-Helix structures. *Journal of the American Chemical Society*, *119*(45), 11116–11117.

Endoh, T., Hnedzko, D., Rozners, E., & Sugimoto, N. (2016). Nucleobase-modified PNA suppresses translation by forming a triple Helix with a hairpin structure in mRNA in vitro and in cells. *Angewandte Chemie, International Edition*, *55*(3), 899–903.

Escude, C., Francois, J. C., Sun, J. S., Ott, G., Sprinzl, M., Garestier, T., et al. (1993). Stability of triple helixes containing RNA and DNA strands: Experimental and molecular modeling studies. *Nucleic Acids Research*, *21*(24), 5547–5553.

Fox, K. R., Brown, T., & Rusling, D. A. (2018). DNA recognition by parallel triplex formation. In M. J. Waring (Ed.), *Vol. 7. Chemical biology no. 7: DNA-targeting molecules as therapeutic agents*: (pp. 1–32): Royal Society of Chemistry.

Griffiths-Jones, S., Saini, H. K., van Dongen, S., & Enright, A. J. (2008). miRBase: Tools for microRNA genomics. *Nucleic Acids Research*, *36*(Database issue), D154–D158.

Gupta, P., Zengeya, T., & Rozners, E. (2011). Triple helical recognition of pyrimidine inversions in polypurine tracts of RNA by nucleobase-modified PNA. *Chemical Communications*, *47*(39), 11125–11127.

Han, H., & Dervan, P. B. (1993). Sequence-specific recognition of double helical RNA and RNA-DNA by triple helix formation. *Proceedings of the National Academy of Sciences of the United States of America*, *90*(9), 3806–3810.

Hansen, M. E., Bentin, T., & Nielsen, P. E. (2009). High-affinity triplex targeting of double stranded DNA using chemically modified peptide nucleic acid oligomers. *Nucleic Acids Research*, *37*(13), 4498–4507.

Hnedzko, D., Cheruiyot, S. K., & Rozners, E. (2014). Using triple-helix-forming peptide nucleic acids for sequence-selective recognition of double-stranded RNA. *Current Protocols in Nucleic Acid Chemistry*, *58*(1), 4.60.1–4.60.23.

Hnedzko, D., McGee, D. W., Karamitas, Y. A., & Rozners, E. (2017). Sequence-selective recognition of double-stranded RNA and enhanced cellular uptake of cationic nucleobase and backbone-modified peptide nucleic acids. *RNA*, *23*(1), 58–69.

Hnedzko, D., McGee, D. W., & Rozners, E. (2016). Synthesis and properties of peptide nucleic acid labeled at the N-terminus with HiLyte Fluor 488 fluorescent dye. *Bioorganic & Medicinal Chemistry*, *24*(18), 4199–4205. http://www.rna.ccbb.utexas.edu/.

Kim, K. T., Chang, D., & Winssinger, N. (2018). Double-stranded RNA-specific Templated reaction with triplex forming PNA. *Helvetica Chimica Acta*, *101*(3) e1700295.

Kotikam, V., Kennedy, S. D., MacKay, J. A., & Rozners, E. (2019). Synthetic, structural, and RNA binding studies on 2-aminopyridine-modified triplex-forming peptide nucleic acids. *Chemistry–A European Journal*, *25*(17), 4367–4372.

Li, M., Zengeya, T., & Rozners, E. (2010). Short peptide nucleic acids bind strongly to Homopurine tract of double helical RNA at pH 5.5. *Journal of the American Chemical Society*, *132*(25), 8676–8681.

Morgan, B. ,. S., Forte, J. ,. E., Hargrove, A. ,. E., & Hargrove, A. ,. E. (2018). Insights into the development of chemical probes for RNA. *Nucleic Acids Research*, *46*(16), 8025–8037.

Mullard, A. (2017). Small molecules against RNA targets attract big backers. *Nat. Rev. Drug. Disc.*, *16*(12), 813–815.

Muse, O., Zengeya, T., Mwaura, J., Hnedzko, D., McGee, D. W., Grewer, C. T., et al. (2013). Sequence selective recognition of double-stranded RNA at physiologically relevant conditions using PNA-peptide conjugates. *ACS Chemical Biology*, *8*(8), 1683–1686.

Roberts, R. W., & Crothers, D. M. (1992). Stability and properties of double and triple helices: Dramatic effects of RNA or DNA backbone composition. *Science*, *258*(5087), 1463–1466.

Sato, T., Sakamoto, N., & Nishizawa, S. (2018). Kinetic and thermodynamic analysis of triplex formation between peptide nucleic acid and double-stranded RNA. *Organic & Biomolecular Chemistry*, *16*(7), 1178–1187.

Shortridge, M. D., & Varani, G. (2015). Structure based approaches for targeting non-coding RNAs with small molecules. *Current Opinion in Structural Biology*, *30*, 79–88.

Tähtinen, V., Granqvist, L., Murtola, M., Strömberg, R., & Virta, P. (2017). 19F NMR spectroscopic analysis of the binding modes in triple-helical peptide nucleic acid (PNA)/MicroRNA complexes. *Chemistry–A European Journal*, *23*(29), 7113–7124.

Thomas, J. R., & Hergenrother, P. J. (2008). Targeting RNA with small molecules. *Chemical Reviews*, *108*(4), 1171–1224.

Velagapudi, S. P., Vummidi, B. R., & Disney, M. D. (2015). Small molecule chemical probes of microRNA function. *Current Opinion in Chemical Biology*, *24*, 97–103.

Warner, K. D., Hajdin, C. E., & Weeks, K. M. (2018). Principles for targeting RNA with drug-like small molecules. *Nat. Rev. Drug. Discov.*, *17*(8), 547–558.

Zengeya, T., Gupta, P., & Rozners, E. (2012). Triple helical recognition of RNA using 2-aminopyridine-modified PNA at physiologically relevant conditions. *Angewandte Chemie, International Edition*, *51*, 12593–12596.

Zhou, Y., Kierzek, E., Loo, Z. P., Antonio, M., Yau, Y. H., Chuah, Y. W., et al. (2013). Recognition of RNA duplexes by chemically modified triplex-forming oligonucleotides. *Nucleic Acids Research*, *41*(13), 6664–6673.

CHAPTER EIGHTEEN

Evaluating riboswitch optimality

Hannah Wayment-Steele[a], Michelle Wu[b], Michael Gotrik[c], Rhiju Das[c,d,*]
[a]Department of Chemistry, Stanford University, Stanford, CA, United States
[b]Program in Biomedical Informatics, Stanford University, Stanford, CA, United States
[c]Department of Biochemistry, Stanford University, Stanford, CA, United States
[d]Department of Physics, Stanford University, Stanford, CA, United States
*Corresponding author: e-mail address: rhiju@stanford.edu

Contents

Abstract

Riboswitches are RNA elements that recognize diverse chemical and biomolecular inputs, and transduce this recognition process to genetic, fluorescent, and other engineered outputs using RNA conformational changes. These systems are pervasive in cellular biology and are a promising biotechnology with applications in genetic regulation and biosensing. Here, we derive a simple expression bounding the activation ratio—the proportion of RNA in the active vs. inactive states—for both ON and OFF riboswitches that operate near thermodynamic equilibrium: $1+[I]/K_d^I$, where $[I]$ is the input ligand concentration and K_d^I is the intrinsic dissociation constant of the aptamer module toward the input ligand. A survey of published studies of natural and synthetic riboswitches confirms that the vast majority of empirically measured activation ratios have remained well below this thermodynamic limit. A few natural and synthetic riboswitches achieve activation ratios close to the limit, and these molecules highlight important principles for achieving high riboswitch performance. For several applications, including "light-up" fluorescent sensors and chemically-controlled CRISPR/Cas complexes, the thermodynamic limit has not yet been achieved, suggesting that

Methods in Enzymology, Volume 623
ISSN 0076-6879
https://doi.org/10.1016/bs.mie.2019.05.028

current tools are operating at suboptimal efficiencies. Future riboswitch studies will benefit from comparing observed activation ratios to this simple expression for the optimal activation ratio. We present experimental and computational suggestions for how to make these quantitative comparisons and suggest new molecular mechanisms that may allow non-equilibrium riboswitches to surpass the derived limit.

1. Introduction

Riboswitches are sophisticated sensors that make use of RNA's ability to bind and recognize a wide variety of substrates. They offer promise as genetic regulators (Sherwood & Henkin, 2016; Vazquez-Anderson & Contreras, 2013), as control elements for precise genetic editing systems via CRISPR/Cas9 (Liu et al., 2016; Tang, Hu, & Liu, 2017), as biosensors (Hallberg, Su, Kitto, & Hammond, 2017; Mehdizadeh Aghdam, Hejazi, & Barzegar, 2016), as devices for disease monitoring and diagnostics (Kellenberger et al., 2015; Seetharaman, Zivarts, Sudarsan, & Breaker, 2001), as metal ion detectors (Wedekind, Dutta, Belashov, & Jenkins, 2017), and as targets for future antibiotic therapies (Deigan & Ferré-D'Amaré, 2011; Lee, Han, & Lee, 2016). As sensors, riboswitches offer many benefits over proteins with comparable functions: RNA molecules can be simpler to synthesize in living systems than proteins (Berens & Suess, 2015) and offer the possibility of modular regulatory platforms that can be readily incorporated into new genes or readouts of interest (Berens, Groher, & Suess, 2015; Ceres, Trausch, & Batey, 2013).

A riboswitch can be defined as an RNA molecule that undergoes a conformational change upon binding a specific "input" ligand by an internal recognition module, or aptamer (blue, Fig. 1). This conformational change affects some form of output, such as altered transcription or translation rates (Suess, Fink, Berens, Stentz, & Hillen, 2004), ribozyme (RNA enzyme) activity (Tang & Breaker, 1997a, 1997b), or fluorescence by controlled capture of a fluorogenic ligand (Kellenberger, Wilson, Sales-Lee, & Hammond, 2013) or labeled protein (Kennedy, Vowles, d'Espaux, & Smolke, 2014). Control of these outputs requires a second RNA segment (red, Fig. 1) that recruits a separate ligand or biomolecule to the riboswitch whose role is to affect the downstream output. If the formation of the input module reduces correct formation and binding by the output segment, the riboswitch is typically called an OFF switch (Fig. 1A).

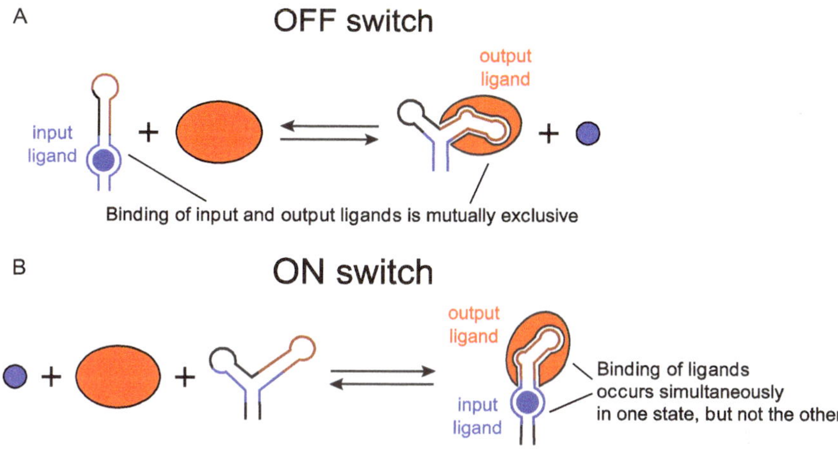

Fig. 1 Schematics for how an RNA conformational change can induce anticooperativity or cooperativity in binding an input and output ligand, corresponding to (A) an OFF-switch and (B) an ON-switch, respectively.

Conversely, if binding of the input ligand promotes formation of the active output segment, it is typically called an ON switch (Fig. 1B).

There has been remarkable progress in riboswitch discovery in recent years, and the field appears poised to create numerous real-world applications. To appreciate this progress, a common figure-of-merit for riboswitches is the "activation ratio" (AR)—the ratio of the output signal in the presence vs. absence of input ligand. While ARs for most natural and synthetic riboswitches have remained around 10-fold or lower, ARs exceeding 10,000 have been reported for riboswitches discovered through in vitro selection (Seetharaman, Zivarts, Sudarsan, & Breaker, 2001). Despite these successes, several questions remain. In studies seeking to design riboswitches that couple new inputs and outputs, AR values of 2- to 10-fold are widespread and often considered satisfactory—with better design, could these values go as high as the AR values in the best riboswitches (>10,000), or are there fundamental laws that limit their performance? To address these questions, biophysical modeling of riboswitches is important but can often become complicated and tied to assumptions particular to the system (Beisel & Smolke, 2009; Espah Borujeni, Mishler, Wang, Huso, & Salis, 2015; Findeiß et al., 2018). Here, inspired by simple analyses of switches in other areas of biophysics (Bintu et al., 2005), we present a thermodynamic model that reveals a simple expression for the maximum activation ratio of a riboswitch. The final expression depends on only two factors: the concentration of the input ligand and the intrinsic dissociation constant of the riboswitch's aptamer module for binding the input ligand. This limit is independent of fine details of the output event, and the limit holds for some systems that are out of thermodynamic equilibrium, including riboswitches deployed in complex biological contexts. Applying the derived expression for the thermodynamic optimum to previously described riboswitch systems helps define limits on the maximal performance achievable by these systems, and we outline a methodology that future studies can take to understand, achieve, and surpass the activation ratios set by the formula.

2. A thermodynamic bound for riboswitch performance

This section derives a simple and quite general expression for the optimal activation ratio of ON and OFF riboswitches that depends only on properties of the riboswitch aptamer, i.e., the RNA's recognition module for the input ligand (blue, Fig. 1). Our derivation follows standard approaches to modeling thermodynamics of biological systems; the reader is referred to an excellent chapter in a previous volume of Methods in Enzymology for an introduction (Garcia, Kondev, Orme, Theriot, & Phillips, 2011).

2.1 Optimal activation ratio for a minimal model

We start where the derivation is simplest, with an OFF switch whose output signal is reduced by reversible binding of an input ligand and effected by reversible recruitment of an output ligand. Fig. 1A shows an example of this mechanism (Kellenberger et al., 2013), in which an RNA conformational change is toggled by the presence of cyclic di-GMP as the input ligand (blue symbol). Binding of this small molecule input to the riboswitch's *c*-di-GMP aptamer segment (blue line) favors a structure that disrupts the formation of the Spinach aptamer (red line), an aptamer that would otherwise recruit the small molecule 3,5-difluoro-4-hydroxybenzylidene imidazolinone (DFHBI, red shape) and induce its fluorescence, generating the output in the ligand-free state. Following (Garcia et al., 2011), we write out the states of the thermodynamic model and their respective weights in Fig. 2A. This model can be expanded to consider the entire set of possible states of an RNA engaging with two ligands; Appendix A gives this complete model and derivation, but, for simplicity, we stick to minimal models, which lead to the same expression.

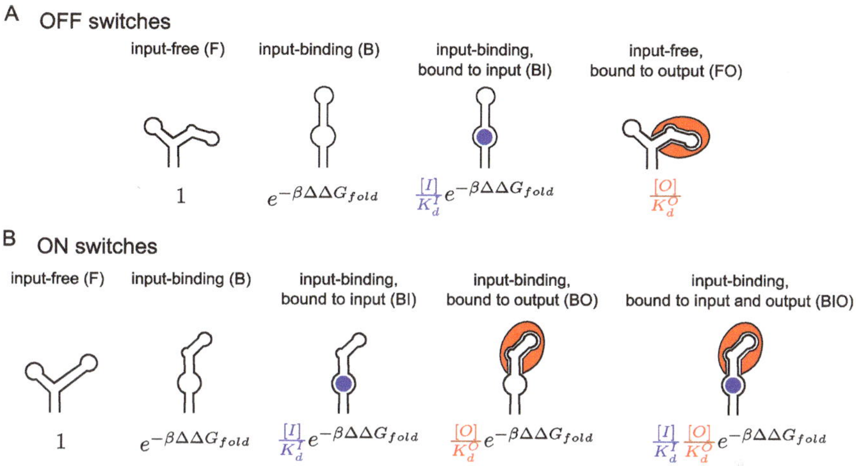

Fig. 2 Minimal thermodynamic model for riboswitch function. Depicted are states and associated Boltzmann weights for (A) an OFF-switch and (B) an ON-switch.

In any riboswitch, the set of RNA conformations can be partitioned into those that present the input-binding aptamer in the correct secondary structure (abbreviated *B*, for binding) and those unable to bind the input (abbreviated *F*, for free). We note that while we show a single structure for each of these states in Fig. 2, each state can involve a panoply of secondary and tertiary structures, and such structural heterogeneity appears to be present for some systems like adenine riboswitches (Tian, Kladwang, & Das, 2018; Warhaut et al., 2017).

In our minimal model, only one state is able to bind the output and thus directly modulate the downstream signal of interest (e.g., binding of the DFHBI fluorophore). Since we are considering an OFF switch, only the *F* state binds the output ligand *O*. Similarly, for OFF switches, only the *B* state is able to bind the input molecule *I*. These considerations result in two additional states with the different ligands bound, which we call *FO* and *BI*. The resulting four-state model is shown in Fig. 2A. We note that unlike typical "short-hand" schematics shown in the literature (e.g., Fig. 1A), it is important to recognize that any RNA conformation that binds a ligand must also exist in the absence of the ligand, and the equilibrium ratio between the two is $[L]/K_d^L$, where $[L]$ is the free concentration of the input ligand and K_d^L is the intrinsic dissociation constant of the riboswitch ligand aptamer (Garcia et al., 2011). It is the presence of these states that are able to bind input ligand, yet remain unbound, that set a useful limit on the activation ratio.

Now, we use the Boltzmann weights for each state (Fig. 2A) to compute the fraction signal observed via the output molecule (O),

$$signal = \frac{e^{-\beta\epsilon_{FO}}}{e^{-\beta\epsilon_F} + e^{-\beta\epsilon_B} + e^{-\beta\epsilon_{BI}} + e^{-\beta\epsilon_{FO}}} = \frac{\frac{[O]}{K_d^O}}{1 + K_{fold} + \frac{[I]}{K_d^I}K_{fold} + \frac{[O]}{K_d^O}} \quad (1)$$

Here, β is $1/k_B T$, where k_B is the Boltzmann constant and T is the temperature. The second expression is rewritten in terms of equilibrium constants: $K_{fold} = e^{-\beta\epsilon_B}/e^{-\beta\epsilon_F}$ is the equilibrium constant between the riboswitch conformations that can bind the input ligand vs. those that cannot. $[I]$ is the input molecule concentration, [O] is the output molecule concentration, and K_d^I and K_d^O are the intrinsic dissociation constants for the correctly folded aptamers toward the input and output ligands, respectively. Rearranging Eq. (1) shows that the signal depends on the concentration of the output ligand, but in a simple form that mimics a standard ligand-binding isotherm with an observed dissociation constant K_d^{obs}:

$$signal = \frac{[O]}{[O] + K_d^{obs}},$$

$$K_d^{obs} = K_d^O\left[1 + K_{fold}\left(1 + \frac{[I]}{K_d^I}\right)\right] \quad (2)$$

That is, the riboswitch has an apparent binding affinity to the output "activator" ligand that becomes weaker (i.e., K_d^{obs} increases) when the input ligand concentration $[I]$ is increased. How much can this signal change with input ligand concentration? The activation ratio is the ratio of the signal in the OFF state (for an OFF riboswitch, the state triggered by ligand) and the ON state (for an OFF riboswitch, the ligand-free state):

$$AR = \frac{\text{signal ON}}{\text{signal OFF}} = \frac{\frac{[O]}{[O] + K_d^{ON}}}{\frac{[O]}{[O] + K_d^{OFF}}} = \frac{[O] + K_d^{OFF}}{[O] + K_d^{ON}} \quad (3)$$

Here, K_d^{OFF} and K_d^{ON} are apparent dissociation constants for the output ligand given by the expression in Eq. (2) evaluated at $[I] = 0$ and at the input ligand concentration used to toggle the switch, respectively. When there is a high concentration of output ligand [O], the AR decreases, as described in Eq. (3). At very large [O], the activation ratio approaches unity, corresponding to no observed switching. The presence of the output ligand pulls the equilibrium of the riboswitch toward the activated state, and high

$[O]$ saturates the riboswitch in the active state. So there is a necessary tradeoff here: the best case scenario (maximal AR) involves using minimal amounts of output ligand, which decreases the magnitude of signal observed. Fig. 3A plots the output signal of the riboswitch as a function of $[O]$, both without and with input ligand. At high $[O]$ the signal saturates as the output ligand "loads" the switch. The maximum activation ratio is attained at lower output concentrations (marked with arrow).

Taking the limit of the activation ratio in Eq. (3) at low $[O]$, we obtain

$$AR < \lim_{[O]\to 0} AR = \frac{K_d^{OFF}}{K_d^{ON}} = \frac{1 + K_{fold}\left(1 + \frac{[I]}{K_d^I}\right)}{1 + K_{fold}} \tag{4}$$

The right-hand side expression is maximized at high K_{fold}. Taking this limit, we have

$$AR < \lim_{K_{fold}\to large} AR = AR_{max} = 1 + \frac{[I]}{K_d^I} \tag{5}$$

Thus, we set a bound on the activation ratio of AR_{max} that depends only on the input ligand concentration $[I]$ and the intrinsic dissociation constant of the riboswitch aptamer segment K_d^I.

Optimal riboswitch performance is achieved in the limit where K_{fold} is large. This corresponds to a situation that, at first glance, may be

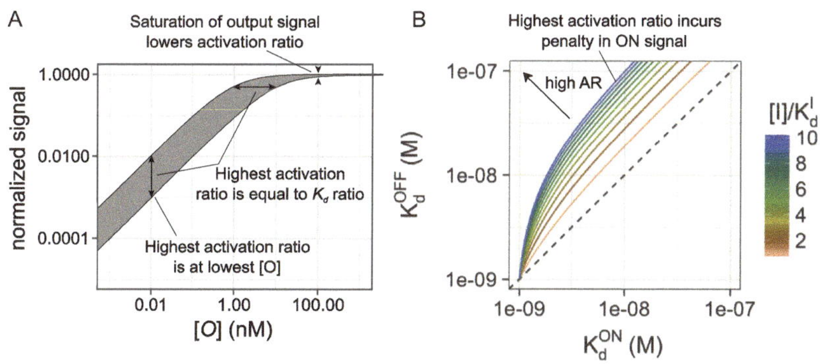

Fig. 3 Modelling illustrates tradeoffs in riboswitch behavior. (A) As the output ligand concentration $[O]$ is changed, there is a tradeoff between achieving high activation ratio and low K_ds to the output, i.e., high activity. (B) The relationship between the achievable K_d^{OFF} and K_d^{ON} is strongly dependent on the ratio of input concentration to input K_d. Within each of these curves, there is another tradeoff between achieving high activation ratio $\frac{K_d^{OFF}}{K_d^{ON}}$ and low K_ds to the output, i.e., high activity.

counterintuitive. The "perfect" OFF riboswitch is achieved when the riboswitch only weakly binds the output ligand in the absence of input and this association becomes even weaker when input ligand is present. Because maximal performance requires minimal [O] (Eq. 3), this additional consideration presents a second tradeoff in which having a large activation ratio necessitates preventing saturated output even in the ligand-free state of the riboswitch.

To illustrate this tradeoff, Fig. 3B shows how the binding of the riboswitch to the output ligand (parameterized by its K_d^{obs}) is toggled at different ligand concentrations, for a wide sweep of the K_{fold} variable. The best possible activation ratio $\frac{K_d^{OFF}}{K_d^{ON}}$ is achieved when K_{fold} is set to give poorer (higher) values of K_d^{ON}—this optimal region corresponds to the part of the curves in Fig. 3B near the top-right of the plot. It is important to note that this observed binding dissociation constant K_d^{ON} is necessarily worse (higher) than the intrinsic dissociation constant of the riboswitch's output-recruiting segment to the output ligand K_d^O (red in Figs. 1 and 2). Indeed, to achieve the best activation ratios, K_d^{ON} may be much worse than the intrinsic dissociation constant for the output K_d^O.

2.2 Optimal activation ratio for an ON switch

The derivation above applies to OFF switches. A minimal thermodynamic model of ON switches is illustrated in Fig. 2B and yields the following expression for the activation ratio,

$$AR < \frac{K_d^{OFF}}{K_d^{ON}} = \frac{\left(1 + K_{fold}\right)\left(1 + \frac{[I]}{K_d^I}\right)}{1 + K_{fold}\left(1 + \frac{[I]}{K_d^I}\right)} \tag{6}$$

This equation is maximized at low K_{fold},

$$AR < AR_{max} = 1 + \frac{[I]}{K_d^I} \tag{7}$$

This AR_{max} for ON switches is the same expression as for OFF switches. Here again, the best activation ratio for an ON switch corresponds to a situation (here, small K_{fold}) in which the riboswitch has highly suppressed activity in the absence of input ligand and then, in the presence of input ligand, gives improved—but not saturating—output activity.

2.3 Generality of the thermodynamic bound

The above analyses of minimal thermodynamic models of OFF and ON riboswitches both lead to the same upper bound on the activation ratio: $AR_{max} = 1 + \frac{[I]}{K_d^I}$ (compare Eqs. 5 and 7). This bound applies even if we relax some assumptions of the analysis. The expressions above have ignored some states; for example, a poorly designed OFF switch may form "leaky" structures that are not bound to input ligand but nonetheless can still bind the output molecule. As described in Appendix A, such states will produce ARs that are even lower than the expressions given above, and the bound $AR_{max} = 1 + \frac{[I]}{K_d^I}$ still holds.

The analyses above involve the free input ligand concentration [*I*]. In typical cases, experimenters set the *total* ligand concentration and not the free concentration, which can be reduced if, for example, the input ligand is sequestered by binding the riboswitch or to other macromolecules in the system (especially in vivo). In those cases, the free ligand concentration will be reduced compared to the input, and the $AR_{max} = 1 + \frac{[I]}{K_d^I}$ bound still holds if [*I*] refers to the total input ligand concentration. Tighter bounds on the activation ratio can be set in models that account for RNA sequestration of input ligand, which can reduce the free concentration relative to total concentration (see, e.g., Espah Borujeni et al., 2015).

Another assumption in the analyses above is that the input aptamer region and the output ligand-recruiting region bind their targets with K_d values that are independent of one another; that is, the riboswitch function is mediated solely through an allosteric RNA conformational change. This assumption appears to hold for the majority of natural and synthetic riboswitches. However, there are notable exceptions, such as a natural ribozyme that binds a glucosamine-6-phosphate input and uses this bound metabolite to help catalyze RNA cleavage. This *glmS* ribozyme is not well described by the models of Fig. 2 and is not considered here.

Many riboswitches create an output through an irreversible event rather than reversible recruitment of an output ligand (red symbols in Fig. 2), but such systems still have limits in which the derived activation ratio should still hold. For example, some synthetic riboswitches effect their output through cleavage of the RNA by an embedded ribozyme (see, e.g., Koizumi, Soukup, Kerr, & Breaker, 1999; Soukup & Breaker, 1999b; Win & Smolke, 2007). If the cleavage rate is slower than RNA conformational changes, the model in Fig. 2 holds with a minor update: the equilibrium constant for recruiting an output $[O]/K_d^O$ to the ON state (red line) needs to change to the rate of cleavage of the ribozyme. The same bound on activation ratio $AR_{max} = 1 + \frac{[I]}{K_d^I}$ still holds as long as the RNA input ligand binding and conformational rearrangements reach equilibrium before cleavage (which has a timescale on the order of minutes). As a more common example of a distinct "output" mechanism, many natural and synthetic riboswitches act during transcription of the riboswitch by bacterial RNA polymerase. These systems effect their output through display of a special terminator hairpin, which signals the RNA polymerase to terminate transcription. In these cases, the RNA polymerase can be considered the output "ligand," and replacing the expression $[O]/K_d^O$ with $k_{termination}$, a rate constant for the terminator hairpin interacting with RNA polymerase, allows derivation of the same bound of $AR_{max} = 1 + \frac{[I]}{K_d^I}$, as long as RNA conformational changes equilibrate faster than the termination event and synthesis of the RNA by polymerase.

As a final note, we emphasize that our expression is based on a thermodynamic model for riboswitch function, but in several riboswitch mechanisms, attaining thermodynamic equilibrium is not guaranteed. Indeed, various riboswitch classes have been experimentally verified to function in the kinetic regime and, among other features, shown to require ligand concentrations significantly above their dissociation constants for activation (Furtig, Nozinovic, Reining, & Schwalbe, 2015; Greenleaf, Frieda, Foster, Woodside, & Block, 2008; Lin, Yoon, Hyeon, & Thirumalai, 2015; Ottink et al., 2007; Sun, Zhao, & Chen, 2018; Wickiser, Cheah, Breaker, & Crothers, 2005). These reports imply that the riboswitch activation process is often a delicate balance between ligand concentration, binding and folding kinetics, and the processing speed of RNA transcription and translation machinery. Modeling and predicting kinetic regulation of riboswitches through cotranscriptional folding and other mechanisms remains an active area of research and discussion. Nevertheless, most out-of-equilibrium effects in riboswitch studies actually work to reduce the activation ratio. For example, if a riboswitch is transcribed so quickly that it does not have time to bind its input ligand before facing the task of activating its output (e.g., transcription termination), its activation ratio will be poor (see, e.g., Ceres, Garst, Marcano-Velazquez, & Batey, 2013; Ceres, Trausch, et al., 2013; Endoh & Sugimoto, 2018; Wickiser et al., 2005). Kinetic mechanisms that might instead boost the activation ratio above the thermodynamic bound of $AR_{max} = 1 + \frac{[I]}{K_d^I}$ are discussed in Section 4.

3. How close are current riboswitches to the thermodynamic optimum?

The bound on the activation ratio $AR_{max} = 1 + \frac{[I]}{K_d^I}$ provides an easily computed metric that the best performing riboswitches should achieve. Do any natural or engineered riboswitches meet this bound? Do any riboswitches surpass it? In this section, we compare observed activation ratios to optimal activation ratios for riboswitch molecules from across literature, harnessing a wide variety of input ligands to a wide variety of outputs. We tabulated experimentally measured AR (AR_{exp}) from these studies and, where possible, we compute the riboswitch's optimality, defined as

$$optimality = AR_{exp} / AR_{max} \tag{8}$$

These calculations were not always possible. On one hand, for experimentally observed activation ratios AR_{exp}, most prior studies of riboswitches directly report—or provide enough information to compute—these values. The output signals (fluorescence of translated GFP, cleavage or transcription termination rate of RNAs, etc.) are expected to be directly proportional to the fraction of the studied riboswitch displaying a correctly folded output-recruiting region (red in Figs. 1 and 2). Other performance measures, such as dynamic range (Ceres, Garst, et al., 2013; Reining et al., 2013), can typically be converted to activation ratio with appropriate additional measurements on, e.g., the maximum output signal of the riboswitch.

On the other hand, for computing the optimal activation ratios AR_{max}, we need the input ligand concentrations $[I]$ and the intrinsic dissociation constant of the aptamer segment for the input ligand K_d^I. Total input ligand concentrations $[I]$ used to toggle the riboswitches are widely reported, although, for in vivo settings, there is uncertainty in what the ligand concentrations are inside cells compared to the concentrations set outside cells. An intracellular concentration higher than the growth media concentration, which can arise due to active transport of metabolites, would result in a higher optimal AR than estimated from the growth media concentration.

Estimating K_d^I can be more challenging. This dissociation constant for the input ligand is *not* the same as the observed midpoint of the riboswitch output signal as the input ligand concentration is varied; in general that midpoint value is higher than the intrinsic dissociation constant for the input ligand K_d^I by a large factor, reflecting the energetic cost of coupling input ligand binding to the riboswitch conformational change. For example, for the OFF switch in Fig. 2A, this inflation factor for the observed dissociation constant for input ligand is $1+K_{fold}$, with K_{fold} being large; see Eq. (2). Measuring the intrinsic dissociation constant for the input ligand K_d^I requires separate measurements on RNA molecules that have been stabilized in a state capable of binding the input ligand via site-directed mutation or careful minimization of the ligand-binding region (state B in Fig. 2). Since these stabilized molecules no longer toggle binding of the output ligand, measurement of K_d^I necessarily requires using some other readout of ligand binding, e.g., nucleotide-resolution chemical mapping of RNA reactivity in the ligand-binding pocket, dialysis of radiolabeled ligand, or competition with fluorophore-labeled ligand analogs. In our survey, these ligand K_d^I values, if not reported in the work itself, were located through referenced works or elsewhere (see Table S1 in the online version at https://doi.org/10.1016/bs.mie.2019.05.028 for sources). Table 1 depicts a sample of collected data from a variety of synthetic riboswitches. In some cases, the solution conditions for these K_d^I determinations were different than in the riboswitch activation ratio measurements, and some of these cases are noted below. We note that riboswitches that bind oligonucleotide ligands through Watson-Crick base pairing are often expected to have intrinsic K_d^I values that are extraordinarily tight (sub-fM) and nucleic acid dissociation rates may be extremely slow (weeks or centuries) compared to other timescales, breaking our assumption of thermodynamic equilibrium; for these reasons, natural and synthetic riboswitches that respond to RNA or DNA (Chappell, Westbrook, Verosloff, & Lucks, 2017; Penchovsky & Breaker, 2005) are not discussed in detail here.

3.1 Optimality of natural riboswitches

Fig. 4 presents the observed activation ratios AR_{exp} and the estimated maximal activation ratios AR_{max} for a range of natural riboswitches, with studies presented in order of appearance in the literature. Fig. 4A depicts the experimental AR in dark bars, and the theoretical maximum AR in light bars; Fig. 4B depicts the optimality of each construct, defined as the ratio of the experimental AR to the theoretical maximum AR (Eq. 8). Each of these natural riboswitches has a clearly defined input and output module in its RNA sequence and therefore should be described well by the modeling of Section 2, at least when operating in the thermodynamic regime. The input ligands span a wide variety of natural metabolites, from the large

Table 1 Example theoretical maximum AR (AR_{max}) and estimated optimality from values reported in literature, including necessary information needed to perform calculation.

Source	Ligand	Switch direction	K_d^I (μM)	AR_{exp}	Source of AR_{exp} estimate	[I] at AR_{max}(μM)	AR_{max}: $1+[I]/K_d^I$	Optimality: AR_{exp}/AR_{max}
Tang and Breaker (1997b)	ATP	OFF	10[a]	180	Ribozyme cleavage products on gel	1000	101	1.78
Soukup and Breaker (1999a)	FMN	ON	0.5[b]	33	Ribozyme cleavage products on gel	20	40	0.83
Suess et al. (2004)	Theophylline	OFF	0.3[c]	8	Beta-galactosidase activity in *B. subtilis*	6000	17,650	0.0005
Wieland, Benz, Klauser, and Hartig (2009)	TPP	ON	0.32[d]	7.5	GFP fluorescence in *E. coli*	100	313	0.02
Kellenberger et al. (2013)	c-AMP-GMP	ON	4.2[e]	6.4	Spinach-DFHBI fluorescence in *E. coli*	100	25	0.26

K_d^I estimates sourced from [a]Sassanfar and Szostak (1993), [b]Burgstaller and Famulok (1994), [c]Jenison, Gill, Pardi, and Polisky (1994), [d]Lang, Rieder, and Micura (2007), [e]Kellenberger et al. (2013), [f]Braselmann et al. (2018). Full data for the optimality estimates presented in Figs. 4 and 5 may be found in Table S1 in the online version at https://doi.org/10.1016/bs.mie.2019.05.028.

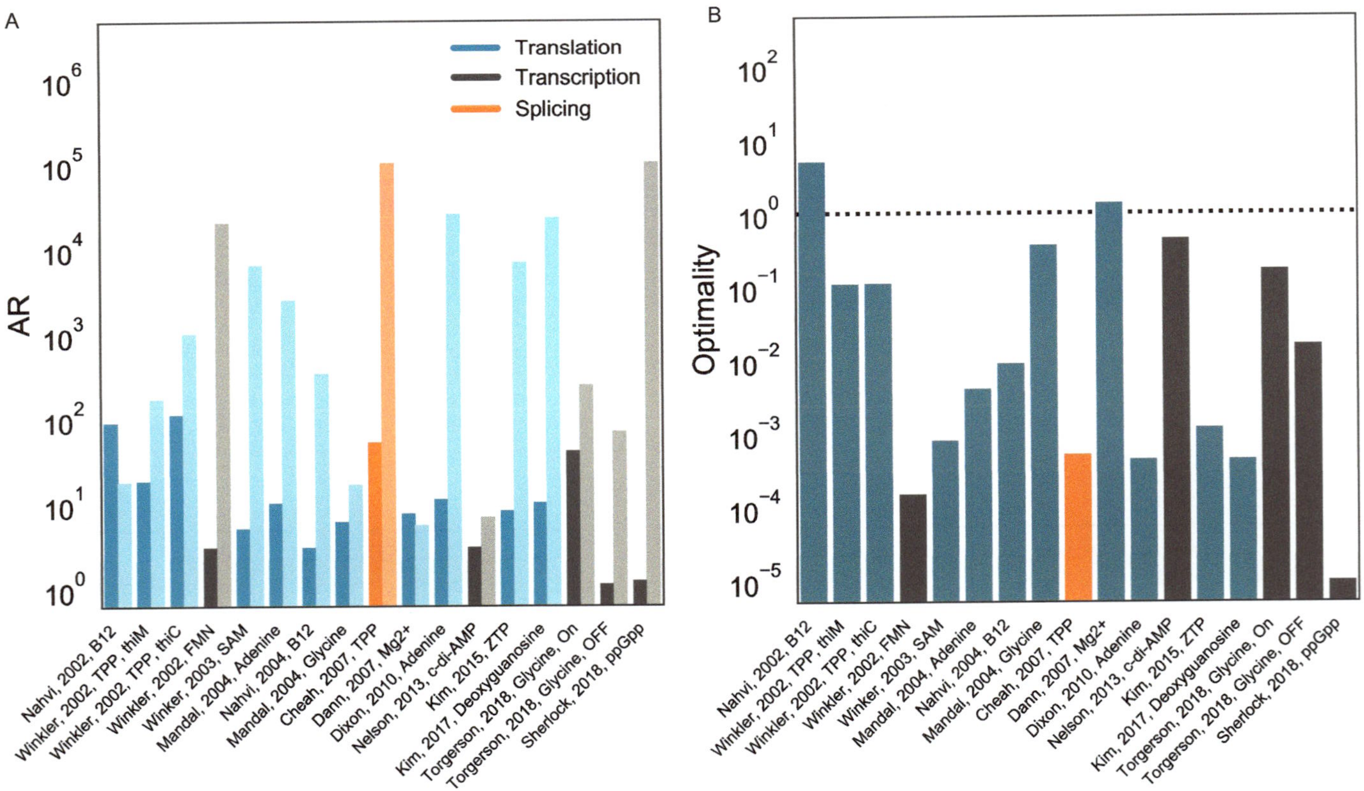

Fig. 4 (A) Comparison of experimental (dark bars) and theoretically optimal ARs (light bars) of natural riboswitches in literature to date, as computed using the thermodynamic model presented here, presented in order of appearance in the literature. (B) Optimalities of natural riboswitches, computed as experimental AR/theoretical AR.

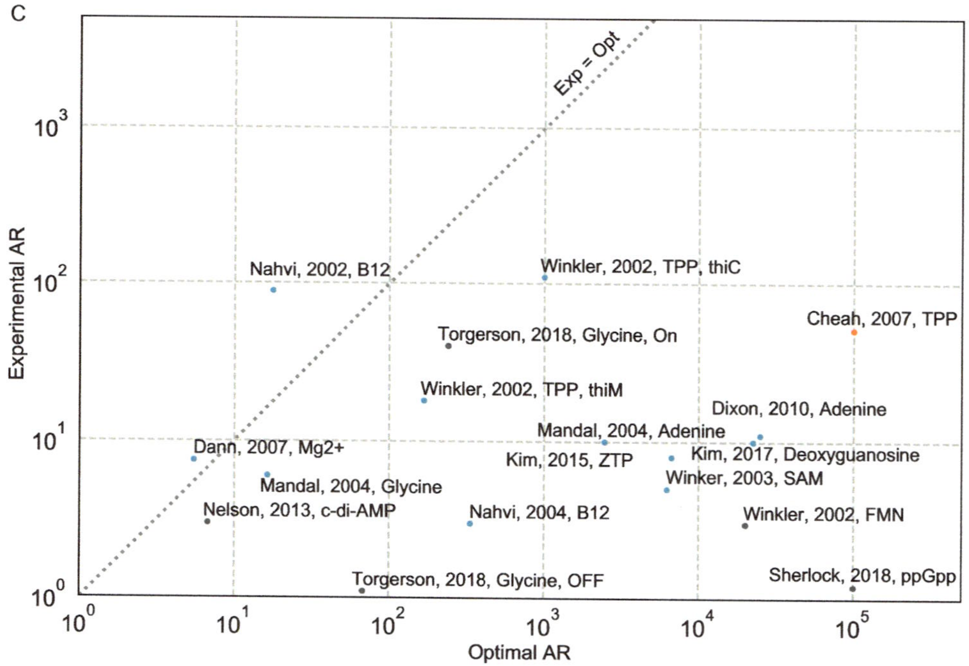

Fig. 4—Cont'd (C) Scatter plot of experimental and theoretical ARs. Values are only shown for cases where previous literature provide enough information to estimate both experimental and theoretically optimal ARs, see main text for caveats.

coenzyme B12 to the small amino acid glycine. In each of the cases described in Fig. 4A, studies isolated and characterized an aptamer from the riboswitch that binds the input ligand. The output mechanisms involve presentation of other segments of the riboswitch in conformations able to recruit other molecular machines. The release of a Shine-Dalgarno region enables recruitment of the bacterial ribosome; the formation of a terminator hairpin trailed by multiple uridine bases initiates termination of the bacterial RNA polymerase. In a few eukaryotic cases, folding or presentation of appropriate pre-mRNA elements toggle recognition by the spliceosome. Data for these respective output mechanisms are colored blue, gray, and orange in Fig. 4. In all of these cases, if ligand-dependent riboswitch RNA rearrangements are fast compared to recruitment of the output, we expect the limit on activation ratio given by $AR_{max} = 1 + \frac{[I]}{K_d^I}$ to hold.

The vast majority of the natural riboswitches we surveyed did give activation ratios less than the optimum set by the input ligand concentration (15 of 17 cases). Indeed, for most of these riboswitches, AR_{exp} was lower than AR_{max} by more than an order of magnitude (13 of 17 cases). In these seemingly sub-optimal cases, the reporter systems used to experimentally characterize activation ratios may have perturbed the riboswitch from achieving peak performance, e.g., by saturating the output (see Eq. 3). For in vivo experiments, the intracellular input ligand concentrations may be smaller than the externally set total ligand concentrations, as noted above. Furthermore, these riboswitches are likely tuned through evolution to achieve the precise activation ratio necessary for their functional roles in natural systems. Indeed, the median AR_{exp} of these switches is 9.0 and is considered respectable for designed synthetic riboswitches (see next section), and natural riboswitches may make a tradeoff to ensure strong activation in their ON states.

For two of the cases we surveyed, the experimentally observed activation ratio exceeded the AR_{max} bound set from thermodynamic considerations. In one of these "super-optimal" cases, a Mg^{2+} sensor (Dann 3rd et al., 2007), both AR_{exp} and AR_{max} values were close to 5.0, and the values of intrinsic dissociation constant and activation ratios came from distinct studies by different groups, so it is unclear if the apparent super-optimality of this riboswitch is significant. The other super-optimal case came from a seminal study by Breaker and colleagues describing the first adenosylcobalamin riboswitch (abbreviated as B12 in Fig. 4), a *btuB* leader element in *E. coli* that toggles translation of a gene for a cobalamin transporter. All numbers needed for the optimality comparison were measured in the same study, using *E. coli* reporter measurements and in vitro equilibrium dialysis as appropriate. The AR_{exp} of this *btuB* leader element was 88.7 with adenosylcobalamin concentration [*I*] of 5 μM. This experimentally observed activation ratio is higher by five-fold than the estimated AR_{max} value of 18, based on dissociation constant K_d^I for the isolated riboswitch aptamer of 0.3 μM. It is possible that the apparent super-optimality of this riboswitch might be due to deviations of the intracellular ligand concentrations due to, e.g., active transport of extracellular adenosylcobalamin into *E. coli*, though it is worth considering possible mechanisms by which a translational riboswitch might surpass the thermodynamic limit (see Section 4).

3.2 Optimality of synthetic riboswitches

Fig. 5 presents observed and estimated maximal activation ratios for synthetic riboswitches, again with studies listed in order of appearance in the literature. As with natural riboswitches, each of these riboswitches contains a readily identifiable aptamer region for the input ligand and a separate region that recruits a separate ligand or molecular machine to activate an output. In terms of input ligands, synthetic riboswitches include aptamer segments for metabolites like thiamine pyrophosphate (abbreviated TPP in Fig. 5), reused from natural riboswitches described above. Synthetic riboswitches also make use of in vitro selected aptamers for a wide variety of small molecules that are not widespread in nature, such as tetracycline. In terms of output mechanisms, synthetic riboswitches have been designed to use the same translational, transcriptional, and splicing output as seen in natural riboswitches (blue, gray, and orange, respectively, in Fig. 5), but also illustrate novel applications that have not yet been seen in Nature. Output mechanisms including the activation of RNA self-cleavage through a hammerhead ribozyme module, binding of fluorescently labeled proteins such as the MS2 bacteriophage coat protein, or small molecules that become fluorescent when bound to RNA, and activation of CRISPR complexes by rearrangement or hammerhead-ribozyme processing of the guide RNA (brown, purple, red, green, and pink, respectively, in Fig. 5) have enabled control of metabolic pathways, visualization of tagged RNAs or cellular metabolites, and chemical control of CRISPR-mediated gene expression (see, e.g., Kellenberger et al., 2015, 2013; Liu et al., 2016; Win & Smolke, 2007).

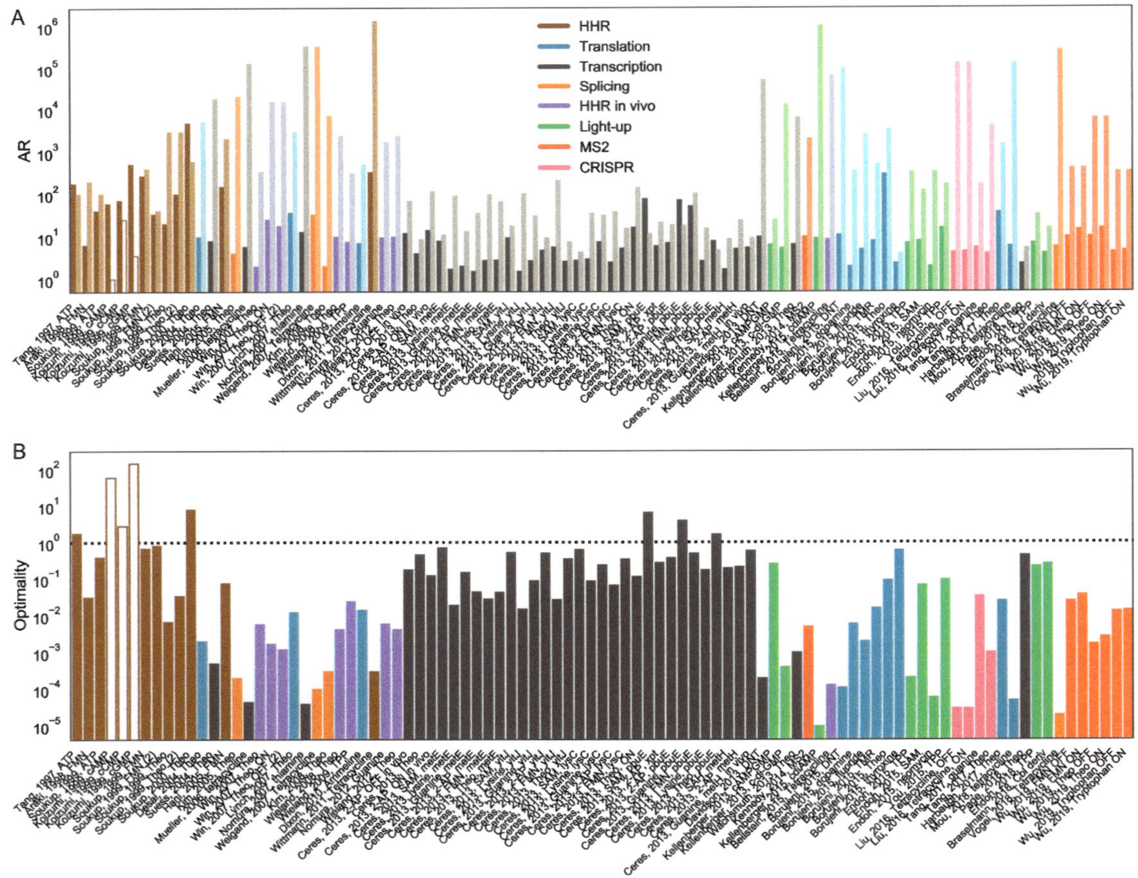

Fig. 5 (A) Comparison of experimental (dark bars) and theoretical ARs (light bars) of synthetic riboswitches in literature to date, presented in order of appearance in the literature. (B) Optimalities of synthetic riboswitches, computed as experimental AR/theoretical AR.

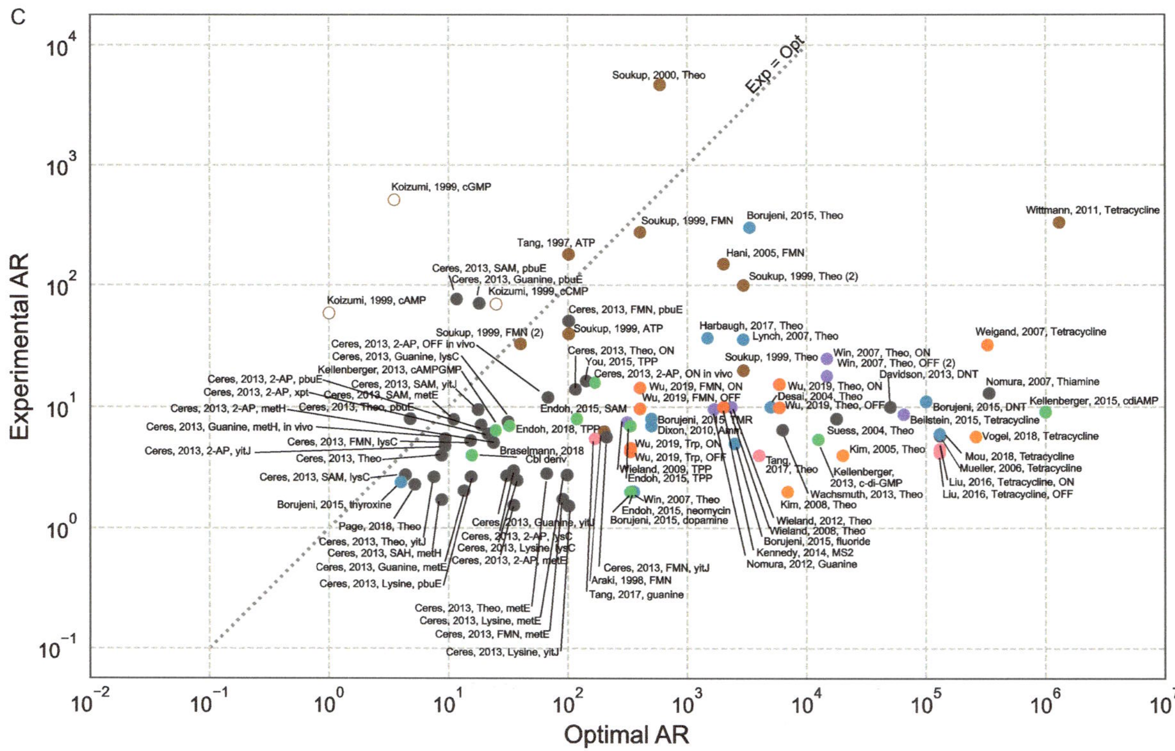

Fig. 5—Cont'd (C) Scatter plot depiction of optimal vs. experimental ARs. Values are only shown for cases where previous literature provide enough information to estimate both experimental and theoretically optimal ARs, see main text for caveats.

As with natural riboswitches, a majority of synthetic riboswitches give experimental activation ratios AR_{exp} that are significantly less than the AR_{max} bound set by $1 + \frac{[I]}{K_d^I}$: 52 of 87 have optimalities below 10%, and 31 of 87 have optimalities below 1%.

It is again instructive to review the riboswitches that did obtain high optimality to try to understand how their mechanisms allow such performance or how caveats in the measurement and prediction of activation ratios might artifactually give rise to such high super-optimality. Synthetic riboswitches with stand-out activation ratios, when compared to their theoretical thermodynamic optimum values, include switches that couple binding of theophylline, flavin mononucleotide (FMN), and ATP to in vitro selected aptamers to hammerhead ribozyme (HHR) cleavage, discovered by Breaker and colleagues (Koizumi et al., 1999; Soukup & Breaker, 1999a, 1999b; Soukup, Emilsson, & Breaker, 2000); the c-AMP-GMP ON switch from (Kellenberger et al., 2013); *S*-adenosyl methionine (SAM), theophylline, 2-aminopyridine (2-AP), and other switches from (Ceres, Garst, et al., 2013; Ceres, Trausch, et al., 2013); the thyroxine OFF-switch from (Espah Borujeni et al., 2015); and a TPP-TAR riboswitch from (Endoh & Sugimoto, 2018). For most of these near-optimal or super-optimal cases, the optimal and experimental activation ratios were each 10-fold or lower and similar in value, and typically inferred from different studies (Endoh & Sugimoto, 2018; Espah Borujeni et al., 2015; Kellenberger et al., 2013); the actual optimality may be somewhat higher or lower depending on experimental uncertainties and it is difficult to comment further. For the HHR-coupled riboswitches reported in Koizumi et al. (1999), plotted with open symbols in Fig. 5, the aptamer segments were selected from random libraries concurrently with selection for ligand-dependent cleavage. It is possible that efforts to characterize the aptamer segment in isolation remain incomplete, or that the actual ligand binding sites involve segments of the hammerhead ribozyme that cannot be cleanly separated from the aptamer (Soukup, DeRose, Koizumi, & Breaker, 2001).

However, the other riboswitches appear exceptional. The general success of Ceres et al.'s modular incorporation of a variety of aptamers into transcription terminator riboswitches (Ceres, Garst, et al., 2013; Ceres, Trausch, et al., 2013), with high optimality for nearly all of their constructs, speaks to the effectiveness of their design principle for mixing and matching natural riboswitch transcriptional units to synthesize new switches. The types of riboswitch that exhibited the highest optimality were hammerhead-based ribozymes designed by Breaker and colleagues. Their success may be tied to the kinetic mechanism of HHR cleavage. Due to the tradeoffs described in Section 2, riboswitches that achieve optimality are unlikely to give strong output signals even under ON conditions, so most riboswitch design studies sacrifice activation ratio in order to yield an observable signal. Use of the hammerhead ribozyme cleavage readout, however, allows for direct characterization of riboswitch output even as cleavage rates change by >10,000-fold, and permits development and in vitro selection of extraordinary riboswitches. Due to the tradeoffs described in Section 2, these switches' active states are not as efficient as constitutive hammerhead ribozymes. Nevertheless, the cleavage rates are still sufficiently fast such that total cleavage is attained in tens of minutes.

Despite these successes—which include several cases from early in the history of in vitro riboswitch engineering—this survey of activation ratios of synthetic riboswitches highlights the general difficulty of achieving optimality. For entire classes of riboswitches, including those that effect changes in pre-mRNA splicing and newer molecules that seek to couple binding of small molecules to binding of fluorophores or toggling of CRISPR-guided gene expression, activation ratios typically remain under 10-fold, and no switches have convincingly achieved the thermodynamic bound in activation ratio derived in Section 2.

4. Going beyond thermodynamic optimality

Deriving the bound on riboswitch activation ratio (Section 2) suggests ways in which the maximum activation ratio $AR_{max} = 1 + \frac{[I]}{K_d^I}$ might be consistently surpassed. These mechanisms might be worth searching for and establishing in natural riboswitches and might also be useful strategies for future studies designing synthetic riboswitches. We describe three mechanisms to go beyond the thermodynamic optimum, each involving riboswitch behaviors beyond the assumptions of our thermodynamic model.

First, the introduction of multiple instances of the aptamer within the target structure, enabling multiple ligands to bind simultaneously, effectively increases the ligand-binding energy. In this case, n copies of the aptamer results in an n-fold increased maximum activation achievable:

$$AR_{max} = 1 + \left(\frac{[I]}{K_d^I}\right)^n \tag{9}$$

Interestingly, some natural riboswitches for glycine, tetrahydrofolate, and cyclic di-AMP bind two copies of their target ligand. Thermodynamic analyses of intrinsic ligand binding to isolated riboswitch aptamers remains incomplete (see, e.g., Kladwang, Chou, & Das, 2012; Sherman et al., 2014), especially because, for some of these cases, the presence of two ligand binding sites only became evident after crystallographic studies that followed biochemical characterizations. However, it may be worthwhile to test if these riboswitches are surpassing the optimal activation ratios achieved with single ligands and to find out how close they get to the higher multiple-ligand AR_{max} bound in Eq. (9). For synthetic riboswitches, metal ion responsive hammerhead ribozymes discovered through in vitro selection give remarkable AR_{exp} values, higher than 10,000 (Seetharaman et al., 2001); biochemical characterization suggests that these RNAs also respond to multiple metal ions, but isolation of the metal ion binding aptamer regions and determination of intrinsic dissociation constants K_d^I has not yet been carried out.

Another way to increase the energy introduced upon ligand binding and, thereby, the activation ratio is to design a riboswitch that utilizes multiple distinct binding sites to coordinate a single input ligand. Such extra contacts beyond the ligand aptamer, potentially including nucleotides in the output region, could increase the affinity of the riboswitch to its ligand, resulting in higher activation ratios than those predicted by the models herein. We predict that such extra contacts may be at play for some of the most effective synthetic riboswitches discovered in the literature, including the cyclic dinucleotide responsive and metal-ion responsive hammerhead ribozymes described by Koizumi et al. (1999) and Seetharaman et al. (2001). 3D structural approaches have not yet been brought to bear on those systems but might reveal contacts of the metal-ion binding regions to the hammerhead ribozyme "output" segment.

A final strategy for increasing activation ratios beyond $AR_{max} = 1 + \frac{[I]}{K_d^I}$ is suggested by natural systems that use out-of-equilibrium kinetic mechanisms to amplify specificities beyond the bounds set by equilibrium thermodynamics. The most famous examples are the kinetic proofreading mechanisms proposed by Ninio and Hopfield and applied to understand numerous molecular biological processes, including translation by the ribosome (Hopfield, 1974; Ninio, 1975). The ribosome is able to achieve extremely low error rates despite the small energy difference between a tRNA binding a correct vs. mismatched codon in the RNA message for a protein. This specificity is realized by requiring multiple pseudo-irreversible steps during the readout mechanism; passing each of these steps requires binding of the correct tRNA. This mechanism can increase optimal activation ratio by the power of n, given an n-step process to reach the downstream readout, as long as each step is separated by an irreversible process. Interestingly, similar to our discussion in Section 2, such mechanisms typically trade output optimality for enhanced specificity. Enforcing such irreversibility requires energy expenditure, and in the case of riboswitches, kinetic amplification of activation ratios can be instantiated through the polymerization of RNA by RNA polymerase or by the translation of ribosomes (R. Das, unpublished results). The riboswitch literature has already pointed out such out-of-equilibrium effects are likely occurring in natural and synthetic riboswitches with potentially positive consequences (see, e.g., Badelt, Hammer, Flamm, & Hofacker, 2015; Espah Borujeni & Salis, 2016; Sun et al., 2018; Wickiser et al., 2005). Establishing and designing "super-optimal" activation ratios in these out-of-equilibrium systems will be important milestones for upcoming riboswitch research.

5. How to evaluate riboswitch optimality

As riboswitch design and discovery efforts become more sophisticated, we propose that future studies should establish the efficacy of new riboswitches by comparing their experimentally observed performance to their maximum expected values. A useful first step in such comparisons is the one carried out herein: comparing experimental observed activation ratios to the best possible value for a riboswitch responding to its input ligand at thermodynamic equilibrium. Comparing the most useful data and experiments that we encountered across previous publications suggests six steps that could help ensure accurate evaluation of riboswitch optimality in future studies:

1. *Measure the observed activation ratio AR_{exp} over a wide variety of input and output ligand concentrations.* A large source of uncertainty in our literature survey of AR_{exp} values involves incomplete characterization in the experimentally measured output signals, which are often presented without estimates of background signal rates (which require negative controls), maximal signal rates (which would require positive controls of ON-stabilized riboswitches), or error estimates. In addition to acquiring this information, riboswitch experiments would benefit from input ligand titrations, rather than a single choice of input ligand concentration. This helps test

for saturation of the riboswitch output, and could provide insights into the mechanisms governing the system. Ceres, Garst, et al. (2013) and Truong, Hsieh, Truong, Jia, and Hammond (2018) provide excellent illustrations of these characterizations. For riboswitches that involve reversible recruitment of output molecules, e.g., small molecule biosensors that bind light-up fluorophores, the concentration of the output ligand ideally would also be varied. Fitting the resulting data to a standard binding isotherm (Eq. 2) would give more precise ARs than current procedures which typically choose a single output ligand concentration.

2. *Estimate input ligand concentrations.* For in vitro applications, the input ligand concentrations ($[I]$ in Section 2) are typically set by the experimentalist and are unambiguous. For in vivo applications, however, intracellular concentrations of the ligand may be different from extracellular concentrations and are typically not well characterized. We believe that an important methodological step in future studies should be to actually measure the total intracellular concentrations of the ligands as a function of extracellular ligand concentration, using quantitative mass spectrometry (Bennett, Yuan, Kimball, & Rabinowitz, 2008).
3. *Estimate* K_d^I. For many synthetic riboswitch design studies, the input aptamer is sourced from previous in vitro selection or biochemical characterization studies, and the intrinsic dissociation constant K_d^I is known, although it is worthwhile repeating that measurement under temperature and solution conditions that match the conditions used to characterize the full riboswitch. For studies of natural riboswitches, mutations or truncations must be introduced that stabilize the natural RNA in the conformation ready to bind the ligand; for riboswitches whose aptamers have been crystallized, such stabilized constructs have already been developed. There are several available methods to estimate dissociation constants of ligands to RNA in vitro in nearly arbitrary solution conditions; perhaps the most general approach involves chemical probing of the aptamer RNA as the ligand concentration is changed (Regulski & Breaker, 2008). Extensions of these chemical mapping approaches to in vivo characterization of dissociation constants would be powerful but have not been established.
4. *Estimate* AR_{max}. Authors should compute $AR_{max} = 1 + \frac{[I]}{K_d^I}$ of their riboswitch system at each input ligand concentration. If the observed activation ratio AR_{exp} is lower than AR_{max}, it is worth considering what

factors might cause this, and, if necessary, how to modulate the riboswitches optimality. Possible factors are alternative conformational states that do not properly couple input to output (see Appendix A), saturation of output, and errors in experimental estimates. If the observed activation ratio AR_{exp} is higher than AR_{max}, the thermodynamic bound has been surpassed, and it is important to understand how that has happened. Section 4 outlines possible scenarios.

5. *Computational prediction of activation ratios.* There are few cases in the literature that computationally predict the activation ratio of a riboswitch given its sequence and the known structures of its input aptamer and output-recruiting module. Sophisticated computational methods have existed for a long time to model riboswitch conformational ensembles and energetics but have typically required software "hacks" to get the information needed to estimate activation ratios. The recent availability of user-defined ligand binding sites and energetic bonuses in the Vienna 2.0 RNA folding package (Findeiß et al., 2018) is therefore strongly welcomed. In Appendix B, we include an example calculation for the activation ratio of a riboswitch that toggles translation with theophylline using this latest Vienna package and find reasonable agreement of the calculation with the experiments.
6. *More detailed modeling.* While we believe that measuring activation ratios and comparing to theoretical optima is a first step in understanding how a riboswitch works, a deeper understanding comes from modeling the complete ligand dependence of the systems, the kinetics of the process, and perhaps idiosyncratic events in how any recruited molecular machines eventually give rise to outputs. For example, some riboswitch studies already provide detailed data sweeping the input ligand concentrations over very wide ranges (e.g., Ceres, Garst, et al., 2013; Ceres, Trausch, et al., 2013). These data strongly constrain models of the riboswitch behavior; for example the K_{fold} parameter in Section 2 can be set based on the midpoint of these input ligand titrations and the known intrinsic aptamer K_d^I. Additional data from these experiments can rigorously test predictions—such as how the maximal activation ratio seen at high input ligand concentrations is set by K_{fold}—and check that mutations of the sequences in switch and aptamer regions behave as predicted from prospective computational methods such as Vienna 2.0. Such quantitative biophysical modeling will be particularly critical for riboswitches designed or discovered to surpass the thermodynamic optimum (Section 4).

In summary, riboswitches are promising genetic regulation elements that can render the exquisite specificity of RNA recognition into a variety of important biological processes, but thus far the optimality of the majority of designed riboswitches has been low. This chapter has presented a method to compute the theoretical maximum activation ratio of a riboswitch operating at thermodynamic equilibrium and highlights existing natural and synthetic riboswitches that have obtained the highest efficiencies. With the method presented here to estimate riboswitch optimality, we hope that researchers will be able to better characterize what factors are currently dampening riboswitch performance and that this will lead to improvements in general riboswitch design.

Appendix A Optimal activation ratio for a general model

We here present the fully generalized nine-state model for riboswitch function. We assume a riboswitch contains two distinct sequence motifs, or aptamers, one of which binds an input molecule (I), and one of which binds an output molecule (O), which may be a ribosome, a fluorogenic molecule, etc. To write an expression for the partition function of the system, we begin by defining the set of states that the riboswitch can adopt. We group all of the possible RNA structures based on the presence of zero, one, or both aptamers. The conversion of each aptamer from a conformation unable to bind a ligand to a conformation that is able to bind a ligand has an associated free energy penalty. Each of these states can bind zero, one, or two molecules (I or O), limited by the presence of the required aptamer motif. There are nine possible states (Fig. A1). Thus, the partition function for our system contains nine terms,

$$\begin{aligned}
Z = {} & \exp\left(-\frac{\Delta G_{fold}^{init}}{k_B T}\right) + \exp\left(-\frac{\Delta G_{fold}^{O}}{k_B T}\right) + \exp\left(-\frac{\Delta G_{fold}^{O} + \Delta G_{bind}^{O}}{k_B T}\right) \\
& + \exp\left(-\frac{\Delta G_{fold}^{I}}{k_B T}\right) + \exp\left(-\frac{\Delta G_{fold}^{I} + \Delta G_{bind}^{I}}{k_B T}\right) + \exp\left(-\frac{\Delta G_{fold}^{I,O}}{k_B T}\right) \\
& + \exp\left(-\frac{\Delta G_{fold}^{I,O} + \Delta G_{bind}^{I}}{k_B T}\right) + \exp\left(-\frac{\Delta G_{fold}^{I,O} + \Delta G_{bind}^{O}}{k_B T}\right) \\
& + \exp\left(-\frac{\Delta G_{fold}^{I,O} + \Delta G_{bind}^{I} + \Delta G_{bind}^{O}}{k_B T}\right),
\end{aligned}$$

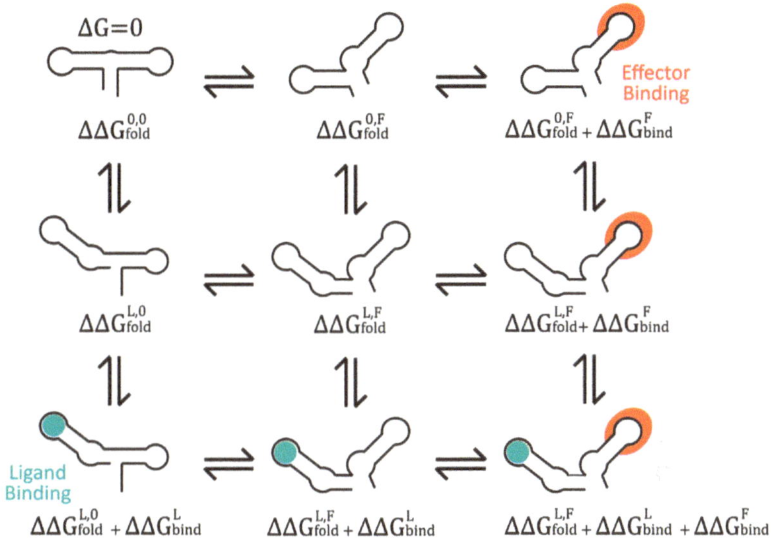

Fig. A1 A depiction of states and weights for a generalized nine-state model of riboswitch function.

where ΔG_{fold} is the Gibbs free energy cost of folding (superscript I and O indicate the presence and absence of the respective aptamer motif in the RNA structure), and ΔG_{bind} is the Gibbs free energy contribution of binding the input or output as indicated (depicted in Fig. A1).

We set the state where neither aptamer has formed, in the absence of ligands I and O, to have an energy of zero, $\Delta G_{fold}^{init}=0$. We substitute $\exp\left(-\frac{\Delta G_{fold}}{k_B T}\right)=S$ for clarity. For both aptamers, we make the substitution

$$\exp\left(-\frac{\Delta\Delta G_{bind}}{k_B T}\right)=\frac{[T]}{K_d^T},$$

where K_d^T is the dissociation constant of the aptamer with molecule T, and $[T]$ is the concentration of the molecule. The partition function above becomes

$$Z=1+S_O\left(1+\frac{[O]}{K_d^O}\right)+S_I\left(1+\frac{[I]}{K_d^I}\right)+S_{IO}\left(1+\frac{[O]}{K_d^O}+\frac{[I]}{K_d^I}+\frac{[O]}{K_d^O}\frac{[I]}{K_d^I}\right).$$

We wish to determine the fraction of RNA that exists in the output-bound state (f_{OB}), as this represents the downstream signal generated by the riboswitch. This can be determined as

$$f_{OB} = \frac{S_O \frac{[O]}{K_d^O} + S_{IO}\left(\frac{[O]}{K_d^O} + \frac{[O]}{K_d^O}\frac{[I]}{K_d^I}\right)}{Z} = \frac{[O]\left(\frac{[O]}{K_d^O} + \frac{S_{IO}}{K_d^O} + \frac{S_{IO}}{K_d^O}\frac{[I]}{K_d^I}\right)}{Z}$$
$$= \frac{[O]\left(\frac{K_d^I S_O + K_d^I S_{IO} + S_{IO}[I]}{K_d^I K_d^O}\right)}{Z}$$

Substituting in the full partition function and simplifying further, we obtain

$$f_{OB} = \frac{[O]}{[O] + K_d^O\left(\frac{K_d^I(1 + S_O + S_I + S_{IO}) + [I](S_I + S_{IO})}{K_d^I(S_O + S_{IO}) + [I]S_{IO}}\right)}.$$

This equation thus describes the magnitude of the riboswitch signal as it relates to the concentrations of reporter and ligand, the binding affinities of the aptamers, and the folding free energies of the different aptamer-containing structures.

We define the activation ratio of a riboswitch as the signal difference in the absence vs. presence of the target ligand. The background signal of the riboswitch is simply f_{OB} evaluated at $[I] = 0$,

$$f_{OB}^{[I]=0} = \frac{[O]}{[O] + K_d^O\left(\frac{1 + S_O + S_I + S_{IO}}{S_O + S_{IO}}\right)}$$

The activation ratio for an ON-switch can then computed as

$$AR = \frac{f_{OB}}{f_{OB}^{[I]=0}} = \frac{[O] + K_d^O\left(\frac{1 + S_O + S_I + S_{IO}}{S_O + S_{IO}}\right)}{[O] + K_d^O\left(\frac{K_d^I(1 + S_O + S_I + S_{IO}) + [I](S_I + S_{IO})}{K_d^I(S_O + S_{IO}) + [I]S_{IO}}\right)}$$

For an OFF-switch, the AR is defined as $\frac{f_{OB}^{[I]=0}}{f_{OB}}$ and gives the same equation. From here, we can identify the properties and conditions of an ideal riboswitch that maximize the activation ratio. First, this ratio is maximized as the concentration of reporter becomes very small (i.e., $[O] \rightarrow 0$):

$$AR^{[O]\rightarrow 0} = \frac{\frac{1 + S_O + S_I + S_{IO}}{S_O + S_{IO}}}{\frac{K_d^I(1 + S_O + S_I + S_{IO}) + [I](S_I + S_{IO})}{K_d^I(S_O + S_{IO}) + [I]S_{IO}}} = \frac{1 + \frac{[I]}{K_d^I}\left(\frac{S_{IO}}{S_O + S_{IO}}\right)}{1 + \frac{[I]}{K_d^I}\frac{S_I + S_{IO}}{1 + S_O + S_I + S_{IO}}}$$

Thus, an optimal riboswitch has very costly transitions into the input and output aptamer-containing structures (described by S_I and S_O, respectively); so S_O, $S_I \ll 1$, and a relatively more favorable (but still costly) transition into the state containing both aptamers (S_{IO}); that is, $S_O + S_I + S_{IO} \approx S_{IO} \ll 1$. Under these conditions, the riboswitch obtains its maximum activation ratio:

$$AR_{max} = 1 + \frac{[I]}{K_d^I}$$

Appendix B Computational prediction of activation ratios

As an example, we walk through the calculation of error rates for a theophylline riboswitch described by Lynch, Desai, Sajja, and Gallivan (2007) (clone 8.1). We assume that translation will be activated proportionally to the prevalence of structures that have the RBS and start codon exposed.

First, we calculate the probability that the sequence will adopt an exposed conformation in the absence of theophylline. We use ViennaRNA 2.4.4 (Lorenz et al., 2011) to compute the ensemble free energies with and without a constraint specifying that the RBS and start codon are unpaired.

```
$ echo -e
"GGUGAUACCAGCAUCGUCUUGAUGCCCUUGGCAGCACCCCGCUGCAAGACAACAAGAUG\n.....
.........................................xxxxxxxxxxxxx" | RNAfold -p -C
    GGUGAUACCAGCAUCGUCUUGAUGCCCUUGGCAGCACCCCGCUGCAAGACAACAAGAUG
(((...))).(((((.....))))).....(((((.....)))))).............. (-15.30)
(((...})).(((((,...,))))).....(((((.....)))))).............. [-15.88]
(((...))).(((((.....))))).....(((((.....)))))..............{-15.30
    d=1.98}
     frequency of mfe structure in ensemble 0.389169; ensemble diversity
    3.22
```

The value −15.88 is the $\Delta G_{constrained}$, the free energy of the RNA in kcal/mol, summed over all structures with the constraint that the RBS and start codon are unpaired.

```
 $ echo "GGUGAUACCAGCAUCGUCUUGAUGCCCUUGGCAGCACCCCGCUGCAAGACAACAA
GAUG" | RNAfold -p
GGUGAUACCAGCAUCGUCUUGAUGCCCUUGGCAGCACCCCGCUGCAAGACAACAAGAUG
(((...))).....(((((((.((..((((.((((.....)))))))).)).)))))))
(-18.40)
```

```
(((,...,).,,..((((((.((..(((,{((((.....)))))))).)).)))))))
[-19.75]
..............(((((((.((..(((.(((((.....)))))))).)).))))))) {-17.10
d=5.39}
 frequency of mfe structure in ensemble 0.112679; ensemble diver-
sity 7.67
```

The value −19.75 is the $\Delta G_{unconstrained}$, the free energy of the RNA in kcal/mol, summed over all structures with no constraints. We then calculate the probability as

$$signal_{OFF} = \frac{\exp(-\Delta G_{constrained}/k_B T)}{\exp(-\Delta G_{unconstrained}/k_B T)} = 1.87 \times 10^{-3}$$

Here, we used $k_B T = 0.616$ kcal/mol.

We perform a similar calculation for signal in the presence of ligand, but with an energetic bonus for the formation of the theophylline aptamer. We compute the energetic bonus as

$$E = -k_B T \ln \frac{[theo]}{K_d^{theo}} = -4.91\,\text{kcal/mol}.$$

where we used $[theo]$ of 1 mM and K_d^{theo} of 340 nM. We then use ViennaRNA to obtain the constrained and unconstrained ensemble free energies, with this ligand bonus.

```
$ echo -e
"GGUGAUACCAGCAUCGUCUUGAUGCCCUUGGCAGCACCCCGCUGCAAGACAACAAGAUG\n...
........................................xxxxxxxxxxxxx" | RNAfold -
motif="GAUACCAG&CCCUUGGCAGC,(...((.(&)....))...),-4.9" -p -C
 GGUGAUACCAGCAUCGUCUUGAUGCCCUUGGCAGCACCCCGCUGCAAGACAACAAGAUG
 ((((...((.(((((.....)))))....))...)))).................... (-16.39)
 ((((...((.(((((,....,)))))....))...)))).................... [-16.84]
 ((((...((.(((((.....)))))....))...)))).................... {-11.50
d=3.10}
  frequency of mfe structure in ensemble 0.48021; ensemble diver-
sity 4.89
$   echo   "GGUGAUACCAGCAUCGUCUUGAUGCCCUUGGCAGCACCCCGCUGCAAGACAACA
AGAUG"| RNAfold -motif="GAUACCAG&CCCUUGGCAGC,(...((.(&)....))...),-
4.9" -p
 GGUGAUACCAGCAUCGUCUUGAUGCCCUUGGCAGCACCCCGCUGCAAGACAACAAGAUG
 (((...))).....(((((((.((..((((.((((.....)))))))).)).))))))) (-18.40)
 (((,...,).,,..((((((.((..(((,{((((.....)))))))).)).))))))) [-19.75]
```

```
..............((((((.((..(((.(((((.....)))))))).)).))))))) {-17.10
d=5.63}
  frequency of mfe structure in ensemble 0.111545; ensemble diver-
sity 8.13
```

This unconstrained free energy is very similar to that without theophylline, due to the relatively low prevalence of the theophylline-aptamer-forming structures in the overall ensemble. We calculate the probability of adopting a translation-compatible conformation as

$$signal_{ON} = \frac{\exp(-\Delta G_{constrained}/k_B T)}{\exp(-\Delta G_{unconstrained}/k_B T)} = 8.88 \times 10^{-3}.$$

Finally, we compute the predicted activation ratio,

$$AR_{predicted} = signal_{ON}/signal_{OFF} = 4.75.$$

This predicts the correct direction of switching. Though the experimentally measured activation ratio (36) is substantially higher than the value above, these calculations suggest that this riboswitch is mostly off, even in the presence of theophylline. This is aligned with the observation that there is a tradeoff between high fold ratio and high signal level (see main text in Section 2).

References

Badelt, S., Hammer, S., Flamm, C., & Hofacker, I. L. (2015). Thermodynamic and kinetic folding of riboswitches. *Methods in Enzymology*, *553*, 193–213. https://doi.org/10.1016/bs.mie.2014.10.060.

Beisel, C. L., & Smolke, C. D. (2009). Design principles for riboswitch function. *PLoS Computational Biology*, *5*(4). e1000363. https://doi.org/10.1371/journal.pcbi.1000363.

Bennett, B. D., Yuan, J., Kimball, E. H., & Rabinowitz, J. D. (2008). Absolute quantitation of intracellular metabolite concentrations by an isotope ratio-based approach. *Nature Protocols*, *3*(8), 1299–1311. https://doi.org/10.1038/nprot.2008.107.

Berens, C., Groher, F., & Suess, B. (2015). RNA aptamers as genetic control devices: The potential of riboswitches as synthetic elements for regulating gene expression. *Biotechnology Journal*, *10*(2), 246–257. https://doi.org/10.1002/biot.201300498.

Berens, C., & Suess, B. (2015). Riboswitch engineering—Making the all-important second and third steps. *Current Opinion in Biotechnology*, *31*, 10–15.

Bintu, L., Buchler, N. E., Garcia, H. G., Gerland, U., Hwa, T., Kondev, J., et al. (2005). Transcriptional regulation by the numbers: Models. *Current Opinion in Genetics & Development*, *15*(2), 116–124.

Braselmann, E., Wierzba, A. J., Polaski, J. T., Chromiński, M., Holmes, Z. E., Hung, S.-T., et al. (2018). A multicolor riboswitch-based platform for imaging of RNA in live mammalian cells. *Nature Chemical Biology*, *14*(10), 964–971.

Burgstaller, P., & Famulok, M. (1994). Isolation of RNA aptamers for biological cofactors by in vitro selection. *Angewandte Chemie International Edition in English*, *33*((10), 1084–1087.

Ceres, P., Garst, A. D., Marcano-Velazquez, J. G., & Batey, R. T. (2013). Modularity of select riboswitch expression platforms enables facile engineering of novel genetic regulatory devices. *ACS Synthetic Biology*, *2*(8), 463–472. https://doi.org/10.1021/sb4000096.

Ceres, P., Trausch, J. J., & Batey, R. T. (2013). Engineering modular 'ON' RNA switches using biological components. *Nucleic Acids Research*, *41*(22), 10449–10461.

Chappell, J., Westbrook, A., Verosloff, M., & Lucks, J. B. (2017). Computational design of small transcription activating RNAs for versatile and dynamic gene regulation. *Nature Communications*, *8*(1), 1051. https://doi.org/10.1038/s41467-017-01082-6.

Dann, C. E., 3rd, Wakeman, C. A., Sieling, C. L., Baker, S. C., Irnov, I., & Winkler, W. C. (2007). Structure and mechanism of a metal-sensing regulatory RNA. *Cell*, *130*(5), 878–892. https://doi.org/10.1016/j.cell.2007.06.051.

Deigan, K. E., & Ferré-D'Amaré, A. R. (2011). Riboswitches: Discovery of drugs that target bacterial gene-regulatory RNAs. *Accounts of Chemical Research*, *44*(12), 1329–1338. https://doi.org/10.1021/ar200039b.

Endoh, T., & Sugimoto, N. (2018). Co-transcriptional molecular assembly results in a kinetically controlled irreversible RNA conformational switch. *Analytical Chemistry*, *90*(19), 11193–11197. https://doi.org/10.1021/acs.analchem.8b03427.

Espah Borujeni, A., Mishler, D. M., Wang, J., Huso, W., & Salis, H. M. (2015). Automated physics-based design of synthetic riboswitches from diverse RNA aptamers. *Nucleic Acids Research*, *44*(1), 1–13.

Espah Borujeni, A., & Salis, H. M. (2016). Translation initiation is controlled by RNA folding kinetics via a ribosome drafting mechanism. *Journal of the American Chemical Society*, *138*(22), 7016–7023. https://doi.org/10.1021/jacs.6b01453.

Findeiß, S., Hammer, S., Wolfinger, M. T., Kühnl, F., Flamm, C., & Hofacker, I. L. (2018). In silico design of ligand triggered RNA switches. *Methods*, *143*, 90–101. https://doi.org/10.1016/j.ymeth.2018.04.003.

Furtig, B., Nozinovic, S., Reining, A., & Schwalbe, H. (2015). Multiple conformational states of riboswitches fine-tune gene regulation. *Current Opinion in Structural Biology*, *30*, 112–124. https://doi.org/10.1016/j.sbi.2015.02.007.

Garcia, H. G., Kondev, J., Orme, N., Theriot, J. A., & Phillips, R. (2011). Thermodynamics of biological processes. *Methods in Enzymology*, *492*, 27–59. https://doi.org/10.1016/B978-0-12-381268-1.00014-8.

Greenleaf, W. J., Frieda, K. L., Foster, D. A. N., Woodside, M. T., & Block, S. M. (2008). Direct observation of hierarchical folding in single riboswitch aptamers. *Science*, *319*(5863), 630–633. https://doi.org/10.1126/science.1151298.

Hallberg, Z. F., Su, Y., Kitto, R. Z., & Hammond, M. C. (2017). Engineering and in vivo applications of riboswitches. *Annual Review of Biochemistry*, *86*, 515–539. https://doi.org/10.1146/annurev-biochem-060815-014628.

Hopfield, J. J. (1974). Kinetic proofreading: A new mechanism for reducing errors in biosynthetic processes requiring high specificity. *Proceedings of the National Academy of Sciences of the United States of America*, *71*(10), 4135–4139.

Jenison, R. D., Gill, S. C., Pardi, A., & Polisky, B. (1994). High-resolution molecular discrimination by RNA. *Science*, *263*(5152), 1425–1429.

Kellenberger, C. A., Sales-Lee, J., Pan, Y., Gassaway, M. M., Herr, A. E., & Hammond, M. C. (2015). A minimalist biosensor: Quantitation of cyclic di-GMP using the conformational change of a riboswitch aptamer. *RNA Biology*, *12*(11), 1189–1197. https://doi.org/10.1080/15476286.2015.1062970.

Kellenberger, C. A., Wilson, S. C., Sales-Lee, J., & Hammond, M. C. (2013). RNA-based fluorescent biosensors for live cell imaging of second messengers cyclic di-GMP and cyclic AMP-GMP. *Journal of the American Chemical Society*, *135*(13), 4906–4909.

Kennedy, A. B., Vowles, J. V., d'Espaux, L., & Smolke, C. D. (2014). Protein-responsive ribozyme switches in eukaryotic cells. *Nucleic Acids Research*, *42*(19), 12306–12321.

Kladwang, W., Chou, F. C., & Das, R. (2012). Automated RNA structure prediction uncovers a kink-turn linker in double glycine riboswitches. *Journal of the American Chemical Society*, *134*(3), 1404–1407. https://doi.org/10.1021/ja2093508.

Koizumi, M., Soukup, G. A., Kerr, J. N. Q., & Breaker, R. R. (1999). Allosteric selection of ribozymes that respond to the second messengers cGMP and cAMP. *Nature Structural & Molecular Biology*, *6*(11), 1062.

Lang, K., Rieder, R., & Micura, R. (2007). Ligand-induced folding of the thiM TPP riboswitch investigated by a structure-based fluorescence spectroscopic approach. *Nucleic Acids Research*, *35*(16), 5370–5378.

Lee, C. H., Han, S. R., & Lee, S.-W. (2016). Therapeutic applications of aptamer-based riboswitches. *Nucleic Acid Therapeutics*, *26*(1), 44–51. https://doi.org/10.1089/nat.2015.0570.

Lin, J. C., Yoon, J., Hyeon, C., & Thirumalai, D. (2015). Using simulations and kinetic network models to reveal the dynamics and functions of riboswitches. In *Methods in enzymology. Vol. 553.* (pp. 235–258). Academic Press.

Liu, Y., Zhan, Y., Chen, Z., He, A., Li, J., Wu, H., et al. (2016). Directing cellular information flow via CRISPR signal conductors. *Nature Methods*, *13*(11), 938–944. https://doi.org/10.1038/nmeth.3994.

Lorenz, R., Bernhart, S. H., Siederdissen, C. H. Z., Tafer, H., Flamm, C., Stadler, P. F., et al. (2011). ViennaRNA Package 2.0. *Algorithms for Molecular Biology*, *6*, 26.

Lynch, S. A., Desai, S. K., Sajja, H. K., & Gallivan, J. P. (2007). A high-throughput screen for synthetic riboswitches reveals mechanistic insights into their function. *Chemistry & Biology*, *14*(2), 173–184.

Mehdizadeh Aghdam, E., Hejazi, M. S., & Barzegar, A. (2016). Riboswitches: From living biosensors to novel targets of antibiotics. *Gene*, *592*(2), 244–259. https://doi.org/10.1016/j.gene.2016.07.035.

Ninio, J. (1975). Kinetic amplification of enzyme discrimination. *Biochimie*, *57*(5), 587–595.

Ottink, O. M., Rampersad, S. M., Tessari, M., Zaman, G. J., Heus, H. A., & Wijmenga, S. S. (2007). Ligand-induced folding of the guanine-sensing riboswitch is controlled by a combined predetermined-induced fit mechanism. *RNA*, *13*(12), 2202–2212.

Penchovsky, R., & Breaker, R. R. (2005). Computational design and experimental validation of oligonucleotide-sensing allosteric ribozymes. *Nature Biotechnology*, *23*(11), 1424–1433. https://doi.org/10.1038/nbt1155.

Regulski, E. E., & Breaker, R. R. (2008). In-line probing analysis of riboswitches. *Methods in Molecular Biology*, *419*, 53–67. https://doi.org/10.1007/978-1-59745-033-1_4.

Reining, A., Nozinovic, S., Schlepckow, K., Buhr, F., Furtig, B., & Schwalbe, H. (2013). Three-state mechanism couples ligand and temperature sensing in riboswitches. *Nature*, *499*(7458), 355–359. https://doi.org/10.1038/nature12378.

Sassanfar, M., & Szostak, J. W. (1993). An RNA motif that binds ATP. *Nature*, *364*(6437), 550.

Seetharaman, S., Zivarts, M., Sudarsan, N., & Breaker, R. R. (2001). Immobilized RNA switches for the analysis of complex chemical and biological mixtures. *Nature Biotechnology*, *19*(4), 336–341. https://doi.org/10.1038/86723.

Sherman, E. M., Elsayed, G., Esquiaqui, J. M., Elsayed, M., Brinda, B., & Ye, J. D. (2014). DNA-rescuable allosteric inhibition of aptamer II ligand affinity by aptamer I element in the shortened Vibrio cholerae glycine riboswitch. *Journal of Biochemistry*, *156*(6), 323–331. https://doi.org/10.1093/jb/mvu048.

Sherwood, A. V., & Henkin, T. M. (2016). Riboswitch-mediated gene regulation: Novel RNA architectures dictate gene expression responses. *Annual Review of Microbiology*, *70*, 361–374. https://doi.org/10.1146/annurev-micro-091014-104306.

Soukup, G. A., & Breaker, R. R. (1999a). Engineering precision RNA molecular switches. *Proceedings of the National Academy of Sciences of the United States of America, 96*(7), 3584–3589.

Soukup, G. A., & Breaker, R. R. (1999b). Nucleic acid molecular switches. *Trends in Biotechnology, 17*(12), 469–476.

Soukup, G. A., DeRose, E. C., Koizumi, M., & Breaker, R. R. (2001). Generating new ligand-binding RNAs by affinity maturation and disintegration of allosteric ribozymes. *RNA*, 7(4), 524–536.

Soukup, G. A., Emilsson, G. A., & Breaker, R. R. (2000). Altering molecular recognition of RNA aptamers by allosteric selection. *Journal of Molecular Biology, 298*(4), 623–632. https://doi.org/10.1006/jmbi.2000.3704.

Suess, B., Fink, B., Berens, C., Stentz, R., & Hillen, W. (2004). A theophylline responsive riboswitch based on helix slipping controls gene expression in vivo. *Nucleic Acids Research, 32*(4), 1610–1614.

Sun, T. T., Zhao, C. H., & Chen, S. J. (2018). Predicting cotranscriptional folding kinetics for riboswitch. *Journal of Physical Chemistry B, 122*(30), 7484–7496. https://doi.org/10.1021/acs.jpcb.8b04249.

Tang, J., & Breaker, R. R. (1997a). Examination of the catalytic fitness of the hammerhead ribozyme by in vitro selection. *RNA, 3*(8), 914–925.

Tang, J., & Breaker, R. R. (1997b). Rational design of allosteric ribozymes. *Chemistry & Biology, 4*(6), 453–459.

Tang, W., Hu, J. H., & Liu, D. R. (2017). Aptazyme-embedded guide RNAs enable ligand-responsive genome editing and transcriptional activation. *Nature Communications, 8*, 15939. https://doi.org/10.1038/ncomms15939.

Tian, S., Kladwang, W., & Das, R. (2018). Allosteric mechanism of the *V. vulnificus* adenine riboswitch resolved by four-dimensional chemical mapping. *eLife*, 7, e29602. https://doi.org/10.7554/eLife.29602.

Truong, J., Hsieh, Y.-F., Truong, L., Jia, G., & Hammond, M. C. (2018). Designing fluorescent biosensors using circular permutations of riboswitches. *Methods, 143*, 102–109. https://doi.org/10.1016/j.ymeth.2018.02.014.

Vazquez-Anderson, J., & Contreras, L. M. (2013). Regulatory RNAs: Charming gene management styles for synthetic biology applications. *RNA Biology, 10*(12), 1778–1797. https://doi.org/10.4161/rna.27102.

Warhaut, S., Mertinkus, K. R., Hollthaler, P., Furtig, B., Heilemann, M., Hengesbach, M., et al. (2017). Ligand-modulated folding of the full-length adenine riboswitch probed by NMR and single-molecule FRET spectroscopy. *Nucleic Acids Research, 45*(9), 5512–5522. https://doi.org/10.1093/nar/gkx110.

Wedekind, J. E., Dutta, D., Belashov, I. A., & Jenkins, J. L. (2017). Metalloriboswitches: RNA-based inorganic ion sensors that regulate genes. *The Journal of Biological Chemistry, 292*(23), 9441–9450. https://doi.org/10.1074/jbc.R117.787713.

Wickiser, J. K., Cheah, M. T., Breaker, R. R., & Crothers, D. M. (2005). The kinetics of ligand binding by an adenine-sensing riboswitch. *Biochemistry, 44*(40), 13404–13414. https://doi.org/10.1021/bi051008u.

Wieland, M., Benz, A., Klauser, B., & Hartig, J. S. (2009). Artificial ribozyme switches containing natural riboswitch aptamer domains. *Angewandte Chemie, International Edition, 48*(15), 2715–2718.

Win, M. N., & Smolke, C. D. (2007). A modular and extensible RNA-based gene-regulatory platform for engineering cellular function. *Proceedings of the National Academy of Sciences of the United States of America, 104*(36), 14283–14288.

Kolibri 360 GmbH
Ferdinand-Jühlke-Straße 7
99095 Erfurt, Deutschland
produktsicherheit@kolibri360.de